SPACE MISSIONS

SPACE MISSIONS

From Sputnik to SpaceShipOne:
The History of Space Flight

Jim Winchester, Editor

THUNDER BAY
P · R · E · S · S

San Diego, California

Thunder Bay Press
An imprint of the Advantage Publishers Group
5880 Oberlin Drive, San Diego, CA 92121-4794
www.thunderbaybooks.com

All notations of errors or omissions should be addressed to Thunder Bay Press, Editorial Department, at the above address. All other
correspondence (author inquiries, permissions) concerning the content of this book should be addressed to Amber Books Ltd., Bradley's Close,
74–77 White Lion Street, London N1 9PF, England, www.amberbooks.co.uk.

ISBN-13: 978-1-59223-580-3
ISBN-10: 1-59223-580-8

Library of Congress Cataloging-in-Publication Data available upon request.

Printed in Singapore

1 2 3 4 5 10 09 08 07 06

CONTENTS

INTRODUCTION

Left: Spacelab was a laboratory module that could be fitted inside the Shuttle's payload bay. First launched in 1983, Spacelab was a great success and heralded the age of space science.

The "Space Age" is less than 50 years old, but in that time mankind has expanded its presence from the Earth's upper atmosphere to the edges of the Solar System, and has landed craft on several planets, moons, comets, and asteroids. Despite all this exploration, there are many mysteries to be solved and each new mission uncovers its share of surprises. Even old certainties such as the number of planets and their moons are constantly challenged by new discoveries made by space probes and telescopes.

With the help of their Lunar Rover, Apollo 16's crew traveled almost 17 miles (27 km) across the lunar surface collecting rock samples for analysis back on Earth.

Artificial satellites have transformed life on Earth, not least for their ability to aid weather prediction and track storms. Radar imaging from space is being used to find minerals, oil and even water under the Earth's surface. The modern global telecommunications and banking systems could not function without satellite relays, and nor could the Internet.

Space stations allow prolonged study of the effects of weightlessness on the human body and on the growth of plants and crystals that may have application in medicine. The record for the longest time in space has invariably been held by Russians—the record for one continuous period currently stands at 438 days, held by Valery Polyakov. A mission to Mars, the nearest planet to Earth, will require several astronauts to share a spacecraft for much longer than this.

Since 1957, Russia/the U.S.S.R. has made well over 2,500 space launches—more than two-and-a-half times the U.S. total. However, the numbers are reversed for space travelers, with over 270 U.S.

Right: Space history is made. Yuri Gagarin's Vostok rocket launches from the Baikonur Cosmodrome in Kazakhstan on 12 April, 1961, making him the first man in space.

Old certainties such as the number of planets and their moons are constantly challenged by new discoveries made by space probes and telescopes.

citizens having reached space compared to 100 from the U.S.S.R. and C.I.S. About 60 astronauts from other nations have flown as part of the two space programs. In all, more than 430 people have flown in space, including 40 women. At time of writing, two individuals have made seven flights in space and four have made six, including John Young, the only person to fly Gemini, Apollo and Shuttle missions. More recently Europe, Japan, China, India, and Israel have joined the "space club." China's Yang Liwei became the first "Taikonaut" in 2003.

Private, commercial spaceflight began in 2004 with the first flights of SpaceShipOne into what is officially defined as space, although orbital flights for paying passengers are still some way in the future.

The U.S. Space Shuttle is the only reusable spacecraft to enter service, although it has fallen far short of its design promise of making over 40 missions per year. Two accidents since the Shuttle's debut in 1981, one on launch and one during re-entry, have cost the lives of 14 astronauts and two of the five flight-capable Shuttles. With the three remaining Shuttles carrying the burden of the U.S. manned space programme until a replacement can be fielded, NASA has become

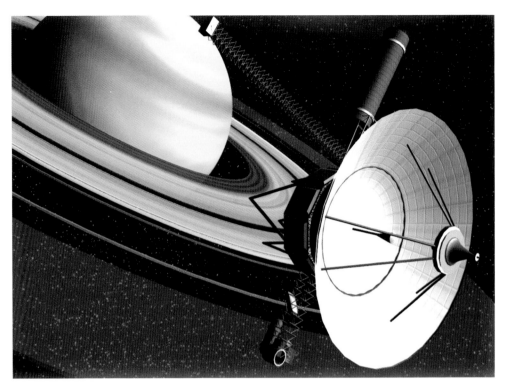

The twin spacecraft Voyagers 1 and 2 have transformed our understanding of the outer solar system. Originally designed to study only Jupiter and Saturn, these two intrepid probes have visited all the gas giants—Jupiter, Saturn, Uranus, and Neptune.

For the first time in 20 years, an American and a Russian spacecraft were linked in Earth orbit on June 29, 1995, when the Shuttle Atlantis *docked with Mir.*

averse to flying "unnecessary" missions. This may spell doom for the Hubble Space Telescope, which requires periodic maintenance that can only be carried out with a space walk. The U.S.S.R.'s Buran shuttle was superficially very similar, but only made a single unmanned flight in 1988.

In 2004 President G.W. Bush announced that America intends to return to the Moon and send astronauts to Mars. A Mars mission is likely to be an international effort due to the many challenges it will pose, not less the long duration of travel and stay needed to await the narrow window available for a return launch. The rockets and spacecraft needed for a true interplanetary mission have yet to be designed, but will no doubt spur the development of as many new technologies as did the Apollo programme of the 1960s.

FIRST STEPS IN SPACE

America's first attempts to catch up with the Soviets' headline-grabbing Sputnik launch were rushed and ended in embarrassing failure on the launch pad. However, by early 1958 Jupiter and Vanguard rockets had successfully carried small science packages into orbit. The full story of Soviet "space shots" is still unclear, because the U.S.S.R. preferred to publicize only those craft which successfully left the atmosphere and for a time in the early 1960s that was only a small proportion of launches. Two explosions of military rockets in 1960 and 1963 caused huge loss of life, including many scientists, at the U.S.S.R.'s Baikonur space centre but were kept secret until 1991. The U.S. suffered the loss of three astronauts in the Apollo 1 launchpad fire in 1967, and four cosmonauts were lost on missions in the 1960s and 1970s, but it is remarkable that the inherently dangerous and experimental nature of spaceflight in its early phase did not result in more disasters. Once the feasibility of single and two-person orbital flight had been proven, the next step (literally) was the "space walk," whereby an astronaut made a tethered excursion outside the spacecraft. This paved the way for many important scientific experiments, missions to launch and repair satellites, and construction of the International Space Station.

On 15 May, 1963, the last of the Mercury astronauts, Gordon Cooper, was launched into orbit atop a Mercury Atlas 9 rocket. Cooper orbited the Earth 22 times and logged more time in space than all five previous Mercury astronauts combined.

SPUTNIK

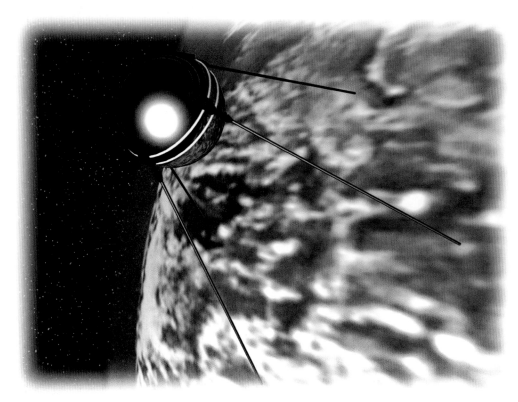

S putnik 1 did not look like the stuff that history is made of. It was a polished aluminum ball, just 22 inches (60 cm) across. But on October 4, 1957, the ball was shot hundreds of miles into space on an adapted Soviet missile, never to return. For the first time ever, an artificial object had reached the 18,000-mph (29,000 km/h) speed required to escape the Earth's atmosphere and go into orbit around our planet. Sputnik, the so-called "Red Moon," amazed the world and panicked the U.S. into accelerating its own rocket program. It also marked the true beginning of the Space Age.

WHAT IF...

...THE U.S. HAD DONE IT FIRST?

A s they were building the rocket that would place the world's first satellite into orbit, designer Sergei Korolev and his team were aware that U.S. engineers were close behind them. In fact, the Korolev design bureau built the Sputnik satellite from scratch in less than a month. They were very concerned that any delay would see the U.S. beat the Soviet Union into space.

Korolev and his team were right to be worried. But the U.S. government may have had political reasons for letting the Soviet Union go first. Any rocket that could launch a satellite could just as easily launch a nuclear weapon. For the U.S. to have sent a satellite over Soviet territory might have been regarded as a hostile act.

Peaceful space exploration depended on "freedom of space": the principle that satellites are entitled to fly over any nation. The U.S. government had already considered ways to establish this principle. One option was to send their first payload around the Equator, well south of Soviet airspace. A second, surer option, was to let the Soviets launch the first satellite so that they could not possibly object to the freedom of space principle.

In 1955, the U.S. and the Soviet Union both announced that they would launch satellites as part of International Geophysical Year, an 18-month period of international science research and cooperation that extended from the poles

The first satellite could have been American, but the Soviet Union might have seen it as a hostile act.

to the depths of the oceans. A year earlier, the German rocket pioneer Wernher von Braun—now employed by the U.S. Army—had proposed Project Orbiter in which the Army's Redstone rocket would be adapted to send a balloon or small instrument package into orbit. But Project Orbiter was rejected because of its links with the Army, for fear of offending the Soviets.

Von Braun's team continued with their tests regardless and could have launched a small satellite during a test flight in September 1956—a year before Sputnik. But von Braun was specifically ordered not to achieve orbit during this launch. Instead, the rival U.S. Navy satellite, Project Vanguard, won official favor. It was more ambitious than von Braun's project—too much so. The first launch attempt, in December 1957, exploded on the pad. The world's press wrote Vanguard off as "flopnik," and it was von Braun's Army team who launched the first U.S. satellite, Explorer 1, in January 1958.

SPUTNIK SPECIFICATIONS

LAUNCH VEHICLE	R-7 SEMIORKA (NATO CODE NAME "SS-6 SAPWOOD")	ORBITAL INCLINATION	65°6'
		APOGEE	583.5 MILES (939 KM)
DIAMETER	22 INCHES (60CM)	PERIGEE	133 MILES (214 KM)
WEIGHT	183 LB (83 KG)	DATE OF LAUNCH	OCTOBER 4, 1957
ON-BOARD EQUIPMENT	2 RADIO TRANSMITTERS	DATE TRANSMISSIONS ENDED	OCTOBER 26, 1957
ORBITAL PERIOD	96.2 MIN	DATE OF REENTRY	JANUARY 4, 1958

BALL IN SPACE

For such an astonishing achievement, the Soviet Union announced the existence of Sputnik 1 in a surprisingly low-key manner. On October 5, 1957, the Earth's first artificial satellite barely made the front page of the government-controlled newspaper Pravda, where it was buried halfway down a side column with the anonymous headline "Tass Report." The satellite's existence was only mentioned in the story's third paragraph. It took the congratulations of the entire world to make the Kremlin realize its propaganda potential. To the Soviet government, Sputnik 1 was merely a sideshow to the all-important ballistic missile program that would maintain military parity with the U.S. in the Cold War.

The Soviet Union's space engineers felt very differently. The real passion of chief missile designer Sergei Pavlovich Korolev was not military superiority, but space exploration. Korolev spent the early 1950s designing and testing missiles with successively longer ranges until, in 1953, he created the A-Series R-7 rocket, which could carry a 5.4-ton payload over a distance of 6,000 miles. It was not long before he suggested modifying the R-7 to place a satellite in orbit, although it it was several years before the Kremlin gave him the final go-ahead.

In July 1955, the Soviet government announced that it would launch an artificial Earth satellite—the day after U.S. President Dwight Eisenhower had made a similar promise. Development of Korolev's satellite went ahead in tandem with flight tests of the R-7. His design bureau built a one-ton probe filled with scientific equipment, but problems with the electronics caused it to be sidelined—although as Sputnik 3, it flew the following year.

SIMPLEST SPUTNIK

In August 1957, Korolev switched to a less advanced design. This consisted of a polished aluminum ball containing a radio, along with some simple scientific instruments: On-board pressure and temperature detectors that were supposed to send data back to the surface by modulating the radio signals. The package was launched in October using an R-7 fitted with additional boosters. It was never given an official name: The Soviet media simply referred to it as "the sputnik," meaning both "satellite" and "traveling companion."

The news electrified the world: Sputnik 1's distinctive "beep, beep" radio signals could be picked up worldwide, and people everywhere searched the night skies, hoping to catch a sight of the satellite. Its limited scientific payload did not work very well, but that hardly mattered. The "Red Moon" had risen. The sky would never seem the same again.

ANATOMY OF SPUTNIK

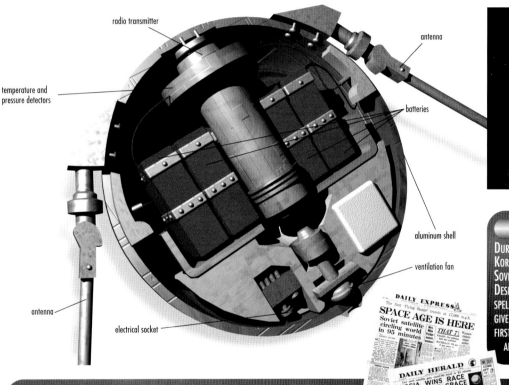

radio transmitter

antenna

temperature and pressure detectors

batteries

aluminum shell

ventilation fan

antenna

electrical socket

DESIGN BRIEF

The sole purpose of Sputnik 1 was to prove that an artificial object could be launched into orbit—and that the Soviet Union could do it first. It was therefore built to be as simple and reliable as possible and to broadcast radio signals on two wavelengths powerful enough to be tracked by amateurs worldwide for two to three weeks. In this way, the Soviet Union made sure that Sputnik could not be hushed up by Western governments.

CHIEF DESIGNER

DURING HIS LIFETIME, SERGEI PAVLOVICH KOROLEV (1906–66) WAS KNOWN TO THE SOVIET PUBLIC ONLY AS "THE CHIEF DESIGNER." KOROLEV (RIGHT) SURVIVED A SPELL IN ONE OF STALIN'S LABOR CAMPS TO GIVE THE SOVIET UNION MANY OF ITS SPACE FIRSTS, INCLUDING SPUTNIK, THE FIRST MAN AND WOMAN IN SPACE AND THE WORLD'S FIRST VIEW OF THE FAR SIDE OF THE MOON.

RELAUNCH

ON THE 1997 ANNIVERSARY OF SPUTNIK 1'S LAUNCH, A REPLICA WAS DEPLOYED FROM THE MIR SPACE STATION BY COSMONAUT PAVEL VINOGRADOV. U.S. ASTRONAUT DAVE WOLF SHARED IN THE MISSION. THE REPLICA ITSELF WAS BUILT JOINTLY BY FRENCH AND RUSSIAN TEENAGERS—A SYMBOL NOT OF THE OLD SPACE RACE, BUT OF A NEW ERA OF COOPERATION.

SPUTNIK IN ORBIT

JULY 30, 1955 THE U.S.S.R. ANNOUNCES ITS INTENTION TO LAUNCH AN ARTIFICIAL SATELLITE TO COINCIDE WITH THE INTERNATIONAL GEOPHYSICAL YEAR.

AUGUST 21, 1957 AFTER FIVE FAILURES, THE R-7 ICBM (ABOVE) PROPELS A DUMMY H-BOMB 2,500 MILES (4,000 KM) ACROSS THE SOVIET UNION.

SEPTEMBER 21, 1957 THE ONE-TON SATELLITE ORIGINALLY PLANNED TO BE FIRST IN ORBIT IS DELAYED BY TECHNICAL PROBLEMS, SO CHIEF DESIGNER SERGEI KOROLEV AND HIS STAFF MANUFACTURE A LESS AMBITIOUS REPLACEMENT INSIDE A MONTH. IT IS KNOWN AS PROSTREISHIY SPUTNIK OR "SIMPLEST SATELLITE". TWO IDENTICAL MODELS ARE PRODUCED.

SEPTEMBER 30, 1957 AN R-7 ROCKET MODIFIED FOR SPACE IS ASSEMBLED IN A BAÏKONUR HANGAR CLOSE TO THE LAUNCHPAD.

OCTOBER 2, 1957 THE R-7 IS ROLLED OUT FOR THE MILE-LONG JOURNEY TO ITS LAUNCHPAD. KOROLEV AND HIS CHIEF DESIGNERS WALK IN FRONT OF THE ROCKET IN SILENCE.

OCTOBER 4, 1957 SPUTNIK 1 IS LAUNCHED INTO AN ELLIPTICAL ORBIT.

OCTOBER 26, 1957 SPUTNIK 1 CEASES ITS DISTINCTIVE "BEEP, BEEP" RADIO TRANSMISSIONS.

JANUARY 4, 1958 SPUTNIK 1'S ORBIT DECAYS AND THE PROBE BURNS UP IN THE EARTH'S ATMOSPHERE. BY THIS TIME, ANOTHER SOVIET SATELLITE HAS BEEN LAUNCHED, SPUTNIK 2, WITH THE SPACE DOG LAIKA INSIDE.

DAILY EXPRESS
The first 'Flying Saucer' travels at 15,000 m.p.h.
SPACE AGE IS HERE
Soviet satellite circling world in 95 minutes
THAT 7!

DAILY HERALD
RUSSIA WINS RACE INTO OUTER SPACE
Night of fear as Warsaw workers riot

ANIMALS IN SPACE

S ince the beginning of the space age in the 1950s, animals of many species have journeyed beyond the Earth's atmosphere. Not all of them have returned alive. Scientists have learned much from their animal helpers, but there is plenty of opposition—from other scientists as well as animal rights activists—to sending animals into space. For some, it is a crime; to others, it is a valuable research resource that helps mankind learn more about the universe.

WHAT IF...

...WE HAD PETS IN SPACE?

The comforts that help to ease the stress of life at home will be even more important in the unnatural confines of a spacecraft. The farther that astronauts venture into the depths of the cosmos, the more they need to reproduce conditions found on Earth. The company of animals as pets on long space voyages could be extremely beneficial to morale.

At present, pets only inhabit spacecraft in science fiction, and the logistics of free-roaming animals on board are far from simple. In the absence of an artificial gravity system, pets would float, disoriented and without control, throughout the vehicle. Unable to understand the reason for the artificial conditions, it would be harder for animals to adapt than for humans. Their distress would make the relationship between humans and pets largely one-sided.

Animals would also be subject to the same space sickness problems encountered by astronauts. The absence of night and day would affect the metabolism of many species, and hygiene could present problems—a kitty-litter tray would be a tricky proposition in zero-gravity conditions.

Some creatures, though, can adapt more readily than others. Tests involving zebra fish bred in a weightless environment have shown that their sense of balance and orientation develop to

Conditions on the spaceship in the film Alien allowed Ripley, played by Sigourney Weaver (above), to keep Jones the cat in comfort. Sadly, a hostile being invaded Jones' body.

suit their unusual situation. In fact, when placed back on Earth, they are unable to cope and exhibit bizarre behavior, such as swimming upside down or twirling aimlessly around.

In the distant future, space colonies may be developed with some form of artificial gravity and enough freedom to allow animals to breathe, eat and exercise healthily. Until it is possible to reproduce the basic conditions of their natural habitat, it will not be practical—or fair—to take animals into space as pets.

ANIMAL FLIGHTS

DATE	ANIMAL	MISSION	STUDY
NOV 1957	DOG LAIKA	SPUTNIK 2	SURVIVABILITY OF SPACE FLIGHT
DEC 1960	RHESUS MONKEY SAM	LITTLE JOE PROJECT	EFFECTS OF HIGH-G ACCELERATION
JAN 1961	CHIMPANZEE HAM	MERCURY REDSTONE 2	SURVIVABILITY OF MANNED SPACE FLIGHT
JUNE 1973	MINNOWS	SKYLAB 3	DISORIENTATION IN THE SPACE ENVIRONMENT
AUG 1973	SPIDERS	SKYLAB 3	ABILITY OF SPIDERS TO ADAPT TO ZERO-G
MAR 1982	MOTHS AND FLIES	SHUTTLE MISSION STS-3	INSECT FLIGHT MOTION STUDY
SEPT 1992	FROGS	SHUTTLE MISSION STS-47	EFFECTS OF WEIGHTLESSNESS ON DEVELOPMENT OF EGGS
OCT 1993	RATS	SHUTTLE MISSION STS-58	EFFECTS OF WEIGHTLESSNESS
MAY 1994	NEWTS AND GOLDFISH	SHUTTLE MISSION STS-65	EFFECTS OF MICROGRAVITY ON EMBRYOS

SPACE BEASTS

Early in November 1957, a dog named Laika ("Little Lemon") became the first living being from the Earth to venture into space and orbit the planet. Laika's one-way mission aboard the Russian spacecraft Sputnik 2 paved the way for manned spaceflight and marked the beginning of animal involvement in the space age.

A year later, two dogs, Belka ("Squirrel") and Strelka ("Little Arrow"), returned safely after a one-day flight on Sputnik 5. Strelka later gave birth to a litter of six puppies, one of which was given to President John F. Kennedy as a gift from the Soviet Union.

At the end of 1960, soon after the Sputnik mission, the U.S. instigated its Little Joe animal-flight program. A rhesus monkey named Sam was sent into space as part of a series of animal flights designed to investigate the effects of high-g acceleration and to test equipment that would later be used in manned missions.

These early animal pioneers proved that there was no danger in space that humans could not face—with a lot of good engineering and a little luck. Later research animals have helped us to understand some of the long-term effects of weightlessness on bodies that have evolved to function in a powerful gravity field.

On short journeys, such effects are minor. But even a week or two in space is enough to affect heart muscles, depress immune systems and distort coordination and balance. A little longer and astronauts begin to suffer from osteoporosis, a weakening of the bones. This weakening is caused when their bodies recycle structural bone material, which is apparently no longer needed because of the absence of gravity.

Although some Russian cosmonauts have lived for many months on the Mir space station, they have needed considerable time to recover upon their return to Earth. Future missions to Mars, or the construction of space colonies, are likely to put the human body under considerable strain.

Spaceborne animals have helped to quantify at least some of the consequences of zero gravity (zero g). But such experiments have their opponents, too. Some scientists argue that the stress of weightlessness itself is enough to invalidate the results of some animal experiments, and that most animal anatomies are not close enough to our own for them to serve as testbeds. Also, creatures such as mice and rats are too short-lived for scientists to gain much long-term data from them. And animal well-being is taken seriously. The Bion program—a long-term collaboration between the U.S. and Russia—insists that animals involved in missions are retired soon after to live out the remainder of their natural lives in comfort.

PIGS CAN FLY

TV CHARACTER MISS PIGGY WAS FEATURED REGULARLY IN A SKETCH ENTITLED "PIGS IN SPACE" ON THE MUPPET SHOW. IN REALITY, NO PIGS HAVE YET TAKEN PART IN SPACE MISSIONS. BUT A FEW HAVE EXPERIENCED FREE-FLOATING IN ZERO GRAVITY ABOARD NASA'S WEIGHTLESS KC-135A TRAINING PLANE, KNOWN AFFECTIONATELY AS THE VOMIT COMET—PROVING THAT PIGS REALLY CAN FLY.

ORBITAL ZOO

AMPHIBIANS
Frogs and newts do not make heavy demands on life-support systems. But they have helped teach scientists how zero gravity affects hearing and balance.

TANGLED WEB

THE SKYLAB MISSIONS OF THE 1970S CARRIED MANY ANIMALS INTO SPACE, INCLUDING A NUMBER OF SPIDERS. WITHOUT GRAVITY, THE SPIDERS LOST THEIR WEB-WEAVING SKILLS. BUT IN TIME THEY LEARNED TO ADAPT TO THEIR NEW SITUATION AND SPUN SUCCESSFULLY IN SPACE.

CHIMPANZEES
The flight of Ham the chimpanzee, aboard Mercury-Redstone 2 in January 1961, paved the way for Alan Shepard's historic Mercury mission four months later.

MONKEYS
The U.S. and Russia have launched 11 unpiloted biosatellites in their joint Bion Program; six have carried Rhesus monkeys. The program aims to increase our understanding of the biological effects of zero gravity.

RODENTS
Long periods in zero gravity often cause osteoporosis, a bone-weakening ailment that also occurs in old age. Mice and rats in orbiting labs have helped doctors to better understand the disease.

LAIKA: SPACE PIONEER

OCTOBER 1957
THE RUSSIAN MONGREL LAIKA UNDERGOES A PROGRAM OF EXTENSIVE TRAINING TO ACCLIMATIZE HER TO CONFINED SPACES, HIGH ACCELERATION FORCES AND ENGINE NOISE THAT SHE WILL EXPERIENCE ON HER VOYAGE.

NOVEMBER 3, 1957
SPUTNIK 2 IS LAUNCHED INTO ORBIT (RIGHT). THE TOP SECTION CONTAINS INSTRUMENTS TO MEASURE RADIATION, THE MIDDLE HOLDS THE RADIO CAPSULE AND BELOW IS THE COMPARTMENT CONTAINING LAIKA.

NOVEMBER 10, 1957
LAIKA DIES AFTER RUNNING OUT OF OXYGEN. SPUTNIK 2 CONTINUES TO ORBIT FOR MORE THAN SIX MONTHS. LAIKA AND THE CRAFT BURN UP ON REENTRY INTO THE ATMOSPHERE.

EXPLORER 1

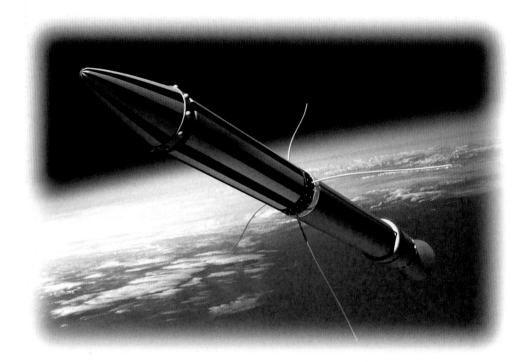

During the late 1950s, the United States and the Soviet Union carried their rivalry into Earth orbit. The Soviets took an early lead by successfully launching two Sputnik satellites. The U.S., uneasy at the thought of Soviet space hardware tracking over North America, redoubled its efforts and responded with Explorer 1 early in 1958. But the first American satellite did more than save face for the U.S. in the Cold War. Once in orbit, the onboard instruments discovered belts of radiation in the Earth's magnetosphere.

WHAT HAPPENED...

...AFTER THE SUCCESS OF EXPLORER 1?

The flight of Explorer 1 was a milestone in American scientific achievement. But just as important was the manner in which it was reached. The night of the launch, at a Pentagon press briefing, the three project managers held up a model of the satellite. The managers were Wernher von Braun, former head of the German rocket engineering team, who had come to work for the U.S. Army; James Van Allen, a researcher in atmospheric physics and satellite development; and William Pickering, director of the Jet Propulsion Laboratory. The three men, despite their vastly different backgrounds, had all served together on the Special Committee on Space Technology set up by the National Advisory Committee for Aeronautics (NACA) in November 1957.

The early years of American space science had seen vigorous competition between different organizations, especially the U.S. Army, Navy and Air Force. During the summer of 1958, this rivalry continued unabated, with each of the three services submitting spaceflight proposals to NACA. But the Special Committee on Space Technology was forging new partnerships between scientists, engineers and the military. New ideas for lunar and planetary exploration were beginning to take shape. It became clear that NACA was changing direction, away from ballistics and aeronautics and toward spaceflight.

Recognition of the fact came on July 29, 1958, when President Eisenhower signed the order to

A triumphant Pickering, Van Allen and von Braun (left, left to right) hold up a replica of Explorer 1 at a press conference at the Pentagon.

The National Advisory Committee for Aeronautics (above) was set up in November 1957 in response to the Soviet launch of the satellite Sputnik 1.

create a new government agency with special responsibility for space—the National Aeronautics and Space Administration, or NASA.

NASA immediately claimed the Vanguard missile program from the Navy and assumed responsibility from the Air Force for the development of the Pioneer family of deep-space probes. Vanguard, unfortunately, was less than successful—eight out of 11 launches between December 1957 and September 1959 ended in failure. But NASA would go on to make some significant scientific breakthroughs with the Pioneer probes. The first successful launch of the series took place on October 11, 1958—just 10 days after NASA's official creation.

EXPLORER 1 SPECIFICATIONS

DIMENSIONS	80 INCHES (203 CM) LONG, 6 INCHES (15CM) IN DIAMETER
WEIGHT	30.66 POUNDS (13.9 KG)
LAUNCH DATE AND TIME	JANUARY 31, 1958, 10:47 P.M. EST
LAUNCH SITE	CAPE CANAVERAL, FLORIDA
LAUNCH VEHICLE	JUPITER-C
ORBITAL INFORMATION	PERIGEE: 1,575 MILES (2534 KM) APOGEE: 224 MILES (360 KM)
	INCLINATION: 33.24 DEGREES PERIOD: 114.9 MINUTES

INTO ORBIT

After the Soviet Union launched Sputnik 1, the world's first artificial satellite, opinion was divided in the United States as to the significance of the event. Did this little metal ball, spinning and beeping its way around the globe, really present a threat to the western world? President Eisenhower, in public, was on the side of the skeptics. But in private, after America's U-2 spy plane was shot down over the Soviet Union, he saw the potential of satellite technology as a safe way of peeking over the Iron Curtain.

Certainly Eisenhower's military advisers saw Sputnik as a potential threat to American security, and their fears were voiced on Capitol Hill by Senator Stuart Symington. "Unless our defense policies are promptly changed, the Soviets will move from superiority to supremacy," Symington warned. The U.S. military wanted more resources for a reconnaissance satellite program.

Meanwhile, the Soviets surged ahead once more. On November 3, 1957, Sputnik 2 was launched carrying Laika the dog—the first living being in space. Sputnik 2 raised the stakes. It was suddenly clear that the Soviet space effort was looking far beyond the obvious military advantages of being able to send missiles into orbit: They wanted to put a man into space—an achievement with extraordinary propaganda value in a war of ideas between competing superpowers. Smaller nations might well be tempted to line up behind the winner of the space race.

The first U.S. attempt to launch a satellite failed on December 6 of that same year, when a Vanguard rocket exploded two seconds into its

flight. But this left the field clear for rocket pioneer Wernher von Braun. On January 31, 1958, a four-stage Jupiter-C lifted off from Cape Canaveral with an upgraded Redstone rocket as the first stage. Inside the fourth stage was Explorer 1, which was launched into an orbit measuring 224 by 1,575 miles (360 by 2,534 km).

Two hours later, Eisenhower told the American people, "The United States has successfully placed a scientific satellite in orbit around the Earth. This is part of our participation in the International Geophysical Year." The International Geophysical Year (IGY) was a global venture, bringing together scientists from 66 countries to investigate Earth's climate and atmosphere.

Explorer 1 made a sensational contribution to the IGY, thanks to the on-board Geiger tube radiation detector that discovered belts of intense radiation girdling the Earth. The instrument was designed by James Van Allen, one of the architects of the IGY. Fittingly, the radiation belts bear his name to this day.

WIRED

THE EXPLORER SERIES CONTINUES TO THIS DAY. EXPLORER 75, THE "WIDE-FIELD INFRARED EXPLORER," OR WIRE (RIGHT), WAS LAUNCHED ON MARCH 4, 1999, AND CARRIED AN INFRARED IMAGING TELESCOPE. UNFORTUNATELY, AN ACCIDENT SHORTLY AFTER LAUNCH MEANT THAT THE SPACECRAFT QUICKLY GREW TOO WARM. THE TELESCOPE COULD NOT BE USED AND THE MISSION WAS DECLARED A LOSS.

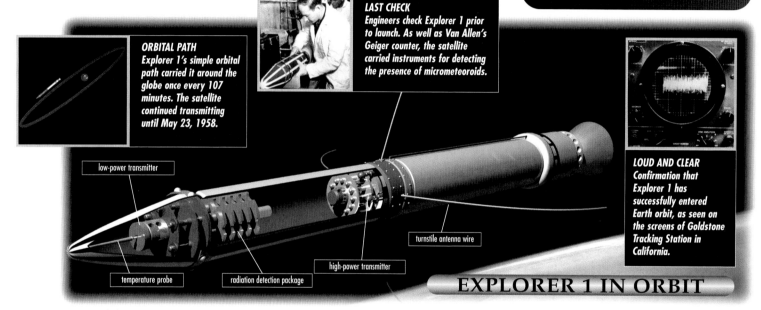

ORBITAL PATH
Explorer 1's simple orbital path carried it around the globe once every 107 minutes. The satellite continued transmitting until May 23, 1958.

LAST CHECK
Engineers check Explorer 1 prior to launch. As well as Van Allen's Geiger counter, the satellite carried instruments for detecting the presence of micrometeoroids.

LOUD AND CLEAR
Confirmation that Explorer 1 has successfully entered Earth orbit, as seen on the screens of Goldstone Tracking Station in California.

low-power transmitter

temperature probe

radiation detection package

high-power transmitter

turnstile antenna wire

EXPLORER 1 IN ORBIT

MEDIA FRENZY

EXPLORER 1 AROUSED INTENSE MEDIA INTEREST. AS VAN ALLEN HIMSELF RECALLED OF ONE PRESS CONFERENCE, "ALTHOUGH IT WAS 1:30 IN THE MORNING, THERE WAS STILL A HUGE CROWD OF REPORTERS WAITING AROUND." BUT THE STORY WAS MORE ABOUT HOW THE U.S. HAD CAUGHT UP WITH THE SOVIETS THAN ABOUT THE SATELLITE'S REMARKABLE SCIENTIFIC ACHIEVEMENTS.

1ST U.S. SATELLITE LAUNCHED, ORBITS IN SPACE WITH SPUTNIK

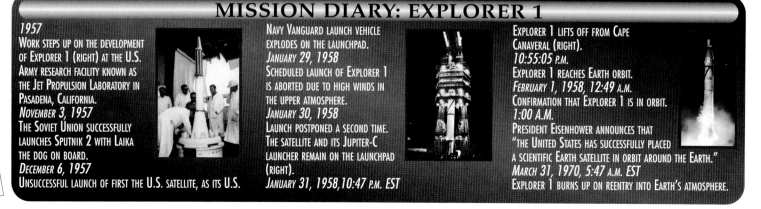

MISSION DIARY: EXPLORER 1

1957
WORK STEPS UP ON THE DEVELOPMENT OF EXPLORER 1 (RIGHT) AT THE U.S. ARMY RESEARCH FACILITY KNOWN AS THE JET PROPULSION LABORATORY IN PASADENA, CALIFORNIA.
NOVEMBER 3, 1957
THE SOVIET UNION SUCCESSFULLY LAUNCHES SPUTNIK 2 WITH LAIKA THE DOG ON BOARD.
DECEMBER 6, 1957
UNSUCCESSFUL LAUNCH OF FIRST THE U.S. SATELLITE, AS ITS U.S.

NAVY VANGUARD LAUNCH VEHICLE EXPLODES ON THE LAUNCHPAD.
JANUARY 29, 1958
SCHEDULED LAUNCH OF EXPLORER 1 IS ABORTED DUE TO HIGH WINDS IN THE UPPER ATMOSPHERE.
JANUARY 30, 1958
LAUNCH POSTPONED A SECOND TIME. THE SATELLITE AND ITS JUPITER-C LAUNCHER REMAIN ON THE LAUNCHPAD (RIGHT).
JANUARY 31, 1958, 10:47 P.M. EST

EXPLORER 1 LIFTS OFF FROM CAPE CANAVERAL (RIGHT).
10:55:05 P.M.
EXPLORER 1 REACHES EARTH ORBIT.
FEBRUARY 1, 1958, 12:49 A.M.
CONFIRMATION THAT EXPLORER 1 IS IN ORBIT.
1:00 A.M.
PRESIDENT EISENHOWER ANNOUNCES THAT "THE UNITED STATES HAS SUCCESSFULLY PLACED A SCIENTIFIC EARTH SATELLITE IN ORBIT AROUND THE EARTH."
MARCH 31, 1970, 5:47 A.M. EST
EXPLORER 1 BURNS UP ON REENTRY INTO EARTH'S ATMOSPHERE.

MERCURY CRAFT

The late 1950s saw the start of the race between the U.S. and the Soviet Union to launch a human being into orbit. In the interests of speed, both nations opted for systems based on a simple, recoverable nose cone that could be launched by an existing intercontinental ballistic missile. In the end, the Soviets won—with Vostok 1 in April, 1961. But America's Mercury capsule, which made its manned debut a month later, was a far more sophisticated craft and flew a total of six manned missions between 1961 and 1963.

WHERE ARE THEY NOW?

THE MERCURY CAPSULES

Anyone looking at one of the surviving Mercury capsules—now on view at various sites around the U.S.—will readily appreciate the courage of the early astronauts who flew in them.

Although the Soviet Union had already launched a successful manned spacecraft, the physical and psychological effects of spaceflight were still largely unknown when the manned Mercury flights began. The success of the Mercury missions proved that manned spaceflight was not as dangerous to the body as many people had thought.

The biggest fears were that weightlessness would result in severe disorientation and that the stress of the g-forces during reentry would cause injury. There were also concerns that space travelers would be psychologically affected.

These fears were largely dispelled by the success of Mercury. Even so, with space technology very much in its infancy, the risks were still enormous. The launch vehicles were no more than lightly modified guided missiles. And the capsules were equally crude by modern standards: The tiny conical section in which the astronaut traveled was just 6 feet 10 inches (2 m) long by 6 feet, 2½ inches (1.9 m) at its widest point.

Of the six manned Mercury capsules, five have been preserved. Alan Shepard's Freedom 7 is normally housed at the National Air and Space Museum, Washington, D.C., as is John Glenn's Friendship 7. Scott Carpenter's craft, Aurora 7, can be seen at the Museum of Science and Industry, Chicago, Illinois, and Wally Schirra's Sigma 7 is in the Astronaut Hall of Fame at the U.S. Space and Rocket Center, Titusville, Florida.

The final craft in the Mercury program, Gordon Cooper's Faith 7, can be seen at Johnson Space Center, Houston, Texas. The one Mercury craft not on public display is Gus Grissom's Liberty Bell 7, which unfortunately sank after splashdown and now lies nearly three miles deep in the Atlantic Ocean some 516 miles (830 km) northwest of Grand Turk Island.

The Mercury capsule at the U.S. Air Force Museum, Dayton, Ohio (above), was flight-rated but never flew. Friendship 7, Alan Shepard's capsule, is on display at the National Air and Space Museum (left).

MERCURY FLIGHT LOG

Name	Mission	Launcher	Launch date	Duration	Crew	Flight
Mercury spacecraft manned flights 1961–3						
Freedom 7	MR-3	Redstone	May 5, 1961	15 min	Shepard	Sub-orbital
Liberty Bell 7	MR-4	Redstone	July 21, 1961	16 min	Grissom	Sub-orbital
Friendship 7	MA-6	Atlas	February 20, 1962	4 hr, 55 min	Glenn	Three orbits
Aurora 7	MA-7	Atlas	May 24, 1962	4 hr, 56 min	Carpenter	Three orbits
Sigma 7	MA-8	Atlas	October 3, 1962	9 hr, 13 min	Schirra	Six orbits
Faith 7	MA-9	Atlas	May 15, 1963	34 hr, 20 min	Cooper	22.5 orbits

MAN IN A CAN

Enclose the driver's seat of a small compact as far as the pedals, put on six layers of clothes, then enter feet-first through the sunroof and you have an idea of what it was like to squeeze into a Mercury capsule. Those who flew it said that you didn't climb into a Mercury, you put it on! To add to the discomfort, the forces on astronauts at launch reached 7 g and during reentry up to 11 g—that is, 11 times their Earth weight.

Mercury's builders could not be sure how astronauts would react in space, so the craft was designed not to rely on them. The chief means of

MERCURY CAPSULE

control was the onboard automatic stabilization and control system, which was monitored from the ground by a rate stabilization and control system. In an emergency, the astronaut could take over some of these functions—a system known as manual proportional control—and there was also a fully manual, fly-by-wire back-up system.

Even so, an astronaut had only limited control over the capsule. He could align it from side to side or up and down, and roll it using 18 hydrogen peroxide-powered thruster nozzles. But if anything major went wrong, there was little that he could do about it.

PRESSURE SUIT
The cabin atmosphere was 100% pure oxygen, and the astronaut wore a pressure suit with a helmet. The helmet's visor could be opened so that the astronaut could take a drink or eat bite-sized chunks of food.

WINDOW ON THE WORLD

The display console included a revolving globe to show the astronaut his position as he orbited the Earth at 17,500 mph (28,160km/h). There was also a periscope system with a screen that displayed a black-and-white image of what lay immediately below. For the first Mercury astronaut, Alan Shepard, this was his only means of viewing the outside world; windows were fitted to all the subsequent capsules in the program.

At the base of the capsule was a pack of three solid-propellant retrorockets that could be manually fired as the astronaut aligned the craft at the correct angle for reentry. The pack was then discarded to expose an ablative—designed-to-melt—heatshield. This absorbed the searing 3,000°F (1,600°C) heat of reentry, which occurred 25 miles (40 km) up, at a speed of 15,000 miles per hour (24,000 km/h).

Because Mercury had a blunt shape and reentered base-down at a slight angle, it generated enough lift to permit some aerodynamic control.

ESCAPE system
The orange-painted nose-mounted launch escape rocket (right) was designed to fire and pull the Mercury capsule to safety in case the launch vehicle malfunctioned either during or immediately after launch.

LUCKY 7

ALAN SHEPARD NAMED HIS SPACECRAFT FREEDOM AND ADDED THE NUMBER 7 BECAUSE IT WAS THE CRAFT'S FACTORY PRODUCTION NUMBER. GUS GRISSOM, WHO FLEW THE NEXT MISSION, NAMED HIS CRAFT LIBERTY BELL. HE, TOO, ADDED THE NUMBER 7, PARTLY BECAUSE SHEPARD HAD, BUT ALSO BECAUSE HE THOUGHT IT WOULD BE A GOOD WAY TO COMMEMORATE THE 7 ASTRONAUTS IN THE MERCURY PROGRAM. THE OTHER ASTRONAUTS FOLLOWED SUIT: FRIENDSHIP 7 (JOHN GLENN), AURORA 7 (SCOTT CARPENTER), SIGMA 7 (WALTER SCHIRRA) AND FAITH 7 (GORDON COOPER). GROUNDED ASTRONAUT DEKE SLAYTON WAS GOING TO NAME HIS CAPSULE DELTA 7.

By using the manual controls the astronaut could vary his flight path and hence the craft's point of splashdown in the ocean.

At an altitude of 10,000 feet (3,048 m) and a speed of 400 miles per hour (644 km/h), a 63-foot (19-m)-diameter chute was deployed to slow the craft to about 20 miles per hour (32 km/h) at sea level. Just before splashdown, an airbag inflated under the heatshield to cushion the impact.

ASTRONAUT WALLY SCHIRRA'S ATLAS 8—SIGMA 7 MERCURY LAUNCH WAS ALMOST ABORTED WHEN HIS ATLAS LAUNCH ROCKET INITIALLY WENT INTO AN ALARMING ROLL. THE RANGE SAFETY OFFICER'S FINGER WAS ON THE ABORT BUTTON, WHICH WOULD HAVE EXPLODED THE ATLAS WHILE SCHIRRA'S CAPSULE WAS PULLED CLEAR BY SIGMA 7'S LAUNCH ESCAPE SYSTEM. FORTUNATELY, THE ROLL WAS CORRECTED.

heatshield

pressurized inner capsule

titanium/nickel alloy outer shell

retrorocket pack

ceramic fiber insulation

control stick

reentry parachutes

control thrusters

periscope

antennae and infrared horizon sensors

escape rocket nozzles

aerodynamic spike

CONTROLS
Aside from limited manual takeover in an emergency, the astronaut had very little control of the craft. Most of the displays were for monitoring.

VOSTOK MISSIONS

The Soviet Union stunned the world in 1957 with the launch of Sputnik 1, the Earth's first artificial satellite. By 1961, they were ready to extend their lead in the space race. Under the brilliant leadership of Sergei Korolev, Soviet engineers built the Vostok craft, designed to take a person into space. And on April 12, it did just that—Yuri Gagarin orbited the Earth in 108 minutes. Five missions later, Valentina Tereshkova became the first woman in orbit. The pioneering age of spaceflight had truly begun.

WHERE ARE THEY NOW?

REMEMBERING NELYUBOV

To the Soviet authorities, the doctoring of photographs was an important weapon in the propaganda war. It was vitally important to promote the image of a fearless team of cosmonaut heroes, building a communist road to the stars. Yuri Gagarin's smile had been one of the factors that had made him "First Cosmonaut." Unfortunately, even the best-laid Soviet plans sometimes went awry. The pair of photographs shown here, or rather the glaring difference between them, illustrates one such story.

Grigori Nelyubov was an outstanding jet pilot. He was also, by all accounts, extremely egotistical—he was good, and he knew it. Nevertheless, his academic and technical skills made him a serious candidate for First Cosmonaut and a certainty for a Vostok flight. Until, that is, one night late in 1961 when Nelyubov, returning to base with two cosmonaut friends after an evening's drinking, got into a fight with guards at a checkpoint. The three pilots were arrested and confined. Nelyubov protested: "You can't do this to me—I'm an important cosmonaut!"

Duly impressed, the officers agreed to let the cosmonauts go—if Nelyubov apologized for his behavior. He refused. The incident was then reported to Nikolai Kamanin, commander of the cosmonaut corps, who resolved to hand down a terrible punishment—Nelyubov was instantly expelled from the corps, and banished to a remote airbase. As he watched first his contemporaries, then his juniors, fly into space,

The cosmonaut who never was. The Soviet propaganda authorities, masters in the art of photo retouching, had no trouble erasing Nelyubov from memory (bottom picture).

Nelyubov fell into a deep depression. On February 18, 1966, drunk and desperate, he took his own life by stepping in front of a train.

At the height of the Cold War, the Soviet Union thought nothing of erasing Nelyubov, and his story, from the records. But with the thaw in East–West relations in the late 1980s, the tragic history of the pilot who fell from grace could finally be told.

VOSTOK MISSION FACTS

Craft	Cosmonaut	Date	Orbits	Duration
V1	Yuri Gagarin	April 12, 1961	1	1 hr 48 min
V2	Gherman Titov	August 6, 1961	17	1 day 1 hr 18 min
V3	Andrian Nikolayev	August 11, 1962	64	3 days 22 hr 22 min
V4	Pavel Popovich	August 12, 1962	48	2 days 22 hr 57 min
V5	Valery Bykovsky	June 14, 1963	81	4 days 23 hr 6 min
V6	Valentina Tereshkova	June 16, 1963	48	2 days 22 hr 50 min

STARS OF THE EAST

Fifteen minutes into Yuri Gagarin's historic flight aboard Vostok 1, a monitoring post off Alaska detected what sounded like a conversation between the Baikonur cosmodrome in Soviet Kazakhstan and a spacecraft in Earth orbit. For once, the usually secretive Soviets were happy to be overheard. Months earlier, U.S. President John F. Kennedy had declared: "If the Soviet Union were first in outer space, that would be the most serious defeat the United States has suffered in many, many years." Now, his words came back to haunt him.

If Vostok 1 was a political victory for the Kremlin, it was a personal triumph for Sergei Korolev and his team at Baikonur. Korolev had survived Stalinist purges, the Nazi invasion of Russia, and many other setbacks in his drive to build a workable spacecraft. As a new decade dawned, it was only a matter of time before one carried a cosmonaut into space. But who would it be?

The two leading contenders were Gagarin and Gherman Titov. Korolev chose the smiling Gagarin—the model of a communist hero. But Titov's Vostok 2 mission was another great leap forward—17 orbits and a full day in space. The American response came in February 1962, with John Glenn's three orbits aboard Friendship 7. Yet within months, the Soviets surged ahead again when Vostoks 3 and 4, launched 24 hours apart, passed within three miles (5 km) of each other in the first space rendezvous.

Vostoks 5 and 6 were also paired in June 1963, giving Korolev the chance to show that docking two spacecraft in orbit was within his reach. But the Soviets' biggest success was in the propaganda war. In Vostok 6 was a woman— Valentina Tereshkova.

MISSION PROFILE

1 LIFTOFF
Vostok blasted off from the Baikonur cosmodrome in Soviet Kazakhstan atop a modified R-7 intercontinental ballistic missile containing no fewer than 32 thrust chambers. Once in orbit, the shields protecting the two-module spacecraft were discarded.

2 ORBIT
Gagarin sat upright in a modified pilot's ejection seat in the spherical crew module. This was supplied with power and a pressurized oxygen/nitrogen mixture by the equipment module. The orbit lasted 89 minutes.

3 REENTRY
Retro-rocket fired to brake the craft, then explosive bolts released the equipment module. The crew module plunged into the atmosphere, protected from the heat of air friction by an ablative shield, which was designed to burn off during reentry. At an altitude of four and a half miles (7 km) above the Soviet Union, Gagarin ejected.

BAD NEWS

It was 3:30 a.m. in Florida when Moscow radio broadcast the news that Gagarin had circled the Earth. When telephoned by reporters for a comment, a shocked and angry Colonel John "Shorty" Powers, press officer for NASA's Mercury program, framed the unfortunate response: "We're all asleep down here!"

SIGN OF HONOR

One of the highlights of a cosmonaut's career is the ceremonial signing of Yuri Gagarin's diary. This tradition was initiated as a tribute to the "First Cosmonaut" after he was killed in a plane crash in 1968. Here, Mir 18 cosmonaut Vladimir Dezhurov signs, watched by guest NASA astronaut Norman Thagard (seated left). The two men had recently returned from Mir aboard the Shuttle Atlantis, after the first Mir–Shuttle docking in 1995.

MISSION DIARY: VOSTOK 1–6

July 1958
Sergei Korolev outlines advantages of Vostok missions in letter to Soviet leaders.
November 1958
Vostok program approved.
March 1960
Twenty cosmonauts (including Gherman Titov, right) begin intensive training for Vostok flights.
August 19, 1960
Test of Vostok launch vehicle with two dogs, Belka and Strelka, aboard. It is successfully recovered.

March 9, 1961
Successful recovery of a Vostok craft with a dog, Chernushka, aboard.
March 25, 1961
Final test launch. Canine passenger Zvezdochka (right) recovered safely.
April 12, 1961
First human spaceflight. Yuri Gagarin orbits Earth aboard Vostok 1.

August 6, 1961
Vostok 2 cosmonaut Titov (right) completes a full day in space.
February 20, 1962
U.S. puts a man in orbit—John Glenn, aboard Friendship 7.
August 11, 1962
Joint mission of Vostoks 3 and 4. Cosmonauts Andrian Nikolayev and Pavel Popovich pass within just over three miles (5 km) of each other in orbit.

August 12
Popovich (right) in Vostok 4 is brought back a day early after ground control misinterprets his comments as being code words for a problem.
June 14–16, 1963
Joint mission of Vostoks 5 and 6. Vostok 5 fails to reach the correct orbit for rendezvous and its mission is cut short. Vostok 6 makes history, as Tereshkova becomes the first woman in space.

MERCURY: THE FIRST STEPS INTO SPACE

NASA's Mercury program and its 7 astronauts had one goal: "To put a manned spacecraft into orbital flight around the Earth." But the program was caught wrong-footed in 1961 when the Russians won the man-into-orbit race. In response, NASA chose Alan Shepard (back row, left) to make a quick, sub-orbital hop. It was a wise decision. Beneath his boyish good looks, Shepard was tough and determined. In later years, he would triumph over an ear disorder to become the oldest man to walk on the Moon.

WHAT HAPPENED...

...TO AMERICA'S FIRST ASTRONAUT?

Alan Shepard returned to Earth not just a national hero, but as the acknowledged senior of the small corps of Mercury astronauts and with a glittering space career ahead of him. He had already been chosen to command the first of the forthcoming Gemini missions when personal disaster struck. Shepard was diagnosed with an inner-ear disturbance that affected his balance—a disastrous complaint for any pilot.

The ailment kept Shepard grounded for the next 6 years. He remained with NASA as chief of the Astronaut Office, responsible for crew selection. But he could only watch in desk-bound frustration as the next generation of astronauts prepared for Project Apollo and the landings on the Moon.

In 1968, Shepard underwent an operation that repaired his damaged eardrum. Once NASA's flight surgeons had checked that his recovery was complete, Shepard put himself forward for the lunar mission that he had always wanted. But he had to wait 3 more years to achieve his goal.

Finally, in 1971, a decade after his historic Mercury flight, Shepard took command of Apollo 14. At the age of 47, he became the oldest man to set foot on the Moon—and the first to

In February 1971, Shepard landed on the Moon (left). Pictured above (center) with his Apollo 14 crew, he was the oldest of the 12 men who reached the lunar surface.

play golf on the lunar surface, to the delight of millions watching back on Earth. He returned to another hero's welcome—and promotion to Admiral in the United States Navy.

After his retirement from NASA, Alan Shepard became one of the charter members of the Astronaut Scholarship Foundation, which provides support to promising science and engineering students. He died on July 22, 1998, at the age of 74.

MERCURY FLIGHT LOG

Mission	Crew	Date	Duration (DAYS, HR, MIN, SEC)
Mercury-Redstone 3	Shepard	May 5, 1961	00:00:15:28
Mercury-Redstone 4	Grissom	July 21, 1961	00:00:15:37
Mercury-Atlas 6	Glenn	February 20, 1962	00:04:55:23
Mercury-Atlas 7	Carpenter	May 24, 1962	00:04:56:05
Mercury-Atlas 8	Schirra	October 3, 1962	00:09:13:11
Mercury-Atlas 9	Cooper	May 15–16, 1963	01:10:19:49

MERCURY'S FIRST PILOT: ALAN SHEPARD

At his 1961 inauguration, the newly elected President John F. Kennedy declared: "This is the new age of exploration; space is our great new frontier." He was confident that the National Aeronautics and Space Administration (NASA), formed less than 3 years before, would ensure that America led the way on the race to his "new frontier."

Barely 3 months later, Kennedy's confidence was badly shaken. On April 12, the Soviet Union launched their first cosmonaut—Yuri Gagarin—into orbit around the Earth. Gagarin deservedly returned as a hero, but Kennedy was embarrassed and humiliated. If America, or its president, were to retain any credibility, they needed to match the U.S.S.R.'s achievement—fast.

NASA's hopes were pinned on Project Mercury, set up in 1959 amid fanfares of publicity to put a man into orbit. Mercury's 7 astronauts-in-training were nationally known figures. But in 1961, the Redstone launch vehicle that NASA was using lacked the power to lift a man all the way into orbit. An improved Atlas booster was on the way, but with Kennedy demanding action ("Are we working 24 hours a day? If not, why not?"), NASA could not wait.

NASA chiefs decided on an immediate sub-orbital flight. Of their 7 astronauts, they chose Alan Shepard to ride a tiny Mercury capsule on a mission that would take him more than 100 miles (160 km) into space before an ocean splashdown 15 minutes later.

COOL LIFTOFF

Shepard could not match Gagarin's full orbit of the Earth. But as compensation, he would have a space "first" of his own. Unlike Gagarin, whose Vostok capsule was under ground control throughout his trip, Shepard would use on-board attitude thrusters to maneuver his Mercury capsule himself.

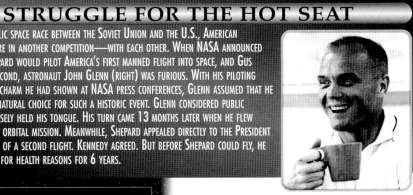

REDSTONE BOOSTER
Shepard blasts off from Cape Canaveral aboard a Redstone launcher—a modified intercontinental ballistic missile.

116 MILES (186 KM) UP
Mercury capsule Freedom 7 reaches peak altitude, then turns around blunt-end-first for reentry.

SPLASHDOWN
Parachutes deploy in the lower atmosphere. Freedom 7 lands gently in the ocean.

SUB-ORBITAL HOP

STRUGGLE FOR THE HOT SEAT

BESIDE THE PUBLIC SPACE RACE BETWEEN THE SOVIET UNION AND THE U.S., AMERICAN ASTRONAUTS WERE IN ANOTHER COMPETITION—WITH EACH OTHER. WHEN NASA ANNOUNCED THAT ALAN SHEPARD WOULD PILOT AMERICA'S FIRST MANNED FLIGHT INTO SPACE, AND GUS GRISSOM THE SECOND, ASTRONAUT JOHN GLENN (RIGHT) WAS FURIOUS. WITH HIS PILOTING SKILLS AND THE CHARM HE HAD SHOWN AT NASA PRESS CONFERENCES, GLENN ASSUMED THAT HE WOULD BE THE NATURAL CHOICE FOR SUCH A HISTORIC EVENT. GLENN CONSIDERED PUBLIC PROTEST, BUT WISELY HELD HIS TONGUE. HIS TURN CAME 13 MONTHS LATER WHEN HE FLEW AMERICA'S FIRST ORBITAL MISSION. MEANWHILE, SHEPARD APPEALED DIRECTLY TO THE PRESIDENT FOR THE CHANCE OF A SECOND FLIGHT. KENNEDY AGREED. BUT BEFORE SHEPARD COULD FLY, HE WAS GROUNDED FOR HEALTH REASONS FOR 6 YEARS.

In true Soviet style, Gagarin's mission had been veiled in secrecy until it was clear that it had succeeded. Alan Shepard would be launched the American way: on live television, in front of millions of people worldwide. On May 5, 1961, just 23 days after Gagarin parachuted back to Earth, Shepard climbed into his tiny capsule, christened Freedom 7, on top of a Redstone rocket. There was a series of upsets during the long, tense hours of the countdown. Shepard was unimpressed. At one point, 2 minutes and 40 seconds before scheduled liftoff, the astronaut calmly told mission control: "I'm a hell of a lot cooler than you guys. Why don't you just fix your little problem and light this candle?"

Mission control fixed their problem. They lit their candle on time, too, and Alan Shepard became the first American in space.

MISSION DIARY: MERCURY REDSTONE 3

LIFTOFF MAY 5, 1961
SPACECRAFT NAME
FREEDOM 7
LAUNCH SITE PAD LC-5,
CAPE CANAVERAL
BOOSTER ROCKET
REDSTONE
CREW ALAN B. SHEPARD,
JR. (RIGHT)

MISSION OBJECTIVE
"TO DETERMINE MAN'S CAPABILITIES IN A SPACE ENVIRONMENT AND IN THOSE ENVIRONMENTS TO WHICH HE WILL BE SUBJECT UPON GOING INTO AND RETURNING FROM SPACE."

9:34 AM EST LIFTOFF FROM THE CAPE (RIGHT).
9:36 AM SHEPARD IS 25 MILES (40 KM) UP, TRAVELING AT 2,700 MPH (4,345 KM/H) AT 6G.
9:37 AM ENGINES SHUT DOWN AND BOOSTER SEPARATES FROM CAPSULE.
9:39 AM FREEDOM 7 REACHES PEAK ALTITUDE. THE CAPSULE AUTOMATICALLY TURNS AROUND, READY

FOR REENTRY.
9:49 AM SPLASHDOWN. A HELICOPTER LIFTS AN ELATED SHEPARD (RIGHT) FROM HIS SCORCHED CAPSULE.
10:30 AM PRESIDENT KENNEDY CALLS SHEPARD FROM WASHINGTON, D.C. A FEW DAYS LATER HE OFFERS SHEPARD HIS CONGRATULATIONS IN PERSON (FAR RIGHT).

MISSION STATS
MAX. ALTITUDE
116.5 MILES (186 KM)
ORBITS COMPLETED 0
DURATION
15 MIN., 28 SEC.
DISTANCE TRAVELED
303 MILES (487 KM)
MAX. VELOCITY 5,134 MPH (8,262 KM/H)
LANDING ATLANTIC OCEAN OFF FLORIDA
27°13.7' N, 75°53' W

GEMINI 1 & 2

Without the Gemini program, there would have been no Apollo. It was NASA's 12 Gemini missions that bridged the gap between single-person and multi-crew spacecraft and grappled with the technical problems of sending astronauts to the Moon. But before the twin-seat Gemini capsule could undertake crewed missions, its technology had to be tested. Geminis 1 and 2 verified that the new spacecraft was safe enough to send two people into orbit—and get them back again.

WHAT IF...

...A GEMINI CAPSULE HAD BEEN SENT TO THE MOON?

By the time the Gemini missions began, NASA was already developing the Apollo spacecraft that would land men on the Moon some seven years later. The Gemini program's task was to test procedures such as orbital rendezvous, docking and spacewalking, all of which would be needed for Apollo. But even before the first Gemini spacecraft flew, some engineers were calling for Gemini itself to fly to the Moon.

Several proposals to take Gemini to the Moon were drawn up by NASA planners and by engineers at McDonnell Douglas, the contractor for the spacecraft. Some of these plans were axed after only a couple of weeks, but others lasted longer. One idea was to send Gemini around the Moon aboard a Centaur rocket booster mated to its existing Titan 2 launcher. Some people within NASA saw this as a backup plan to beat the Russians to the Moon if the Apollo program was delayed.

During 1961, a group of engineers hatched an even more ambitious plan—to land a Gemini capsule on the Moon. The landing module was envisioned as a lightweight, open-cockpit spacecraft that would allow a single spacesuited astronaut to make a daring descent to the surface.

Later, following the 1967 Apollo 1 launchpad fire that killed three astronauts, the safety of the Apollo spacecraft came into question. NASA

The Gemini-derived lunar rescue module was one option NASA planned to use to recover stranded Apollo astronauts.

rated the chances of a crew being stranded in lunar orbit or on the surface of the Moon as dangerously high, and they began to study rescue schemes built around their tried-and-tested Gemini capsule.

In one plan, a Gemini spacecraft would be blasted into lunar orbit to rendezvous with the disabled Apollo. The stranded astronauts would then spacewalk to Gemini and crowd in together for the return to Earth.

Gemini was also proposed as a survival shelter that could be dispatched uncrewed to the Moon in an emergency. If an Apollo crew found themselves unable to lift off the surface, the two astronauts would walk to the Gemini shelter and await rescue by another Apollo craft. Another plan called for an enlarged Gemini that would land on the Moon, pick up the stranded astronauts and return them to Earth.

Gemini never did fly to the Moon, mainly due to lack of money and NASA's commitment to Apollo. But in hindsight, a Gemini Moonshot could have been achieved faster, at far less cost, than the Apollo program.

GEMINI DATA

	GEMINI 1	GEMINI 2
CREW	NONE	NONE
LAUNCH VEHICLE	TITAN 2	TITAN 2
LAUNCH WEIGHT	3.51 TONS	3.43 TONS
LAUNCH DATE	APRIL 8, 1964	JANUARY 19, 1965
LAUNCH COMPLEX	PAD 19, CAPE KENNEDY	PAD 19, CAPE KENNEDY
MISSION	THREE PLANNED ORBITS (BUT LASTED FOR 64)	18-MINUTE SUB-ORBITAL FLIGHT
TEST GOALS	LAUNCHER, LAUNCHER/SPACECRAFT COMPATIBILITY	ALL SYSTEMS, HEAT SHIELD, RECOVERY
HIGHEST ALTITUDE	200 MILES (321 KM)	107 MILES (172 KM)
RECOVERY DATE	NOT RECOVERED; BURNED UP APRIL 12, 1964	JANUARY 19, 1964, IN SOUTH ATLANTIC

MACHINE BEFORE MAN

The first two Gemini flights thundered off the launchpad in April 1964 and January 1965 without crew. In their place were instruments to record temperature, vibration, g force and other factors—and help make sure future launches were safe for astronauts.

Gemini 1, sitting atop its equally new two-stage Titan 2 launcher, reached space safely and was tracked as it orbited the Earth. Four days and 64 orbits after leaving Cape Kennedy, it burned up as planned over the South Atlantic Ocean. Now NASA knew that their new spacecraft could fly. Gemini 2's job was to determine whether it could return astronauts in one piece.

Gemini 2's flight was planned as a short, sub-orbital mission—a ballistic hop, similar to Alan Shepard's Freedom 7 flight, that would test all systems from launch to splashdown. After four weather-related delays and one false start, Gemini 2 was finally launched successfully on its 18-minute flight. The first and second stages of the Titan launcher separated without a hitch, and soon the capsule was firing its retro-rockets to begin the return to Earth. Gemini 2 was programmed to shoot through the atmosphere at much higher speeds than later flights to test its heat shield. It survived and splashed into the ocean, suspended from its single parachute.

The way was now clear for more complex crewed missions. Two months after Gemini 2 splashed down, astronauts Gus Grissom and John Young entered Earth orbit aboard Gemini 3 to begin practicing the spaceflight techniques that would one day be used to land men on the Moon.

TWIN TEST MISSIONS

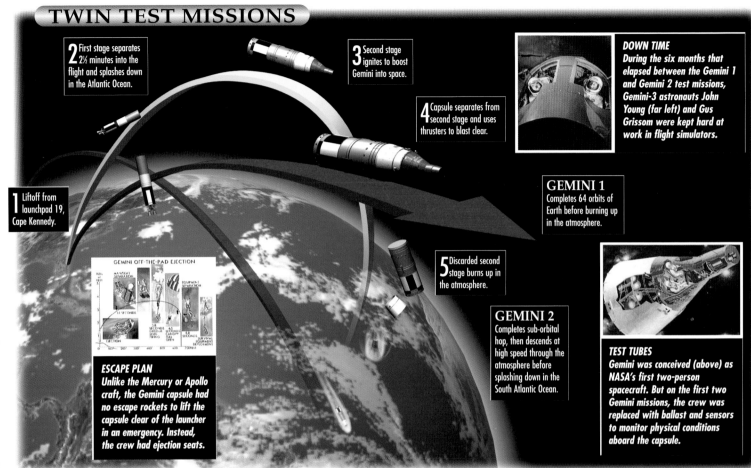

2 First stage separates 2½ minutes into the flight and splashes down in the Atlantic Ocean.

3 Second stage ignites to boost Gemini into space.

4 Capsule separates from second stage and uses thrusters to blast clear.

1 Liftoff from launchpad 19, Cape Kennedy.

5 Discarded second stage burns up in the atmosphere.

DOWN TIME
During the six months that elapsed between the Gemini 1 and Gemini 2 test missions, Gemini-3 astronauts John Young (far left) and Gus Grissom were kept hard at work in flight simulators.

GEMINI 1
Completes 64 orbits of Earth before burning up in the atmosphere.

GEMINI 2
Completes sub-orbital hop, then descends at high speed through the atmosphere before splashing down in the South Atlantic Ocean.

GEMINI OFF-THE-PAD EJECTION

ESCAPE PLAN
Unlike the Mercury or Apollo craft, the Gemini capsule had no escape rockets to lift the capsule clear of the launcher in an emergency. Instead, the crew had ejection seats.

TEST TUBES
Gemini was conceived (above) as NASA's first two-person spacecraft. But on the first two Gemini missions, the crew was replaced with ballast and sensors to monitor physical conditions aboard the capsule.

SAVING THE DAY

GEMINI 2 (RIGHT) WAS ORIGINALLY SCHEDULED FOR A DECEMBER 9, 1964, LAUNCH. THE COUNTDOWN RAN SMOOTHLY AND AT 11:41 THE TITAN LAUNCHER'S FIRST STAGE ENGINES BURST INTO LIFE— ONLY TO SHUT DOWN THREE SECONDS LATER. THE AUTOMATIC FAULT DETECTION SYSTEM HAD LOCATED A GLITCH IN ONE OF THE TITAN'S THRUSTER STEERING CHAMBERS AND SHUT DOWN THE ENGINES FASTER THAN ANY HUMAN COULD HAVE REACTED. GEMINI 2 WAS SAVED FOR ANOTHER DAY.

MISSION DIARY: GEMINI 1 & 2

GEMINI 1
MAY 21, 1963
GEMINI'S BRAND-NEW TITAN LAUNCHER IS COMPLETED.
JULY 5, 1963
TESTING OF GEMINI 1 BEGINS.
OCTOBER 4, 1963
GEMINI 1 IS DELIVERED TO CAPE KENNEDY IN FLORIDA.
MARCH 3, 1964
GEMINI 1 IS PLACED ON LAUNCHPAD.
APRIL 8, 1964
GEMINI 1 IS LAUNCHED (ABOVE). THE SPACECRAFT COMPLETES

THREE EARTH ORBITS AS PLANNED.
APRIL 12, 1964
AFTER A FURTHER 61 ORBITS, GEMINI 1 BURNS UP IN THE EARTH'S UPPER ATMOSPHERE.

GEMINI 2
AUGUST-SEPTEMBER 1964
HURRICANES DELAY GEMINI 2'S LAUNCH. ITS TITAN LAUNCHER IS REMOVED FROM THE LAUNCHPAD.
SEPTEMBER 21, 1964
THE GEMINI 2 CAPSULE ARRIVES AT CAPE KENNEDY. THE LAUNCH IS RESCHEDULED FOR LATE FALL.

OCTOBER 18, 1964
GEMINI 2 IS DISPATCHED TO THE LAUNCHPAD.
NOVEMBER 28, 1964
FINAL TESTS ON THE SPACECRAFT AND LAUNCHER ARE COMPLETED.
DECEMBER 9, 1964
SECONDS AFTER THE TITAN'S LAUNCHER'S ENGINES IGNITE, THE LAUNCH IS AUTOMATICALLY HALTED.
JANUARY 19, 1965
ON ITS SECOND ATTEMPT, GEMINI 2 LAUNCHES (LEFT). DURING ITS 2,000-MILE (3,218 KM) JOURNEY, AN ONBOARD CAMERA BELOW ITS THRUSTERS TAKES THE PICTURE OF EARTH (ABOVE).

VOSKHOD 1

October 13, 1964: Newspapers worldwide reported the flight of Voskhod 1, the first "space passenger ship." Only one-man spaceflights had taken place before Voskhod ("Sunrise") carried its three cosmonauts. But "space passenger ship" was not an accurate description. The Soviet Union had to upstage the forthcoming U.S. Gemini flights at almost any cost, and the three brave men were squeezed into a stripped-down single-seat Vostok capsule. They risked their lives for a mission whose sole purpose was propaganda.

WHAT IF...

...VOSKHOD WAS SCRAPPED?

Without Soviet leader Khrushchev's interference in his country's crewed space program in 1963 and his liking for space spectaculars, the Voskhod flights might never have taken place. The irony is that without such interference, it could have been a Soviet cosmonaut who claimed the ultimate goal of the space race: to take mankind's first steps on the surface of the Moon.

Khrushchev was obsessed with the public image of Soviet superiority in space, and in 1963, he saw the upcoming American Gemini flights as a threat to Soviet space leadership. Gemini had two objectives that worried Khrushchev: Each Gemini would have a crew of two astronauts, and on an early flight, one of the crewmembers would walk in space. Khrushchev wanted both of these objectives to be achieved first by the Soviet Union—at any cost.

Soviet space engineers worked hard during late 1963 and 1964 to ensure that Voskhod flew before Gemini, but in doing so, they diverted their efforts from the development of the new Soyuz—which was no propaganda vehicle but a real multi-person spacecraft. A key element in the Soviet lunar program, Soyuz eventually become the Soviet space workhorse and is still in use today.

The two Voskhod flights—the first in October 1964 and the second in March 1965—achieved their public relations purpose. The second mission even saw the world's first spacewalk, by cosmonaut Alexei Leonov. The first U.S. Gemini mission did not

March 23, 1965: The first Gemini launch. Had the Soviets not tried to beat Gemini into space with their own multi-seater craft, they could have pushed ahead with the Soyuz program— the equivalent of Apollo.

take place until after the Voskhod 2 crew was back on Earth. But the Soyuz program paid the price for the propaganda coup.

Without Voskhod, Gemini would have been the first space multi-seater. The Soviet leadership would have suffered a modest propaganda setback, but the Soviet space program would have been in better shape. Soyuz would have been very close behind Gemini, with a first flight some time in 1965. As things were, the initial Soyuz mission did not take place until April, 1967. Even then, the spacecraft was plagued by design and construction flaws that cost the life of the first Soyuz cosmonaut, Vladimir Komarov.

In real life, Komarov had commanded the Voskhod 1 mission. If Soviet resources had not been wasted on that publicity stunt, he might have made his first flight on a successful Soyuz. The Soviet Union would still have been in the race to the Moon.

By the time the Soyuz design had been debugged, the U.S. Apollo program was well underway. The Soviets lost their momentum, and it was American astronauts who first reached the surface of the Moon in 1969.

VOSKHOD 1 STATISTICS

MISSION	VOSKHOD 1 (CODE-NAME RUBY), 13TH CREWED SPACEFLIGHT
CREW	VLADIMIR KOMAROV (COMMANDER; RIGHT, CENTER)
	KONSTANTIN FEOKTISTOV (ENGINEER; NEAR RIGHT)
	BORIS YEGOROV (DOCTOR; FAR RIGHT)
LAUNCH	OCTOBER 12, 1964, 7:30 A.M. GMT ABOARD AN A-2 ROCKET
LAUNCH SITE	BAIKONUR LAUNCH COMPLEX 1
SPACECRAFT DIMENSIONS	16.4 FEET (5 M) LONG , 7.9 FEET (2.4 M) IN DIAMETER
MASS	11,731 LB (5,321 KG)
MAXIMUM ALTITUDE	208 MILES (335 KM)
NUMBER OF ORBITS	16

MISSION DURATION	1 DAY 17 MINUTES
RECOVERY	OCTOBER 13, 1964, 7:47 A.M. GMT

CLOSE QUARTERS

The Soviet Vostok craft had put the first men into space and firmly established the Soviet lead in the space race. When Vostok flights ended in 1963, the Soviets hoped to follow it with a larger spacecraft called Soyuz, capable of carrying two or more cosmonauts. But Soyuz development was behind schedule. Its first flight was not expected until 1965 at the very earliest.

Meanwhile, in the U.S., the Gemini program—which would place a 2-man crew in orbit—was forging ahead. Soviet leader Nikita Khrushchev was appalled by the idea of an American "first." He ordered his space officials to ensure that three cosmonauts would fly in space before the first Gemini liftoff.

Because Soyuz was nowhere near ready, Soviet space engineers had no choice but to modify the obsolete, single-seater Vostok. Without its ejection seat and reserve parachute system, there was room for three cosmonauts with two days' supplies to squeeze in sideways—but without spacesuits. The "new" spacecraft, in reality a dangerously overloaded Vostok, was called Voskhod.

On October 12, 1964, Voskhod 1 headed toward space from the same launch pad used by Yuri Gagarin on his historic flight three years earlier. Mission commander Komarov sat in the right-hand seat in front of the capsule's controls. The mission engineer, Konstantin Feoktistov, was in the left-hand seat and sitting between the two men, raised a few inches above them, was Boris Yegorov, the mission's doctor.

BARD NEWS

THE VOSKHOD CREW WAS BROUGHT DOWN BY THE WORDS OF WILLIAM SHAKESPEARE (RIGHT). WHEN THEY ASKED TO STAY IN SPACE FOR MORE THAN A DAY, CHIEF DESIGNER SERGEI KOROLEV QUOTED HAMLET: "THERE ARE MORE THINGS IN HEAVEN AND EARTH, HORATIO, THAN ARE DREAMT OF IN YOUR PHILOSOPHY." POSSIBLY A REFERENCE TO THE OVERTHROW OF PARTY SECRETARY KHRUSHCHEV, IT MAY HAVE BEEN KOROLEV'S WAY OF TELLING THEM THAT THE MYSTERIES OF SPACE MUST AWAIT FUTURE FLIGHTS.

THREE MEN IN A CAN

communication antenna

The hurriedly improvised Voskhod spacecraft was not built for comfort or safety. The three crewmembers had to squeeze into a space designed for one. And in order to fit in, the crew did not wear spacesuits and had no ejection seats. If anything had gone wrong, they would have had no means of escape.

command antenna

modified Vostok capsule, with three seats instead of one

external oxygen tanks

KONSTANTIN FEOKTISTOV
Technical scientist Feoktistov gained invaluable experience during his only spaceflight. He went on to help design spacecraft and space stations.

VLADIMIR KOMAROV
Flying his first space mission on Voskhod 1, Komarov was the only trained cosmonaut in the crew. He was killed in 1967 when Soyuz 1's parachutes failed to open for landing.

BORIS YEGOROV
The 27-year-old doctor was the youngest person aboard Voskhod 1. Yegorov returned to medicine and a successful career. He died of a heart attack in September 1994.

MISSION DIARY: VOSKHOD 1

MAY 1963 SPACE ENGINEERS BEGIN WORK ON SUCCESSOR CRAFT TO THE SINGLE-SEAT VOSTOK.

SEPTEMBER 1963 RAPID DEVELOPMENT OF THE VOSKHOD SPACECRAFT BEGINS (RIGHT).

MARCH 1964 A GROUP OF COSMONAUTS BEGIN THEIR TRAINING FOR THE VOSKHOD 1 MISSION.

JUNE 1964 YEGOROV, A DOCTOR, AND FEOKTISTOV, AN ENGINEER, JOIN THE CREW AS "SPECIALIST COSMONAUTS" JUST FOUR MONTHS BEFORE LAUNCH.

OCTOBER 6, 1964 LESS THAN A WEEK BEFORE THE ACTUAL FLIGHT ,THE UNMANNED COSMOS 47 MISSION TESTS A VOSKHOD SPACECRAFT.

OCTOBER 12, 1964 VOSKHOD 1 IS LAUNCHED (RIGHT) FROM BAIKONUR COSMODROME. NIKITA KHRUSHCHEV SPEAKS TO THE CREW WHILE ON VACATION AT A BLACK SEA RESORT. THIS WILL BE KHRUSHCHEV'S LAST PUBLIC STATEMENT: HE IS REMOVED FROM POWER DURING THE FLIGHT. THE CREW SENDS GREETINGS TO PARTICIPANTS AT THE TOKYO OLYMPIC GAMES. CREWMEMBERS YEGOROV AND FEOKTISTOV SUFFER FROM SPACESICKNESS.

OCTOBER 13, 1964 THE VOSKHOD 1 SPACECRAFT REENTERS THE ATMOSPHERE AND PARACHUTES BACK TO THE SURFACE, RETURNING THE CREW SAFELY TO EARTH.

DAY TRIPPERS

Since three cosmonauts were aboard what was really no more than a one-man spacecraft, the experiments carried out during the flight of Voskhod 1 were very basic. Yegorov carried out some simple medical examinations on his two colleagues. Komarov tested the spacecraft's control system and new ion thrusters, while Feoktistov used a special horizon sensor to test orbital navigation techniques. The three cosmonauts weren't alone in the crowded capsule—fruit flies and plants were carried to examine the effects of zero g on these life-forms. As Voskhod 1 passed over the Soviet Union, a television camera on board beamed pictures of the crew into Russian homes, and the cosmonauts used a hand-held camera to take snapshots of Earth's surface 208 miles (335 km) below them.

After a day in orbit, the spacecraft was positioned for its return home. The inexperienced Yegorov was frightened by the blaze created as Voskhod began to plunge through the atmosphere: He thought the craft had caught fire. Deceleration increased to a crushing 8 g. But at 16,000 feet (4,900 m), Voskhod's parachutes opened, and just before touchdown, small retro-rockets reduced the landing speed so effectively that the crew didn't feel touchdown. They only knew that they had landed because they could hear wheat stubble rustling against the hull of the capsule.

The cosmonauts had only been in space a day, but in that time Khrushchev had been ousted from power. When they returned to Moscow it was a new Soviet leader—Leonid Brezhnev—who welcomed them.

Although officially the mission's purpose was to test a new spacecraft and carry out scientific and medical experiments, in reality, Voskhod 1 was not much more than a propaganda mission. Its sole purpose was to beat Gemini into space, and allow the Soviet Union to maintain a psychological advantage over the Americans.

THE FIRST SPACEWALK

On March 18, 1965, Voskhod 2 cosmonaut Alexei Leonov crawled into an 8-foot (2.4 m) airlock tunnel far above the Earth. He floated out of the other end into open space, to hurtle around the planet as continents and clouds passed below. The spacewalk was the last Soviet space spectacular, carried out in haste to upstage the U.S. Gemini missions. Leonov's orbital adventure was risky and he was lucky to get back into his spacecraft alive. But survive he did—to go down in history as the first human to walk in space.

WHAT HAPPENED TO...

ALEXEI LEONOV

Alexei Leonov returned to the Earth aboard Voskhod 2 as a space hero. As the first human to step into orbit outside a spacecraft, he was made a Hero of the Soviet Union by the government and also received the Order of Lenin, common practice for all returning cosmonauts.

Two years after his spacewalk Leonov had another zero-gravity trip in his sights—a mission to the Moon. Although these plans were abandoned, Leonov did not give up hope of a Moon landing, and as Neil Armstrong trained on simulated lunar surfaces in the U.S., Alexei Leonov did the same in the Soviet Union.

But the Soviet government's enthusiasm for landing people on the Moon fizzled when the crew of Apollo 11 marked out their territory with a U.S. flag in 1969. With this particular round of the space race over, Leonov's last chance of reaching the Moon vanished. He was recruited to work on the Salyut space station program instead.

Leonov was due to fly to Salyut 1 aboard the Soyuz 11 spacecraft, but only days before launch, one of the other crew members became ill and the main crew, including Leonov, was replaced by backups. As it turned out, Leonov had a lucky escape—the backup crew died when an air valve failed in their spacecraft as they returned to Earth.

Leonov himself had earned a reputation as a survivor—he cheated death twice during his spacewalk and survived an assassination attempt on Premier Brezhnev when bullets showered their car. Despite the dangers involved, Leonov

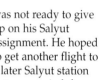
Between attending space shows around the world, Leonov has written books and cowritten the script for a Soviet science fiction movie.

was not ready to give up on his Salyut assignment. He hoped to get another flight to a later Salyut station but was again left grounded by technical difficulties.

The world had changed a lot by the time Leonov finally went into orbit again. Cooperation had replaced competition on the U.S. and Soviet Union space agenda, and Leonov was assigned commander of the Soviet half of the Apollo-Soyuz Test Project. This high-profile U.S. and Soviet docking mission launched in 1975. Leonov spent six days in space aboard Soyuz 19, which linked to a U.S. Apollo spacecraft for two days to carry out joint experiments and transfers. Leonov trained for the flight in Houston, where he became popular among U.S. astronauts thanks to his quick wit.

Soyuz 19 was to be Leonov's last spaceflight. After the mission, he rose through the ranks of cosmonaut training and was deputy director of the Gagarin Cosmonaut Training Center until his retirement in 1991. Leonov's diverse interests in art and writing still focus on space, and he will be happy to be remembered for his most dangerous moment—floating 300 miles (500 km) above the Earth, with a clear view from Africa to the Urals.

VOSKHOD 2 MISSION

SPACECRAFT	VOSKHOD 2, THE WORLD'S 14TH CREWED SPACEFLIGHT AND THE 8TH FROM THE SOVIET UNION	AIRLOCK DIMENSIONS	4 FT X 8 FT 4 IN (1.2 x 2.5M)
		AIRLOCK MASS	551 POUNDS (250 KG)
		SPACEWALK DURATION	24 MINUTES OUTSIDE CAPSULE, WITH 12 MINUTES SPENT OUTSIDE AIRLOCK IN OPEN SPACE
LAUNCH	MARCH 18, 1965, FROM THE BAIKONUR COSMODROME		
CREW	PAVEL BELYAYEV AND ALEXEI LEONOV	MISSION DURATION	1 DAY 2 HR 2 MIN
MAXIMUM ALTITUDE	310 MILES (499 KM)	LANDING	MARCH 19, 1965, IN NORTHERN RUSSIA
SPEED	17,400 MPH (28,000 KM/H)		

RISKY VENTURE

Cold War officials in the Soviet Union made cosmonaut Alexei Leonov's first historic steps in space sound like a walk in the park. A government news agency reported that Leonov "felt well" during his swim in space and on his return to the Voskhod capsule. Leonov apparently enjoyed good control over his movements thanks to a 50-foot (15-m) tether. To the U.S. government and the world at large, the first spacewalk was presented as an easy triumph for Soviet engineering.

In reality, Leonov had been doing anything but enjoying the view during his 24-minute space adventure on March 18, 1965. He nearly suffered heat stroke as he somersaulted around space, and spent several fraught minutes struggling to reenter Voskhod 2. Leonov recently revealed that he carried a suicide pill in case commander Pavel Belyayev was forced to leave him in orbit.

The politics of the Cold War made it vital to paint Leonov's maneuver as a total success. With the Gemini program, the U.S. threatened to seize the lead in the space race, and the Soviets were determined to meet or beat Gemini's goals, one of which was the first spacewalk. At short notice and with Soyuz still under development, Soviet engineers had to work with what they had. In the end, Leonov exited a modified Vostok capsule based on the design that took Yuri Gagarin to another first—first person in orbit—in 1961. Voskhod 2 was fitted with a Volga inflatable airlock to maintain cabin pressures and prepare Leonov for his spacewalk.

ART IN SPACE

ALEXEI LEONOV IS A TALENTED SELF-TAUGHT ARTIST AND TOOK COLORED PENCILS AND PAPER ABOARD THE VOSKHOD 2 MISSION. HIS UNIQUE SKETCHES OF THE COSMOS HAVE APPEARED IN A NUMBER OF EXHIBITIONS AND BOOKS, ALONG WITH MORE TRADITIONAL SCENES OF RUSSIAN CHURCHES AND SIBERIAN SNOW SCENES. ONE OF LEONOV'S MOST FAMOUS PAINTINGS IS A SELF-PORTRAIT OF THE FIRST SPACEWALK (ABOVE).

THE WALK OF LEONOV'S LIFE

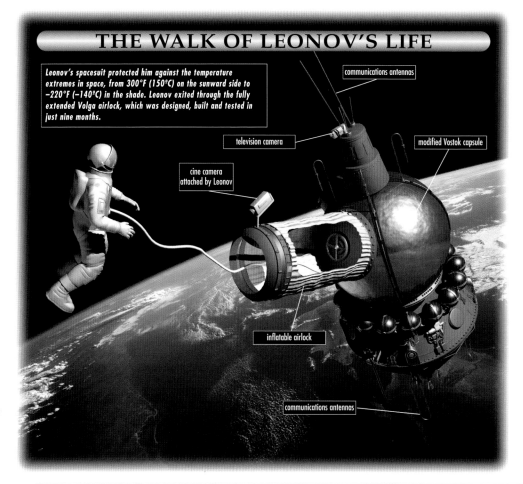

Leonov's spacesuit protected him against the temperature extremes in space, from 300°F (150°C) on the sunward side to −220°F (−140°C) in the shade. Leonov exited through the fully extended Volga airlock, which was designed, built and tested in just nine months.

communications antennas

television camera

modified Vostok capsule

cine camera attached by Leonov

inflatable airlock

communications antennas

MISSION DIARY: THE LONG MARCH

APRIL 1964
BELYAYEV AND LEONOV ARE SELECTED AS THE CREW FOR THE VOSKHOD 2 MISSION AND BEGIN TRAINING (RIGHT).
JUNE 1964
ENGINEERS BEGIN THE DESIGN OF AN AIRLOCK FOR VOSKHOD 2.
FEBRUARY 1965
COSMOS 57 IS LAUNCHED TO TEST THE VOSKHOD 2 MISSION BUT DISINTEGRATES IN ORBIT.
MARCH 18, 1965, 2:00 A.M. EST

VOSKHOD 2 IS LAUNCHED FROM THE BAIKONUR COSMODROME IN KAZAKHSTAN.
MARCH 18, 2:30 A.M.
THE INFLATABLE AIRLOCK IS EXTENDED FROM THE SIDE OF VOSKHOD 2.
MARCH 18, 2:45 A.M.
LEONOV ENTERS THE AIRLOCK FROM HIS SEAT IN VOSKHOD 2.
MARCH 18, 3:30 A.M.
AFTER THE AIRLOCK HAS BEEN PRESSURIZED, THE OUTER HATCH IS OPENED AND LEONOV FLOATS INTO OPEN SPACE.
MARCH 18, 3:42 A.M.

LEONOV BEGINS TO REENTER THE AIRLOCK (RIGHT) WITH DIFFICULTY.
MARCH 18, 3:49 A.M.
LEONOV IS SAFELY BACK INSIDE VOSKHOD 2.
MARCH 19, 1965
VOSKHOD 2 RETURNS TO EARTH BUT OVERSHOOTS THE LANDING AREA, ENDING UP IN A REMOTE SNOW-COVERED FOREST IN NORTHERN RUSSIA. THE CREW HAS A LENGTHY, COLD WAIT BEFORE THEY ARE RESCUED.

HUMAN SATELLITE

Leonov entered the airlock tunnel through Voskhod's hatch 300 miles (500 km) above the Pacific Ocean. There was a short wait while the airlock pressure was lowered to the space vacuum—if the pressures had not been equalized Leonov would have shot out like a cork from a bottle. As the airlock opened, Leonov floated headfirst into space.

The conditions inside Leonov's spacesuit recreated a little Earth atmosphere, with steady temperature and the correct levels of breathing gases. But without the luxury of gravity, Leonov found it hard to control the cord that tied him to Voskhod. After a poorly timed pull, Leonov crashed into the spacecraft, rocking his comrade.

Leonov spent 12 minutes floating freely in space before he was instructed to reenter the airlock. But with his spacesuit inflated like a balloon, Leonov could not bend his legs enough to climb back into the airlock. After several attempts, Leonov was forced to bleed some of the suit's air to make it less rigid. In doing so, he risked a dangerous attack of the bends. With his pulse racing, Leonov finally squeezed back into the airlock and closed the outer hatch. He was close to suffering heat stroke and reported that he was "up to the knees" in sweat.

More trouble awaited on Earth. A malfunction in Voskhod's systems forced Belyayev to make a manually controlled descent. The spacecraft overshot its designated landing site by over 700 miles (1,100 km), to land in a remote, snowy forest in northern Russia. The cosmonauts shivered for over two hours before a rescue helicopter arrived.

Whatever the difficulties, Leonov had made a landmark achievement—and most importantly, beaten U.S. astronaut Ed White to it by three months. Although the U.S. regained supremacy with the Apollo Moon landing and Neil Armstrong's giant leap, the first steps belong to Leonov.

GEMINI 8–12

W hen the Gemini program began in 1965, no one knew if humans could survive in space long enough to get to the Moon and back. No one had done an in-orbit rendezvous or a docking between two vehicles. And no one had made an EVA—a spacewalk. But to achieve a lunar landing, these feats would have to be routine. Early Gemini missions started to tackle the issue, but the last five got results. Geminis 8 through 12 saw the first dockings and first successful spacewalk—crucial steps in reaching the Moon.

INSIDE STORY

THE GEMINI ASTRONAUTS

T he last five Gemini missions tested the skills of the 10 astronauts on board to their limits. By the end of these missions, NASA had a pretty good idea which of the men were merely brilliant in space, and which of them could perform flawlessly. They all proved highly capable—so they all found themselves on the crews of the later Apollo missions.

Three of them ended up on the crew of Apollo 11. Neil Armstrong and Buzz Aldrin need little introduction. Michael Collins was originally chosen as Command Module pilot on Apollo 8, but a problem with his spine required surgery and kept him from the first mission to circle the Moon. Instead, he was appointed to the first Moon landing, and was unlucky enough to be the only member of the crew required to stay in orbit while his colleagues descended to the surface. As for Tom Stafford, he became Commander of Apollo 10, with Gene Cernan and John Young to accompany him. Young and Cernan went on to command Apollos 16 and 17. Meanwhile, Dave Scott flew on Apollo 9, and then commanded Apollo 15. Pete Conrad and Dick Gordon, the crew of Gemini 11, again flew together on Apollo 12. They had been best friends since they were at test pilot school together, so no one wanted to split them up. When Conrad was given the position of

Gemini 11 astronauts Pete Conrad (far left) and Dick Gordon later formed the crew for Apollo 12. The two had been best friends since they met at U.S. Navy test pilot school. Conrad went on to join one of the Skylab missions. Sadly, in the same month that Gordon was attending the 30th anniversary gala of Apollo 11 (July 1999), Conrad—the third man to walk on the moon—died in a motorcycle crash on his way to Monterey, California.

Commander there seemed no question that Gordon should go along too. He was assigned the position of Command Module pilot.

Not long after Apollo 12, Jim Lovell became almost as famous as Armstrong and Aldrin. As the Commander of Apollo 13, the whole world watched as he battled to bring his crew home in their damaged spacecraft. But he himself considered the highlight of his career to be Apollo 8, when he was one of the first three men to orbit the Moon and see the far side. Luck seemed to be ever present throughout Lovell's career: He was the one who replaced the incapacitated Collins. In total, seven of the 11 Apollo missions were commanded by veterans of the last five Gemini missions. That NASA had faith in its Gemini astronauts was beyond doubt.

GEMINI MISSIONS 8–12

	GEMINI 8	GEMINI 9	GEMINI 10	GEMINI 11	GEMINI 12
CREW	NEIL ARMSTRONG, DAVE SCOTT	TOM STAFFORD, GENE CERNAN	JOHN YOUNG, MICHAEL COLLINS	PETE CONRAD, DICK GORDON	JIM LOVELL, BUZZ ALDRIN
LAUNCH	11:41:02 A.M. EST, MAR. 16, 1966	8:39:33 A.M. EST, JUNE 3, 1966	5:20:26 P.M. EST, JULY 18, 1966	9:42:26 A.M. EST, SEPT. 12, 1966	3:46:33 P.M. EST, NOV. 11, 1966
MISSION HIGHLIGHT	FIRST DOCKING OF TWO VEHICLES IN SPACE.	CERNAN PERFORMED AN EVA.	DOCKED WITH AGENA ROCKET; COLLINS MADE A SPACEWALK.	SET WORLD ALTITUDE RECORD; PERFORMED A FULL ORBIT OF EARTH.	ALDRIN MADE THE MOST SUCCESSFUL SPACEWALK OF THE GEMINI PROGRAM.

WHAT'S UP, DOCK?

Neil Armstrong and Dave Scott were blasted into space aboard Gemini 8 on March 16, 1966. Their assignment was a space first—to dock with an Agena rocket circling the Earth. But the mission nearly became one of the worst disasters in U.S. space history.

Armstrong soon caught up with the target rocket and performed a flawless docking. But after flying around the world for 30 minutes attached to the rocket, the spacecraft began rolling and yawing simultaneously. Suspecting it was a problem with the Agena, Armstrong hurriedly undocked. The spacecraft, though, spun even faster. The Sun and Earth flashed dizzyingly past the window. If the problem wasn't fixed soon, Armstrong and Scott would lose consciousness and Gemini would break apart. The fault was in the reaction control system, but Armstrong had no way of knowing. Still, he was resourceful and cool-headed. Switching off the thrusters, he used the reentry control system to gain control of the spacecraft. Mission control instructed him to perform a deorbit burn. Without the main system, the flight would have to be aborted. So just 10 hours and 41 minutes after launch, Gemini 8 splashed into the Pacific.

Gemini 9 fared little better. This time the Agena failed on launch. An alternative docking target was launched. As Tom Stafford and Gene Cernan closed in, they saw that it, too, was faulty. The maneuver was abandoned. Cernan made his spacewalk as planned, connected to Gemini by nothing more than a 30-foot (9 m) umbilical cord. But with little training, he floundered around, quickly becoming exhausted, and the spacewalk was cut short.

SUCCESS AT LAST

The first spacecraft to make a completely successful docking was Gemini 10. Michael Collins and John Young circled the Earth for 39 hours attached to their Agena vehicle. A second difficult spacewalk was made by Collins. Gemini 11 also docked successfully with its Agena. Dick

HOOKING UP
The Agena Target Docking vehicle, photographed from Gemini 8 on March 16, 1966, is two feet (60 cm) away from the spacecraft's nose (bottom left). The Agena's instrument panel can be seen at the center of this picture.

IN COMMAND
Command Pilot John Young photographed from the Gemini 10 spacecraft. In Young's capable hands, this mission was the first to complete an entirely successful docking with the Agena vehicle.

SPACE STEPS
Buzz Aldrin during his EVA (extravehicular activity), the most successful spacewalk of the Gemini program, as snapped by the onboard Gemini 12 camera.

TUCKING IN
Technicians prepare to close the hatches on astronauts Tom Stafford (left) and Gene Cernan (right) before Gemini 9's liftoff. Backup astronauts Jim Lovell and Buzz Aldrin taped a humorous message (top center) to the spacecraft.

ORIGINAL CREW

THE ORIGINAL CREW FOR THE GEMINI 9 MISSION WERE CHARLIE BASSETT (LEFT) AND ELLIOT SEE (RIGHT). BUT BOTH WERE KILLED ON FEBRUARY 28, 1966, WHEN A COMBINATION OF BAD WEATHER AND PILOT ERROR SENT THEIR T-38 JET CRASHING INTO THE ROOF OF THE BUILDING THAT HOUSED THEIR SPACECRAFT. THEY WERE REPLACED ON GEMINI 9 BY BACKUP ASTRONAUTS TOM STAFFORD AND GENE CERNAN.

Gordon climbed out of the capsule and attempted the third spacewalk. But like all his predecessors, he kept floating away from the capsule. He was struggling so much that he overloaded the cooling system in his suit and his heart rate reached 180 beats per minute. So the spacewalk was cut short again. After a not-very-restful sleep period, the tired astronauts opened the hatch a second time. Gordon stood up to take some photographs of the Earth and its clouds. No one was in a hurry. Before he realized it, Conrad had fallen asleep. Waking up with a start, he yelled to Gordon, "Hey, Dick, would you believe I fell asleep?" For a reply he received a "Huh? What?" To his amazement, Gordon had fallen asleep, too, with his head sticking out of the capsule into outer space, as the Gemini hurtled around the Earth at 17,000 miles an hour (27,500 km/h).

Gemini 12 was the final mission in the program, and it suffered from a faulty radar. So while Jim Lovell flew the spacecraft, Buzz Aldrin dug out a set of rendezvous charts and manually calculated the maneuvers needed to dock with the Agena. It was an impressive display of skill, and Aldrin was later to make the Gemini program's most successful spacewalk. Admittedly, NASA had learned to train him astronauts underwater and to provide them with plenty of handrails and tethers in space. But Aldrin nevertheless demonstrated real ability. Few were surprised when he and the cool-headed Armstrong were chosen for the first Moon landing. Gemini had not only given NASA the chance to test its technical expertise, but it had helped the agency try out its astronauts, too.

MISSION DIARY: GEMINI 8–12

FEBRUARY 28, 1966
ELLIOT SEE AND CHARLIE BASSETT, THE TWO ASTRONAUTS ASSIGNED TO GEMINI 9, ARE KILLED WHEN THEIR T-38 JET CRASHES.
MARCH 16, 1966, 11:41:02 A.M. EST GEMINI 8 IS LAUNCHED FROM CAPE CANAVERAL (RIGHT).
MARCH 17, 1966
GEMINI 8 DOCKS WITH AN AGENA TARGET VEHICLE 185 MILES (298 KM) ABOVE THE EARTH, BUT AFTER HALF AN HOUR A THRUSTER ON THE GEMINI CAPSULE MALFUNCTIONS AND THE SPACECRAFT SPINS OUT OF CONTROL. AFTER REGAINING CONTROL, ARMSTRONG AND SCOTT BRING THE MISSION TO A PREMATURE END JUST 10 HOURS 41 MINUTES AFTER LAUNCH.

JUNE 3, 1966, 8:39:33 A.M. EST GEMINI 9 IS LAUNCHED.
JUNE 5, 1966
CERNAN MAKES AN EVA (RIGHT).
JUNE 6, 1966
GEMINI 9 SPLASHES DOWN.
JULY 18, 1966, 5:20:26 P.M. EST GEMINI 10 IS LAUNCHED.
JULY 21, 1966 GEMINI 10 SPLASHES DOWN.
SEPTEMBER 12, 1966, 9:42:26 A.M. EST GEMINI 11 IS LAUNCHED. THE SPACECRAFT DOCKS WITH THE AGENA.
SEPTEMBER 15, 1966 GEMINI 11 SPLASHES DOWN.
NOVEMBER 11, 1966, 3:46:33 P.M. EST GEMINI 12 IS LAUNCHED. NOVEMBER 15, 1966 GEMINI 12 SPLASHES DOWN.

THE PIONEERS

The first space travelers were dogs and chimpanzees. The very first living creature in orbit was the U.S.S.R.'s Laika, a husky cross found straying on a Moscow street. She survived four days aboard Sputnik 2, only the second craft to enter orbit, before running out of oxygen. The first human in space was cosmonaut Yuri Gagarin, who made just one orbit of the earth in April 1961. American efforts to catch up began more cautiously, with Alan Shepard making a sub-orbital flight in May 1961 before John Glenn became the first U.S. astronaut in orbit in February 1962. The so-called "Mercury 7" astronauts all flew in space solo before the more ambitious two-crew Gemini programme began, followed by the three-man Apollo missions that culminated in the Moon landings and Neil Armstrong's first steps on another world.

Although the Space Shuttle may seem to be a mature, rather than a pioneering space vehicle, it requires more piloting skills than have previous spacecraft. U.S. astronauts have traditionally come from military test pilot backgrounds, until recently ruling out female commanders. Eileen Collins became the first woman to command a US space mission in 1999 and was chosen to pilot the first Shuttle to fly following the 2003 Columbia disaster, confirming that the "Right Stuff" is not just a male attribute.

John Glenn clambers aboard Friendship 7 on 20 February, 1962. Previous launches had been postponed due to technical problems and bad weather, but this time Glenn finally reached space, becoming the first American to orbit the earth.

YURI GAGARIN

My God, he's got two daughters, how did he decide to do that? He must be crazy!" On April 12, 1961, these were the words with which Zoya Gagarin greeted the news that her brother Yuri had become the first man in space. Elsewhere, the reaction was more positive. Idolized by the world press, Yuri Gagarin became an icon of Soviet achievement. Gagarin died in 1968, but he still remains respected as one of the great pioneers of space exploration.

INSIDE STORY

THE FLIGHT OF VOSTOK 1

At 9:15 a.m. Moscow time on April 12, 1961, the Vostok 1 capsule burst into space with one man on board: Yuri Gagarin. Behind him, he left a plume of rocket exhaust plus the husks of four drop-away boosters. These boosters had helped get his Vostok rocket—basically a modified intercontinental ballistic missile—off the ground and on course for space. From Earth to orbit, the journey had taken just nine minutes.

Gagarin's surroundings were uncomfortable. Facing him in the cramped Vostok capsule were a camera, a small locker of food and a keypad whereby he could tap in the code for manual control of the spacecraft. The camera got in his way; the food was not needed; and the keypad was redundant—the flight was preprogrammed and only in the direst circumstance would manual control have been necessary.

The supposed silence of space was disturbed by the roar of machinery and by static from the open radio channel. But the view was magnificent. As he sped around the globe at over 17,000 miles (27,500 km/h) per hour, Gagarin peered through his tiny porthole. Down below was the Earth. "Am carrying out observations," he reported. "Visibility good. I can see the clouds. I can see everything. It's beautiful!"

Gagarin completed almost a full orbit of the Earth before his engines kicked him back into the atmosphere at 10:25 a.m. Here the trouble started. The capsule was weighted so that, once

The Vostok with Yuri Gagarin on board rockets into space—and history—that April morning in 1961 from the Baikonur Cosmodrome in Kazakhstan. The name Vostok means East.

its engines broke free, it would swing around to hit the atmosphere heat shield first. But the engines did not break free. They hung on, linked by stubborn cables, and the capsule began to spin. As the capsule spun, it presented its unprotected side to the firestorm of reentry. The temperature rose, and Gagarin could hear an ominous crackling from outside. Then, to Gagarin's great relief, the troublesome cables finally burned through and Vostok was able to right itself.

By now, however, the capsule had been twisted onto an entirely different flight path and was spinning so sickeningly that Gagarin nearly blacked out. But when he recovered, he could see blue skies through the porthole. He was almost home.

At 4.3 miles (7 km) above the Earth, the hatch blew and Gagarin was automatically ejected from the capsule. While the Vostok careened on its course, Gagarin parachuted safely down to Soviet soil and landed at Smelovaka, near Saratov. The trip took exactly 108 minutes and made him not only the first man in space but also the most famous person on Earth.

LIFE LINES: YURI GAGARIN

FULL NAME	YURI ALEXEYEVICH GAGARIN	**FAMILY**	MARRIED VALYA (VALENTINA) GORYACHEVA 1957; TWO DAUGHTERS
DATE OF BIRTH	MARCH 9, 1934		
PLACE OF BIRTH	KLUSHINO, RUSSIA	**CAREER**	TRAINEE ORENBURG PILOT'S SCHOOL
EDUCATION	PRIMARY AND SECONDARY SCHOOLS IN KLUSHINO AND GZHATSK; APPRENTICE AT MOSCOW'S LYUBERTSY STEEL PLANT 1950–1; STUDENT OF TRACTOR CONSTRUCTION AT THE SARATOV TECHNICAL SCHOOL 1951–5		1955–7; FIGHTER PILOT, BASED NEAR MURMANSK 1957–9; COSMONAUT 1961; DEPUTY DIRECTOR OF THE COSMONAUT TRAINING CENTER AT STAR CITY 1963–7; TEST PILOT 1968

PEOPLE'S HERO

On October 4, 1957, the Soviet Union launched the world's first artificial satellite. Two years later the Soviets decided to replicate their triumph—but this time the satellite would carry a human being. From a list of 2,200 eager applicants for the honor of becoming the first cosmonaut, a shortlist of 20 was selected for further training. Among them was a 25-year-old fighter pilot called Yuri Gagarin.

Born on March 9, 1934, Yuri Gagarin was not an obvious choice of cosmonaut. A farmworker's son from Smolensk, he had survived the German occupation of World War II and had trained as a foundry apprentice before joining tractor school in 1951. On graduation he had progressed to the Soviet air force, where he proved himself an enthusiastic MiG pilot with a taste for heavy g-forces.

Gagarin was shorter than average—he needed to sit on a cushion to get a clear view over the nose of his MiG—and had open, cheerful features. He married young and by the age of 25 he was already a father, living with his new family in standard Soviet military housing. Gagarin was unassuming, uncomplaining and optimistic. In short, he was a textbook example of a Soviet citizen.

Gagarin's flying experience had little to do with his inclusion on the shortlist. He had spent 252 hours and 21 minutes in the air. Of these, only 75 hours had been as a solo pilot. Other candidates had been airborne for some 1,500 hours. Instead, the selectors were swayed by his solid background, by his determination and, most prosaically, by his size—the Vostok capsule in which the successful cosmonaut would travel had very little legroom.

Gagarin's flight will stir people's imaginations as long as the Earth exists.
SOVIET WRITER KONSTANTIN PAUSTOVKSI

Gagarin was the first man in space (above) on April 12, 1961. His capsule, the Vostok (center), was designed by Sergei Korolev and his team. After his record flight, he was treated as a hero, awarded medals (right) and feted by Soviet leader Nikita Khrushchev (top right).

Gagarin and his family became icons of Soviet progress (above center) and his untimely death sent millions into mourning. His state funeral (above) was one of the biggest ever seen in Moscow.

ROCKET RIDE TO FAME

By the end of 1960, the 20-man squad had been whittled down to six. And by the following April there were two favorites: Gagarin and an equally short pilot named Gherman Titov. After an examination of each man's abilities and his ideological soundness—Titov sported a suspiciously bourgeois hairstyle—the decision was reached on April 7, 1961. Five days later, Gagarin hurtled into space and completed a 108-minute orbit of the Earth.

On his return he was showered with honors. A publicist's dream, he smiled photogenically, talked simply but expressively, and charmed heads of state around the world. Such was his popularity that a special department was created to handle his fan mail.

Fame took its toll, however. On October 3, 1961, he incurred severe head injuries by jumping from a second-floor window in a Crimean resort. Officially, he had hurt himself rescuing his daughter from the Black Sea. In fact, his wife had discovered him with another woman.

Gagarin continued to work on the space program, but his advisory capacities were nullified by rapid leaps in technology. And in 1967, when he criticized the disastrous launch of a Soyuz spacecraft—he had been back-up to the man who died aboard it—he was relieved of duties. He returned to the air force, and died in a plane crash in 1968. Ironically, he had been flying with an instructor. He left behind him a widow and two daughters.

SPY OR ALIEN?

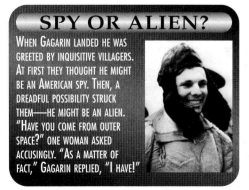

When Gagarin landed he was greeted by inquisitive villagers. At first they thought he might be an American spy. Then, a dreadful possibility struck them—he might be an alien. "Have you come from outer space?" one woman asked accusingly. "As a matter of fact," Gagarin replied, "I have!"

CAREER TIMELINE

1934 Born in the village of Klushino located 100 miles (160 km) west of Moscow.
1951 After a year's apprenticeship at Moscow's Lyubertsy Steel Plant, enrolls in the Saratov Technical School to study tractor construction.
1955 On graduation joins the Orenburg Pilot's School, where he makes his first MiG solo flight on March 26, 1957.
1957 Marries Valya Goryacheva. The following month graduates from Orenburg and is posted to Nikel airbase near Murmansk.

1960 Along with 19 other would-be cosmonauts, Gagarin is transferred to the Star City space base, 30 miles (48 km) northeast of Moscow.
April 12, 1961 Chosen as the world's first cosmonaut only five days previously, Gagarin becomes the first man in space. His colleague, Gherman Titov, makes the second Soviet spaceflight a few months later.
April 14, 1961 A rapturous reception from Nikita Khrushchev inaugurates several years as a national hero.
1963 To keep Gagarin out of trouble, he is appointed

Deputy Director of the Cosmonaut Training Center at Star City.
1967 Criticizes authorities when a Soyuz spacecraft crashes, killing its occupant, and is dismissed from the space team.
1968 Not having flown for five months, takes off with an instructor in a MiG-15. The plane crashes in poor weather. Gagarin is identified by a birthmark on a scrap of flesh from the back of his neck, and then given a hero's funeral.

GUS GRISSOM

Raised in small-town Indiana, Virgil "Gus" Grissom was a man of few words. One of NASA's "Original Seven" pioneering astronauts, his flying record spoke for itself. In July 1961, he made the second Mercury flight, a 15-minute sub-orbital trip. He then commanded the first crewed Gemini mission, and his performance on that flight marked him out as a future key player in the Apollo missions to the Moon. Appointed commander of the first Apollo mission, Grissom died along with his crew in a tragic training accident.

INSIDE STORY

ESCAPE FROM LIBERTY

With his flight aboard the Mercury capsule Liberty Bell 7, Gus Grissom became the third person in space, after Yuri Gagarin and Alan Shepard. But he also won the dubious honor of being the first space traveler to encounter an emergency. The flight itself was flawless, as Grissom soared to an altitude of 118 miles (190 km), experiencing over five minutes of weightlessness. Splashdown in the Atlantic was just three miles (5 km) away from the predicted coordinates, 300 miles (482 km) out from the Florida coast. Grissom had already radioed the recovery helicopter crews to stand by while he completed his instrument checks. When he was done, he gave the signal for the helicopters to approach.

Procedure called for the lead helicopter to attach a line to the capsule, and lift it slightly out of the water. Then the astronaut in the capsule would press a plunger to activate a small explosive charge, blowing the hatch. As the recovery crew approached, Grissom pulled the safety catch on the plunger, and waited. "I was lying there minding my own business," he said later, "when I heard a dull thud." The hatch had blown, and sea water began pouring in. Grissom quickly realized he was in danger of going down with his ship. He wrenched off his helmet, hauled himself out of the capsule and swam away as fast as he could. But his troubles were still not over. In

A rescue helicopter strains against the weight of Grissom's flooded Mercury capsule. The pilot, Jim Lewis, risked stalling his helicopter in the effort to rescue the stricken Liberty Bell.

the heat of the moment, he had left a valve open in his spacesuit and was losing buoyancy. The circling helicopters were churning up the water, making it difficult for Grissom to keep his head above the water. After much frantic waving, he was winched up by one of the helicopters. Liberty Bell was not so lucky. The lead helicopter had managed to attach a line to the vessel, but the sinking capsule was now far too heavy to lift. As their helicopter threatened to stall under the strain, the crew had to detach the line and watch as Liberty Bell sank beneath the waves.

Grissom was cleared of any blame, but the incident was never explained satisfactorily. In July 1999, a salvage team led by explorer Curt Newport managed to raise Liberty Bell from the Atlantic floor, three miles (5 km) down. But while the recovery and restoration of the capsule has provided space historians with many valuable artifacts, Newport's efforts could shed no light on the mystery of the blown hatch. Theories abound, but there is still no convincing explanation.

LIFE LINES

FULL NAME	VIRGIL IVAN "GUS" GRISSOM
DATE OF BIRTH	APRIL 3, 1926
PLACE OF BIRTH	MITCHELL, INDIANA
EDUCATION	MITCHELL HIGH SCHOOL CLASS OF 1944; BS, MECHANICAL ENGINEERING, PURDUE UNIVERSITY, 1950
FAMILY	MARRIED BETTY MOORE, JULY 1945; TWO SONS, SCOTT AND MARK
CAREER	FIGHTER PILOT, KOREAN WAR, 1951–2; U.S.A.F. TEST PILOT, WRIGHT-PATTERSON AIR FORCE BASE, OHIO, 1957–9; SELECTED FOR ASTRONAUT TRAINING, 1959; SECOND AMERICAN IN SPACE, ABOARD MERCURY FLIGHT LIBERTY BELL 7, JULY 1961; COMMANDER, FIRST MANNED GEMINI MISSION, MARCH 1965; COMMANDER, APOLLO 1, 1967
DATE OF DEATH	JANUARY 27, 1967, IN APOLLO 1 FIRE

FRONTIER MARTYR

At the end of World War II, Gus Grissom was discharged from the U.S. Air Force. At the age of 20, he found himself without qualifications and unemployed. He had married high-school sweetheart Betty Moore in the summer, while on leave, and money was tight. But Grissom had a burning ambition to be a fighter pilot. He earned a degree in engineering at Purdue University in Indiana, funded by his job in a diner, Betty's income, and a small government grant. By 1952, Grissom had achieved his ambition, and had flown 100 missions in the Korean War.

By 1959, Grissom was a respected test pilot, qualified to instruct. But then came the telegram from Washington, labeled "Top Secret," that changed everything. NASA, the new government agency in charge of advanced aviation and space projects, was looking for recruits. After two months of exhaustive physical and psychological testing, Grissom and six others made the final cut and were quickly dubbed the "Mercury Seven."

When the Soviet Union launched Yuri Gagarin into Earth orbit in April 1961, U.S. pride was on the line. Grissom served as backup with John Glenn for Alan Shepard's sub-orbital flight on May 5, 1961. Eleven weeks later, Grissom become the second American in space. After a 15-minute excursion beyond the atmosphere, his Liberty Bell 7 capsule splashed down in the Atlantic. Grissom's life was suddenly imperiled when a hatch unexpectedly blew and water flooded in, but he managed to swim to safety as the capsule sank.

The conquest of space is worth the risk of life.
GUS GRISSOM

Gus Grissom (right) was one of the first seven American astronauts (below) selected for the Mercury project. His second flight was with the first Gemini mission, which launched March 25, 1965 (left). He lost his life January 27, 1967, with the crew of Apollo 1 (mission patch, below left) when fire broke out in the training capsule (after the accident, below right).

DURING THE 5-HOUR FLIGHT OF GEMINI 3, ASTRONAUTS GRISSOM AND YOUNG (RIGHT) HAD SEVERAL EXPERIMENTS TO PERFORM. GRISSOM WAS DREADING ONE IN PARTICULAR: TESTING THE LATEST SPACE MEALS. DISTRIBUTION OF THE DEHYDRATED FOOD WAS YOUNG'S JOB, BUT WHEN THE TIME CAME, HE HAD A SURPRISE FOR HIS COMMANDER. "I WAS CONCENTRATING ON OUR SPACECRAFT'S PERFORMANCE," GRISSOM LATER SAID, "WHEN SUDDENLY JOHN ASKED, 'CARE FOR A CORNED-BEEF SANDWICH, SKIPPER?'" YOUNG HAD SMUGGLED HIS COMMANDER'S FAVORITE SNACK INTO SPACE, TO THE DISMAY OF MISSION CONTROL. BUT A GRATEFUL GRISSOM RATED THE SANDWICH "ONE OF THE HIGHLIGHTS OF OUR FLIGHT."

RETURN TO SPACE

When Alan Shepard was grounded and John Glenn left the space program, Grissom was penciled in to pilot the next generation spacecraft, the twin-seated Gemini. He commanded its first crewed outing, with John Young as his co-pilot. With the loss of Liberty Bell 7 still on his mind, Grissom named the Gemini craft Molly Brown, after the "unsinkable" survivor of the Titanic. Grissom became the first human to fly in space twice.

In March 1966, NASA confirmed Grissom's status as "first among equals" in the astronaut corps, by giving him command of the first Apollo mission. He began training with Gemini pilot Ed White and rookie Roger Chaffee for an Earth-orbit test of the craft that was designed to take Americans to the Moon. On January 27, 1967, with launch just weeks away, fire broke out during a command module test at the Kennedy Space Center. Sealed into the mock capsule by a hatch that was difficult to open from the inside, Grissom, White and Chaffee were dead within seconds. At Grissom's funeral at Arlington National Cemetery, his casket was borne by the six surviving members of the Mercury Seven.

CAREER TIMELINE

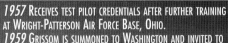

1926 BORN IN MITCHELL, INDIANA.
1944–5 SERVES IN THE U.S. AIR FORCE, BUT IS DISCHARGED AT THE END OF WORLD WAR II BEFORE RECEIVING FLIGHT TRAINING.
1946–50 STUDIES FOR A DEGREE IN MECHANICAL ENGINEERING AT PURDUE UNIVERSITY, INDIANA, AND REENLISTS IN THE U.S.A.F. AFTER GRADUATION.
1951 GRISSOM WINS HIS AIR FORCE PILOT WINGS (ABOVE). FLIES 100 COMBAT MISSIONS IN THE KOREAN WAR WITH THE 334TH FIGHTER-INTERCEPTOR SQUADRON. AWARDED DISTINGUISHED FLYING CROSS AND AIR MEDAL WITH CLUSTER.
1952–5 SERVES AS FLIGHT INSTRUCTOR TO AIR FORCE CADETS.

1957 RECEIVES TEST PILOT CREDENTIALS AFTER FURTHER TRAINING AT WRIGHT-PATTERSON AIR FORCE BASE, OHIO.
1959 GRISSOM IS SUMMONED TO WASHINGTON AND INVITED TO APPLY FOR ASTRONAUT TRAINING.
APRIL 9, 1959 AFTER INTENSIVE TESTING, GRISSOM BECOMES ONE OF THE FINAL SEVEN MERCURY ASTRONAUTS.
MAY 5, 1961 ALAN SHEPARD TAKES THE FIRST MERCURY MISSION, A 15-MINUTE SUB-ORBITAL FLIGHT.
JULY 21, 1961 GRISSOM IS LAUNCHED ATOP A REDSTONE ROCKET FOR THE SECOND AND FINAL SUB-ORBITAL MERCURY FLIGHT. THE MISSION GOES ACCORDING TO PLAN UNTIL SPLASHDOWN IN THE ATLANTIC OCEAN. GRISSOM HAS TO SWIM FOR SAFETY AFTER A CAPSULE HATCH BLOWS OPEN.

MARCH 1965 GRISSOM COMMANDS THE FIRST CREWED FLIGHT OF NASA'S 2-MAN GEMINI CAPSULE.
MARCH 1966 NAMED COMMANDER OF THE FIRST APOLLO MISSION.
JANUARY 27, 1967 GRISSOM AND CREWMATES ED WHITE AND ROGER CHAFFEE ARE KILLED IN A CAPSULE FIRE DURING TRAINING AT CAPE KENNEDY.
JANUARY 30, 1967 GRISSOM IS BURIED AT ARLINGTON NATIONAL CEMETERY, ARLINGTON, VIRGINIA (ABOVE).

JOHN GLENN

I n 1962, wartime fighter pilot John Glenn became the first American to orbit the Earth. His dramatic mission made him an international hero, but his new-found fame put a stop to his career as an astronaut. President Kennedy grounded him, believing that if Glenn died, the public would turn against the U.S. space program. But Glenn always believed that one day he would travel into space again, and 36 years later his belief was vindicated. In 1998, at age 77, he became the oldest man in space.

INSIDE STORY

THE FLIGHT OF FRIENDSHIP 7

Glenn's 9 ft-7 in by 6 ft-3 in capsule (left) returns to Cape Canaveral.

The launch countdown at Cape Canaveral had been repeatedly delayed for more than two hours, but John Glenn kept a cool head the morning of February 20, 1962. The astronaut was suited up and strapped into the Mercury space capsule he had named Friendship 7, and the sensors monitoring him picked up a relaxed pulse of 60 to 80 heartbeats per minute. Glenn was already familiar with the craft since this was the 11th launch attempt in two months.

But this attempt was different—it was successful. At 9:47 a.m., the mission announcer shouted, "Glenn reports all spacecraft systems go! Mercury Control is go!" The Atlas rocket eased Glenn and Friendship 7 off the ground, and as it accelerated toward orbit, the g-force pinned Glenn to the seat at six times Earth's gravity. Minutes later, he was weightless. To his relief, he found that he did not suffer from vertigo as had been predicted by some scientists.

When Glenn swung around the Earth and into his first orbital sunrise, he saw that the capsule was surrounded by thousands of particles glittering in sunlight refracted through the Earth's atmosphere. They were probably flecks of frozen coolant expelled from the capsule, but Glenn thought they looked like "little stars."

Meanwhile, at Mission Control, a sensor indicated that the capsule's heatshield was loose. If the shield fell off, Friendship 7 would burn up when it plunged back into the Earth's atmosphere. The shield was sandwiched between a rocket pack and a shock-reducing landing bag on the bottom of the capsule. Because the bag had come unlocked, only its straps held it and the shield to the capsule. The rocket pack's task was to slow Friendship 7 so that it reentered the atmosphere; the pack would then be jettisoned. NASA figured that if he kept the pack, it might just hold the heat shield in place.

As Glenn hurtled to Earth, he saw burning chunks of metal blowing by his small porthole. "That's a real fireball out there," Cape Canaveral heard over the radio. But what, everyone wondered, was burning: rocket pack or heat shield? Fortunately, it was only the rocket pack, and Glenn splashed down safely moments later. As he clambered out of the tiny capsule after a flight of under five hours, he could never have imagined it would take him more than three decades to get back into a spacecraft again.

LIFE LINES

Full Name	John Herschel Glenn, Jr.
Date of Birth	July 18, 1921
Place of Birth	Cambridge, Ohio
Education	Primary and secondary schools in New Concord, Ohio; Muskingum College, New Concord. Bachelor of Science in Engineering, plus nine honorary doctoral degrees
Family	Married Anna Margaret Castor, 1943; two children, John David and Carolyn Ann, and two grandchildren
Career	U.S. Marines pilot and instructor 1943–1956; test pilot 1956–59; NASA astronaut and spacecraft engineer 1959–1964; business executive 1965–1974; elected U.S. Senator 1974–1998; payload specialist, Shuttle mission STS-95 October 1998

SPACE SENATER

By the winter of 1961, the United States was losing the Space Race—the bitterly contested battle for space supremacy being fought between the U.S. and the Soviet Union. That spring, Soviet cosmonaut Yuri Gagarin had become the first person to orbit the Earth. Several months later, after a triumphant 17-orbit space flight by Soviet cosmonaut Gherman Titov, the score was: Soviet Union 18 crewed orbits, U.S. none.

These Soviet successes sent NASA scrambling to catch up, and on February 20, 1962, millions of people watched the television coverage of a 95-foot (29-m) Mercury-Atlas launch vehicle shooting into the sky and out of sight. The rocket's Mercury space capsule, Friendship 7, circled the Earth three times before reentering the atmosphere and landing in the Atlantic Ocean near Grand Turk Island, West Indies.

The occupant of the tiny capsule was Lieutenant-Colonel John H. Glenn, Jr., a U.S. Marine Corps pilot who had joined the NASA Space Task Group in 1959. Glenn, born in Cambridge, Ohio, on July 18, 1921, saw action as a fighter pilot in both World War II and the Korean War and went on to became a test pilot.

Scheduled for no further space flights after his historic Mercury mission, Glenn left NASA in 1964 and spent most of the next decade as a businessman. In 1974, he was elected to represent Ohio in the Senate. On Capitol Hill, Glenn took a strong interest in environmental issues and his technical expertise was often called upon when he served on committees, including the Senate Armed Forces Committee. He was elected to the Senate a record fourth consecutive time in 1992.

> *Too often people set their lives by the calendar. It takes the fun out of life.*
> JOHN GLENN

Friendship 7 (top) lifts off with John Glenn on board. He returned to Earth a national hero, which helped him launch a career as a senator (top left). The Shuttle Discovery (top center) blasts off in October 1998. On board, Glenn takes a picture of Earth (center). Before takeoff, his wife Annie snaps a photo of Glenn (above).

U.S. President Bill Clinton samples some of Glenn's space food (left). Glenn and his wife Annie (above) pose after his historic second spaceflight.

FIT FOR WORK

DRIVEN NO DOUBT BY THE DESIRE TO GET BACK INTO SPACE, JOHN GLENN, AT THE AGE OF 77, WAS STILL LIFTING WEIGHTS AND SPEED-WALKING TWO MILES EVERY DAY BEFORE HIS 1998 SPACEFLIGHT. THE FLIGHT ENDED A WAIT OF 36 YEARS. IT'S NO SURPRISE THAT HE WAS ALSO THE OLDEST ASTRONAUT EVER, BEATING STORY MUSGRAVE (WHO LAST FLEW IN 1996) BY 16 YEARS. WHILE GROUNDED, GLENN KEPT FIT, SERVED IN CONGRESS, AND UNDERWENT REGULAR CHECKS WITH NASA DOCTORS. THAT PERSISTENCE MAKES HIM THE LONGEST-STUDIED ASTRONAUT. NASA'S MEDICAL FILE ON HIM GOES BACK 42 YEARS, MORE THAN HALF HIS LIFE.

A 36-YEAR WAIT

For years, Senator Glenn lobbied for his own return to space and in 1998 his persistence finally paid off. NASA administrator Dan Goldin called Glenn "the most tenacious man alive" when he signed up the 77-year-old for the Shuttle mission STS-95 that fall.

On board the Shuttle Discovery, Glenn found that NASA technology had come a long way since the 1960s. During takeoff, he experienced less than half the gravitational force that he had on Friendship 7, and he had far more room in which to move around.

As the oldest astronaut ever, Glenn once again served as an orbiting guinea pig. Weightlessness affects the human body much as old age does, for example, with a loss of bone mass. Glenn, now going through both at once, wore an electrode cap while asleep to monitor his brainwaves, and had blood samples taken by fellow astronauts. The data collected during the mission was used in an ongoing study by the National Institutes of Health into the effects of aging.

But in one way, at least, 36 years haven't aged the hero. A day into the flight, Shuttle commander Curtis Brown observed, "Let the record show that John has a smile on his face and it goes from one ear to the other one, and we haven't been able to remove it yet." Glenn and the other crewmembers landed safely eight days later.

CAREER TIMELINE

1942 RECEIVES DEGREE IN ENGINEERING FROM MUSKINGUM COLLEGE, NEW CONCORD, OHIO.
1942 ENTERS NAVAL AVIATION CADET PROGRAM.
1943 ENLISTS IN THE MARINE CORPS.
1943–5 DURING WW2, GLENN FLIES 59 COMBAT MISSIONS IN F-4U FIGHTER PLANES.
1948–1950 ADVANCED FLIGHT INSTRUCTOR IN CORPUS CHRISTI, TEXAS.
1950–3 IN THE KOREAN WAR, GLENN FLIES A FURTHER 90 COMBAT MISSIONS. GLENN HAS SIX DISTINGUISHED FLYING CROSSES AND THE AIR MEDAL WITH 18 CLUSTERS FOR HIS SERVICE IN TWO WARS.

1957 FLIES AN F-8U CRUSADER FROM LOS ANGELES TO NEW YORK IN A RECORD TIME OF 3 HOURS 23 MINUTES. THIS IS THE FIRST TRANSCONTINENTAL FLIGHT TO AVERAGE ABOVE THE SPEED OF SOUND.
1956–9 HELPS DESIGN FIGHTER PLANES AT U.S. NAVY BUREAU OF AERONAUTICS.
1959 CHOSEN WITH SIX OTHERS (CARPENTER, COOPER, GRISSOM, SCHIRRA, SHEPARD AND SLAYTON) AS A MERCURY ASTRONAUT.
FEBRUARY 20, 1962 GLENN FLIES ON MERCURY 6, MAKING THE FIRST AMERICAN ORBITAL FLIGHT AND BECOMING THE FIFTH PERSON IN SPACE. HE MAKES THREE ORBITS DURING THE FIVE-HOUR FLIGHT.

1964 RESIGNS FROM SPACE PROGRAM.
1974 BECOMES U.S. SENATOR FOR OHIO.
OCTOBER 29, 1998 RETURNS TO SPACE ON BOARD THE SHUTTLE DISCOVERY AS A PAYLOAD SPECIALIST.
NOVEMBER 7, 1998 RETURNS TO EARTH ON DISCOVERY (RIGHT) AFTER A MISSION LASTING NEARLY NINE DAYS.

VALENTINA TERESHKOVA

A MARRIAGE MADE FOR THE MEDIA?

Nobody could deny Tereshkova's achievement —or courage. But her flight had more to do with sexist propaganda than with feminism. Given 1960s attitudes, the Soviets intended to show the world they were so far ahead of the Americans that they could put "even" a woman into space, and the Americans were just as irked as Soviet leader Nikita Krushchev had hoped.

Valentina Tereshkova, here flanked by Nikita Krushchev (right) and fiancé Andrian Nikolayev, was a propaganda triumph for the Soviet Union.

After Tereshkova's return, Krushchev saw yet another chance for publicity. When he learned that she was dating cosmonaut Andrian Nikolayev, the so-called "Iron Man" who had flown Vostok 3, he pressured the couple into marriage. On November 3, 1963, they obliged. The ceremony was held at the Moscow Wedding Palace and was followed by a lavish reception presided over by General Secretary Krushchev himself.

On June 8, 1964, Valentina gave birth to a daughter, but shortly afterward the "space family" fell apart. Rumors spread that the marriage was not merely a propaganda exercise but had a darker purpose—to see what effects space travel had on human reproductive capability. Whatever the truth, Tereshkova's daughter was perfectly healthy and showed no ill effects from her parents' exposure to cosmic radiation.

Following her one mission, Valentina Tereshkova was often in the public eye. She was made a Hero of the Soviet Union, received the Order of Lenin and hailed as an icon of feminism. In 1977 she was awarded the UN Gold Medal of Peace and 10 years later was given the Order of the Red Banner of Labor. She rose to the rank of Colonel-Engineer in the air force and later became a member of the Supreme Soviet.

But she never went back to space. Although a three-woman mission was discussed in 1965, it did not materialize. Instead, four years later, the female cosmonaut division was quietly disbanded. Tereshkova's flight was a token, a fact made clear by the March 1980 edition of Soviet Weekly which declared that "Life aboard spacecraft is extremely wearing, demanding and physically difficult...we feel the time is not yet right to impose such a strain on a woman." Two years later, though, when the U.S. announced its intention to send a woman into space, the Soviets quickly pre-empted with the first female cosmonaut since Tereshkova: Svetlana Savitskaya. In 1985, Savitskaya went on to become the first woman to walk in space.

On June 19, 1963, in the wilds of the Soviet republic of Kazakhstan, herdsmen on horseback found a young Russian woman calmly waiting. Despite her curious appearance—she wore a white helmet and bright orange overalls—the herdsmen hospitably offered bread, cheese and fermented mare's milk. She was eating heartily when a recovery helicopter arrived. Valentina Tereshkova had just spent almost three days in orbit: The first woman in space.

LIFE LINES

FULL NAME	VALENTINA VLADIMIROVNA TERESHKOVA
DATE OF BIRTH	MARCH 6, 1937
PLACE OF BIRTH	MASLENNIKOVO, YAROSLAVL, former U.S.S.R.
EDUCATION	GRADUATED SPINNING TECHNOLOGIST 1961; graduate ZHUKOVSKY AIR FORCE ENGINEERING ACADEMY 1969; CANDIDATE OF TECHNICAL SCIENCES 1976
FAMILY	MARRIED FELLOW COSMONAUT ANDRIAN NIKOLAYEV 1963 (DIVORCED 1982); ONE DAUGHTER, YELENA ANDRIANOVA
CAREER	TEXTILE WORKER 1955–62; JOINED THE COMMUNIST PARTY 1962; COSMONAUT 1963; MEMBER OF SUPREME SOVIET 1966; MEMBER OF THE SOVIET CENTRAL COMMITTEE 1974; ELECTED TO CONGRESS OF PEOPLE'S DEPUTIES 1989

STAR WOMAN

Born on March 6, 1937, in the village of Maslennikovo in the Yaroslavl Region, Valentina Tereshkova had a hard childhood. Her father, a tractor driver on a local collective, was killed during World War II. His widow brought up three children on her own. The family moved to the nearby city of Yaroslavl where Valentina quit school at the age of 16 to supplement the family income, first with a job in a tire factory and then, from 1955, as a skilled loom operator in the Krasnyi Perekop Cotton Mill.

She might have spent her life as a factory worker—had it not been for Yuri Gagarin. In 1961 he became the first man in space and inspired thousands of young Soviet citizens to apply for the chance of cosmonaut training. Valentina Tereshkova was one of them.

She never expected to succeed. True, she was fit, strong and a keen parachutist, but a diploma in cotton-spinning was not exactly a qualification for space. Luckily for her, qualifications mattered little to Soviet leader Nikita Khruschev. He had beaten the Americans by putting the first man in space. Now, he wanted to beat them again.

COSMONAUT CRASH COURSE

The Soviet Vostok capsule did not have to be flown; ground control attended to that. Vostok was really no more than an endurance test, and a woman could endure it at least as well as any man. The only part of the trip that

I have invaded their little playground.
VALENTINA TERESHKOVA ON BECOMING A TRAINEE COSMONAUT

After her flight, Tereshkova became an international celebrity. She was given numerous awards including a gold medallion from the British Interplanetary Society in 1964 (above). Just a few months later, she gave birth to a baby girl (below), an event of some interest to scientists investigating the physiological effects of space travel.

After months of training (above left), Valentina Tereshkova became the first woman in space in 1963. Even years after her historic flight, Tereshkova was given VIP treatment. On a visit to England, for example, she became one of the first women to sit at the controls of the supersonic Concorde airliner (top).

CAREER TIMELINE

MARCH 6, 1937 BORN IN THE RUSSIAN VILLAGE OF MASLENNIKOVO.
1953 LEAVES SCHOOL AND BEGINS WORK AT A TIRE FACTORY, BUT CONTINUES TO STUDY BY CORRESPONDENCE COURSE. TWO YEARS LATER SHE BECOMES A LOOM OPERATOR AT A COTTON MILL.
MAY 21, 1959 MAKES FIRST PARACHUTE JUMP AT YAROSLAVL AVIATION CLUB. LATER FORMS THE TEXTILE MILL WORKERS PARACHUTE CLUB.
1961 GRADUATES AS A COTTON-SPINNING TECHNOLOGIST AND BECOMES SECRETARY OF THE LOCAL KOMSOMOL (YOUNG COMMUNIST LEAGUE).
FEBRUARY 16, 1962 SELECTED AS ONE OF FIVE

TRAINEE FEMALE COSMONAUTS (RIGHT).
JUNE 16, 1963 BECOMES THE FIRST WOMAN IN SPACE ON BOARD VOSTOK 6.
NOVEMBER 3, 1963 MARRIES FELLOW COSMONAUT ANDRIAN NIKOLAYEV AND SEVEN MONTHS LATER, ON JUNE 8, 1964, GIVES BIRTH TO A DAUGHTER, YELENA ANDRIANOVA.
1964 ENTERS ZHUKOVSKY MILITARY AIR ACADEMY TO COMPLETE HER EDUCATION.
OCTOBER 1969 BECOMES A STAFF MEMBER AT THE

YURI GAGARIN TRAINING SCHOOL FOR COSMONAUTS. THE FEMALE COSMONAUT DETACHMENT IS DISBANDED.
1974 ELECTED TO THE PRESIDIUM OF THE SUPREME SOVIET AND BECOMES A GOVERNMENT REPRESENTATIVE, APPEARING AT NUMEROUS INTERNATIONAL EVENTS.
1982 AFTER A PROLONGED SEPARATION SHE IS FINALLY DIVORCED FROM ANDRIAN NIKOLAYEV.
1989 ELECTED TO CONGRESS OF PEOPLE'S DEPUTIES.
1990 AFTER THE COLLAPSE OF COMMUNISM TERESHKOVA FADES FROM PUBLIC LIFE. BUT SHE STILL TRAVELS AND LECTURES, VISITING ENGLAND IN 1995.

RIGHT STUFF

WELL BEFORE TERESHKOVA'S FLIGHT, 13 AMERICAN WOMEN HAD PASSED THROUGH THE SAME TESTS AND TRAINING AS NASA'S "MERCURY SEVEN." MOST (INCLUDING PILOT JERRI COBB, RIGHT) SCORED HIGHER THAN THE MEN. BUT BEFORE THE GROUP WENT FOR FINAL JET-TRAINING, NASA CHANGED ASTRONAUT REQUIREMENTS TO INCLUDE JET-PILOT EXPERIENCE. SINCE THE MILITARY PROHIBITED WOMEN FROM FLYING JETS, THAT WAS THE END OF THE PROGRAM. IN 1963 SENATE HEARINGS, ASTRONAUT JOHN GLENN TESTIFIED THAT IT "IS A FACT OF OUR SOCIAL ORDER" THAT WOMEN COULD NOT GO INTO SPACE. THAT SAME YEAR, TERESHKOVA REACHED ORBIT.

required experience was the parachute descent after reentry. And Valentina Tereshkova had made more than 100 jumps. Even better, she was also a sound communist.

In February 1962, she was invited to join four other women at the Cosmonaut Training Center. Tereshkova's progress was impressive. Within a year she had a grasp of navigation, geophysics and spacecraft construction, had passed rigorous physical tests and could even fly a jet.

She lifted off from Baikonur cosmodrome at noon on June 16, 1963, aboard Vostok 6 and orbited the Earth 48 times in tandem with another capsule, Vostok 5, carrying Valeri Bykovsky. The two cosmonauts came within three miles (5 km) of each other and regularly beamed TV pictures to the planet below. Tereshkova returned to Earth after two days, 22 hours and 50 minutes. Bykovsky followed her 2½ hours later. As Khruschev had intended, her flight stung American pride. Not only was she the first woman in space— she had been out there longer than all the U.S. astronauts put together.

JOHN YOUNG

More than 35 years after he partnered Gus Grissom on the first Gemini crew in 1965, John Young was still listed by NASA as an active astronaut. Following Gemini, Young took part in the final dress rehearsal for the first lunar landing, served as command module pilot on Apollo 10 and went back to walk on the Moon as commander of Apollo 16. As Chief of the Astronaut Office in the 1970s, he prepared the astronaut corps for Space Shuttle operations and piloted the Shuttle's first orbital test in 1981.

INSIDE STORY

CRATER CRAWLING

John Young is one of a select group of three astronauts who have traveled to the Moon twice. His first lunar voyage was as command module pilot on Apollo 10, the final test of the hardware and procedures designed to put Neil Armstrong on the Moon just two months later.

Apollo 10 lifted off from Kennedy Space Center on May 18, 1969, crewed by Young and fellow Gemini veterans Tom Stafford and Gene Cernan. Their mission was to rehearse every detail of a full lunar landing attempt, except for the final descent to the surface. The course of Apollo 10 was plotted to take the craft right over the projected target area for Apollo 11.

After a 3-day translunar flight, Apollo 10 settled into a roughly circular orbit, 68 miles (109 km) above the Moon's surface. Upon separation from the command module Charlie Brown, the Apollo 10 lunar module Snoopy became the first lunar lander to fly on its own around the Moon. Stafford and Cernan took Snoopy to within 10 miles (16 km) of the lunar surface. The Apollo 11 countdown had begun.

After the historic triumph of the first landing, John Young served as commander on the backup crew of Apollo 13, with Ken Mattingly and Charlie Duke. The established sequence of crew rotation meant that these three astronauts then became the prime crew of Apollo 16.

John Young did go back to the Moon, and this

During the Apollo 16 lunar landing in 1972, John Young (left) and Charlie Duke set up scientific equipment on the Moon's surface and drove more than 16 miles (26 km) on three separate geological traverses.

time he was going all the way. He took his first step in the Descartes Highlands of the Moon at 11:59 a.m. EST, April 21, 1972. Young and Duke suffered an early setback in their first period of extravehicular activity (EVA): John Young stumbled over a cable and damaged an experiment. But their excursion to Flag Crater on the Cayley Plains in the lunar roving vehicle was trouble-free. Young observed that the Plains had been misnamed, and were in fact "just craters on top of craters."

During the second EVA, Young and Duke's itinerary took them to a sampling area on Stone Mountain, where the astronauts found themselves more than 500 feet (150 m) above the valley floor, at the highest point reached by any Apollo moonwalker. The third, and longest, EVA took Young and Duke to North Ray Crater and "Shadow Rock." By the end of the mission, they had collected 208 lb (95 kg) of sample rocks, and spent more than 20 hours working outside the lunar module.

Young and Duke finally lifted off from the lunar surface on the evening of April 26, and rejoined Ken Mattingly in the command module Casper for the journey back to Earth.

LIFE LINES

FULL NAME	JOHN WATTS YOUNG
DATE OF BIRTH	SEPTEMBER 24, 1930
PLACE OF BIRTH	SAN FRANCISCO, CALIFORNIA
FAMILY	MARRIED BARBARA WHITE 1955 (DIVORCED); MARRIED SUZY FELDMAN 1972; TWO CHILDREN FROM FIRST MARRIAGE
EDUCATION	ORLANDO HIGH SCHOOL, ORLANDO, FLORIDA;. BS IN AERONAUTICAL ENGINEERING, GEORGIA INSTITUTE OF TECHNOLOGY, GRADUATED 1952
CAREER	SELECTED FOR ASTRONAUT TRAINING 1962; SIX SPACEFLIGHTS: GEMINI 3 (1965, FIRST CREWED FLIGHT OF GEMINI CRAFT), GEMINI 10 (1966), APOLLO 10 (1969), APOLLO 16 (1972, FIFTH LUNAR LANDING), STS-1 (1981, FIRST CREWED FLIGHT OF THE SPACE SHUTTLE), STS-9 (1983); CHIEF OF NASA ASTRONAUT OFFICE, 1974–87; SPECIAL ASSISTANT TO DIRECTOR OF JOHNSON SPACE CENTER, 1987–96; ASSOCIATE DIRECTOR (TECHNICAL), JSC, 1996 TO DATE

ROUND TRIP

The Mercury program was in full swing when John Young reported for astronaut selection tests in 1962. But NASA was looking much farther ahead: President Kennedy had set the U.S. the goal of a Moon landing before the end of the decade.

Nineteen candidates competed for a place in the U.S. aviation elite alongside the "Original Seven" astronauts. And when nine successful applicants were chosen, their names were broadcast: Armstrong, Borman, Conrad, Lovell, McDivitt, See, Stafford, White, Young. One of their number, Elliott See, was killed before his first spaceflight. But the names of the remaining eight are part of space history. Neil Armstrong was the first person to step onto the Moon. Jim Lovell flew two Gemini missions and two Apollo missions to lunar orbit. But John Young was the only member of the "New Nine" to fly two Gemini flights, walk on the Moon, and then pilot the next generation of spacecraft, the Space Shuttle.

Young's background—as a Navy test pilot and a holder of climb-to-altitude records flying Phantom jets—earned him a great deal of respect, and he was chosen to ride with Gus Grissom on the first orbital flight of Project Gemini in 1965. After three orbits, the crew made the first manually controlled reentry and splashed down in the Atlantic. The following year, Young accompanied Michael Collins on Gemini 10. Each

John Young's career in space began with the first Gemini mission (right) and ended with the Space Shuttle. But the highlight was the 1972 Apollo 16 mission (left), culminating in the lunar landing (below).

I don't do science. I just fly.
JOHN YOUNG

mission in the Gemini series was more technically challenging than the previous one. Collins and Young spent a busy three days in orbit, rehearsing maneuvers and docking procedures.

LUNAR LANDING

By May 1969, the Moon was within reach. Young traveled to lunar orbit as command module pilot of Apollo 10. His crewmates Tom Stafford and Gene Cernan took the lunar module Snoopy to reconnoiter Apollo 11 landing sites, while Young remained aboard Charlie Brown, circling the Moon at a distance of 68 miles (109 km). As commander of Apollo 16, three years later, Young made the round trip to the Moon a second time. But on this occasion, he was able to go the extra 68 miles. He spent nearly three days on the lunar surface, collecting scientific samples and data in three exhausting moonwalks. Apollo missions ended that year. But while many of his colleagues sought careers in business or other challenges, Young stayed at NASA.

Young became Chief of the Astronaut Office in 1974, and supervised the Apollo-Soyuz Test Project and the development of Space Shuttle crew procedures and training. As a tribute to his exceptional talents and his unwavering commitment to spaceflight, NASA gave John Young command of the maiden flight of the Space Shuttle, on April 12, 1981—the 20th anniversary of Yuri Gagarin's historic first journey into Earth orbit.

CAREER TIMELINE

1952 YOUNG JOINS THE U.S. NAVY.
1959 COMPLETES NAVY TEST PILOT TRAINING.
1962 AS A PILOT IN THE PHANTOM FIGHTER (RIGHT), YOUNG SETS WORLD RECORDS FOR THE FASTEST ASCENT TO 3,000 M (9,850 FEET) AND 25,000 M (82,000 FEET). SELECTED FOR ASTRONAUT TRAINING, WHICH INCLUDED SURVIVAL EXERCISES IN THE DESERT (FAR RIGHT).
MARCH 1965 YOUNG FLIES TO ORBIT WITH GUS GRISSOM ON THE FIRST CREWED GEMINI FLIGHT.
DECEMBER 1965 SERVES WITH GRISSOM AS BACKUP CREW FOR

GEMINI 6.
1966 FLIES HIS SECOND GEMINI MISSION, WITH MICHAEL COLLINS. GEMINI 10 SUCCESSFULLY DOCKS WITH ITS AGENA TARGET.
1968 MEMBER OF BACKUP CREW FOR APOLLO 7.
1969 YOUNG, TOM STAFFORD AND GENE CERNAN TRAVEL IN APOLLO 10 INTO LUNAR ORBIT. STAFFORD AND CERNAN TAKE THE LUNAR MODULE SNOOPY TO WITHIN 10 MILES (16 KM) OF THE MOON'S SURFACE.
1970 BACKUP FOR APOLLO 13.
APRIL 1972 COMMANDER OF FIFTH MOON LANDING MISSION. YOUNG AND CHARLIE DUKE LAND THE LUNAR MODULE ORION AND

SPEND ALMOST THREE DAYS ON THE SURFACE.
DECEMBER 1972 BACKUP FOR APOLLO 17.
1973 APPOINTED HEAD OF THE SPACE SHUTTLE BRANCH OF THE ASTRONAUT OFFICE.
1974–87 CHIEF OF THE ASTRONAUT OFFICE. YOUNG OVERSEES THE APOLLO-SOYUZ TEST PROJECT, AND 25 SHUTTLE MISSIONS.
1981 YOUNG COMMANDS THE SPACE SHUTTLE'S INAUGURAL FLIGHT.
1983 YOUNG'S SIXTH SPACEFLIGHT, AS COMMANDER OF STS-9.

EDWARD WHITE

T he son of a pioneer aviator, Ed White was a gifted and ambitious pilot who wanted to fly spacecraft. He joined the Gemini program in 1962, alongside Neil Armstrong and Jim Lovell. His skill and dedication soon made him one of NASA's brightest stars. In June 1965, he flew on the Gemini 4 mission and became a national hero when he executed the first spacewalk by an American. But White would never fly in space again: He lost his life in the horrific fire that engulfed the Apollo 1 capsule in January 1967.

INSIDE STORY

ED SKYWALKER

O n March 18, 1965, Soviet cosmonaut Alexei Leonov left his Voskhod capsule and floated in space for 10 minutes, attached to a 10-foot (3-m) tether. It hardly mattered that he almost didn't make it back into his ship: He had walked in space—becoming the first human to do so—and survived to return to Earth as yet another symbol of Soviet dominance in the space race.

The American response was to hurriedly revise the flight plan for the upcoming Gemini 4 flight, scheduled for June. Ed White was to perform the first extravehicular activity (EVA) by an American astronaut. In the weeks leading to the Gemini 4 launch, preparations were feverish. White's EVA suit and the hand-held maneuvering unit he would take with him into the vacuum were pulled off the drawing board and into production. White himself underwent intensive training for his new mission objective, spending hours wearing spacesuits in pressure chambers.

By the launch date, June 3, everything was set. The first American spacewalk was scheduled for Gemini 4's second Earth orbit, but when the time came, White and his commander Jim McDivitt asked for a delay. They had found themselves rushing through the EVA checklist, and felt they needed more time to prepare.

On the third orbit, the two astronauts pronounced themselves ready. After spacecraft

White uses a newly produced self-propulsion unit (gripped in his right hand) to maneuver during his 21-minute sojourn in the vacuum of space. He left the capsule over Hawaii, and was able to see all of Florida, Cuba and Puerto Rico by the time his spacewalk was over.

decompression, Ed White triple-checked his camera. "I wanted to make sure I didn't leave the lens cap on," he later said. Then he eased himself out of the hatch, experiencing no sensation of speed, despite traveling at a velocity of some 17,500 mph (28,000 km/h).

Tethered by an umbilical oxygen cord 25 feet (7.5 m) long, White maneuvered using the hand-held self-propulsion unit, but this quickly ran out of fuel, and movement became difficult. This caused White to collide gently with the Gemini craft, and McDivitt called out, "You smeared up my windshield, you dirty dog!"

In contrast to Leonov's hair-raising 10 minutes in space, the first American spacewalk played to an audience of millions tuning in to their radios and televisions—and the astronauts put on a great show. White summed it up by saying, "I'm very thankful in having the experience to be first...this is fun!"

LIFE LINES

FULL NAME	EDWARD HIGGINS WHITE II	FAMILY	MARRIED PATRICIA EILEEN FINEGAN; TWO CHILDREN, EDWARD AND BONNIE LYNN
DATE OF BIRTH	NOVEMBER 14, 1930		
PLACE OF BIRTH	SAN ANTONIO, TEXAS	CAREER	SELECTED BY NASA FOR ASTRONAUT TRAINING, 1962; PILOT, GEMINI 4, JUNE 1965; BACKUP COMMANDER, GEMINI 7, DECEMBER 1965; SENIOR PILOT, APOLLO 1
EDUCATION	WESTERN HIGH SCHOOL, WASHINGTON D.C., GRADUATED 1948; BACHELOR OF SCIENCE DEGREE, WEST POINT MILITARY ACADEMY 1948–52; AWARDED MASTER OF SCIENCE DEGREE (AERONAUTICAL ENGINEERING), UNIVERSITY OF MICHIGAN, 1959		
		DATE OF DEATH	JANUARY 27, 1967

BORN TO FLY

After his first and only spaceflight, in 1965, Ed White was interviewed for Life magazine. He articulated the simple motivation behind the imminent Apollo program: "A lot of us here on Earth are getting pretty curious about what the Moon's made of, and you'll never satisfy man's curiosity until a man goes himself." The quote illustrates the direct, can-do attitude that White inherited from his pilot father, who had flown balloons for the U.S. Army before powered flight became widely used. The family philosophy had long been to set personal goals and then to strive hard to achieve them. In order to achieve the goal of flying in space, Ed White did whatever he had to do.

After serving with the U.S. Air Force in Germany for three and a half years, White got an advanced degree in aeronautics to improve his credentials. In 1959, after he learned that the "Original Seven" Mercury astronauts were all test pilots, he joined the Air Force Test Pilot School at Edwards Air Force Base. After graduating, he was posted to Ohio, where he had his first contact with the space program. He later recalled, "I flew the big Air Force cargo planes through weightless maneuvers to test what happens to a pilot in zero gravity. Two of my passengers were John Glenn and Deke Slayton...other passengers of mine were Ham and Enos, the chimps that went up before the astronauts." In 1962, NASA selected nine astronauts—the "New Nine"— to join the Original Seven and fly the Gemini missions and, ultimately, the Apollo Moon missions. Ed White was among them.

Ed White's moment of glory, the flight of Gemini 4, (top center and right) was the culmination of years of training, including even geology field trips (center).

...I look forward a great deal to making the first flight. There's a great deal of pride involved in making a first flight.
EDWARD WHITE

The Apollo 1 crew members (below right and above, left to right) were Virgil "Gus" Grissom, an experienced Mercury astronaut; Ed White; and Roger Chaffee, a promising rookie. The report on the craft's burned-out shell (left) called for numerous changes.

PLASTERED

EACH OF THE GEMINI ASTRONAUTS HAD TO UNDERGO WEEKS OF RIGOROUS TESTS AND SIMULATIONS TO PREPARE FOR SPACEFLIGHT. BUT ONE OF THE MORE UNUSUAL, AND LESS STRENUOUS, OF THE PREPARATIONS WAS THE CREATION OF A PLASTER CAST OF THE ASTRONAUT'S HEAD TO MAKE SURE HIS HELMET LINING WOULD FIT PERFECTLY. ED WHITE IS SHOWN HERE, GETTING "PLASTERED." A CLOSE, COMFORTABLE FIT WOULD BE ESPECIALLY IMPORTANT FOR WHITE AS HE MANEUVERED HIMSELF IN THE VACUUM OUTSIDE THE GEMINI SPACECRAFT DURING HIS SPACEWALK.

STEPPING OUT

In June 1965, White got the flight he wanted—four days in Earth orbit aboard Gemini 4. With just a week to go before launch, White received confirmation of a change to the flight plan—he would perform an EVA, the first extravehicular activity by an American in space.

On Gemini 4's third Earth orbit, Ed White ventured outside the spacecraft and took photographs of the planet as he floated free above it at 17,500 miles an hour (28,000 km/h). On their return, Ed White and crewmate Jim McDivitt were awarded the Exceptional Service Medal by President Johnson. McDivitt went on to command Apollo 9, the first Earth orbit flight of the entire Apollo spacecraft.

But by then, Ed White had been lost. He had perished at the Kennedy Space Center in Florida, in the flash fire that began as an electrical spark inside an Apollo command module on January 27, 1967. In the sealed capsule, the fire was ferocious, fueled by coolant fumes and combustible foam padding. Ed White, Gus Grissom and Roger Chaffee were the first fatalities of the American space program. And in White, Apollo had lost one of the best men on the team.

CAREER TIMELINES

1952 UPON GRADUATION FROM WEST POINT MILITARY ACADEMY, ED WHITE JOINS THE U.S. AIR FORCE.
1953 AFTER FLIGHT TRAINING IN FLORIDA AND TEXAS, WHITE EARNS HIS PILOT'S WINGS.
1954–8 POSTED TO GERMANY, FLYING F-86 AND F-100 FIGHTER JETS. SUCCESSFULLY COMPLETES AIR FORCE SURVIVAL COURSE IN BAD TOLZ, GERMANY.
1959 RECEIVES MASTER'S DEGREE IN AERONAUTICAL ENGINEERING FROM UNIVERSITY OF MICHIGAN. ATTENDS TEST PILOT SCHOOL AT EDWARDS AIR FORCE BASE IN CALIFORNIA, GAINING TEST PILOT CREDENTIALS IN THE SAME YEAR.
1959–62 EXPERIMENTAL TEST PILOT AT WRIGHT-PATTERSON AIR FORCE BASE, OHIO.
1962 SELECTED BY NASA FOR ASTRONAUT TRAINING, AS

ONE OF THE GROUP NICKNAMED THE "NEW NINE."
JUNE 3, 1965 SPENDS FOUR DAYS IN EARTH ORBIT ABOARD GEMINI 4, ACCOMPANIED BY JIM McDIVITT (RIGHT, LEADING WHITE TO PAD 19, FROM WHICH GEMINI 4 WAS LAUNCHED). EXECUTES A 20 MINUTE SPACEWALK, THE FIRST BY AN AMERICAN.
JUNE 11, 1965 McDIVITT AND WHITE MEET PRESIDENT JOHNSON (RIGHT), AND PRESENT HIM WITH A PHOTOGRAPH OF WHITE'S HISTORIC SPACEWALK.

JANUARY 27, 1967 TRAINING FOR WHAT WAS TO BE THE FIRST CREWED APOLLO COMMAND AND SERVICE MODULE, WHITE AND HIS CREWMATES GUS GRISSOM AND ROGER CHAFFEE ARE KILLED IN AN ACCIDENT. SEALED INSIDE THE COMMAND MODULE, THEY ARE POWERLESS TO CONTAIN A FIRE STARTED BY AN ELECTRICAL FAULT AND FED ON THE HIGH-PRESSURE, OXYGEN-RICH MODULE ATMOSPHERE.
APRIL 5, 1967 WHITE'S WIFE PATRICIA ACCEPTS THE HALEY ASTRONAUTICS AWARD ON BEHALF OF HER LATE HUSBAND.

JAMES LOVELL

During his career as an astronaut, Jim Lovell scored some spectacular firsts. His Gemini 7 flight saw the first rendezvous between two spacecraft in Earth orbit. As Command Module pilot on the crew of Apollo 8, Lovell shared with Frank Borman and Bill Anders the distinction of being the first humans to leave Earth for another world. But Lovell holds one record he never wanted. As Commander of the aborted Apollo 13 lunar mission, Lovell became the only astronaut to travel to the Moon twice without landing on it.

INSIDE STORY

NASA GOES TO HOLLYWOOD

Tom Hanks played Lovell in the movie of the Apollo 13 story. Inspired to take up the cause of space exploration, Tom Hanks sent a letter to Congress appealing for funds for NASA programs.

Co-written by Jim Lovell and Jeffrey Kluger, Lost Moon was published in 1994, Apollo 11's 25th anniversary year. The first lunar landing, mankind's "giant leap," was justly celebrated in the text. But Apollo 13, was also great copy. It was a true tale of human courage in the face of impossible odds, with all the spectacular trappings of spaceflight, complete with a happy ending. Hollywood couldn't pass it up, and studios bid huge sums for the rights to Jim Lovell's account.

Imagine Films won the battle for the rights to the movie of the Apollo 13 story, with director Ron Howard in charge. But for Howard, the story really only came to life when he sat down with Jim Lovell in an office in Los Angeles and heard it from the man himself. Howard later wrote, "I realized the story would need no Hollywoodizing. Its power...lay in the detailed truth of this amazing adventure." In order to capture this "detailed truth" on film, Howard, his production team and his actors sought out the advice and recollections of many Apollo veterans. Apollo 15 commander Dave Scott was technical adviser on the set. Set designers used parts from the actual Apollo 13 Command Module, which had been disassembled and stored after the mission. And Howard even took his actors aboard NASA's weightlessness training plane, known to astronauts as the Vomit Comet.

Twenty-five years after Jim Lovell had hung up his spacesuit for the last time, he took part in a reenactment of his last and most dramatic homecoming from space. Howard had asked Lovell if he would take a cameo part in the last scene of Apollo 13. He found himself on the movie set, acting the role of Rear Admiral Donald C. Davis of the recovery ship *Iwo Jima*. It was the Admiral who had welcomed Lovell back to Earth in 1970. When director Ron Howard called "Action!" Lovell shook the hand of his younger self, played by Tom Hanks. The scene, and the film itself, was an affectionate tribute to an American hero. But Lovell didn't let himself get carried away. Eagle-eyed moviegoers will have noticed that, though playing a Rear Admiral, Lovell wore a Captain's uniform, the rank he held when he retired from the United States Navy in 1973—and so was entitled to wear.

Apollo 13 (the mission) was called "a successful failure"; Apollo 13 (the movie) was a hit. Jim Lovell may have lost the Moon, but his story continues to inspire yet another generation of young Americans to reach for the Moon—and beyond.

LIFE LINES

FULL NAME	JAMES ARTHUR LOVELL JR.
DATE OF BIRTH	MARCH 25, 1928
PLACE OF BIRTH	CLEVELAND, OHIO
EDUCATION	UNIVERSITY OF WISCONSIN 1946–8; U.S. NAVAL ACADEMY, BACHELOR OF SCIENCE 1952; TEST PILOT SCHOOL, PATUXENT RIVER, MARYLAND, 1958; AVIATION SAFETY SCHOOL, UNIVERSITY OF SOUTHERN CALIFORNIA 1961; ADVANCED MANAGEMENT PROGRAM, HARVARD BUSINESS SCHOOL 1971
FAMILY	MARRIED MARILYN GERLACH, 1952. CHILDREN: BARBARA, 1953; JAMES, 1955; SUSAN, 1958; JEFFREY, 1966
CAREER	SELECTED FOR ASTRONAUT TRAINING 1962; GEMINI 4 BACKUP CREW, JUNE 1965; GEMINI 7 CREW, DECEMBER 1965; GEMINI 9 BACKUP CREW, JUNE 1966; GEMINI 12 CREW, NOVEMBER 1966; APOLLO 8 COMMAND MODULE PILOT, DECEMBER 1968; APOLLO 11 BACKUP COMMANDER, JULY 1969; APOLLO 13 COMMANDER, APRIL 1970

MOON OR BUST

As a student in 1940s Milwaukee, Jim Lovell was obsessed with space. Inspired by the Moon voyages in the novels of Jules Verne, he experimented with model rockets propelled by gunpowder. Lovell attended the University of Wisconsin and then graduated from the U.S. Naval Academy at Annapolis in 1952. After four years at the Naval Test Pilot School in Maryland, he found himself on the shortlist for the U.S. space program.

NASA was recruiting test pilots to support and succeed the original seven Mercury astronauts. In 1962, Lovell was selected for the second astronaut group, or the "New Nine." His first spaceflight came in December 1965, alongside Frank Borman in Gemini 7. The two astronauts endured two weeks aboard a craft dubbed "a flying men's room" due to its unsanitary conditions. The flight set a new spaceflight duration record, but its real achievement came 11 days into the mission. Gemini 6, piloted by Walter Schirra and Tom Stafford, matched orbit with Gemini 7 and approached within two feet. The crews waved at each other across the vacuum of space.

When astronauts Elliot See and Charlie Bassett were killed in a jet crash, Lovell inherited the last Gemini flight with rookie Buzz Aldrin. Their Gemini 12 mission in November 1966 was most notable for Aldrin's successful 5.5-hour EVAs solving problems that had dogged

Lovell was one of NASA's most experienced astronauts. He flew the record-breaking Gemini 7 mission (near left) and the final Gemini, Gemini 12 (below, center). Ironically, he is remembered for Apollo 13 (launch, far left), his only failure. With Fred Haise and Jack Swigert (bottom center), Lovell regained control of the craft (left) and returned to Earth (right). Later, he read about his ordeal in the papers (below, right) and was greeted as a hero by President Nixon (bottom right).

Apollo 13 is a significant addition to the knowledge of what human beings are capable of.
JAMES LOVELL

previous spacewalkers. By the end of 1968, NASA was under pressure to send a crew to circumnavigate the Moon before the Soviets. Lovell and Bill Anders, with Frank Borman commanding, got the job.

SO NEAR, SO FAR

Apollo 8 was launched atop the first Saturn V rocket to carry passengers. Lovell, Anders and Borman braked into lunar orbit on Christmas Eve. As millions watched on television, the astronauts read passages from Genesis while the stark moonscape rolled below. Looking back at the home planet, Lovell said, "the Earth from here is a grand oasis in the big vastness of space." On Christmas Day, after 10 Moon orbits, Apollo 8 headed home.

Jim Lovell would leave his "grand oasis" once again on April 11, 1970. Lovell had been Neil Armstrong's backup for the first lunar landing. Now he had his own mission, with rookies Fred Haise and Jack Swigert for company. But after an oxygen tank in the Service Module exploded, Apollo 13's primary mission objective became survival. Using the Lunar Module Aquarius as a lifeboat, Lovell and his crew swung around the Moon on a "free return" path. After a cold trip home, the exhausted astronauts splashed down to a hero's welcome. Lovell has claimed he wouldn't change a thing about his career, even the events of April 1970. "Apollo 13 was a test pilot's mission," he said. He came through the test with distinction.

SHADOWING

In April 1967, Astronaut Office chief Deke Slayton gathered 18 astronauts and told them, "The guys who are gonna fly the first lunar missions are...in this room." When Jim Lovell was assigned to Apollo 8 (right), the first circumlunar mission, he knew his backup crew—Neil Armstrong, Buzz Aldrin and Fred Haise—were in line for the first landing. But Haise had to make way for Michael Collins on Apollo 11, joining Lovell and Bill Anders as backup. As commander of the crew-in-waiting, Lovell, like Armstrong, trained to be the first man on the Moon.

CAREER TIMELINE

1958–62 Spends four years as a test pilot at the Naval Air Test Center, Patuxent River, Maryland. Serves as development program manager for the F4H Phantom jet fighter.

1962 After a year at the Aviation Safety School, Lovell is selected for astronaut training as a member of the "New Nine."

June 1965 First crew assignment, teaming up with Frank Borman to back up Gemini 4 astronauts Ed White and Jim McDivitt.

December 1965 Lovell and Borman (above right) orbit Earth for 14 days on Gemini 7. The mission sees the first-ever spacecraft rendezvous, as Walter Schirra and Tom

Stafford close to within two feet aboard Gemini 7.

June 1966 Another backup assignment, this time paired with Buzz Aldrin, on Gemini 9.

November 1966 Second spaceflight, lasting just four days, with Buzz Aldrin on Gemini 12. Lovell's first command.

December 1968 Lovell's third flight makes history. First crewed launch of the Saturn V rocket sends Apollo 8 on course for a rendezvous with the Moon. Lovell spends Christmas in lunar orbit with Frank Borman and Bill Anders.

July 1969 As backup to Neil Armstrong, commander of the first Moon landing, Lovell trains to lead his own mission to the lunar surface.

April 1970 Apollo 13 lifts off from Florida, bound for the Fra Mauro highlands of the Moon. An oxygen tank explosion 200,000 miles (322,000 km) into the flight forces the crew to abandon the landing. They use the Moon's gravity to swing their stricken ship back to Earth.

1973 Retires from Navy as a Captain.

1994 Publishes a book about the Apollo 13 drama, Lost Moon. The following year, it becomes the Hollywood movie, Apollo 13.

NEIL ARMSTRONG

A short hop off the bottom rung of his lunar lander's ladder, and Neil Armstrong was firmly on the Moon's surface—the first human ever to stand upon another world. "That's one small step for man, one giant leap for mankind," Armstrong radioed to Mission Control. The words were his own, not the product of NASA's public relations office. Armstrong messed up his delivery a little—he had meant to say "a man." But the line typified an astronaut who said less so that he could think more.

INSIDE STORY

HOW THE EAGLE LANDED

The final landing of the lunar module Eagle was so dramatic that it nearly didn't happen at all. Neil Armstrong and fellow astronaut Buzz Aldrin had to work around a balky computer, a pessimistic fuel gauge, and immense pressure to succeed.

In this two-minute excerpt from NASA transcripts, Eagle has just passed through the point of no return. Armstrong pilots the module, Aldrin reads altitude and vertical and horizontal speed in feet per second from the navigational computer, and capsule communicator (Capcom) Charles Duke at mission control in Houston, Texas, reads off the remaining fuel in seconds.

Armstrong: Okay, how's the fuel?
Aldrin: Take it down.
Armstrong: Okay. Here's a…looks like a good area here.
Aldrin: I got the shadow [of the lunar module on Moon surface] out there. […]
Armstrong: Going to be right over that crater.
Aldrin: Two hundred feet [altitude]. Four and one half [feet per second] down.

Aldrin: Five and one half down.
Armstrong: I got a good spot.
Aldrin: One hundred sixty feet, six and one half down.
Five and one half down, nine [feet per second] forward. You're looking good.
One hundred twenty feet.
One hundred feet, three and one half down, nine forward.
Five percent [of fuel left]. [Low fuel] Quantity light.
Okay. Seventy-five feet. It's looking good. Down a half, six forward.
Duke: Sixty seconds [of fuel remaining].
Aldrin: [Fuel] Light's on.
Sixty feet, down two and one half.
Two forward. Two forward. That's good.
Forty feet, down two and one half. Picking up some dust.
Thirty feet, two and one half down.
Four forward. Four forward. Drifting right a little. Twenty feet [altitude], down a half.
Duke: Thirty seconds.
Aldrin: Drifting forward just a little; that's good.
Contact light.
Armstrong: Shutdown.
Aldrin: Okay. Engine stop. […]
Duke: We copy you down, Eagle.
Armstrong: Engine arm off.
Houston, Tranquillity Base here. The Eagle has landed.

The astronauts spent 2½ hours outside Eagle on the lunar surface, where they collected rock samples and set up scientific experiments.

LIFE LINES

FULL NAME	NEIL ALDEN ARMSTRONG
DATE OF BIRTH	AUGUST 5, 1930
PLACE OF BIRTH	ANGLAIZE COUNTY, OHIO
EDUCATION	STUDENT AT PURDUE UNIVERSITY 1947–8, 1952–5; BS IN AERONAUTICAL ENGINEERING 1955; MS IN AEROSPACE ENGINEERING, UNIVERSITY OF SOUTHERN CALIFORNIA 1970; SEVERAL HONORARY DOCTORATES
FAMILY	MARRIES JANET SHEARON; TWO CHILDREN
CAREER	U.S. NAVY PILOT 1948–52; TEST PILOT FOR THE NATIONAL ADVISORY COMMITTEE ON AERONAUTICS, LATER PART OF NASA 1955–62; NASA ASTRONAUT AND AERONAUTICS ADMINISTRATOR 1962–71; PROFESSOR OF AERONAUTICAL ENGINEERING, UNIVERSITY OF CINCINNATI 1971–9; CURRENTLY CHAIRMAN OF AIL SYSTEMS

LUNAR LANDER

Even among a team of astronauts selected for their courage, fast reactions and coolness under pressure, the first man to set foot on the Moon stood out. It seemed Neil Armstrong had no sense of fear.

Once, Armstrong and co-pilot Edwin "Buzz" Aldrin were practicing the Moon landing in the Lunar Module simulator. The lander began to spin out of control. As it hurtled toward the virtual Moon surface, Armstrong never punched the abort button. They crashed.

Aldrin thought Armstrong had frozen. Aware of NASA's fact-tallying mentality, he worried that the "crash" would be a strike against the two of them. Later, Armstrong said his decision was deliberate. He wanted to test the reactions of ground control—and himself.

From childhood, Armstrong set high standards for himself, and reached them. Like many boys growing up in the 1940s, he built model airplanes. But he tested his in a homemade wind tunnel. Like many teenagers, he got his first learner's permit at 16, but his was for an airplane, not the family car.

SUCCESS IN SPACE

Armstrong flew fighters in the Korean War, and later the X-15 rocket plane, before NASA selected him for astronaut training in 1962. He was in space less than four years later.

That first mission was a dramatic success. Before mission control had even given the go-ahead, Armstrong and his co-pilot performed the first-ever space docking. Half an hour later, they managed to stop their now out-of-control capsule from shaking them to pieces.

After the mission, Armstrong rejoined the astronaut duty roster. His name, along with that of Aldrin and Michael Collins, came up next on the list for the Apollo 11 mission— originally scheduled to be the second landing trip, with Armstrong due to be the third man on the Moon. But when NASA realized that the lunar module would not be ready in time, the plans were moved back. Apollo 10

> *It suddenly struck me that that tiny pea, pretty and blue, was the Earth.*
> NEIL ARMSTRONG

The Saturn 5 rocket blasted Apollo 11 into space on the morning of July 16, 1969 (far left). Four days later, the lunar module Eagle (left) carried Neil Armstrong (top left) and Buzz Aldrin down to the Moon's surface, where they planted the American flag (center). The command module Columbia returned safely to Earth on July 24 (below).

The Apollo 11 crew were put into quarantine (above) when they returned to Earth. Their ticker-tape parade through New York City was the biggest in history (right).

ARMSTRONG DESCRIBED THE SURFACE OF THE MOON AS "FINE AND POWDERY." HE CONTINUED, "I CAN KICK IT UP LOOSELY WITH MY TOE... IT DOES ADHERE IN FINE LAYERS, LIKE POWDERED CHARCOAL, TO THE SOLE AND SIDES OF MY BOOTS. I ONLY GO IN [TO THE SURFACE] A SMALL FRACTION OF AN INCH, MAYBE AN EIGHTH OF AN INCH, BUT I CAN SEE THE FOOTPRINTS OF MY BOOTS AND THE TREADS IN THE FINE, SANDY PARTICLES."

MOON PLAQUE

THE DESCENT STAGE OF THE APOLLO 11 LUNAR MODULE, WHICH REMAINED BEHIND ON THE MOON, CARRIED A COMMEMORATIVE PLAQUE. THE PLAQUE SAID, "HERE MEN FROM THE PLANET EARTH FIRST SET FOOT UPON THE MOON. JULY 1969 A.D. WE CAME IN PEACE FOR ALL MANKIND." IT ALSO BORE THE SIGNATURES OF NEIL ARMSTRONG, HIS FELLOW CREW MEMBERS MICHAEL COLLINS AND BUZZ ALDRIN, AND U.S. PRESIDENT RICHARD NIXON.

rehearsed every part of the complex Moon mission except the landing itself.

Its smooth success cleared the way for Apollo 11. On July 20, 1969, Armstrong fulfilled U.S. President John F. Kennedy's eight-year-old, $26 billion pledge to put a man on the Moon. Modestly, the astronaut insisted that the thousands of Apollo support personnel be credited, too: "It's their success more than ours."

Two years later, Armstrong left NASA to become a professor of aerospace engineering, and in the late '70s he went into business. Today, he lives in semi-retirement on his Ohio farm.

CAREER TIMELINE

1948–52 U.S. NAVY PILOT. FLIES **78** MISSIONS DURING THE KOREAN WAR. EJECTS TO SAFETY WHEN HIS PLANE STRIKES A TRAP WIRE STRETCHED OVER A VALLEY. RECEIVES THREE MEDALS FOR BRAVERY.
1955 RECEIVES BACHELOR'S DEGREE IN AERONAUTICAL ENGINEERING, PURDUE UNIVERSITY.
1955 MADE TEST PILOT, NATIONAL ADVISORY COUNCIL OF AERONAUTICS, EDWARDS AIR FORCE BASE, CALIFORNIA.

1956 MARRIES JANET SHEARON.
MARCH 16, 1966 GEMINI 8 LAUNCHES. ARMSTRONG MAKES FIRST-EVER DOCKING IN SPACE, AND STABILIZES CRAFT AFTER FAULTY THRUSTER PUTS CAPSULE IN DANGER.
SEPTEMBER 12–15, 1967 SERVES AS BACKUP PILOT FOR GEMINI 11 (ON EARTH).
1968 BAILS OUT OF LUNAR LANDING TRAINING VEHICLE SECONDS BEFORE IT CRASHES.
JULY 16, 1969 LAUNCH OF APOLLO 11 MISSION, WHICH TAKES ARMSTRONG, MICHAEL

COLLINS AND BUZZ ALDRIN TO THE MOON.
JULY 20, 1969 BECOMES FIRST PERSON TO WALK ON THE MOON AFTER LANDING LUNAR MODULE WITH BUZZ ALDRIN.
JULY 24, 1969 APOLLO 11 RETURNS TO EARTH. CREW IS KEPT IN A MOBILE QUARANTINE FACILITY UNTIL 21 DAYS AFTER LUNAR LIFTOFF.
1971–9 PROFESSOR AT UNIVERSITY OF CINCINNATI.
1986 SERVES ON PRESIDENTIAL COMMISSION ON SPACE SHUTTLE CHALLENGER ACCIDENT.

GENE CERNAN

A s commander of the Apollo 17 Moon landing, Gene Cernan took humanity's last small steps on the lunar surface. The last Apollo landing was also the last flight of Cernan's distinguished career as an astronaut. He had flown a Gemini mission in 1966, was America's second spacewalker, took the Apollo 10 lunar module to within 10 miles (16 km) of the Moon and, on Apollo 17, assisted geologist Harrison Schmitt in three epic moonwalks, gathering more scientific data than any other mission.

APOLLO 17

When Gene Cernan climbed down the ladder of the Lunar Module Challenger and onto the surface of the Moon in December 1972, he had good reason to feel a great sense of personal satisfaction. On his first two spaceflights, Gemini 9 and Apollo 10, Cernan was the junior partner of Tom Stafford. Determined to get his own command, Cernan had turned down the chance to walk on the Moon with John Young on Apollo 16, as lunar module pilot. Cernan staked everything on getting command of Apollo 17, the last lunar landing—and won. Cernan landed the Lunar Module just 200 feet (60 m) from the target area—a spot that had been carefully picked out months earlier by NASA.

Apollo 17 was a bumper mission: three full days on the surface, including three 7-hour excursions in the valley of Taurus-Littrow, with an expert geologist on hand to direct the scientific side of things. Harrison Schmitt was the first scientist-pilot to fly in space, and his mission made him the envy of geologists the world over. On the surface, Schmitt called the lunar landscape "a geologist's paradise." It was Cernan's job to assist Schmitt in gathering as much information about this paradise as possible.

During the first work session on the surface, Cernan hit trouble immediately when he tried to obtain a deep core sample of rock. Brute force was required to extract a core, and the effort ate into his oxygen reserves. This meant that the first geology

Just before placing his size 10 1/2 boot on the Moon, Cernan said, "I'd like to dedicate the first steps of Apollo 17 to all those who made it possible. Oh my golly. Unbelievable."

field trip in the Lunar Rover had to be shortened. And when they did get underway, a broken fender on the Rover meant that the two astronauts were caked in dirt by the time they returned to Challenger. After a night's rest, Cernan and Schmitt embarked on the longest lunar drive to date—an hour-long journey taking them five miles from base, and into the shadow of the Taurus Mountains. They spent 63 minutes in the foothills of the South Massif, collecting samples and reporting visual impressions back to the scientists at Mission Control.

By the end of their third excursion, Cernan and Schmitt had covered 19 miles (30.5 km) of territory and collected a record 220 lb (100 kg) of rock samples of high quality and variety. They had made hundreds of observations and taken dozens of photographs and readings. Finally, Cernan unveiled a plaque on the Lunar Rover, which was left on the Moon. The plaque reads, "Here Man completed his first explorations of the Moon, December 1972 A.D. May the spirit of peace in which we came be reflected in the lives of all mankind."

LIFE LINES

FULL NAME	EUGENE ANDREW CERNAN	EDUCATION	GRADUATED FROM PROVISO TOWNSHIP HIGH SCHOOL, MAYWOOD, ILLINOIS, 1952; BS IN ELECTRICAL ENGINEERING, PURDUE UNIVERSITY, 1952–6; MS AERONAUTICAL ENGINEERING; U.S. NAVAL POSTGRADUATE SCHOOL, MONTEREY, 1964
DATE OF BIRTH	MARCH 14, 1934		
PLACE OF BIRTH	CHICAGO, ILLINOIS		
FAMILY MARRIED	BARBARA ATCHLEY 1961 (DIVORCED); ONE DAUGHTER, TERESA, BORN 1963; MARRIED JAN NANNA 1987	CAREER	ENTERED U.S. NAVY FLIGHT SCHOOL 1956; SELECTED FOR NASA ASTRONAUT GROUP 1963; GEMINI 9 PILOT, JUNE 1966; APOLLO 10 LUNAR MODULE PILOT, MAY 1969; APOLLO 17 COMMANDER, DECEMBER 1972; SPECIAL ASSISTANT ON APOLLO-SOYUZ TEST PROJECT 1973–5

TO THE MOON

Early in 1966, NASA's Gemini program was in full swing. Neil Armstrong and David Scott were preparing for a March liftoff for Gemini 8, and Gemini 9 would follow in June, crewed by Elliot See and Charlie Bassett. On February 28, See and Bassett were killed when their T-38 jet crashed at the McDonnell facility in Missouri. For the astronaut corps, the tragic accident was a stark reminder of the dangers they faced daily. But for Gemini 9 backups Gene Cernan and Tom Stafford, it meant unexpected promotion to a prime crew.

Captain Gene Cernan of the U.S. Navy had been selected for the Gemini program in October 1963. His background in electrical engineering and aeronautics stood him in good stead for his initial NASA assignment: working on rocket booster development. After serving as communications chief for the twinned Gemini missions 6 and 7, Cernan trained for his first spaceflight.

The Gemini 9 flight was beset with problems. A docking target failed to reach orbit—its replacement made it to space, but failed to deploy properly. The highlight of Cernan's 3-day mission was to be a spacewalk, America's second. Cernan experienced many problems during the EVA—his visor became fogged and the lack of handholds left him with no leverage. The EVA was eventually cut short when Mission Control agreed that he could not safely perform all his scheduled tasks.

But the difficulties faced by Gemini astronauts provided invaluable experience, and when NASA drew up training rosters for flights to the Moon, Cernan made the list.

THREE DAYS ON THE MOON

Cernan served as backup Lunar Module pilot for Apollo 7. And in May 1969 he was aboard Apollo 10, the dress rehearsal for the first lunar landing attempt in July. Tom Stafford commanded the mission, with John Young as Command Module pilot. Four days after leaving Earth, Stafford and Cernan took their lunar module, Snoopy, into an eccentric orbit around the Moon. The trajectory brought

We just landed on another world somewhere in this universe.
GENE CERNAN

In three separate spaceflights, Gene Cernan logged 566 hours 15 minutes in space. He made America's second spacewalk (top left, glimpsed from inside the spacecraft) on Gemini 9. On Apollo 17 (launch, left), Cernan piloted the Lunar Module (above), explored the surface (far left) with Harrison Schmitt, and was the last person to walk on the Moon (below).

NEGOTIATOR

AFTER COMMANDING APOLLO 17, GENE CERNAN HELPED PLAN THE APOLLO-SOYUZ TEST PROJECT, THE FIRST CREWED INTERNATIONAL SPACE MISSION. HE OVERSAW THE TRAINING OF AMERICAN ASTRONAUTS AND, AS THE SENIOR U.S. NEGOTIATOR WITH THE SOVIET SPACE AUTHORITIES, HE HAMMERED OUT THE DETAILS FOR THE LINKUP IN ORBIT OF AN APOLLO COMMAND MODULE AND A SOYUZ CRAFT ON JULY 17, 1975.

the craft within 10 miles (16 km) of the surface. As their altitude fell, they began to see mountains, cliffs and boulders. "We is go and we is down among 'em, Charlie!" Cernan said to the Capcom, unable to hide his excitement.

The success of Apollo 10 paved the way for the first Moon landings, and Cernan was rewarded with command of his own mission. This time, he would go the extra 10 miles and walk on the Moon. Apollo 17 was the only Moon landing to feature a scientist. Cernan accompanied geologist Harrison Schmitt on three grueling but highly productive EVAs, gathering a record haul of samples. Then, after three days on an alien world, Gene Cernan, the last man on the Moon, closed the hatch of the lunar module. The Apollo adventure was over.

CAREER TIMELINE

1956 JOINS THE U.S. NAVY.
1957 QUALIFIES AS A NAVY PILOT.
1961 ENTERS U.S. NAVAL POSTGRADUATE SCHOOL AT MONTEREY NAVAL STATION, CALIFORNIA (RIGHT).
1963 CERNAN IS SELECTED AS ONE OF 14 TRAINEE ASTRONAUTS FOR THE GEMINI AND APOLLO PROGRAMS.
1965 APPOINTED BACKUP PILOT FOR GEMINI 9. IN DECEMBER, SERVES AS CAPCOM ON THE ORBITAL RENDEZVOUS OF GEMINIS 6 AND 7.
FEBRUARY 1966 ELLIOT SEE AND CHARLIE BASSETT ARE KILLED IN A JET CRASH. CERNAN AND TOM STAFFORD TAKE OVER THEIR MISSION.

JUNE 3–6 CERNAN AND STAFFORD SPEND THREE DAYS ON GEMINI 9 (RIGHT). THEIR DOCKING TARGET BLOWS UP ON THE WAY TO SPACE. A REPLACEMENT TARGET ALSO FAILS. CERNAN MAKES A RECORD SPACEWALK OF 2 HOURS 10 MINUTES.
1968 CERNAN IS BACKUP LUNAR MODULE PILOT ON APOLLO 7.
MAY 18–26, 1969 CERNAN, STAFFORD AND JOHN YOUNG TAKE THE APOLLO 10 SPACECRAFT THROUGH A DRESS REHEARSAL FOR THE FIRST LUNAR LANDING. CERNAN PILOTS LUNAR MODULE

SNOOPY TO WITHIN 10 MILES (16 KM) OF THE MOON'S SURFACE.
1971 CERNAN SERVES AS BACKUP COMMANDER ON APOLLO 14.
DECEMBER 7–11, 1972 CERNAN COMMANDS THE LAST APOLLO MISSION TO THE MOON AND IS THE LAST TO LEAVE THE MOON'S SURFACE.
1973–5 CERNAN (LEFT) SERVES AS SPECIAL ADVISOR TO THE APOLLO-SOYUZ TEST PROJECT—THE FIRST RUSSIAN-AMERICAN SPACE RENDEZVOUS.
1976 RETIRES FROM THE NAVY AND FROM NASA.
1981 STARTS CERNAN CORPORATION.

BUZZ ALDRIN

On July 20, 1969, minutes after Neil Armstrong had stepped out on to the lunar surface, Buzz Aldrin became the second human to walk on the Moon. It was the high point of an impressive career. In the preceding years, Aldrin had not only demonstrated his ability as an astronaut, but had also been instrumental in helping the NASA ground crew solve the technical problems involved in sending people to the Moon. He was both intelligent and practical—just what was needed on the Moon.

BRAINS AND BRAWN

The instant Buzz Aldrin stepped onto the Moon, his name became famous around the world. But his moonwalk with Neil Armstrong was not necessarily the greatest contribution he made to the Apollo program. To walk on the Moon, NASA first had to figure out the best method of getting there. And that was by no means obvious. At first, the space agency favored a direct ascent approach, in which a single spacecraft flies to the Moon and back propelled by a huge rocket. The problem was that in order to carry enough fuel for the round trip, the rocket would have to be truly enormous. Such a craft would have cost a fortune, and would have taken too long to develop.

The only other option involved launching several smaller craft capable of performing a rendezvous in space. Aldrin favored this approach long before many NASA personnel had given it any serious attention. NASA was uncertain about the viability of the space rendezvous method. It was thought to be highly dangerous. Rendezvousing in lunar orbit, as opposed to Earth orbit, was deemed particularly risky—if anything went wrong, the crew would be too far away to be rescued.

Aldrin believed so strongly in the rendezvous method that he wrote his doctoral thesis on the subject. This required him to calculate highly technical spacecraft

Aldrin studied for his doctorate in aeronautics at the Massachusetts Institute of Technology (left). His unconventional approach to becoming an astronaut made him a valuable jack-of-all-trades at NASA.

maneuvers. Flying in orbit, he noted, is totally unlike flying an aircraft within Earth's atmosphere. In orbit, speed and altitude are inseparable—the faster a spacecraft travels, the lower its orbit must be. So if the Lunar Module wanted to catch up with the Command Module in order to dock, simply firing its rocket would not do the job. The Lunar Module would have to approach at a carefully planned trajectory from below its target.

Among the Apollo astronauts, Aldrin was the unquestioned expert on rendezvous trajectories. He could often see solutions that no one else spotted. For example, before the first lunar rendezvous mission, Apollo 9, NASA officials were concerned that the Sun would blind the pilot. Aldrin suggested that the Lunar Module should simply fly upside down—it would make little difference in zero g. NASA officials at first thought he was joking, but soon realized that he was right—it was the obvious solution. Insights like these quickly earned Aldrin the nickname "Dr. Rendezvous" among his colleagues.

LIFE LINES

FULL NAME	BUZZ ALDRIN (ORIGINALLY EDWIN EUGENE ALDRIN JR.; CHANGED TO BUZZ IN 1988)	FAMILY	MARRIED TO LOIS DRIGGS CANON (THIRD WIFE); 3 CHILDREN FROM PREVIOUS MARRIAGE
DATE OF BIRTH	JANUARY 20, 1930	CAREER	ACTIVE SERVICE IN KOREA, 1951–3; AERIAL GUNNER INSTRUCTOR, NELLIS AIR FORCE BASE, NEVADA, 1953; AIDE TO FACULTY DEAN AT THE AIR FORCE ACADEMY, 1953–6; PILOT IN 36TH FIGHTER DAY WING IN GERMANY, 1956; TRAINEE ASTRONAUT, 1963; COMMANDER OF THE TEST PILOTS SCHOOL, EDWARDS AIR FORCE BASE, 1971–2
PLACE OF BIRTH	MONTCLAIR, NEW JERSEY		
EDUCATION	GRADUATED FROM MONTCLAIR HIGH SCHOOL, 1947; WEST POINT MILITARY ACADEMY, 1948–51; MASSACHUSETTS INSTITUTE OF TECHNOLOGY, 1959–63		

WHAT'S THE BUZZ?

Like nearly all his Apollo colleagues, Buzz Aldrin was a fighter pilot before he became an astronaut. He flew 66 missions in the Korean War and shot down two MiG-15s. But he was not one of the test pilots—the elite group from which the vast majority of early astronauts were drawn. He chose a more academic route into space. In 1959, instead of going to the Edwards Air Force Test Pilot School as he had previously planned, Aldrin went to Massachusetts Institute of Technology. His doctoral thesis, titled Guidance for Manned Orbital Rendezvous and written in 1963, contributed directly to the Apollo program. Aldrin applied to NASA, and in October 1963 was taken on as a trainee astronaut.

He waited for three years before his first mission, but on November 11, 1966, he finally entered orbit aboard Gemini 12. He and his Commander Jim Lovell were going to put into practice what Aldrin had spent so long studying—a docking in space. Previous Gemini missions had proved it was possible, but it needed to be perfected if NASA was to send men to the Moon.

Lovell was in the pilot's seat as they approached the target, an Agena rocket stage. But then disaster struck. The radar malfunctioned, meaning that the spacecraft's computer was deprived of the vital data it needed to calculate the trajectories. Undaunted, Aldrin made the necessary calculations on the spot. Lovell guided the spacecraft to a successful docking based on Aldrin's figures. Aldrin then took the controls and put all his theorizing into practice to carry out a flawless docking.

The significance [of Apollo 11] was the reaction of the people watching it...It changed lives.
BUZZ ALDRIN

The most important week in Aldrin's life began on July 16, 1969, as he prepared for the Moon (far right). The historic event itself (above left) was followed by three weeks in quarantine, which included a meeting with President Nixon (bottom right). Then the crew got a hero's welcome around the world (New York ticker-tape parade, top right.)

A LONG JOURNEY HOME

After Gemini 12, Aldrin's next spaceflight was to make him the second human on the Moon. With Neil Armstrong he spent over two hours walking on the lunar surface, collecting rocks and performing simple experiments. It was an incredible achievement, but for Aldrin, the hardest part was coming home. He, Armstrong and Command Module pilot Michael Collins went on a goodwill tour of the world on their return. It wasn't long, though, before life returned to normal, and Aldrin suffered a bout of depression. After walking on the Moon, he found it difficult to adjust to a relatively ordinary existence.

Aldrin worked for a time as commander of the Test Pilot School at Edwards Air Force Base before retiring from the Air Force in 1972. He wrote several books, including a frank autobiography, Return to Earth, that detailed his experience with depression. Today, though, he has managed to put his life in order. He travels the world to lecture and promote his own vision of the future of space exploration. In doing so, Aldrin continues to inspire new generations of space travelers—even those too young to remember the Moon landings.

CAREER TIMELINE

1951 GRADUATES WITH A BACHELOR OF SCIENCE DEGREE FROM THE WEST POINT MILITARY ACADEMY, NEW YORK. HE IS THIRD IN HIS CLASS. ASSIGNED TO 51ST FIGHTER WING IN KOREA FLYING F-86S.
1953 POSTED TO NELLIS AIR FORCE BASE, NEVADA, AS AN AERIAL GUNNER INSTRUCTOR; BECOMES AIDE TO A FACULTY DEAN AT THE U.S.A.F. ACADEMY, COLORADO SPRINGS, COLORADO.
1956 F-100 PILOT BASED AT BITBURG, GERMANY.
1959 GOES TO MASSACHUSETTS INSTITUTE OF TECHNOLOGY (MIT) TO STUDY ASTRONAUTICS.
1963 GAINS HIS DOCTORATE WITH A THESIS CALLED GUIDANCE FOR MANNED ORBITAL RENDEZVOUS.

OCTOBER 1963 JOINS NASA ASTRONAUT TEAM.
AUGUST 1965 CAPCOM (SPEAKING TO THE ASTRONAUTS FROM MISSION CONTROL) FOR GEMINI 5.
JULY 1966 CAPCOM FOR GEMINI 10.
NOVEMBER 11, 1966 LAUNCHED INTO SPACE WITH ASTRONAUT JIM LOVELL ABOARD GEMINI 12.
NOVEMBER 15, 1966 SPLASHES DOWN IN GEMINI 12.
JULY 16, 1969 LIFTS OFF ABOARD APOLLO 11 WITH NEIL ARMSTRONG AND MICHAEL COLLINS.
JULY 20, 1969 BECOMES THE SECOND MAN TO WALK ON THE MOON (RIGHT).
JULY 24, 1969 RETURNS TO EARTH ABOARD APOLLO 11 COMMAND MODULE.

1971 BECOMES COMMANDER OF THE TEST PILOTS SCHOOL AT EDWARDS AIR FORCE BASE, CALIFORNIA.
1972 RETIRES FROM THE AIR FORCE AND GOES INTO BUSINESS.
1973 WRITES HIS AUTOBIOGRAPHY, RETURN TO EARTH.
1989 WRITES MEN FROM EARTH, DESCRIBING HIS APOLLO MISSION AND HIS VISION OF AMERICA'S FUTURE IN SPACE, WITH MALCOLM MCCONELL.
1996 ENCOUNTER WITH TIBER, A SCIENCE FICTION NOVEL WRITTEN WITH JOHN BARNES, IS PUBLISHED.

STORY MUSGRAVE

During his 30 years as a career astronaut, Story Musgrave accumulated several spaceflight records. He flew on the maiden voyage of Space Shuttle Challenger in 1983, and went on to fly on all five orbiters. On his last Shuttle flight, in 1996, he became the oldest person to have flown in space. But for all his technical expertise, Musgrave always went in to orbit with a deep sense of wonder. He even earned a degree in literature so that he could write about his experiences and "bring space back to people."

INSIDE STORY

HUBBLE TOIL AND TROUBLE

In December 1993, Story Musgrave played a leading role in the 11-day mission to repair and service the Hubble Space Telescope. The HST was deployed in April 1990, by the crew of Shuttle Discovery, and was designed to provide astronomers with their clearest view yet of the wonders of the universe. All was not well, however. The telescope's primary mirror was found to suffer from a "spherical aberration"—it was the wrong shape.

With the credibility of NASA on the line, Story Musgrave was chosen as payload commander for the crucial HST repair. His extensive experience was required for three of the five spacewalks necessary to complete the mission. The telescope was to be fitted with a device called the Corrective Optics Space Telescope Axial Replacement (COSTAR). The size of a phone booth, COSTAR contained a series of 10 tiny mirrors, and had to be delicately maneuvered from the payload bay of Shuttle Endeavour. Once in place, the device dramatically improved the quality of data from the Faint Object Camera, the Faint Object Spectrograph, and the Goddard High Resolution Spectrograph. Partnered by Jeff Hoffman, another experienced astronaut, Musgrave made the first of the EVAs, replacing gyroscopes on the HST and preparing Hubble for the COSTAR installation. At

Story Musgrave rides the Shuttle's Remote Manipulator System arm on the last of his three spacewalks. He performed the final Hubble upgrades before the telescope was deployed into orbit.

7 hours 54 minutes, that first work session at the Hubble was the second-longest spacewalk in NASA history. But the repair went like clockwork, and two extra spacewalks allowed as a contingency were not needed. Musgrave later said, "It may have been the most well-rehearsed mission since the first Moon-walk."

Other elements of the mission were in fact scheduled upgrades. The Wide Field/Planetary Camera was replaced with a next-generation version, and the telescope's onboard computer received a new coprocessor. The two solar arrays, provided by the European Space Agency, were replaced. But the key question remained—would Hubble now do the job it was designed to do? When the first images came through from the rejuvenated telescope, the answer was a resounding yes.

Musgrave summarized Hubble's significance by saying that "anything that powerful tends to bridge the gap between cosmology, theology and philosophy." As one of NASA's philosophers, it was fitting that he took the lead in making Hubble work.

LIFE LINES

FULL NAME	FRANKLIN STORY MUSGRAVE
DATE OF BIRTH	AUGUST 19, 1935
PLACE OF BIRTH	BOSTON, MASSACHUSETTS
EDUCATION	ST. MARK'S SCHOOL, SOUTHBOROUGH, MASSACHUSETTS, GRADUATED 1953; BS IN MATH AND STATISTICS, SYRACUSE UNIVERSITY, 1958; MBA, UNIVERSITY OF CALIFORNIA, 1959, SPECIALIZING IN OPERATIONS ANALYSIS AND COMPUTER PROGRAMMING; BA IN CHEMISTRY, MARIETTA COLLEGE, 1960; MD IN MEDICINE FROM COLUMBIA UNIVERSITY, 1964; MS IN PHYSIOLOGY AND BIOPHYSICS, UNIVERSITY OF

KENTUCKY, 1966; MA IN LITERATURE, UNIVERSITY OF HOUSTON, 1987

CAREER: AVIATION TECHNICIAN, U.S. MARINE CORPS, 1953–8; SELECTED AS A NASA SCIENTIST-ASTRONAUT 1967; BACKUP SCIENCE PILOT, SKYLAB 2, 1973; CAPSULE COMMUNICATOR, SKYLABS 3 AND 4, 1973–4; TEST AND VERIFICATION PILOT, SPACE SHUTTLE PROGRAM, 1979–84; SIX SHUTTLE FLIGHTS BETWEEN 1983 AND 1996: STS-6, STS-51F/SPACELAB-2, STS-33, STS-44, STS-61, STS-80

SHUTTLE EXPERT

In an interview in 1994, Story Musgrave talked about an unusual object he saw and photographed in space on two separate shuttle flights: "What I call 'the snake,' like a 7-foot [2-m] eel swimming out there. It may be an uncritical rubber seal from the main engines...it has its own internal waves like it's swimming...At zero g, things have an incredible freedom." The freedom of spaceflight and its unique perspective inspired Story Musgrave throughout his 30-year career as an astronaut. And he stands out among the astronaut corps, not only for his considerable achievements as a scientist and pilot, but for his gift for communicating his experiences in space.

Born in Massachusetts in 1935, the young Musgrave escaped an unhappy family life by sneaking off at night to gaze at the stars. He joined the U.S. Marine Corps in 1953, straight out of high school, working as a technician for flight crews in Korea, Japan and Hawaii. By 1967, he had acquired degrees in math and statistics, business administration, chemistry and physiology, as well as a doctorate in medicine. That year, NASA was looking to recruit scientists to train as astronauts, and Musgrave's application was approved.

He became an astronaut at the height of the Apollo program. In 1967, NASA wanted to go to the Moon, build a space station, build a lunar base and go to Mars. So did Musgrave. He was hoping to get a berth on the Mars expedition as flight physician. In reality, Musgrave had to wait 16 years for his first space flight.

> *Eventually we'll start seeing ourselves as universal creatures, and won't even refer to other living creatures out there as aliens.*
> STORY MUSGRAVE

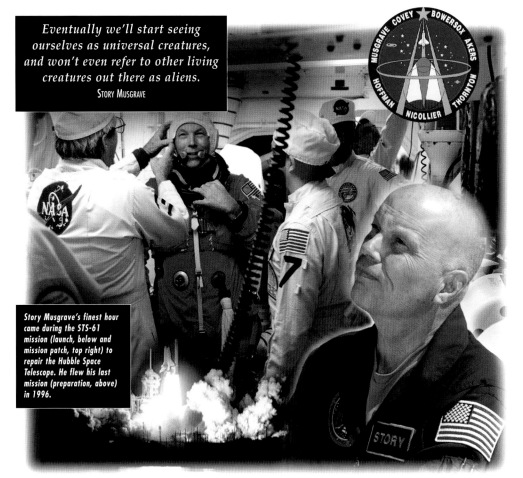

Story Musgrave's finest hour came during the STS-61 mission (launch, below and mission patch, top right) to repair the Hubble Space Telescope. He flew his last mission (preparation, above) in 1996.

TREASURE

SHUTTLE ASTRONAUTS ARE ALLOWED TO TAKE OBJECTS WITH THEM INTO ORBIT. THESE ARE SOMETIMES SIGNIFICANT OBJECTS FLOWN FOR AN INSTITUTION, OR SIMPLY FOR THE ASTRONAUTS TO GIVE AS GIFTS WHEN THEY GET BACK. THE ALLOWANCE IS 12 ITEMS, WEIGHING NO MORE THAN AN OUNCE EACH. STORY MUSGRAVE HAS MADE THE MOST OF THIS CONCESSION, TAKING DINOSAUR BONES, FAVORITE BOOKS AND HIS SON'S TOY CARS INTO SPACE. HIS INTEREST IN HISTORY LED HIM TO TAKE INTO ORBIT A FRAGMENT OF THE ANCIENT BRITISH MONUMENT STONEHENGE (ABOVE), AS WELL AS THE FOSSILIZED REMAINS OF CREATURES THAT LIVED ON EARTH 350 MILLION YEARS AGO.

HOME SWEET HOME

Musgrave finally lifted off aboard the 1983 inaugural flight of the Space Shuttle Challenger. "I was a long-term investor," he said. Musgrave's investment brought him another five flights between 1985 and 1996, including one of the most important Shuttle missions to date, the repair of the Hubble Space Telescope in December 1993. In 1996, Musgrave flew into orbit for the last time, and achieved a NASA first as the first astronaut to fly on all five Shuttles. But it was Challenger that gave the veteran pilot the biggest fright of his career. Five minutes after the July 1985 launch, one of the ill-fated orbiter's engines shut down—the only such failure to occur in the Shuttle program so far. The orbiter had to dump the unused fuel and failed to reach its target altitude.

For Story Musgrave, the occasional bumpy ride was a small price to pay for the rich variety of experiences offered to the space traveler. When asked if he ever wished a Shuttle flight could go on longer, he replied: "Oh, yeah, months longer. I'd like to move out there." Dick Covey, mission commander for the Hubble repair, said of Musgrave, "He never stops learning."

CAREER TIMELINE

1953–8 MUSGRAVE WORKS AS AVIATION TECHNICIAN FOR U.S. MARINE CORPS. DUTY TOURS IN JAPAN, KOREA AND HAWAII, INCLUDING A STINT ON U.S.S. WASP (FAR RIGHT).

1964–7 AFTER RECEIVING HIS DEGREE IN MEDICINE, MUSGRAVE TRAINS AS A SURGEON AT THE UNIVERSITY OF KENTUCKY MEDICAL CENTER.

1967 SELECTED BY NASA AS ONE OF 11 SCIENTISTS CHOSEN TO TRAIN AS ASTRONAUTS.

1973 BACKUP SCIENCE PILOT FOR THE FIRST CREWED MISSION TO SKYLAB, THE UNITED STATES' FIRST SPACE STATION. CAPSULE COMMUNICATOR, OR CAPCOM, FOR THE SECOND AND THIRD (LEFT) SKYLAB MISSIONS.

1974 AWARDED NASA'S EXCEPTIONAL SERVICE MEDAL FOR HIS CONTRIBUTION TO THE SKYLAB PROGRAM.

1979–84 WORKS AT THE SHUTTLE AVIONICS INTEGRATION LABORATORY AT JOHNSON SPACE CENTER, AS A TEST AND VERIFICATION PILOT FOR THE SPACE SHUTTLE DEVELOPMENT PROGRAM.

APRIL 4, 1983 FIRST SPACEFLIGHT, AS MISSION SPECIALIST ON CHALLENGER, STS-6. PERFORMS THE FIRST SPACEWALK FROM THE SHUTTLE, WITH DON PETERSON.

JULY 29, 1985 SECOND SHUTTLE FLIGHT, ON CHALLENGER, STS-51F.

NOVEMBER 22, 1989 MISSION SPECIALIST ON DISCOVERY, STS-33. NOVEMBER 24, 1991 MUSGRAVE'S FOURTH FLIGHT, ABOARD SHUTTLE ATLANTIS.

DECEMBER 2, 1993 ON ENDEAVOUR, FLIGHT STS-61, MUSGRAVE HELPS REPAIR THE HUBBLE TELESCOPE.

NOVEMBER 19, 1996 ON HIS FINAL SPACEFLIGHT, ABOARD SHUTTLE COLUMBIA, MUSGRAVE BECOMES THE FIRST ASTRONAUT TO FLY INTO SPACE ON ALL FIVE ORBITERS.

EILEEN COLLINS

When the Space Shuttle Columbia launched on July 23, 1999, it was the first time that a woman had sat in the Commander's seat. Air Force Lieutenant Colonel Eileen Collins was on her third space flight and in command of a crucial mission that would deploy the multi-billion-dollar Chandra X-Ray Observatory. On the return flight she would make history again—as the first woman to guide the Shuttle to touchdown. It was a great moment—for NASA, for Eileen Collins, and for women around the world.

INSIDE STORY

THE SHUTTLE'S FIRST WOMAN COMMANDER

On March 5, 1998, Hillary Clinton strode into the Roosevelt Room at the White House to let the world world know that, for the first time, a woman would command the Space Shuttle. Lieutenant Colonel Eileen Collins would take Columbia into space in December with a crew of four on mission STS-93. The Shuttle would be carrying an extremely important payload—the $2.78-billion Chandra X-Ray Observatory.

But it soon emerged that there was a technical problem with the observatory. NASA, unwilling to repeat the embarrassing Hubble episode (the orbiting space telescope was launched in 1990 with a crippling flaw in its lens), delayed the mission until July 23, 1999. On that day, Collins and her crew finally climbed into Columbia and were launched into space just after midnight. At the time, the Chandra Observatory and the rocket booster that would blast it into its own orbit made up the heaviest payload ever carried by the Shuttle.

Collins was leading a seasoned crew: Her pilot, Navy Captain Jeff Ashby, was the only rookie astronaut. They were joined by French air force Colonel Michel Tognini, making his second trip into space. In charge of Chandra were Air Force Lieutenant Colonel Catherine Coleman, on her second space flight, and Stephen Hawley, who had deployed the Hubble Telescope and a veteran of 40 other flights.

Columbia lifts off for mission STS-93. Collins and her crew deployed the Chandra X-Ray Observatory, which now peers into hitherto invisible regions of the cosmos.

By 7:47 a.m., Chandra was ready to go. Coleman and Tognini donned their spacesuits and ventured out into the cargo bay to guide the 25-ton (22.5-tonne) load into space. Hawley directed the two astronauts from inside, while Collins stayed at the controls of the Shuttle. With the observatory successfully deployed, Collins guided the Shuttle to a safe distance before Chandra's rocket booster fired, carrying the observatory into Earth orbit. The crew could now set to work on their secondary missions.

On day five, it was time to come home. Although Collins had been Pilot of the Shuttle twice before, in practice it is the Commander who actually controls the spacecraft. Landing the Shuttle is the most challenging part of a Shuttle flight: Crews note that in the atmosphere, it flies like a brick. To make things even more interesting, STS-93 was to be a night landing. At 9:19 p.m. EST, Collins activated the deorbit burn, and at 9:49 p.m. Columbia entered Earth's atmosphere. At 10:20:35 p.m. the Shuttle touched down on runway 33 at the Kennedy Space Center. Collins' historic flight was complete.

LIFE LINES

FULL NAME	EILEEN COLLINS	FAMILY	DAUGHTER OF JIM AND ROSE COLLINS; MARRIED TO PAT YOUNGS; ONE DAUGHTER
DATE OF BIRTH	NOVEMBER 19, 1956		
PLACE OF BIRTH	ELMIRA, NEW YORK	CAREER	T-38 INSTRUCTOR, VANCE AIR FORCE BASE, OKLAHOMA, 1979–82; C-141 AIRCRAFT COMMANDER AND INSTRUCTOR, TRAVIS AIR FORCE BASE, CALIFORNIA, 1983–5; ASSISTANT PROFESSOR IN MATH AND T-41 INSTRUCTOR PILOT, U.S.A.F. ACADEMY, COLORADO, 1986–9; JOINED NASA 1990; STS-63, PILOT; STS-84, PILOT, STS-93, COMMANDER
EDUCATION	GRADUATED FROM ELMIRA FREE ACADEMY, 1974; ASSOCIATE DEGREE IN MATH AND SCIENCE, CORNING COMMUNITY COLLEGE, 1976; BA IN MATH AND ECONOMICS, SYRACUSE UNIVERSITY, 1978; MS IN OPERATIONS RESEARCH, STANFORD UNIVERSITY, 1986; MA IN SPACE SYSTEM MANAGEMENT, WEBSTER UNIVERSITY, 1989		

GIRL POWER

E ileen Collins was born on November 19, 1956, just as the space race was entering its most exciting phase. In 1969, when she was 12, she watched with millions as Neil Armstrong and Buzz Aldrin walked on the Moon. But the figures who inspired her most were the early women aviators. Their pioneering example showed that women had as much of a role to play in aviation as men—despite what America's aerospace establishment thought.

After attending Corning Community College, Collins won a scholarship to Syracuse University, where she earned a degree in mathematics and economics. During her time at Syracuse, she also managed to find the time to obtain her pilot's license. Collins now had the qualifications she needed to join the U.S. Air Force and, in 1978, armed with a letter of recommendation from her flying instructor, she was accepted on the pilot training program at Vance Air Force Base in Oklahoma.

So successful was Collins that she stayed on at Vance after her graduation in 1979 to be an instructor. As her career blossomed in the following years, she conceived her boldest ambition yet: She would become an astronaut.

Collins began to build up the experience and qualifications that would make her a good astronaut candidate. It paid off. In 1989, she was selected for the Test Pilot School at Edwards Air Force Base, California. This was her chance: Edwards is the traditional selection center for trainee astronauts. It was not long before Collins' talents were recognized. After a year of training at Vance, she started her career with NASA in 1991.

FLYING INTO HISTORY

B efore Collins could enjoy the glory of spaceflight, she had to do her share of ground jobs. She worked as a Shuttle engineer and in mission control as CapCom—the person who maintains voice contact with Shuttle crews.

In 1995, she finally got the job she really wanted. Assigned to the crew of STS-63, she was to be the Shuttle's first woman pilot. On

Eileen Collins (left, as Commander of the historic STS-93 Shuttle mission) had flown a number of aircraft before the Shuttle, including the massive C-141 Starlifter (below). She visited Mir with STS-84 in 1997 (mission patch, below left) and deployed the Chandra X-Ray Observatory (bottom) when she commanded STS-93 in 1999.

When I was a child, I dreamed about space—I admired pilots, astronauts, and I've admired explorers of all kinds. It was only a dream that I would someday be one of them.
EILEEN COLLINS

February 2, 1995, Collins began her first space flight on board Discovery. STS-63 was scheduled to perform the first Shuttle rendezvous with the Mir space station. The crew had to contend with a leaky Reaction Control System (RCS) thruster that was threatening to prevent the maneuver. In the end, Discovery came within 37 feet (11 m) of Mir. Both the Americans and Russians were delighted. The mission paved the way for the first Shuttle-Mir docking, STS-71.

Two years later, Collins was back in space. As pilot on board STS-84, she got to see the inside of Mir when the Shuttle docked with the space station. But her most famous mission was yet to come. As commander of STS-93, Collins would break new ground for women, just like the early aviators she so admired.

CAREER TIMELINE

NOVEMBER 19, 1956 BORN IN ELMIRA, NEW YORK.
1979 GRADUATES FROM AIR FORCE UNDERGRADUATE PILOT TRAINING, VANCE AIR FORCE BASE, OKLAHOMA.
1979–82 T-38 JET INSTRUCTOR PILOT AT VANCE AIR FORCE BASE.
1983–5 C-141 AIRCRAFT COMMANDER AND INSTRUCTOR PILOT AT TRAVIS AIR FORCE BASE, CALIFORNIA.
1986–9 ASSISTANT MATH PROFESSOR AND T-41 INSTRUCTOR PILOT AT THE AIR FORCE ACADEMY IN COLORADO.
1990 GRADUATES FROM AIR FORCE TEST PILOT SCHOOL, EDWARDS AIR FORCE BASE, CALIFORNIA. SELECTED BY NASA TO JOIN THE ASTRONAUT TRAINING PROGRAM.

JULY 1990 RETURNS TO VANCE FOR ASTRONAUT TRAINING.
JULY 1991 BEGINS WORK AT KENNEDY SPACE CENTER, FLORIDA, AS PART OF THE SHUTTLE ORBITER ENGINEERING SUPPORT TEAM.
FEBRUARY 2–11, 1995 AFTER FOUR YEARS OF GROUND DUTIES, COLLINS FINALLY GOES TO SPACE AS PILOT OF STS-63. SHE AND THE CREW ENGAGE IN THE FIRST SHUTTLE RENDEZVOUS WITH THE MIR SPACE STATION.
MAY 15–24, 1997 RETURNS TO SPACE AS PILOT OF STS-84. THIS TIME SHE ACTUALLY BOARDS MIR AS PART OF A 9-DAY MISSION.
MARCH 5, 1998 GOES TO THE WHITE HOUSE TO MEET

PRESIDENT CLINTON AND HILLARY CLINTON. MRS. CLINTON ANNOUNCES COLLINS AS COMMANDER OF SHUTTLE MISSION STS-93.
JULY 22–27, 1999 BECOMES THE FIRST FEMALE COMMANDER OF A SHUTTLE MISSION, STS-93 (RIGHT). SHE OVERSEES THE SUCCESSFUL LAUNCH OF THE CHANDRA X-RAY OBSERVATORY, AND BECOMES THE FIRST WOMAN TO LAND THE SHUTTLE.

ROCKETS AND LAUNCHERS

The first practical payload-carrying rockets were the A4 or V-2 rockets designed by Wernher von Braun and used by Nazi Germany in 1944–45. After World War II, V-2s and German scientists fell into Allied hands and V-2 production continued in the U.S. and U.S.S.R., where the rockets were used for upper atmosphere science as well as military missile research. Intermediate-range missiles such as the U.S. Redstone followed, and these proved adaptable to carrying manned capsules such as Mercury. Using converted missiles for manned flight purposes was a calculated risk, as they had a worrying tendency to blow up upon launch. Fortunately this never occurred with astronauts aboard. The U.S.S.R.'s first rockets were hardly more successful, particularly the modified SS-6 "Sapwood" missiles lifting early Sputnik and Venera missions, many of which failed to reach orbit. By the mid-1960s purpose-built civilian space launchers such as the Saturn series were replacing military designs, and were followed by the Titan, Delta, and Europe's Ariane series. In addition to the U.S., Russia and Europe, India, China, and Japan have flown rockets of their own designs and are likely to be joined by further nations, particularly in Asia.

An Atlas-Agena rocket is launched on 12 September, 1966, as part of the Gemini 11 mission. Astronauts Pete Conrad and Richard Gordon would maneuver their craft to rendezvous with the Agena rocket in orbit, as preparation for the planned Apollo moon shots.

FROM V-2 TO EXPLORER 1

W hen the world's first ballistic missiles—Adolf Hitler's V-2 rockets—began raining down on London in September 1944, it heralded the violent birth of the space age. But little could the V-2's designer, the brilliant Wernher von Braun, have imagined that just 14 years later he would have his finger on the ignition button of the rocket carrying America's first satellite, Explorer 1, into orbit around the Earth.

WHAT IF...

...THE U.S. HAD BEEN FIRST?

E xplorer 1 should not have been the first American satellite. Not only that, but Juno should not have been the rocket that launched it, since officially the Army Ballistic Missile Agency (ABMA) had no business building satellite launch vehicles at all.

The official U.S. entrant in the space race was Project Vanguard—a Navy endeavor. But Vanguard's first attempt to reach orbit—after the Soviet Union had launched both the world's first- and second-ever satellites—died in flames on the launchpad.

The Jupiter rocket that took its place, thinly disguised as a Juno, was a direct descendant of Project Orbiter, the brainchild of ABMA research chief Wernher von Braun. Orbiter had been officially canceled some two and a half years earlier in favor of Vanguard, because the Navy project was based on the Viking scientific research rocket. Orbiter, by contrast, was a by-product of the U.S. ballistic missile program, and likely to be seen by the rest of the world as a flagrant escalation of the nuclear arms race.

Yet despite repeated warnings from the Department of Defense not to develop a rocket with the ability to launch satellites, the ABMA continued to do just that. On September 20, 1956, the Army rocketeers proved their point emphatically with the first launch of their Jupiter C rocket—derived, like the proposed Project Orbiter vehicle, from the Redstone ballistic missile. The Jupiter C reached an altitude of 682 miles (1,098 km) and could

Kennedy: He saw a successful U.S. space program as a means of galvanizing the nation in the face of the perceived threat from the Soviet Union. His 1961 pledge to land a man on the Moon by the end of the decade turned national humiliation into technological triumph.

Eisenhower: His staff was embarrassed by the links between space research and the U.S. ballistic missile program.

have put its fourth stage into orbit. The only thing that stopped von Braun's team was government policy. It was more than a year before the Soviet Union launched Sputnik 1.

The U.S. could have been first, but if it had won the first round of the space race, would it have had so much to prove later? Less than four years after Sputnik 1 and Explorer 1 were launched, President John F. Kennedy announced the goal of placing a man on the Moon by the end of the decade. It was an ambitious move. But had the U.S. not felt so humiliated at coming in second to the Soviet Union, perhaps we would still be waiting for a manned lunar landing.

EARLY US ROCKETS

ROCKET	STAGES	LENGTH (FT/M)	MASS (LB/KG)	LIFTOFF THRUST (LBF/KN)
V-2 (A-4)	ONE	40/12	28,230/128,804	59,547/265
BUMPER	TWO	54/16	28,930/13,122	59,547/265
VANGUARD	THREE	65/20	22,123/10,034	27,833/124
REDSTONE	ONE	60/18	62,700/28,440	82,600/367
JUNO (JUPITER C)	FOUR	65/20	64,066/29,059	82,969/369

WEAPONS FOR SCIENCE

America gained 103 unusual new citizens on April 14, 1955, in a ceremony at Huntsville High School in Alabama. Ten years earlier, many of them had been among Adolf Hitler's top weapons scientists. Their leader, Wernher von Braun, was the brilliant engineer who had designed the V-2, the Nazis' long-range missile that terrorized London in the final months of World War II.

At the end of the war, von Braun and 125 of his staff gave themselves up to the U.S. Army. They were shipped back to America and put to work designing a long-range missile that could deliver a nuclear warhead. The first fruits of this research came in 1953 with the Redstone missile. But space exploration was never far from the scientists' minds: In 1954, von Braun put forward his first practical proposal for launching a satellite. The following January, the idea was submitted to the U.S. Department of Defense as a joint Army-Navy endeavor called Project Orbiter.

Six months later, President Dwight D. Eisenhower announced that the U.S. would launch an artificial satellite into orbit before December 1958. But Project Orbiter was rejected in favor of Project Vanguard, which was based on the Navy's Viking research rocket.

But von Braun's Army engineers continued with unauthorized tests of their own. So when Vanguard 3 exploded on the launchpad in December 1957—after two successful Soviet satellite launches—von Braun's team was able to produce a new rocket. In less than two months, they rescued U.S. pride by launching Explorer 1.

FAMILY TREE

BUMPER
This was the first rocket to reach space. Launched from White Sands Proving Ground, New Mexico, in 1949, it broke a world record by reaching a speed of 5,510 mph (8,867 km/h). Bumper was a hybrid rocket that used a German V-2 captured at the end of World War II to boost a U.S.-built WAC Corporal upper stage to a maximum altitude of 244 miles (393 km).

VANGUARD
This was the rocket that was intended to launch the first U.S. satellite. It had liquid propellant engines in the first and second stages and a solid propellant third stage. Vanguard was derived from the Navy's Viking and Aerobee research rockets.

PROJECT ORBITER
First the U.S. and then the Soviet Union announced intentions to launch a scientific satellite into orbit as part of the 1957–8 International Geophysical Year, a worldwide effort to study the Earth. The Army proposed Project Orbiter to launch the U.S. satellite, using a modified Redstone missile, which in turn had been derived from the German V-2. The government eventually vetoed this plan in favor of the politically more acceptable Vanguard, which was descended from research rockets.

JUNO
America at last managed to put the satellite Explorer 1 into orbit using Juno, a renamed Jupiter C rocket. Designed by the von Braun team and built by the Army, the Jupiter C was a modified Redstone ballistic missile with a top stage designed for reentry. The rocket had a liquid-propellant main-stage engine, while the upper stages used solid propellant.

EXPLORER 1
Explorer 1, the first U.S. probe into orbit, occupied the fourth stage of the Juno rocket and was built by the Army Ballistic Missile Agency and the Jet Propulsion Laboratory. Instruments on board Explorer 1, designed by scientist James van Allen, revealed the Earth's radiation belts, which were named Van Allen Belts in his honor. Explorer weighed 10.5 pounds (4.75 kg) and measured just six inches (15 cm) across and less than 40 inches (1 m) long. After circling the Earth more than 58,000 times, it reentered the atmosphere in March 1970 and burned up.

LOSING FACE

On December 6, 1957, only two months after the U.S.S.R. launched Sputnik 2, America tried to catch up. A huge Vanguard rocket stood on the launchpad, ready to fire the first U.S. satellite into space. But Vanguard reached an altitude of only four feet (1.2 m) before it exploded in a fireball. The tiny satellite that it had been carrying managed to survive the blast and rolled into a nearby bush, where it began broadcasting radio signals. It was a bad day for the U.S. space program.

ROCKETS INTO SPACE

October 3, 1942 First successful launch of the V-2 by the Nazis.

February 24, 1949 U.S. Army launches Bumper 5, which becomes the first artificial object in space.

August 20, 1953 A prototype Redstone missile is launched.

June 25, 1954 Wernher von Braun proposes a Redstone-based rocket for launching satellites.

July 29, 1955 President Dwight D. Eisenhower announces that the U.S. will send a satellite into orbit before the end of 1958.

September 20, 1956 First launch of the U.S. Army Jupiter C rocket.

December 8, 1956 The first test rocket in the official U.S. satellite program, a single-stage Viking, attains an altitude of 126 miles (203 km).

August 8, 1957 The nose cone of a Jupiter C rocket is recovered 1,333 miles (2,145 km) from the launchpad, after reaching a height of 260 miles (418 km). It is the first object ever retrieved from space.

October 4, 1957 The Soviet Union launches Sputnik 1, the first satellite into orbit.

November 3, 1957 The Soviet Union launches Sputnik 2.

December 6, 1957 The first Vanguard rocket with three live stages, carrying a satellite, explodes on the launchpad.

January 31, 1958 A Juno rocket (above) successfully launches Explorer 1.

MERCURY REDSTONE

At the end of the 1950s, the U.S. and the Soviet Union were in the grip of the Cold War. Space was a new frontier where East and West could compete for prestige and find another use for some of their vast arsenals of missiles. The Soviet Union struck first, when a modified R-7 intercontinental missile put Sputnik 1 into orbit in 1957. An upgraded Redstone missile allowed the U.S. to catch up. Further modifications created the Redstone-Mercury rocket that would loft America's first astronaut into space.

WHAT IF...

...MISSILES ARE USED TO LAUNCH SATELLITES?

A final check of the flight console confirms the RSM-54 missile is ready for launch. The Russian submarine commander is given a last briefing on the imminent departure of the missile—all systems are go. A call to the Russian air force confirms their acknowledgement of the missile's trajectory. Back on board the submarine, the final seconds of the countdown echo through the hull. To the sounds of cheers, the missile streaks away from the sub, rips out of the ocean and heads skyward.

Eavesdropping on such an event at the beginning of 21st century, you could be forgiven for thinking you're witnessing the start of World War III. In fact, it's the thawing of East-West relations that has allowed this launch to occur. With the end of the Cold War, both Russia and the United States are using their stockpiles of intercontinental ballistic missiles (ICBMs) as a new and cheap way of reaching low Earth orbit—a technique that may be common in the future.

Since the signing of the two Strategic Arms Reduction Treaties—START 1 and 2—both East and West have found themselves with a huge surplus of obsolete ICBMs. Today, these missiles are being modified into launch vehicles to serve the international demand for launch services into low-Earth orbit. The lead comes from the Khrunichev State Research and Production Space Center, home of the joint Russian-German venture called Rockot.

A U.S. Pershing missile soars spaceward. Pershings were designed to drop nuclear warheads on an enemy. But after some modifications, most Cold War missiles could carry peaceful payloads into orbit.

The Rockot satellite launch vehicle is a modified RS-18 ICBM. The 2-stage RS-18 missiles were manufactured in the hundreds by the Soviet Union during the 1970s and 1980s. By giving the missile a third stage, known as Briz, the Rockot launch vehicle is capable of carrying approximately 4,000 pounds (1,800 kg) into a 250-mile (400-km) orbit. While the Rockot booster is designed for launch from the Plesetsk space center in Russia, another modified missile, the RSM-54, can be launched from Delfin-class submarines. Unlike a fixed launch pad, the submarine can travel to a near-ideal position for any particular orbital requirements.

The U.S. is also hoping to cash in on the use of Cold War hardware now surplus to requirements. After a few modifications, the Minuteman 2 missile should be able to lift a payload into low-Earth orbit from the Vandenberg Air Force Base in California. While the first Minuteman launches will be mainly to test the technology, it is probable that in the future, the U.S., like Russia, will use its surplus ICBMs to meet the ever-increasing demand for reliable launch vehicles to place satellites into space.

MERCURY REDSTONE ROCKET

NOVEMBER 21, 1960 MERCURY-MR1 LAUNCH FAILURE, ENGINE CUTS OUT 1 SECOND AFTER IGNITION	**MAY 5, 1961** ALAN SHEPARD BECOMES AMERICA'S FIRST ASTRONAUT, RIDING ON MERCURY MR-3
DECEMBER 19, 1960 SUCCESSFUL LAUNCH OF MERCURY-MR1A	**JULY 21, 1961** ASTRONAUT GUS GRISSOM IS LAUNCHED ON MERCURY MR-4; LAST MANNED FLIGHT OF THE MERCURY-REDSTONE ROCKET
JANUARY 31, 1961 SUCCESSFUL LAUNCH OF MERCURY-MR2; CARRIED HAM, A CHIMPANZEE	
MARCH 24, 1961 SUCCESSFUL LAUNCH OF MERCURY-MR BOILERPLATE; CARRIED MERCURY TEST CAPSULE	**OCTOBER 30, 1964** MERCURY-REDSTONE ROCKET IS RETIRED FROM MILITARY SERVICE

MAN LIFTER

The story of the Mercury-Redstone rocket began at the end of World War II, when German rocket pioneer Wernher von Braun and his colleagues surrendered to the advancing U.S. Army in 1945. Von Braun had developed Germany's V-2 rocket, the world's first true ballistic missile. Post-war, his technological know-how combined with American skills led to creation of the Redstone, the first operational U.S. missile, by 1953.

The Redstone was designed to throw a 6,000-pound (2,700-kg) nuclear warhead 200 miles (320 km). With a small upper stage in place of a bomb, the rocket would be powerful enough to put a small satellite into orbit—which had been von Braun's ambition for decades. But in 1956, the U.S. government vetoed von Braun's request for a satellite launch. Instead, the first U.S. satellite would be lofted by the specially designed Vanguard rocket, a Navy project with civilian funding that President Eisenhower hoped would "demilitarize" the nascent space program.

On October 4, 1957, though, the Soviets won the satellite race when Sputnik 1 reached orbit on a modified R-7 missile. A month later, the half-ton Sputnik II took a live dog into space aboard the same booster. In December, Vanguard failed spectacularly in a launchpad explosion. Shaken into action, American Secretary of Defense Neil McElroy gave von Braun the go-ahead that the rocket scientist had sought for more than a year. With a new upper stage, the trusty Redstone metamorphosed into the Jupiter-C rocket, which in February 1958 launched Explorer I—America's first satellite.

HERO'S RETURN

AFTER A FLIGHT THAT LASTED JUST OVER 15 MINUTES, ALAN SHEPARD (LEFT) WAS HAULED ABOARD A HELICOPTER FROM THE RECOVERY VESSEL U.S.S. LAKE CHAMPLAIN. MERCURY 3 SPLASHED DOWN IN THE ATLANTIC OCEAN 300 MILES (482 KM) EAST OF CAPE CANAVERAL, WHERE THE HISTORIC MISSION BEGAN.

ROCKET SCIENCE
Technicians check out the Rocketdyne A-6 engine used on Mercury 3 and Mercury 4. The liquid-fueled motor burned a mixture of alcohol and liquid oxygen to yield 78,000 pounds of thrust (347 kN) for a burn time of 155 seconds.

MODIFIED MISSILE

Just like its wartime V-2 ancestor, the Redstone burned alcohol with liquid oxygen. The Redstone-Mercury version had an upgraded motor that provided more thrust and a longer burn time—not enough to put an astronaut into orbit, but sufficient to lift the Mercury capsule briefly to a height of 116 miles (187 km).

emergency escape rocket and tower
Mercury capsule
instrument compartment
alcohol fuel
liquid oxygen
fins and rudders
Rocketdyne A-6 motor

OLD RELIABLE
A Redstone rocket makes a successful test launch. The Redstone relied on just one well-tried engine to supply the thrust needed to launch the Explorer 1 satellite and—in Mercury-Redstone configuration—carry the first two Mercury astronauts on their sub-orbital flights in May and July 1961.

HAPPY DAY

ON THE DAY COMMANDER ALAN SHEPARD BECAME THE FIRST U.S. ASTRONAUT IN SPACE, THE REDSTONE ROCKET TEAM IN HUNTSVILLE, ALABAMA, STAGED A RALLY IN CELEBRATION. PRINCIPAL SPEAKERS INCLUDED THE REDSTONE-MERCURY'S CHIEF CREATOR, WERNHER VON BRAUN (ABOVE, LEFT). VON BRAUN HAD BY THEN BECOME THE DIRECTOR OF THE MARSHALL SPACE FLIGHT CENTER.

The modifications for manned spaceflight began in 1959. Included in the redesign was the more reliable Rocketdyne engine, which mixed liquid oxygen and kerosene to provide 78,000 pounds of thrust (347 kN) at launch. To prolong the burn time of the engine, the Redstone's fuel tank was extended by six feet to achieve the increased speed and altitude needed to carry an astronaut into space. By 1960, the Redstone rocket had metamorphosed once more—into the Mercury-Redstone launch vehicle.

The first Mercury-Redstone launch attempt took place in November 1960. But the rocket lifted only an inch or two above the pad before the engine shut down. Faulty circuitry on the ground was the culprit. Luckily, success followed with further launches. During the next few months, three Mercury-Redstones took to the skies, one of them carrying a full-size dummy Mercury capsule.

To the chagrin of America's rocketeers, the Soviets beat them once more: Their sturdy R-7 put cosmonaut Yuri Gagarin into orbit on April 12, 1961. But Mercury-Redstone was not far behind. After passing its final tests, the rocket stood ready to launch its first human cargo into space. On the morning of May 5, 1961, Mercury astronaut Alan Shepherd soared 116 miles (187 km) skyward, remaining weightless for 4 minutes and 45 seconds. The sub-orbital hop was trivial by the standards of later missions, but the flight of Mercury-Redstone 3 had carried the first American into space.

THE NEXT ROUND

Now, both nations raced for the next goal: a man in orbit. The newly formed National Aeronautics and Space Administration (NASA) turned again to the Redstone rocket to use as a launch vehicle for project Mercury—the project designed to place the American agency's first astronauts into orbit. But before the Redstone could qualify as a manned launch vehicle, approximately 800 engineering changes were needed.

AGENA ROCKETS

Every major undertaking has an unsung hero that does most of the work. For the U.S. Space Program, that hero was the Agena rocket. Lockheed first developed Agena in 1955 for the U.S. Air Force. It became the most reliable upper-stage rocket ever built, and was used on almost half of all U.S. space missions. Agenas have been used on some 362 launches, including those of classified military missions and probes for the exploration of Venus, Mars and the Moon—and a new version is currently in development.

WHAT IF...

...AGENA IS IMPROVED?

The last Agena mission was launched on February 12, 1987, from Vandenburg Air Force Base, California, atop a Titan 34B rocket. But that may not have been the end of the Agena's space career. Before being sold to Aerojet in 2003, Atlantic Research Corporation (ARC) in Gainsville, Virginia, began work on an updated Agena rocket, designated Agena 2000. The project is currently inactive, but it is possible that an updated Agena rocket may eventually appear. ARC planned to use modern fabrication techniques to build a stronger yet lighter frame and fuel storage tanks, and new electronics were to make the improved Agena even more precise in delivering payloads to exact orbits. Additionally, the engine was to burn UDMH (unsymmetrical dimethyl hydrazine) with nitrogen tetroxide instead of with IRFNA (inhibited red fuming nitric acid), greatly increasing the thrust of the engine.

All of these improvements would have increased the delivery payload of the Agena 2000, which was was put forward as a possible upper stage for the new Evolved Expendable Launch Vehicle being built by Lockheed Martin for the U.S. space program. The EELV project is designed to cut costs and provide enhance lift capability into the 2020s.

The Agena 2000 was to have significantly improved performance over the last version of the Agena, the Agena-D, but all of the proposed improvements were just upgrades that use

Lockheed's Evolved Expendable Launch Vehicle (EELV) is one of a new generation of launchers. Such launchers might one day carry improved Agena upper stages.

1990s technology instead of 1960s technology. Could the performance have been increased even more? There are several additional new technologies that would provide large improvements over the Agena-D. The exhaust nozzle of the Agena is a bell nozzle. It could be replaced by a wedge-shaped Aerospike nozzle, increasing the engine burn efficiency. Another change could be to use different fuels that would generate greater thrust. Finally, the Agena could achieve significant weight savings if it was fabricated from lightweight metal composites, such as the aluminum-lithium alloy that is being used in the construction of the new Space Shuttle External Tanks.

But whether or not an updated version of Agena appears, the list of missions it has already served reads like a history of the U.S. space program. The Agena will never lose the distinction of having played a key role in the beginning of humanity's exploration of space.

AGENA SYSTEM STATISTICS

	AGENA-A	AGENA-B	AGENA-D
LENGTH	19 FT 4 IN (5.88 M)	23 FT 7 IN (7.18 M)	23 FT 7 IN (7.18 M)
DIAMETER	4 FT 11 IN (1.49 M)	4 FT 11 IN (1.49 M)	4 FT 11 IN (1.49 M)
THRUST	15,200 LB (67.6 KN)	16,000 LB (71.1 KN)	16,000 LB RESTARTABLE
WEIGHT	1,700 LB (771 KG)	2,300 LB (1.043 KG)	VARIED
PROPELLANTS	JP-4/IRFNA	UDMH/IRFNA	NITROGEN TETROXIDE/UDMH
PAYLOAD (LOW ORBIT)	5,000 LB (2268 KG)	5,780 LB (2,622 KG)	5,980 LB (2,712 KG)

SERVICE RECORD

By the time the Apollo missions were putting men on the Moon, in 1969, the Agena upper stage had been launched 257 times, carrying over 80% of all of the U.S. space program's payloads into orbit. Over the course of the Agena's lifetime of 362 launches, it was responsible for only six mission failures. No other upper stage can boast such a long and successful service life.

A large part of the Agena's success came from its excellent yet simple design. By using storable hypergolic propellants—fuels and oxidizers that ignite when they come into contact with each other—the Agena didn't need any special containment for its propellants, just simple pumps to bring the fuel and oxidizer together in the rocket nozzle. Originally, the Agena engines burned JP-4 jet fuel—also used in jet airplanes—and UDMH (unsymmetrical dimethyl hydrazine), but design improvements changed that to IRFNA (inhibited red fuming nitric acid) and UDMH.

The workhorse of America's early space program was the combination of a Thor launcher and an Agena upper stage. Some six different combinations of Thor and Agena accounted for nearly 180 launches out of Cape Canaveral between 1958 and 1972. And although the majority of these launches carried secret military payloads, NASA used the Thor-Agena to launch the Echo, Nimbus and Oscar communications satellites, as well as to conduct research on various satellite and reentry vehicle designs.

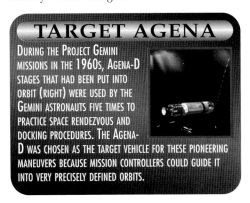

TARGET AGENA
DURING THE PROJECT GEMINI MISSIONS IN THE 1960S, AGENA-D STAGES THAT HAD BEEN PUT INTO ORBIT (RIGHT) WERE USED BY THE GEMINI ASTRONAUTS FIVE TIMES TO PRACTICE SPACE RENDEZVOUS AND DOCKING PROCEDURES. THE AGENA-D WAS CHOSEN AS THE TARGET VEHICLE FOR THESE PIONEERING MANEUVERS BECAUSE MISSION CONTROLLERS COULD GUIDE IT INTO VERY PRECISELY DEFINED ORBITS.

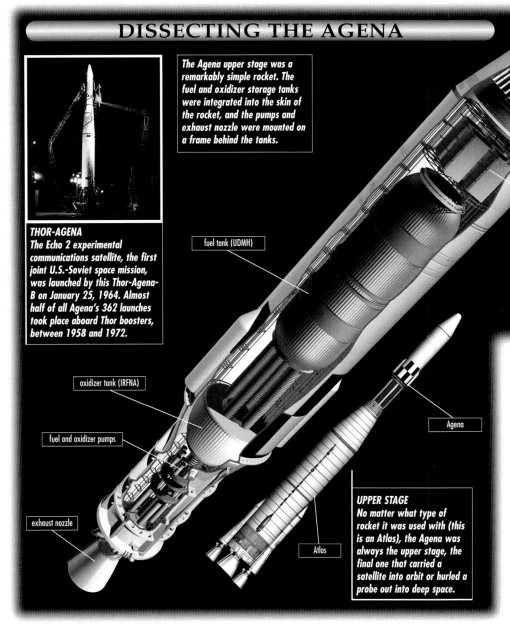

DISSECTING THE AGENA

The Agena upper stage was a remarkably simple rocket. The fuel and oxidizer storage tanks were integrated into the skin of the rocket, and the pumps and exhaust nozzle were mounted on a frame behind the tanks.

THOR-AGENA
The Echo 2 experimental communications satellite, the first joint U.S.-Soviet space mission, was launched by this Thor-Agena-B on January 25, 1964. Almost half of all Agena's 362 launches took place aboard Thor boosters, between 1958 and 1972.

payload fairing

payload

guidance, control and communications equipment

fuel tank (UDMH)

oxidizer tank (IRFNA)

fuel and oxidizer pumps

exhaust nozzle

Agena

Atlas

ATLAS-AGENA
An Atlas-Agena-B lifts off in 1962 carrying Ranger 3, a Moon probe. Three Ranger probes failed due to Agena malfunctions—1 and 2 did not reach their intended high Earth orbits because their Agenas failed to restart, and Ranger 3 missed the Moon because it was given excess velocity.

UPPER STAGE
No matter what type of rocket it was used with (this is an Atlas), the Agena was always the upper stage, the final one that carried a satellite into orbit or hurled a probe out into deep space.

SUCCESSFUL COMBOS

The Agena was also paired with Atlas and Thor launch vehicles. A total of 117 Atlas-Agena combinations were launched between 1959 and 1978, when the Atlas-Agena program was ended. Of those 117 launches, only three mission failures could be attributed to the Agena upper stage. Over time, technological improvements to both Atlas and Agena increased the size of the combination from about 100 feet in height to 115 feet (30 to 35 m), and the payload it could insert into a geosynchronous transfer trajectory grew from 1,800 to 2,240 pounds (816 to 1,016 kg).

But as the sophistication of satellites grew, so did their size. Eventually, some payloads grew too large for the Atlas-Agena to launch and more a capable rocket was needed. The Titan 3B rocket with an Agena upper stage stood over 150 feet (46 m) tall and could place an impressive 7,800-pound (3,538-kg) payload into low Earth orbit.

A total of 70 Titan-Agena vehicles were launched, all carrying military payloads. All of them were launched from Vandenburg Air Force Base in California—no Titan-Agena was ever launched from Cape Canaveral.

SOVIET N-1

There was more than one horse in the race to land a man on the Moon during the 1960s. The Soviet Union had plans to blast two cosmonauts into lunar orbit, one of whom would touch down on the surface. But the project faced impossible hurdles. It was started too late, in an atmosphere of political infighting, and it lacked support from the military who ran the Soviet space program. Even worse, the N-1 rocket—the U.S.S.R.'s answer to NASA's mighty Saturn 5—failed on all four of its launch attempts.

...THE SOVIET UNION HAD GOTTEN TO THE MOON FIRST?

July 20, 1969, is a day that millions around the world will never forget—the day that astronaut Neil Armstrong steered the lunar module Eagle unsteadily to its final resting place in the Sea of Tranquility and took one "giant leap for mankind." Apollo 11's landing was arguably the greatest live TV event that the world had ever seen, with unforgettable images not only of the Moon, but of the relieved Mission Controllers who had steered the Apollo craft to a successful landing.

Judging by the secrecy and lack of media cooperation that surrounded other Soviet space triumphs of the 1960s, things would have been very different if a red flag had been run up the mast in place of Old Glory. NASA would have certainly been hard-pressed to find praise for the Russians' achievement, given that they had been narrowly edged out and had failed to reach the goal set them by President John F. Kennedy in 1961.

Perhaps history would have recorded the event something like this:

"There was no official news of Babushka I until after the launch. But once the K1's fourth stage had fired successfully to propel the tiny craft on its planned lunar trajectory, the Soviet propaganda machine went into full swing. Pravda and Tass gave the mission headline coverage, claiming that it was a flight to test a spacecraft "near the Moon." There was no mention of a landing, either in the press, or in the onboard TV clips that were released to western networks the following day

Beating the U.S. to the Moon would have been a great propaganda coup for the Soviets—if their lone cosmonaut had returned safely.

(having first been vetted by the authorities).

Three days later, while NASA scientists pondered the meaning of it all, Flight Control Center, Baikonur, received confirmation that a lone cosmonaut had set foot on the lunar surface. Twelve hours after that, the first grainy TV footage of the Babushka's lunar landing, shot from an automatic camera above the exit hatch, crackled across the world's television screens. Viewers watched bemused as cosmonaut Valentin Sadoyevesky paused to fix a device like a hula hoop to his spacesuit (since he was alone, he could not afford to trip and be unable to regain his feet). Sadoyevesky spent two hours deploying scientific experiments and collecting rock samples before returning to the lunar lander. In a telephone linkup, Soviet leader Leonid Brezhnev congratulated the cosmonaut warmly. Brezhnev went on to proclaim the landing "yet another triumph for socialism"and called on third-world nations to note which country had reached the Moon first.

Meanwhile, in the United States, Congress called for increased spending on the Apollo program. Construction of a lunar base was given top priority, since CIA reports indicated that the Soviet government had approved plans for advanced N-1M rockets to help build a Soviet lunar base. The Cold War was spreading to the Moon."

N-1 LAUNCHER STATS

PAYLOAD (INTO 125-MILE/200-KM EARTH ORBIT)	99 TONS (90 TONNES)	1ST STAGE (BLOCK A)	30 x NK-33 ENGINES,
TOTAL WEIGHT (AT LAUNCH)	3,086 TONS (2,800 TONNES)		EACH OF 170 TONS-FORCE (1,512 KN)
PROPELLANT MASS (OXYGEN)	1,906 TONS (1,729 TONNES)	2ND STAGE (BLOCK B)	8 x NK-43 ENGINES,
PROPELLANT MASS (KEROSENE)	750 TONS (680 TONNES)		EACH OF 198 TONS-FORCE (1,760 KN)
TOTAL HEIGHT	345 FT (105 M)	3RD STAGE (BLOCK V)	4 x NK-3 ENGINES,
BASE WIDTH	55 FT (16 M)		EACH OF 45 TONS-FORCE (400 KN)

CATALOG OF DISASTER

The N-1 rocket was developed by the Soviet Union during the 1960s and early 1970s as a launch vehicle to carry two men to the Moon. The plan was similar to the Apollo program, in that it employed the concept of Lunar Orbit Rendezvous. This involved firing a pair of spacecraft with a two-person crew toward the Moon on a giant rocket (the N-1). The two craft would separate, leaving a lunar module (code-named LK) to land on the surface. Part of the LK would then lift off to rejoin its sister craft (code-named LOK) in lunar orbit and the crew would fire the LOK's main engine to head home.

The N-1 rocket, however, was very different from the Apollo program's Saturn 5 launch vehicle. Soviet chief designer Sergei Korolev, under pressure from his masters in Moscow, was aware that his design team simply didn't have the time to perfect the super-efficient cryogenic liquid hydrogen rocket engines being developed by NASA. To stay in the race, he had to rely on the tried-and-tested liquid oxygen/kerosene engines that had blasted previous Soviet rockets into space.

Korolev knew that the pound-for-pound performance of these engines was considerably less than that of the American rockets. To compensate, the first stage of the N-1 rocket had no less than 30 engines—compared with just five in Saturn 5. This presented major construction difficulties that Korolev hoped to get around by designing the propellant tanks as huge spheres and then building the walls of the rocket around them.

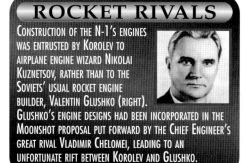

ROCKET RIVALS

CONSTRUCTION OF THE N-1'S ENGINES WAS ENTRUSTED BY KOROLEV TO AIRPLANE ENGINE WIZARD NIKOLAI KUZNETSOV, RATHER THAN TO THE SOVIETS' USUAL ROCKET ENGINE BUILDER, VALENTIN GLUSHKO (RIGHT). GLUSHKO'S ENGINE DESIGNS HAD BEEN INCORPORATED IN THE MOONSHOT PROPOSAL PUT FORWARD BY THE CHIEF ENGINEER'S GREAT RIVAL VLADIMIR CHELOMEI, LEADING TO AN UNFORTUNATE RIFT BETWEEN KOROLEV AND GLUSHKO.

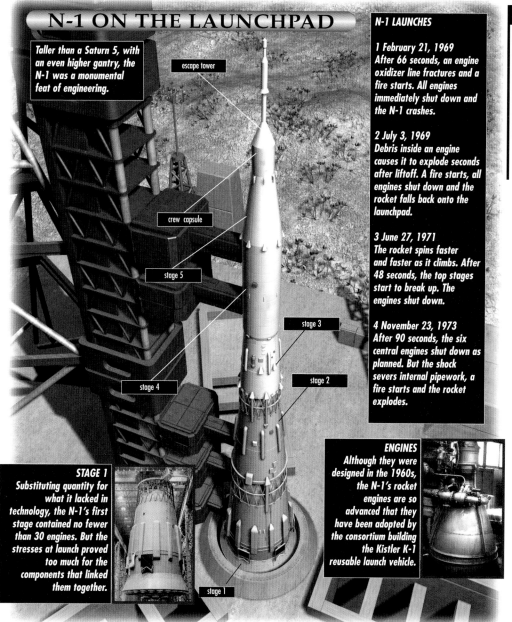

N-1 ON THE LAUNCHPAD

Taller than a Saturn 5, with an even higher gantry, the N-1 was a monumental feat of engineering.

escape tower

crew capsule

stage 5

stage 3

stage 4

stage 2

stage 1

STAGE 1
Substituting quantity for what it lacked in technology, the N-1's first stage contained no fewer than 30 engines. But the stresses at launch proved too much for the components that linked them together.

ENGINES
Although they were designed in the 1960s, the N-1's rocket engines are so advanced that they have been adopted by the consortium building the Kistler K-1 reusable launch vehicle.

N-1 LAUNCHES

1 February 21, 1969
After 66 seconds, an engine oxidizer line fractures and a fire starts. All engines immediately shut down and the N-1 crashes.

2 July 3, 1969
Debris inside an engine causes it to explode seconds after liftoff. A fire starts, all engines shut down and the rocket falls back onto the launchpad.

3 June 27, 1971
The rocket spins faster and faster as it climbs. After 48 seconds, the top stages start to break up. The engines shut down.

4 November 23, 1973
After 90 seconds, the six central engines shut down as planned. But the shock severs internal pipework, a fire starts and the rocket explodes.

TOO HEAVY BY FAR

Unfortunately, Korolev's design solution made for a heavier, even less efficient rocket. The N-1 was so big that it had to be built panel by panel at the launchpad using parts shipped in by rail. And, as it turned out, all four test launches failed while the first stage was firing due to component failures.

Above the N-1's problematic first stage

HARD EVIDENCE

BIG BLAST
The second N-1 launch failure ended with the rocket falling back onto launchpad 110R, completely destroying it and showering the neighboring pad, where another N-1 rocket was being readied for launch, with debris. Rebuilding the 110R took two years. The blast damage and scars (right) were clearly visible on contemporary U.S. spy satellite photos.

were conventional second and third stages, similar to those on the Saturn 5. Beyond these were two "extra" stages—specially designed to burn in Earth orbit—that would give the LOK/LK craft its final kick toward the Moon. The last stage, called "Block D" was originally developed in the mid-1960s under the code name "Sputnik." It is still used today on the Proton rocket.

At the very top of the N-1 was an escape tower to carry the crew capsule away from a launchpad disaster. The tower is known to have fired during three of the four launch failures, carrying real, but unoccupied, crew capsules to safety. Rumors persist, however, of a manned escape tower failure.

Despite the problems, development of the N-1 continued as various engine design bureaus sought to incorporate more powerful liquid hydrogen/oxygen engines into the original design concept. Pairs of these rockets, designated N-1M, were intended to launch large lunar landers that could be assembled in Earth orbit.

The N-1 program was finally canceled in 1974, by which time two more rockets had been prepared for launch. Its successor—the brainchild of chief designer Korolev's old rival, Valentin Glushko—was a rocket called Vulkan, later to become the Energia. Today, reminders of the once-mighty N-1 litter the remote plains of Baikonur Cosmodrome. Ever short of funds, the resourceful Russians have put the old hardware to good use—as water tanks, a bandstand and even a children's play area in the nearby town of Leninsk.

SATURN SERIES

The 364 ft (111 m) -tall monster that carried men to the Moon was the last and largest variant of an entire family of Saturn designs. German rocket designer Wernher von Braun and his team of engineers configured their Saturn launcher in a variety of different ways, to serve all of America's potential needs in space. In the end, only three members of the Saturn series ever exchanged the drawing board for the launchpad—the Saturn 1, the upgraded and crew-rated Saturn 1B and the Moon-bound Saturn 5.

WHAT IF...

...A SATURN 5 LAUNCHED ASTRONAUTS TO MARS?

The Saturn 5 was the largest, most powerful rocket ever built. More than a quarter of a century after its final flight, no space agency possesses a launcher that can carry a crewed spacecraft beyond low Earth orbit, let alone boost it onto a lunar or interplanetary trajectory. So would it be possible to resurrect the Saturn 5 design for the next giant leap in space exploration—a crewed mission to Mars?

In theory, yes, although the practical difficulties would be considerable. The heavy plant used to build the Saturns was disposed of after 15 of the massive rockets had been delivered, and no one knows for sure if even the factory blueprints are still intact. Added to this would be the difficulty of sourcing the launcher's thousands of individual parts—many of them now obsolete—plus the skilled labor to assemble them.

In the unlikely event that these problems could be overcome for less than the cost of developing a new launcher, what sort of Mars missions would the Saturn 5 make possible? As with the lunar program, mission planners would have two basic choices.

The most straightforward, but expensive, option would be to use a series of Saturn 5s to

For the lone astronaut aboard NASA's cut-price Saturn-launched mission to Mars, planting Old Glory on the planet's surface would be bittersweet: After a stopover of just 30 days, he or she would have to blast off again on the 12-month trip home.

place separate modules in Earth orbit. These modules could then be combined into a Mars ship big enough to permit a leisurely stay of 18 months on the planet's surface before the Earth and Mars were once again optimally aligned for the voyage home.

Cheaper, but riskier, would be to launch a single Saturn 5 toward the Red Planet, perhaps with a high-energy Centaur or nuclear-powered third stage for the Earth orbit escape burn. There still would not be enough power to transport an Apollo-style spacecraft—plus the supplies and life support systems needed to support three astronauts for a return journey lasting a year or more. A single hardy astronaut might just make it, provided the mission allowed only a token 30-day sojourn on the Martian surface. But to travel for so long, in such cramped conditions and for such a short stay raises the question of whether it would be worth going at all.

THE SATURN STORY

DECEMBER 30, 1957	WERNHER VON BRAUN PRODUCES A "PROPOSAL FOR A NATIONAL INTEGRATED MISSILE AND SPACE VEHICLE DEVELOPMENT PLAN," WHICH PROPOSES DEVELOPMENT OF THE SATURN 1
JULY 29, 1958	SATURN 1 PROJECT CONTRACT ISSUED BY ARPA
NOVEMBER 2, 1959	TRANSFER OF SATURN 1 PROJECT FROM THE U.S. ARMY TO NASA ANNOUNCED
OCTOBER 27, 1961	FIRST SATURN 1 FLIGHT
JANUARY 5, 1962	NASA ANNOUNCES DEVELOPMENT OF THE SATURN 5 LAUNCH VEHICLE FOR APOLLO
OCTOBER 30, 1963	MANNED SATURN 1 FLIGHTS CANCELED, TO BE REPLACED BY SATURN 1B FLIGHTS
FEBRUARY 26, 1966	FIRST SATURN 1B FLIGHT
NOVEMBER 9, 1967	FIRST SATURN 5 FLIGHT (APOLLO 4)
OCTOBER 11, 1968	FIRST MANNED SATURN 1B FLIGHT (APOLLO 7)
DECEMBER 21, 1968	FIRST MANNED SATURN 5 FLIGHT (APOLLO 8)
JULY 16, 1969	SATURN 5 LIFTS OFF WITH CREW OF APOLLO 11
MAY 14, 1973	FINAL SATURN 5 LAUNCH CARRIES SKYLAB INTO ORBIT IN PLACE OF ITS THIRD STAGE

GROWING FAMILY

The Saturn launchers were the first rockets designed solely to transport men into space. Earlier designs, such as the Redstone, the Atlas and the Russian R-7, were ballistic missiles modified to carry crew capsules in place of warheads. The Saturns' peaceful origin is perhaps surprising, given that when designer Wernher von Braun and his team began work on the rockets in 1958, they were still employed by the military.

After the U.S. Army plucked the rocket builders out of defeated Nazi Germany in 1945, they worked at the Army Ballistic Missile Agency (ABMA), based in Huntsville, Alabama. But von Braun had always dreamed of the peaceful exploration of space. He proposed a 1.5-million-pound thrust (66,700 kN) booster, 10 times more powerful than the existing Jupiter, to be used for the purpose. In the aftermath of Russia's surprise launch of Sputnik 1, the Pentagon responded positively to von Braun's proposal. The ABMA began work on a first-stage Saturn booster while also studying possible upper-stage designs.

CLUSTERS, QUICK!

To save time and money, the kerosene and liquid oxygen first stage was built by clustering together eight Rocketdyne H-1 engines—previously used one at a time in the Thor and Jupiter missiles. This raised concerns about whether so many boosters could work together reliably: Skeptics dubbed the Saturn 1

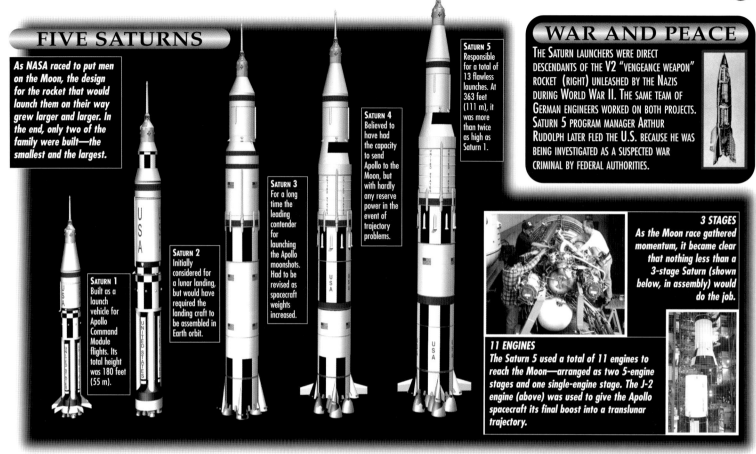

FIVE SATURNS

As NASA raced to put men on the Moon, the design for the rocket that would launch them on their way grew larger and larger. In the end, only two of the family were built—the smallest and the largest.

SATURN 1 Built as a launch vehicle for Apollo Command Module flights. Its total height was 180 feet (55 m).

SATURN 2 Initially considered for a lunar landing, but would have required the landing craft to be assembled in Earth orbit.

SATURN 3 For a long time the leading contender for launching the Apollo moonshots. Had to be revised as spacecraft weights increased.

SATURN 4 Believed to have had the capacity to send Apollo to the Moon, but with hardly any reserve power in the event of trajectory problems.

SATURN 5 Responsible for a total of 13 flawless launches. At 363 feet (111 m), it was more than twice as high as Saturn 1.

WAR AND PEACE

THE SATURN LAUNCHERS WERE DIRECT DESCENDANTS OF THE V2 "VENGEANCE WEAPON" ROCKET (RIGHT) UNLEASHED BY THE NAZIS DURING WORLD WAR II. THE SAME TEAM OF GERMAN ENGINEERS WORKED ON BOTH PROJECTS. SATURN 5 PROGRAM MANAGER ARTHUR RUDOLPH LATER FLED THE U.S. BECAUSE HE WAS BEING INVESTIGATED AS A SUSPECTED WAR CRIMINAL BY FEDERAL AUTHORITIES.

3 STAGES As the Moon race gathered momentum, it became clear that nothing less than a 3-stage Saturn (shown below, in assembly) would do the job.

11 ENGINES The Saturn 5 used a total of 11 engines to reach the Moon—arranged as two 5-engine stages and one single-engine stage. The J-2 engine (above) was used to give the Apollo spacecraft its final boost into a translunar trajectory.

"Cluster's Last Stand." In response, the infant civilian space agency NASA contracted Rocketdyne to build a new engine—the F-1—with enough thrust to match all eight H-1s.

PRESSED FOR TIME

In November 1959, NASA inherited the Saturn program, along with the ABMA itself, from the Army. During the next 18 months, combination tests of H-1 engines showed cluster fears to be misplaced, and work proceeded on a liquid hydrogen and oxygen upper stage for the Saturn.

The first 2-stage Saturn 1 flew from Cape Canaveral on a suborbital test flight on October 27, 1961. Initially, von Braun believed that it would be sufficient to achieve his dream

of men on the Moon—the plan was to launch up to 15 Saturn 1s, to assemble a moonship in Earth orbit. But President Kennedy's pledge to place a man on the Moon by the end of the decade put a strict deadline on the project that the Saturn 1 plan could not meet.

To reach the Moon in a single launch, a much more powerful multi-stage rocket would be needed. Von Braun's answer was the 3-stage Saturn 5, so-called because it had no less than five mighty F-1 engines clustered together to form the first stage.

As work proceeded on the Saturn 5, the Saturn 1 was upgraded with improved first-stage engines, increased automation with a computerized Instrumentation Unit, and a liquid hydrogen and oxygen upper stage that also served as the Saturn 5 third stage.

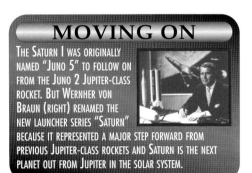

MOVING ON

THE SATURN 1 WAS ORIGINALLY NAMED "JUNO 5" TO FOLLOW ON FROM THE JUNO 2 JUPITER-CLASS ROCKET. BUT WERNHER VON BRAUN (RIGHT) RENAMED THE NEW LAUNCHER SERIES "SATURN" BECAUSE IT REPRESENTED A MAJOR STEP FORWARD FROM PREVIOUS JUPITER-CLASS ROCKETS AND SATURN IS THE NEXT PLANET OUT FROM JUPITER IN THE SOLAR SYSTEM.

This upgraded model, known as the Saturn 1B, had double the payload capacity of its predecessor, enabling it to be used for flight tests of Apollo hardware. The first uncrewed Saturn 1B flew in February 1966.

The first Saturn 5 followed in November 1967. Just under a year later, a crewed Saturn 1B successfully flew as Apollo 7, followed in December by a crewed Saturn 5 that sent Apollo 8 around the Moon.

For the next four years, Saturn 5s carried men to the lunar surface, and in 1973 the final Saturn 5 hoisted the Skylab space station into low Earth orbit. Meanwhile, Saturn 1Bs carried three separate crews up to Skylab and, in 1975, lifted the Apollo Command Module that docked with a Russian Soyuz spacecraft in orbit. It was a perfect flight record.

SATURN 5

S aturn 5, the launcher that lifted the Apollo crews to the Moon, was the mightiest rocket the United States has ever built. Taller than the Statue of Liberty, heavier than a Navy destroyer and containing almost three million working parts, this massively powerful machine never failed once in its short lifetime. The giant Saturn 5 first flew in 1967, and its final mission—to launch the Skylab space station—came just six years later. Only 15 Saturn 5s were built and, because of budget cuts, two of these never left the ground.

WHAT HAPPENED TO...

THE SATURN 5 ROCKETS?

When President Kennedy announced in 1961 that the U.S. aimed to land a man on the Moon "before the end of the decade," NASA calculated that 15 Saturn 5 launches would be needed to achieve Kennedy's goal. In fact, just six Saturn launches put the first men on the Moon. But after Apollo's success, public and political support for the space program waned. Congress refused to fund any Saturn 5s beyond the 15 originally ordered.

Of these, two were used for uncrewed test launches (Apollos 4 and 6), 10 lifted the manned Apollo missions (Apollos 8 through 17) and one was used to orbit the Skylab space station. Two Saturn 5s stayed on the ground: NASA could no longer afford to launch them.

Stages from these remaining two rockets are now on public display, along with parts of other versions used for static testing. You can see them at the Johnson Space Center, Texas (Space Center Houston), the Kennedy Space Center, Florida (Spaceport USA), and the Marshall Space Flight Center, Alabama (U.S. Space & Rocket Center). The planned payload for one of the unused Saturns, a backup Skylab, is on exhibition at the National Air & Space Museum in Washington, D.C.

Though none was ever built, NASA studied

Saturn 5 stages (above) attract visitors to the Johnson Space Center in Houston, Texas. These and other surviving stages exhibited across America have been described, somewhat unfairly, as "the most expensive lawn ornaments in history."

improved designs of the Saturn 5. They included a version with strap-on solid rocket boosters to lift more payload; another with aerospike engines like those on the X-33; and a variant with parachute systems that would allow the controlled descent and splashdown of the second and third stages for recovery and reuse. There were even plans for a Saturn 5 with a nuclear-powered third stage that could boost a crewed mission to Mars.

The future for NASA turned out to be a reusable rocket—the Space Shuttle—which in theory was cheaper than a Saturn 5, even though it could never reach the Moon. In fact, the Shuttle ended up costing far more than Saturn. And it still cannot reach the Moon.

THE SATURN 5 ROCKET

	First Stage (S-1C)	Second Stage (S-2)	Third Stage (S-4B)
Prime contractor:	Boeing	North American	Douglas Aircraft
Height:	138 ft/42 m	81.5 ft (25 m)	58.6 ft (18 m)
Diameter:	33 ft/10 m	33 ft (10 m)	21.7 ft (6.6 m)
Total thrust:	7.9m lbf/35,124 kN (liftoff)	1.15m lbf/5,113 kN (average)	200,000 lbf/889 kN
Engines:	5 x Rocketdyne F-1	5 x Rocketdyne J-2	1 x Rocketdyne J-2
	liquid oxygen/kerosene	liquid oxygen/liquid hydrogen	liquid oxygen/liquid hydrogen

General Statistics			
Launches:	13 (all successful)	Total weight:	400,000 lb/181,500 kg (unfueled),
Cost:	$200 million each (1960s prices)		6.2m lb/2.8m kg (fueled)
Combined height:	363 ft (111 m)	Payload:	250,000 lb (113,400 kg) to Earth orbit
			100,000 lb (45,359 kg) to Lunar orbit

MOON ROCKETS

On July 16, 1969, Neil Armstrong, Buzz Aldrin and Michael Collins lifted off from Kennedy Space Center (KSC), Florida, at the start of the first mission to land men on the Moon. The rocket that carried them into space was a Saturn 5, a three-stage vehicle over 360 feet (110 m) high, 33 feet (10 m) in diameter at its widest and weighing over 3,000 tons (3,306 tonnes). At liftoff, its 5 Rocketdyne F-1 rocket engines generated a total thrust of 7.65 million pounds (34,000 kN). Later versions of these engines, the most powerful ever built, produced up to 7.9 million pounds thrust (35,000 kN).

The three stages of the Saturn 5 were themselves all rockets of the Saturn family—the first stage was a Saturn 1C, the second a Saturn 2, and the third a Saturn 4B. During a launch, the engines of the first stage burned for two minutes 50 seconds, taking the vehicle to an altitude of about 38 miles (61 km). Then explosive bolts were fired to separate it from

the two upper stages, and it fell into the Atlantic Ocean about 375 miles (600 km) downrange from KSC.

If there had been an emergency during liftoff—such as a launch pad explosion—or any life-threatening problem during the first three minutes of the flight, an escape rocket on top of the Saturn 5 would have fired to haul the command/service module and its crew clear of the vehicle. Later in their mission, the crew could have separated their command module from the Saturn 5 without the assistance of the escape rocket.

After first-stage separation, the five Rocketdyne J-2 second-stage engines were ignited. These burned for another 6½ minutes until at an altitude of 115 miles (185 km), more explosive bolts fired and the second stage, too, fell back into the ocean far below.

THIRD STAGE

The single J-2 engine of the third stage burned for about 2½ minutes to put the spacecraft into Earth orbit. After one orbit, Apollo was redirected toward the Moon by a second, five-minute firing of the same engine.

Finally, the third stage separated from the rest of the spacecraft, leaving thrusters to steer the module safely out of harm's way. What remained of the once-mighty rocket was left to crash on the Moon.

SATURN 5

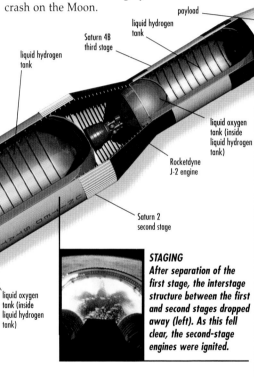

payload

liquid hydrogen tank

Saturn 4B third stage

liquid hydrogen tank

liquid oxygen tank (inside liquid hydrogen tank)

Rocketdyne J-2 engine

Saturn 2 second stage

liquid oxygen tank (inside liquid hydrogen tank)

five Rocketdyne J-2 engines

liquid oxygen tank

RP-1 (kerosene) tank

Saturn 1C first stage

five Rocketdyne F-1 engines

ENGINE NOZZLES
The nozzle of each F-1 engine in the first stage of the Saturn 5 was big enough to hold a small car. The first stage burned up to 15 tons (13.5 tonnes) of fuel per second, and some of the fuel lines were big enough for a man to crawl through.

STAGING
After separation of the first stage, the interstage structure between the first and second stages dropped away (left). As this fell clear, the second-stage engines were ignited.

CAMERA CAPSULE

ROCKET ENGINEERS WERE CONCERNED ABOUT THE RELIABILITY OF SATURN 5 STAGE SEPARATIONS, SO EARLY ROCKETS CARRIED MOVIE CAMERA CAPSULES TO FILM THESE CRITICAL MOMENTS OF THE FLIGHT. THE CAMERAS WERE EJECTED INTO THE OCEAN, WHERE THEY WERE RECOVERED BY DIVERS WHO REACHED THE SCENE BY PARACHUTE.

lens

"paraballoon" for descent braking

quartz window

camera   stabilizing fin

BLOCKBUSTER

THE ROAR OF THE FIRST SATURN 5 LAUNCH, ON NOVEMBER 9, 1967, WAS LOUD ENOUGH TO ROCK BUILDINGS MILES AWAY. AT THE PRESS CENTER THREE MILES (5 KM) FROM THE LAUNCH PAD, STAFF IN THE CBS TELEVISION BOOTH HAD TO HOLD THE WALLS AND WINDOWS TO STOP THE WHOLE BOOTH FROM COLLAPSING. VETERAN BROADCASTER WALTER CRONKITE (RIGHT) JUST KEPT ON TALKING.

PROTON LAUNCHERS

The workhorse of the Soviet and Russian space programs, the Proton (or D-Series) rocket was developed as a direct result of a Soviet Central Committee decree in the early 1960s. Originally, it was designed as a family of rockets that could serve dual use as missiles and space launchers, but changes in the Soviet political world resulted in development for space launch only. Since then, the Proton has been constantly upgraded and improved and has become one of the most reliable and efficient launch vehicles in the world.

WHAT IF...

...PROTON ROCKETS HAD GONE TO THE MOON?

At the beginning of the 1960s, when President Kennedy announced America's intention to put a man on the Moon, rocket designers in the U.S.S.R., such as Vladimir Chelomei and Sergei Korolev, decided to get a Soviet cosmonaut there first.

Vladimir Chelomei's Universal Rocket family included the UR-700, which Chelomei had designed for direct crewed flight to the surface of the Moon. The UR-700 was to use UR-500 Proton core stages as the basis of its modules and upper stages.

The basic configuration of the UR-500, including the modular design that would allow it to be used to build the UR-700, was selected in January 1962. In April 1962, the Soviet government approved development of the UR-500, but not of the UR-700. The UR-700 had fallen prey to political infighting between Chelomei and Korolev, with Korolev's massive N-1 booster coming out on top.

But that wasn't the end of Proton's lunar ambitions. Three years later, Korolev, although originally opposed to the Proton, actually suggested using the UR-500K to fly around the Moon. He proposed to use the Block D stage from his own N1-L3 lunar rocket stack as a fourth stage on the Proton to launch the 7K-L1 manned lunar module onto a translunar trajectory.

The government bought the idea, setting up

Oleg Makarov (left) and Alexei Leonov would have been among the first cosmonauts to fly around the Moon, and perhaps even the first to land on it, if the Soviet crewed lunar program had been successful.

an 18 UR-500K rocket program called Zond that would send L1 spacecraft around the Moon, at first uncrewed, then with a crew. Unfortunately, malfunctions in the Zond craft were not ironed out until 1970, by which time American astronauts had not only flown around the Moon but also landed on it.

When Apollo 11 landed on the Moon, Soviet plans to land one of their cosmonauts there evaporated. And though Chelomei's bureau studied a number of different ways of clustering the basic modules for a lunar attempt, the project was finally and definitively suppressed in 1974.

The hardware was hidden away, and official Soviet policy was to deny they had ever even considered going to the Moon. Details of Soviet lunar designs came to light only at the end of the 20th century, and today, much of the equipment is finding its way into space museums around the world.

PROTON LAUNCHERS

FIRST LAUNCH	JULY 16, 1965		PAYLOAD	48,400 LB (22,000 KG) LOW-EARTH ORBIT	
LIFTOFF THRUST	APPROXIMATELY 2,025,000 LB (9000 KN)			5,500 LB (2,500 KG) LUNAR TRANSFER ORBIT	

	STAGE 1	STAGE 2	STAGE 3	STAGE 4
LENGTH	69 FT (21 M)	46 FT (14 M)	21 FT, 4 IN (6.5 M)	18–28 FT (5.5–8.5 M)
DIAMETER	24 FT (7M)	13 FT (4 M)	13 FT (4 M)	11.5–13 FT (3.5–4 M)
GROSS MASS	904,000 LB (410,000 KG)	365,000 LB (165,500 KG)	123,000 LB (55,800 KG)	38.9–44,000 LB (17,690–20,000 KG)
ENGINES	6	4	1 PLUS 4 VERNIERS	1
PROPELLANT	N204/UDMH	N204/UDMH	N204/UDMH	LOX/RP-1

PREMIER BOOSTER

The Proton booster, also known as the D-Series, was the first Soviet booster that was not based on an existing intercontinental ballistic missile design. It was entirely new and went through many configurations before settling on the excellent design used today. The original design concept included three models that could serve as launchers for either spacecraft or ballistic missiles—the medium-lift UR-200, heavy lift UR-500, and a lunar-launch-capable UR-700 (the "UR" in the designations stood for "Universal Rocket").

But the Soviets canceled the UR-200, concentrated their efforts on developing the UR-500 and UR-700 as space launch vehicles only, and ended up with the Proton, one of the most successful series of launchers ever built. The reason for the Proton's success is its flexibility. It can be launched in either 3- or 4-stage configurations and has several different upper stages that can be used as the fourth stage.

Although it was first launched in 1965, the Proton was not considered fully operational until 1970, after its 61st launch. In fact, in 1969, the Proton failed in seven out of nine launch attempts, but Soviet engineers isolated and fixed the problems and the vehicle went on to become a highly reliable launcher. The

SPACE TRADE

A U.S.-Russian consortium, International Launch Services (ILS), now offers commercial launch facilities using Proton boosters. The consortium, formed in 1995, consists of Lockheed Martin and Russian rocket makers Khrunichev and RSC Energia. The first Western satellite to be launched on an ILS Proton, in 1996, was the Astra 1F communications satellite—the launch (right) is the Astra 1H satellite, on June 18, 1999.

PROTON PARTS

OUT TO THE PAD
A completed Proton K/Block DM rocket is transported by rail to its Baikonur launch pad (left). This 4-stage rocket successfully launched the Inmarsat 3 communications satellite on September 6, 1996.

ON THE RISE
The Proton carrying Inmarsat 3 (left) is lifted into its launch position at Baikonur. The satellite was the second Western payload to be launched by a Proton rocket operated by International Launch Services.

The Proton 8K82K (or D-1e) was first launched in 1967. It has an all-aluminum-alloy structure covered by an external thermal finishing paint.

payload

payload shroud

3rd stage—one single-chamber liquid propellant rocket engine

4th stage—one liquid propellant rocket engine

shroud

2nd stage—four single-chamber liquid propellant rocket engines

UR VARIANTS
Some of the many variants of the UR series of launchers, the development of which led to today's very successful Proton vehicles. Those shown here are the UR-200 (far left) plus several UR-500 models including the massive UR-700K (far right).

1st stage—six single-chamber liquid propellant rocket engines

annual number of Proton launches grew from six in 1970 to a peak of 13 in 1985. Between 1983 and 1986, Proton rockets achieved a string of 43 consecutive successes, a record not yet matched by any other launcher.

ROCKETS BY RAIL

Sections of the Proton are manufactured in the Khrunichev factory outside Moscow, which can produce up to 16 rockets per year

and needs to make and sell at least six a year to stay in business. The finished sections are shipped by rail and water barge to the Baikonur Cosmodrome in Kazakhstan, where they are assembled horizontally in the Integration and Testing Facility.

Completed Protons are transported by rail from the assembly building out to one of four operational pads. Once erected, the launch vehicle is fastened directly onto the launch pad by its tail, and the payload is lifted vertically and joined to the rocket. The vehicle is launched within three to five days of its arrival at the pad. At the moment of liftoff, the tower actually stays connected to the rocket for a short time instead of completely releasing it, as happens with other rockets. A safety mechanism on the tower rises with the rocket, tracking the first few moments of flight to ensure everything is going correctly.

Proton rockets have launched more scientific probes than any other vehicle in the world, having boosted the Phobos missions to Mars, Venera probes to Venus, Luna series to the Moon, and the Vega probe to Halley's Comet. And as well as launching probes, they are now being used to lift sections of the International Space Station into orbit and to carry out commercial satellite launches for Western customers.

SOYUZ SERIES

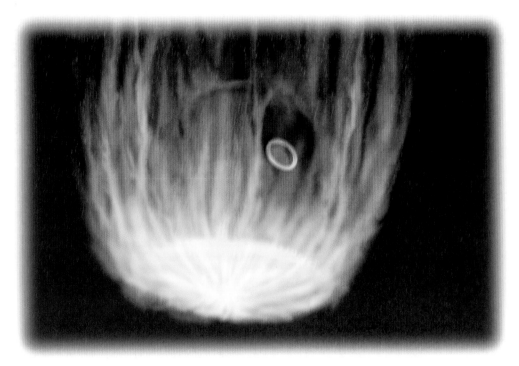

The orbiting workhorse of the Soviet and Russian space programs was first proposed in 1962 as a 2-man capsule for space rendezvous and docking that could also be used on a Moon mission. But as with the disaster that killed three Apollo astronauts in January 1967, catastrophe forced the redesign of the craft. The Soyuz ("Union") capsule that emerged proved phenomenally successful, flying more than 100 missions. In modified form, it will be used well into this century—over 40 years after its first flight.

WHAT IF...

...SOYUZ STAYS IN SERVICE?

The name "Soyuz" actually refers to both the spacecraft and its launch vehicle, the only crew-rated Russian booster. This highly successful launch vehicle, the most frequently flown rocket in the world, is a member of the R-7 series of rockets. It was first used in 1960, to launch Mars probes, and since 1964 it has launched every Russian crewed space mission.

Both spacecraft and launcher have roles to play in the creation of the International Space Station (ISS), a 16-nation venture which was conceived as a US project during the Cold War and scheduled for completion in 2004. The Space Station project was plagued by delays and problems, and finally ended up becoming a cooperative project between many nations – some of them enemies at the time of the station's design. Originally 37 of the 46 launches required for the assembly were to be U.S. Shuttles and nine of the launches would use Russian vehicles. The first two of these, using Proton rockets, delivered the control and service modules.

The Space Shuttle Colombia disaster and subsequent problems with the shuttle fleet cast doubt on the future of the already-imperilled ISS. In the interim, Soyuz shouldered the burden of keeping the station alive. Beginning In November 2000, Soyuz spacecraft began delivering crewmembers and returning them safely to Earth. Cabin space limitations mean that two personnel are delivered to the ISS at a time instead of the intended three, but by stepping into the breach Soyuz has ensured that the station can operate while the Shuttle service is interrupted.

While the station is supplied by uncrewed Progress rockets, it is Soyuz that carries humans and stands by as the station's 'lifeboat'. In the event of a catastrophe threatening the crew's quarters—such as a fire, or a meteoroid strike that knocks out the power and makes the station uninhabitable—the crew will use Soyuz craft to make their escape. At least one Soyuz TMA craft, capable of carrying three astronauts, remains docked at the station for use in any life-threatening emergency. There is usually also a Progress vehicle present.

If the station has to be abandoned, its crew will board the Soyuz craft and fly safely back to Earth. 'lifeboats' are rotated each time a new crew arrives; the old crew returning home aboard their lifeboat and the new arrival becoming the escape vessel for the current crew. This avoids any danger of the lifeboat developing faults due to excessive time in space.

Once the Shuttle returns to full service, the ISS will likely be brought up to a full crew complement of 6. Two Soyuz TMA craft will be needed to provide the necessary escape capability, and these will be rotated every six months.

SOYUZ TM AND PROGRESS M

	Length	Max. Diameter	Habitable volume	Weight
Soyuz TM Spacecraft	24.6 ft (7.5 m)	8.9 ft (2.7 m)	318 cubic ft (9 m³)	15,984 lb (7,250 kg)
Descent Module	7.2 ft (2.2 m)	7.2 ft (2.2 m)	141 cubic ft (4 m³)	6,614 lb (3,000 kg)
Orbital Module	9.8 ft (3.0 m)	7.5 ft (2.3 m)	177 cubic ft (5 m³)	2,866 lb (1,300 kg)
Service Module	7.5 ft (2.3 m)	8.9 ft (2.7 m)	n/a	6,504 lb (2,950 kg)

	Length	Max. Diameter	Overall Mass	Payload
Progress M Spacecraft	23.6 ft (7.2 m)	8.9 ft (2.7 m)	16,424 lb (7,450 kg)	5,600 lb (2,540 kg)
Cargo Module	9.8 ft (3 m)	7.5 ft (2.3 m)	5,555 lb (2,519 lb)	2,954 lb (1,340 kg)
Refueling Module	7.2 ft (2.2 m)	7.2 ft (2.2 m)	4,365 lb (1,980 lb)	2,646 lb (1,200 kg)
Service Module	6.8 ft (2.1 m)	8.9 ft (2.7 m)	6,504 lb (2,950 lb)	n/a

SPACE CARRIER

Soyuz is the most successful series of spacecraft yet built, but its early days were dogged by disaster. The first Soyuz flight, in April 1967, ended in tragedy: Its pilot, Vladimir Komarov, was killed when the craft crashed after reentry. It took 18 months for the program to recover, with a successful orbital near-docking of Soyuz 2 and Soyuz 3.

Then in June 1971, another catastrophe struck. Three cosmonauts were sent aloft in Soyuz 11 and transferred to a new 20-ton space station, Salyut. This new Soviet "first" drew tremendous publicity, but when Soyuz 11 returned to Earth, on June 30, its occupants were found dead in their capsule. A valve had been jolted open during reentry, releasing all the capsule's air, and the crew had suffocated.

But the Soviets continued developing Soyuz and produced a thoroughly reliable design. Their faith in the craft was publicly demonstrated in 1975, when a Soyuz docked successfully in orbit with a U.S. Apollo spacecraft. This Apollo-Soyuz linkup led nowhere, because the Apollo program was almost over. But Soyuz soldiered on. After 34 missions, it was updated to the T (for "transport") version and ferried crews to and from the Salyut space stations. These space station missions were supported with supplies sent up in Soyuz's uncrewed version, Progress. And in a third incarnation, the TM series, some two dozen Soyuz missions enabled cosmonauts to build the first giant space station, Mir.

HARD EVIDENCE

SOFT LANDING
When Yuri Gagarin returned to Earth after making the first human spaceflight, he ejected from his Vostok capsule after reentry and parachuted to the ground. With Soyuz, which holds two or three cosmonauts, ejection is impossible, so it descends by parachute (left). Just before landing, rocket engines cut in to slow its fall, allowing it to land gently at a mere 2 mph (3 km/h). On several occasions these rockets have failed, making touchdown bumpy but survivable.

SOYUZ SPACECRAFT

The Soyuz spacecraft, built by the RSC Energia company, is about 25 feet long and carries a crew of three. The latest version, the Soyuz TM, first flew in 1986.

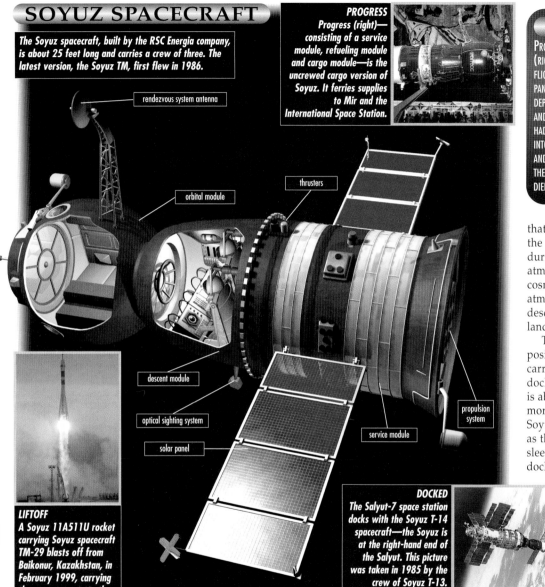

rendezvous system antenna

orbital module

thrusters

descent module

optical sighting system

solar panel

service module

propulsion system

PROGRESS
Progress (right)—consisting of a service module, refueling module and cargo module—is the uncrewed cargo version of Soyuz. It ferries supplies to Mir and the International Space Station.

LIFTOFF
A Soyuz 11A511U rocket carrying Soyuz spacecraft TM-29 blasts off from Baikonur, Kazakhstan, in February 1999, carrying three cosmonauts to the Mir space station.

DOCKED
The Salyut-7 space station docks with the Soyuz T-14 spacecraft—the Soyuz is at the right-hand end of the Salyut. This picture was taken in 1985 by the crew of Soyuz T-13.

SOYUZ TM

The latest crew-carrying version of Soyuz, the TM, is a 3-part vehicle consisting of a descent module, an orbital module and a service module. The bell-shaped descent module, about seven feet long and seven feet in diameter, is where the crew of three stays during launch, orbital maneuvers and reentry, and contains the spacecraft's main control systems. It sits between the orbital and service modules, and is the only part of the spacecraft

that returns to Earth at the end of a mission—the orbital and service modules are jettisoned during reentry and are left to burn up in the atmosphere. The descent module carries the cosmonauts back down through the atmosphere, using parachutes to slow its descent and retrorockets to ensure a soft landing.

The near-spherical orbital module, positioned in front of the descent module, carries life-support and rendezvous and docking systems. Its habitable internal volume is about 177 cubic feet (5 m³)—about 25% more than that of the descent module. When Soyuz is coasting in orbit, the module serves as the cosmonauts' working, recreation and sleeping quarters, and when the craft is docked to the Mir space station or the International Space Station, it functions as an airlock.

The service module, to the rear of the descent module, is over seven feet long with a maximum diameter of nearly nine feet (2.7 m). It contains the orbital flight systems, including propulsion and maneuvering engines, and a pair of wing-like solar panels with a span of about 35 feet (10.6 m). These panels are stowed away during the launch phase of a mission and unfurl when the craft is in orbit. They have a total area of about 108 square feet (10 m²) and generate 600 watts of electricity for the spacecraft, which also carries batteries.

ARIANE

WHAT IF...

...ARIANE GOT BIGGER?

The Ariane 5, which first flew in 1997, can carry larger payloads than most other rockets in the world, but it will face competition from other new heavy-lift launchers, such as Boeing's Delta 4 and Lockheed Martin's Atlas 5. Fortunately for the Ariane 5, a new main engine is under development, and because the upper stage is undersize relative to the main stage, there is plenty of scope for improvements in its design.

The Vulcain main engine is under continuous development. Vulcain 2, with nearly 20% more thrust, will be used in the Ariane 5 ES ATV when it makes its maiden flight in 2006. This engine, combined with upgrades to the solid booster rockets that will increase their thrust by around 4%, will greatly increase the launcher's payload capacity. The new boosters will be lighter, too, because their bolted joints will be replaced by welds.

Ariane's upper stage is also under development. The first upgrade was the introduction of Versatile, an improved version of the existing stage using a new model of the Aestus upper stage engine capable of multiple restarts. This increased maximum payload capacity to about 17,640 lb (6,000 kg) and allowed Ariane to place satellites into orbits that the original Ariane 5 could not reach. The Ariane 5 ECA variant, which was designed to replace the Aestus upper stage, is powered by the HM-

Ariane 5 will undergo a continuous program of upgrades. These will bring improvements to its boosters and the introduction of new and more powerful main and upper stage engines.

7B cryogenic (liquid oxygen/liquid hydrogen) engine used for the third stage of the Ariane 4. It first flew in February 2005.

The final planned upgrade, Ariane 5 ESC-B, is to use an entirely new upper stage engine, the Vinci cryogenic motor. The Vinci motor will produce over five times the thrust of the Aestus motor and will raise the geosynchronous transfer orbit cargo capacity of the Ariane 5 to an amazing 13.2 tons (12 tonnes). Another version of the Ariane 5 under development is Ariane 5 EATV, due in service by 2006 and able to lift over 22 tons (20 tonnes) into low Earth orbit. It will carry the Automated Transfer Vehicle (ATV) that will be used to resupply the International Space Station (ISS) and periodically reboost the station to a higher orbit.

But Ariane 5 could also be used for crewed flights—it was originally designed to launch the Hermes spaceplane. So perhaps, at some time in the future, crewed missions to the Moon or Mars will be launched by Ariane rockets from French Guiana.

Many countries around the world have developed the capability to launch machines and people into space, mostly in the interest of national security. But space is also big business. The space launch industry spends approximately $20 billion a year putting satellites into orbit, and one company, Arianespace, has led that market for over a decade. Arianespace has developed a series of rockets that have become some of the cheapest and most successful launchers on the planet, the Ariane 4 and Ariane 5.

ARIANE 4 AND ARIANE 5

	ARIANE 42P	ARIANE 42L	ARIANE 44P	ARIANE 44L	ARIANE 5
LENGTH	191.6 FT (58.4 M)	191.6 FT (58.4 M)	191.6 FT (58.4 M)	191.6 FT (58.4 M)	191.6 FT (58.4 M)
MAX DIAMETER	12.5 FT (3.8 M)	12.5 FT (3.8 M)	12.5 FT (3.8 M)	12.5 FT (3.8 M)	37.7 FT* (11.5 M)
WEIGHT	373 TONS (338 TONNES)	441 TONS (400 TONNES)	394 TONS (357 TONNES)	520 TONS (471 TONNES)	785 TONS (712 TONNES)
LIFTOFF THRUST	445 TONS (3,964 kN)	455 TONS (4,047 kN)	587 TONS (5,223 kN)	606 TONS (5,388 kN)	1,404 TONS (12,492 kN)
GTO PAYLOAD	6,523 LB (2,958 KG)	7,691 LB (3,488 KG)	7,625 LB (3,458 KG)	10,800 LB (4,898 KG)	15,000 LB (6,803 KG)
COST	$70 MILLION	$80 MILLION	$90 MILLION	$105 MILLION	$120 MILLION
PROPULSION	N204/UDMH	N204/UDMH	N204/UDMH	N204/UDMH	LOX/LH2
	PLUS 2 SOLIDS	PLUS 2 LIQUIDS	PLUS 4 SOLIDS	PLUS 4 LIQUIDS	PLUS 2 SOLIDS

*INCLUDING SOLID BOOSTERS

SPACE BUSINESS

Almost since its inception, Arianespace has been the most successful space launch company in the world, consistently cornering 50% or more of the commercial launch market. Arianespace is actually the European Space Agency's (ESA) launch services operator, and funding for all ESA projects, including the development of the Ariane 5 launcher, comes from the 12-nation consortium that makes up the ESA.

Responding to the needs of a European market, the ESA began development of a European space launcher in the late 1970s. France's space agency, the Centre National d'Etudes Spatiales (CNES), quickly led the development effort and the Ariane 1, a modification of the French-designed Diamant rocket, made its debut in 1979.

Over the years, ESA, CNES, and Arianespace have all made tremendous advances in their technology and technical abilities. Their latest effort, the Ariane 5, demonstrates the high degree of skill and knowledge ESA's engineers have developed. A multibillion-dollar program, the Ariane 5 took more than 10 years to go from the drawing board to the launch pad. The vehicle was completely new in design, having a cryogenic main stage with only one engine and twin solid-fueled boosters. Despite the totally new design, Arianespace expects the Ariane 5 to end up being cheaper, safer, and easier to operate than its predecessors.

LUXURY LAUNCH

ARIANE ROCKETS ARE LAUNCHED FROM THE EUROPEAN SPACE AGENCY (ESA) CENTER NEAR KOUROU ON THE NORTHEASTERN COAST OF FRENCH GUIANA. THE ESA OWNS THE CENTER (RIGHT) AND ALL FACILITIES THERE, WHICH ARE OPERATED ON ITS BEHALF BY ARIANESPACE. THESE FACILITIES WERE DEVELOPED WITH COMMERCIAL CLIENTS IN MIND—THE CITY OF KOUROU NOW OFFERS FINE HOTELS, ENTERTAINMENT AND TRAVEL FACILITIES.

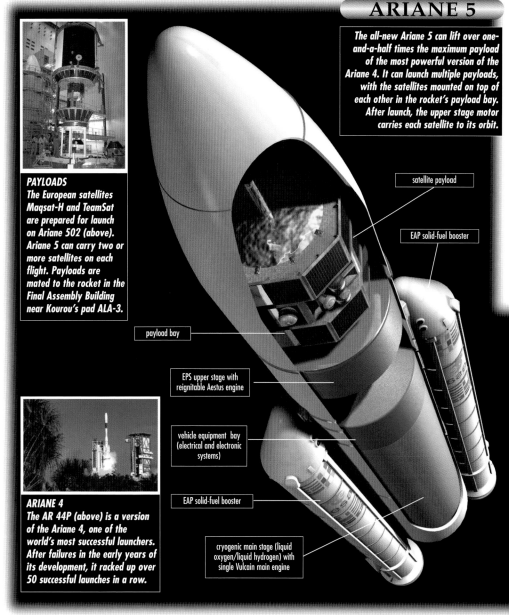

ARIANE 5

The all-new Ariane 5 can lift over one-and-a-half times the maximum payload of the most powerful version of the Ariane 4. It can launch multiple payloads, with the satellites mounted on top of each other in the rocket's payload bay. After launch, the upper stage motor carries each satellite to its orbit.

PAYLOADS
The European satellites Maqsat-H and TeamSat are prepared for launch on Ariane 502 (above). Ariane 5 can carry two or more satellites on each flight. Payloads are mated to the rocket in the Final Assembly Building near Kourou's pad ALA-3.

ARIANE 4
The AR 44P (above) is a version of the Ariane 4, one of the world's most successful launchers. After failures in the early years of its development, it racked up over 50 successful launches in a row.

satellite payload

EAP solid-fuel booster

payload bay

EPS upper stage with reignitable Aestus engine

vehicle equipment bay (electrical and electronic systems)

EAP solid-fuel booster

cryogenic main stage (liquid oxygen/liquid hydrogen) with single Vulcain main engine

HIGH FIVE

The 177-ft (54-m) high Ariane 5 launcher consists of a single-engined main stage, a single-engined upper stage, and two strap-on, solid-fueled boosters. These solid booster rockets provide over 90% of the Ariane 5's thrust at liftoff, and after burning for just over 2 minutes, they separate from the main stage and fall into the Atlantic Ocean. The propellant in the solid boosters is a mix of 68% ammonium perchlorate (oxidizer), 18% aluminum (fuel) and 14% polybutadiene (binder). The boosters hold about 262 tons (237 tonnes) of propellant each.

The main stage, a cryogenic rocket, is 100 feet (30.4 m) long and holds some 146 tons (132 tonnes) of liquid oxygen and 28 tons (25 tonnes) of liquid hydrogen to feed its main Vulcain engine. The Vulcain ignites on the launch pad and continues to burn for about eight minutes after the boosters have separated, and then it too falls into the ocean, leaving the upper stage to deliver the payload to orbit.

The upper stage is powered by a single Aestus engine, fueled by MMH with nitrous oxide as the oxidizer, and the fuel and oxidizer are forced into the engine by pressurized helium. It can be shut down and reignited to deliver multiple payloads to different orbits, and has a total burn time of 1,100 seconds.

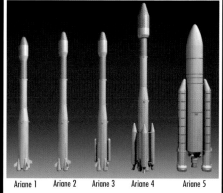

Ariane 1   Ariane 2   Ariane 3   Ariane 4   Ariane 5

SPACE FAMILY ARIANE
The Ariane family first developed as a series of upgrades and expansions from Ariane 1 to Ariane 4. Ariane 1 was a redesign of the Diamant B launch vehicle, 2 and 3 had stretched engines and strap-on boosters, and 4 was a much more powerful model with triple the payload capacity. The Ariane 5 is an entirely new design from the ground up.

CHINESE LAUNCHERS

The world's most populous nation has moved on from a rice-bowl economy and entered the space age. China sent its first satellite into orbit in April 1970—in the words of premier Zhou Enlai, "…through our own unaided efforts." Briefly, between 1956 and 1960, China relied on help from the Soviet Union. But, since then, it has been on its own. Over the last four decades, its space program has survived several political crises and a number of disasters, and emerged as a commercial and scientific force on the international stage.

WHAT IF...

...CHINA TOOK THE LEAD IN THE SPACE RACE?

One of the great mysteries of China's space program was whether it would ever put an astronaut into orbit. China has always been secretive about its space programme, but observers were able to detect continued interest in manned space flight over the years.

In the 1970s, there were photographs in newspapers of astronauts in training. China could also have used its recoverable (FSW) satellites to put someone in space, as the U.S. Mercury missions had done in the 1960s. Long March 2E, the most massive of China's launchers, can lift 11 tons (10 tonnes) into orbit— a heavier load than the 3-person Russian Soyuz spacecraft—yet there were no manned missions despite a more than adequate capability.

The indications were that China simply was not interested in crewed spaceflight. But there were hints that this might not be true. In 1996 they Chinese bought a spacecraft life support system and a docking system devised for the Soyuz ferry craft. Two trainee cosmonauts started an extended stay at Star City, the Russian space training centre, perhaps preparing to train the first generation of Chinese astronauts.

In 1992, a research paper outlined a new generation of rockets that would outperform the existing Long March designs. These more powerful rockets could be used to launch manned flights but instead were used for satellite launches. China made this new capability available to the open market but there were few takers until 1986, when three U.S. rockets blew up, as did Europe's Ariane. Suddenly, Long March looked attractive. Telecommunications companies in Hong Kong, Pakistan and Sweden all commissioned Long March launches.

Then came several disasters: Of 15 commercial launches in 1990–7, seven failed. But after stringent checks, Long March has re-established its reputation, leading to long-term contracts with two U.S. corporations, Motorola and Hughes. By 1998, China had launched 58 satellites, 14 of them commercial launches. Long March was pounding on.

And then, in October 2003, a Shenzou ('Divine Vessel') spacecraft made a manned flight lasting 14 orbits with 'Taikonaut' Yang Liwei aboard. A second mission, this time with two crewmembers, is planned for late 2005. The question has been answered – China does indeed want to put people into space.

China has also expressed interest in its own shuttle, an orbiter that would be launched from a 150-ft (45-m) reusable rocket plane. This concept would not rival the U.S. Shuttle in total load—it could put only three people or about six tons into orbit—but it would probably be cheaper.

THE LONG MARCH (CZ)-B3

Length	172 feet (52.4 m)	**Diameter**	11 feet (3.3 m)
Capability	5,000 lb (2,267 kg) to geostationary orbit	**Weight at liftoff**	417 tons (378 tonnes)

	Strap-ons	1st Stage	2nd Stage	3rd Stage
Length	52.5 feet (16 m)	76 feet (23 m)	34 feet (10.3 m)	29 feet (8.8 m)
Weight	45 tons (41 tonnes)	197 tons (178 tonnes)	44 tons (40 tonnes)	22.7 tons (20.5 tonnes)
Engines	YF-20B	4 x YF-20B	YF-25/23	2 x YF-75
Fuel	Liquid	Liquid	Liquid	Liquid
Thrust	292 tons (2,596 kN)	366 tons (3,254 kN)	93 tons (827 kN)	17.6 tons (157 kN)
Burn time	128 seconds	155 seconds	135 seconds	47 seconds

ONWARD & UPWARD

By 1960, the two greatest communist nations—China and the Soviet Union—had abandoned their past cooperation. If the Chinese wanted a satellite, they had to build their own rocket. They did: Chang Zheng (Long March), named for an epic communist fighting retreat during the Chinese civil war in the 1930s.

Long March had two lower stages, using liquid fuels. But these were not enough to lift a satellite into orbit. Work started on a 13-foot (4-m) solid-fuel third stage, and somehow, despite Mao Zedong's brutal anti-intellectual Cultural Revolution, managed to survive the late 1960s.

Finally, on April 24, 1970, the 82-ton (74-tonne) Long March 1 streaked off. Thirteen minutes later, China's first satellite, Dong Fang Hong (The East is Red), was in orbit, broadcasting the anthem after which it was named.

Two new launchers followed. One, Feng Bao, orbited the first of a series of scientific satellites, and three other secret ones, possibly used for surveillance. The other, an updated Long March, delivered the first of nine recoverable satellites into orbit.

These satellites, known as FSW, for Fanhui Shi Weixing, or Recoverable Experimental Satellite, were a technical leap forward. They had a service module with a retro-rocket and a recoverable capsule. The launch vehicle, Long March 2, was more powerful than anything used before, with four engines that could swivel to steer the rocket. Special computer systems were designed from scratch to meet the launcher's new demands.

COMRADES

TSIEN HSUE-SHEN IS THE ACKNOWLEDGED FATHER OF CHINESE ROCKETRY. IRONICALLY, HE GAINED THIS POSITION AS A RESULT OF BEING EXPELLED FROM THE U.S. LIKE MOST CHINESE ROCKET SCIENTISTS, TSIEN WENT TO THE U.S. BEFORE THE COMMUNISTS SEIZED POWER, IN 1949. DURING THE McCARTHY ANTI-COMMUNIST WITCHHUNT OF THE 1950'S, THE U.S. EXPELLED TSIEN WITH 93 OTHER CHINESE SCIENTISTS. HE RETURNED HOME AND USED HIS U.S. TRAINING TO FOUND CHINA'S SPACE PROGRAM.

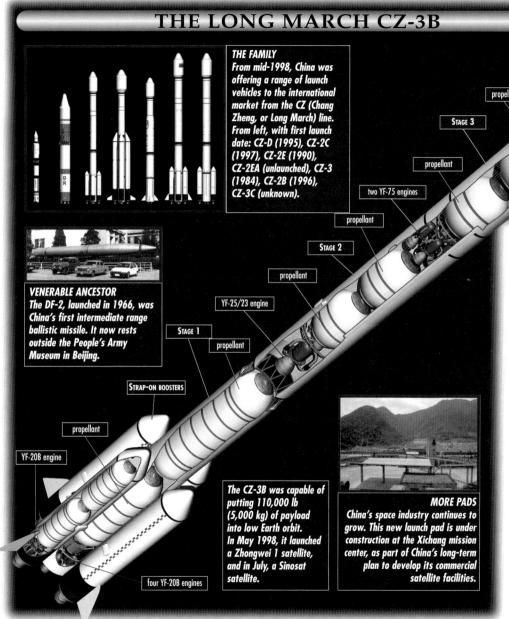

THE LONG MARCH CZ-3B

THE FAMILY
From mid-1998, China was offering a range of launch vehicles to the international market from the CZ (Chang Zheng, or Long March) line. From left, with first launch date: CZ-D (1995), CZ-2C (1997), CZ-2E (1990), CZ-2EA (unlaunched), CZ-3 (1984), CZ-2B (1996), CZ-3C (unknown).

propellant

STAGE 3

propellant

satellite

two YF-75 engines

propellant

STAGE 2

propellant

VENERABLE ANCESTOR
The DF-2, launched in 1966, was China's first intermediate range ballistic missile. It now rests outside the People's Army Museum in Beijing.

YF-25/23 engine

STAGE 1

propellant

STRAP-ON BOOSTERS

propellant

YF-20B engine

The CZ-3B was capable of putting 110,000 lb (5,000 kg) of payload into low Earth orbit. In May 1998, it launched a Zhongwei 1 satellite, and in July, a Sinosat satellite.

MORE PADS
China's space industry continues to grow. This new launch pad is under construction at the Xichang mission center, as part of China's long-term plan to develop its commercial satellite facilities.

four YF-20B engines

LONG MARCH TO SUCCESS

After one explosion on takeoff in 1974, Long March launched the first FSW in November 1975. Forty-seven orbits later, it returned to Earth and placed China in the same league as the U.S. and U.S.S.R., the only two nations so far to have recovered a satellite from orbit.

Now China set its sights higher. It planned to place communications satellites 22,500 miles (36,210 km) above fixed points on Earth, in geostationary orbits. With one of these, it could communicate across the nation; with three, across the world. But to reach that orbit a more powerful rocket with more sophisticated maneuvering was needed.

At first, the satellite program suffered setbacks. In 1978, a third-stage motor exploded, killing several people. And when Long March 3 blasted off on January 29, 1984, the third stage failed, leaving the satellite stranded in a low orbit. Four months later, all went well. China had its first comsat, servicing 200 phone lines and 15 channels.

Now Long March was available in versions of varying power to anyone ready to pay. There were few takers until 1986, when three U.S. rockets blew up, as did Europe's Ariane. Suddenly, Long March looked attractive. Telecommunications companies in Hong Kong, Pakistan and Sweden all commissioned Long March launches.

Then came several disasters: Of 15 commercial launches in 1990–7, seven failed. But after stringent checks, Long March has reestablished its reputation, leading to long-term contracts with two U.S. corporations, Motorola and Hughes. By 1998, China had launched 58 satellites, 14 of them commercial launches. Long March was pounding on.

COMMERCIAL LAUNCH VEHICLES

Cheap Charlie's

Established 2000

GET 'EM WHILE THEY'RE CHEAP
ALL MAJOR CREDIT CARDS WELCOME

MASTISA   Carris

TODAY'S SPECIAL: ATHENA II

N o one could ever call the U.S. space program profitable—the Apollo program cost about $25 billion, and the International Space Station could cost NASA up to four times that amount. But now, a new generation of privately developed, relatively low-cost launch vehicles has begun to enter service, launching communication satellites for profit. They are hardly the most powerful rockets ever built, but these commercial launch vehicles get the job done.

WHAT IF...

...ALL SPACEFLIGHT BECOMES COMMERCIALIZED?

B ig, government-sponsored spaceflight programs could be nearly over. Now that private companies are launching private satellites, the future of spaceflight is bound to change.

In the short term, we will see more and more satellites placed in orbit around the Earth. The rapidly growing demand for communication systems that drives satellite production and launch seems nearly inexhaustible, given the 6-billion-strong audience for their products. And as long as there is demand from customers and competition among the suppliers, the launchers and the satellites will continue to improve.

Science will also benefit, because as prices fall, research probes that are built with commercial-satellite technology will become cheaper. We will be able to learn more and more about space, and our knowledge will come at a lower and lower cost.

With steadily falling prices, some people will undoubtedly want to experience space themselves. And commercial space vehicles will make achieving that dream far easier. The first ambitious projects would be to send tourists into space for a few hours in low-Earth orbit. Already, the Athena-2 can lift 4,400 lb (2,000 kg) into space—1,400 lb (635 kg) more than the launch weight of the Mercury capsule Friendship 7 that carried John Glenn, the first U.S. astronaut to orbit the Earth, in 1962.

One day in the not-too-distant future, commercial space exploration will extend to interplanetary space and private spacecraft, perhaps powered by solar sails, will carry paying passengers to other planets.

The sequence that is occurring may follow a pre-existing pattern. In the 15th century, European explorers sailed the seas with their monarchs' patronage—effectively, for their governments. With time, large corporations more or less independent of the government, such as Britain's East India Company, began gaining access to these new corridors of exploration. Gradually, more and more explorers bought their own ships and started carrying fare-paying passengers, and each step toward fully commercial voyages increased the rate of exploration.

The same rule may hold in space travel. In the beginning, astronauts selected and trained by the government took the first steps outside Earth's atmosphere. Now we are at the point at which private companies are broadening access to space for anyone in the marketplace.

At the moment, only governments and communications giants are able to afford space access. But a new era of cheap space exploration may be just around the corner.

COMMERCIAL LAUNCHERS

NAME	HEIGHT	WEIGHT (FUELED)	STAGES	FUEL TYPE	COST/FLIGHT (IN $)	LOW-EARTH ORBIT PAYLOAD WEIGHT
ATHENA-1	49.2 FT (15 M)	146,264 LB (66,344 KG)	2	SOLID	$17.5 MILLION	1,760 LB (800 LB)
ATHENA-2	85.3 FT (26 M)	265,000 LB (120,200 KG)	3	SOLID	$23 MILLION	4,400 LB (2,000 KG)
TAURUS	89 FT (27 M)	161,000 LB (73,000 KG)	4	SOLID	$22 MILLION	3,000 LB (1,360 KG)
CONESTOGA	50 FT (15 M)	192,000 LB (87,089 KG)	4	SOLID	$19.8 MILLION	1,960 LB (889 KG)

BOOSTER BARGAINS

Today, the lightweight launch vehicle industry is booming, thanks to the growing demand for communications satellites (comsats). Modern global communications networks are based on constellations of 20 to 30 satellites in low-Earth orbits (LEOs), and the telecom companies that operate these networks pay private firms generously to put the satellites into orbit.

What keeps the commercial launch vehicle industry in business is a slight advantage that low-orbit satellites have over the geostationary satellites that were launched in the 1970s and 1980s. Geostationary satellites orbit the Earth at the same rate as the Earth spins, so they appear stationary in the sky. These satellites travel in very high orbits—22,400 miles (36,000 km) above Earth—and so they require large launchers. At that distance, it takes a signal about a quarter of a second to travel from the ground to the satellite and back, a time lag large enough to create significant transmission problems. Low-Earth orbit satellites, only about 600 miles (968 km) above Earth, benefit from much smaller time lags.

Small comsats and their launchers are relatively cheap because they are mass produced from off-the-shelf parts, and if something goes wrong—if a satellite burns out, or a rocket explodes on launch—it is easy to send up another. In contrast, large geostationary communication satellites and their correspondingly large launch vehicles tend to be expensive, and it takes a great deal of time and money to replace a failure.

SIMPLE SOLID FUELS

Production-line economics also explain why most commercial launch vehicles burn relatively low-performance solid fuels, instead of higher-thrust liquid fuels. Solid fuels are easy to handle and need no complex gas tanks, valves or pumps. And solid-fuel rocket engines are cheaper, lighter and more reliable than their liquid-fuel counterparts because they have fewer moving parts. Rocket makers can also save money by using solid-fuel

engines reclaimed from decommissioned missiles.

One of the most successful commercial launch vehicles today is Lockheed Martin's

Athena rocket, originally called the Lockheed Martin Launch Vehicle (LMLV). Although there were problems with some of the early launches, two versions of the Athena have sent

up two satellites—Lewis and ROCSAT-1—and the Lunar Prospector space probe, which used a Star kick motor to power its way from low-Earth orbit to the Moon.

SHUTTLE

In 1989, NASA turned over responsibility for commercial satellite launches to the rocket manufacturers, so the Space Shuttle no longer launches them. Today, NASA employs a fleet of launch vehicles for government satellites, and only a few scientific and military satellites—and some space probes such as the Galileo Jupiter probe (above)—leave Earth on the Shuttle.

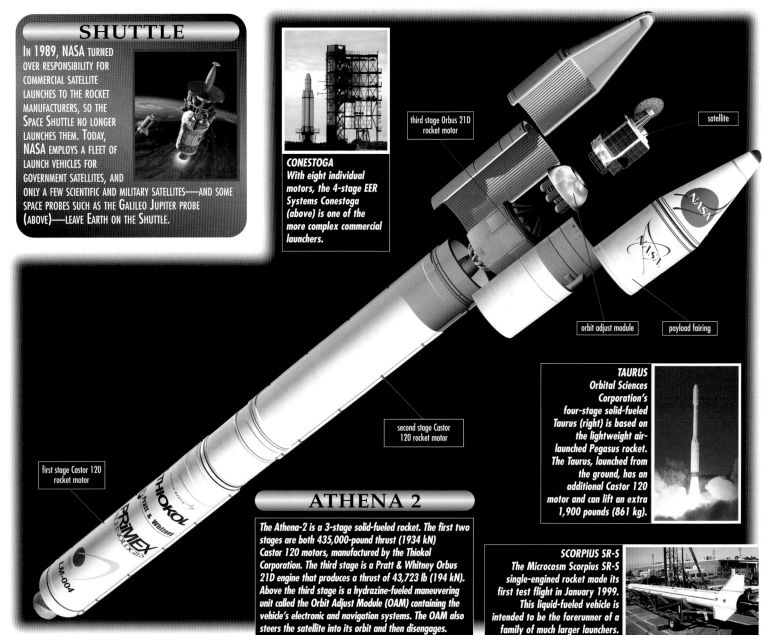

CONESTOGA
With eight individual motors, the 4-stage EER Systems Conestoga (above) is one of the more complex commercial launchers.

third stage Orbus 21D rocket motor

satellite

orbit adjust module

payload fairing

second stage Castor 120 rocket motor

first stage Castor 120 rocket motor

TAURUS
Orbital Sciences Corporation's four-stage solid-fueled Taurus (right) is based on the lightweight air-launched Pegasus rocket. The Taurus, launched from the ground, has an additional Castor 120 motor and can lift an extra 1,900 pounds (861 kg).

ATHENA 2

The Athena-2 is a 3-stage solid-fueled rocket. The first two stages are both 435,000-pound thrust (1934 kN) Castor 120 motors, manufactured by the Thiokol Corporation. The third stage is a Pratt & Whitney Orbus 21D engine that produces a thrust of 43,723 lb (194 kN). Above the third stage is a hydrazine-fueled maneuvering unit called the Orbit Adjust Module (OAM) containing the vehicle's electronic and navigation systems. The OAM also steers the satellite into its orbit and then disengages.

SCORPIUS SR-S
The Microcosm Scorpius SR-S single-engined rocket made its first test flight in January 1999. This liquid-fueled vehicle is intended to be the forerunner of a family of much larger launchers.

PEGASUS

The launch of a rocket may be a spectacular sight, but to space administrators it is money going up in smoke. Economics—not to mention the environment—demand that we find more efficient ways to escape the grasp of the Earth's gravity, which is why many of the best brains in rocketry are now busily searching for less wasteful launch systems that will cut the cost of hardware and energy. The first major success story is a modest launch vehicle named Pegasus. With a secondhand airliner for a partner, this little rocket is pointing the way to the future of space travel.

WHAT IF...

...AIR-LAUNCHING TAKES OFF?

For decades, spacecraft designers have had to contend with the wastefulness of single-shot rockets. Their goal has always been to build a fully reusable launch vehicle—and thanks to air-launching, they may be about to reach it at last.

Orbital Sciences Corporation (OSC), the U.S. company that flies Pegasus, has proven the concept of air-launching. Pegasus rockets have launched over 70 satellites from various bases and broken several records in the process, including becoming the first winged vehicle to exceed Mach 8.

Pegasus was the first privately developed space launch vehicle and has demonstrated immense flexibility, operating from six different sites in the US, Europe and the Marshall Islands. The reduction in launch costs that comes with a reusable air-breathing first stage (the launch aircraft) make commercial launches of small satellites much more economically viable.

Since its first flight in 1990, Pegasus has continued development. For many years its designers at OSC worked with NASA on the next generation of air-launched vehicles. The first was to be an experimental hypersonic rocketplane called X-34 that was carried aloft by a jet aircraft in much the same manner as Pegasus and could land on a conventional runway.

The X-34 concept combined the best of the Space Shuttle and Pegasus. Like the Shuttle, the X-34 was to glide back to Earth to land on a runway and fly another day; like Pegasus, the airplane "booster" is fully reusable and can take off from any suitable air base. Together, these advantages create a further economy—a single base would serve the X-34 and its Tristar partner for both takeoff and landing. The Shuttle program, by contrast, ties up a Boeing 747 simply to ferry the orbiters back from their touchdown airstrips to Cape Canaveral for relaunch.

However, X-34 was beset by technical problems. Costs spiralled and time frames lengthened, and eventually NASA decided to 'pull the plug' on the project and direct its funds into more beneficial programmes. At present, the short-term goal is a two-stage-to-orbit vehicle to replace the Space Shuttle fleet.

A workable single-stage-to-orbit vehicle for now remains a goal that will be difficult to attain, though the economic and environmental benefits are such that the concept will not be abandoned. Indeed Lockheed Martin, which was heavily involved with X-34, is proceeding with its own single-stage-to-orbit craft, VentureStar. Lessons learned from X-34 may provide critical insight. As a private venture, VentureStar is a risky project for Lockheed Martin, but Pegasus, which proved the air-launch concept, was also privately developed.

PEGASUS XL STATISTICS

STAGE	MODEL	TYPE	DIAMETER	LENGTH	THRUST
1	ALLIANT TECHSYSTEMS ORION 50S	SOLID FUEL	4.2 FT (1.2 M)	29.1 FT (8.7 M)	132,410 LB (588 KN)
2	ALLIANT TECHSYSTEMS ORION 50XL	SOLID FUEL	4.2 FT (1.2 M)	11.7 FT (3.6 M)	34,515 LB (153 KN)
3	ALLIANT TECHSYSTEMS ORION 38	SOLID FUEL	3.2 FT (1 M)	4.4 FT (1.34 M)	7,435 LB (33 KN)
4 (OPTIONAL)	HAPS (HYDRAZINE AUXILIARY PROPULSION SYSTEM, FOR PRECISION INJECTION INTO ORBIT)	HYDRAZINE	3.2 FT (1 M)	7.3 FT (2.22 M)	172 LB (0.8 KN)
4 (OPTIONAL)	STAR 27 (FOR GOING BEYOND EARTH ORBIT)	SOLID FUEL	3.2 FT (1 M)	4.2 FT (1.28 M)	6,079 LB (27 KN)

LAUNCH COST	ABOUT $12 MILLION	TOTAL LENGTH	55 FT (16 M)
MAXIMUM PAYLOAD	1,100 LB (500 KG) TO LOW EARTH ORBIT; 300 LB (136 KG) TO ESCAPE EARTH	WINGSPAN	22 FT (6.7 M)
		WEIGHT (EXCLUDING PAYLOAD)	55,000 LB (24,950 KG)

SPACE HORSE

The ancient Greeks had a myth about a winged horse called Pegasus that carried its rider high into the air. The modern Pegasus is a winged rocket that rides an airplane. The idea is simple: Ground-launched rockets expend a lot of effort and fuel forcing their way through the dense lower layers of the atmosphere; Pegasus gets a lift on a conventional jet airplane seven miles (11 km) up into the stratosphere, where the air is thinner.

According to its operators, the Pegasus system cuts in half the cost of launching small satellites. Another advantage is that there is no need for a specialized launch complex. Not only can the partner airplane take off from any large air base, but launches are also theoretically safer because Pegasus is always released over the ocean, far from populated areas.

Even so, we have yet to see the last of ground launches. Pegasus is only capable of lifting relatively small payloads (400–1,100 lb [180–450 kg] to orbit; 300 lb [135 kg] to escape Earth's gravity altogether). At the moment, no carrier aircraft exists to safely lift the much heavier rockets needed to launch larger payloads.

BALLOONS AND COMPUTERS

The idea of air-launching goes back to 1950s' rockoons—rockets launched from high-altitude balloons—but the first commercial use of the concept came only in 1987. That was when Orbital Sciences Corporation (OSC) won a U.S. Defense Department contract for an air-launched rocket capable of quickly launching light satellites—possibly in secret.

The result was Pegasus, America's first new satellite launch vehicle in 20 years, and the first rocket to enter service without a single test flight—or even any wind tunnel tryouts. Pegasus was tested by computer simulation alone, and on its first mission in 1990, the computers were proved right: Two satellites were successfully delivered into orbit.

To keep the design simple, the original Pegasus used proven technology. A flight computer from an Israeli tank and a guidance system from a U.S. torpedo. Later models carried a Global Positioning System (GPS) navigation receiver. When OSC acquired its own airplane, it was not a high-tech, custom design, but a converted Air Canada Tristar airliner.

To date, Pegasus has launched satellites to study subjects as diverse as the ozone layer, plankton, the aurorae and lightning, with the governments of Spain, Brazil and the U.S. among OSC's customers. This least dramatic of rockets is making a big contribution to our presence in space.

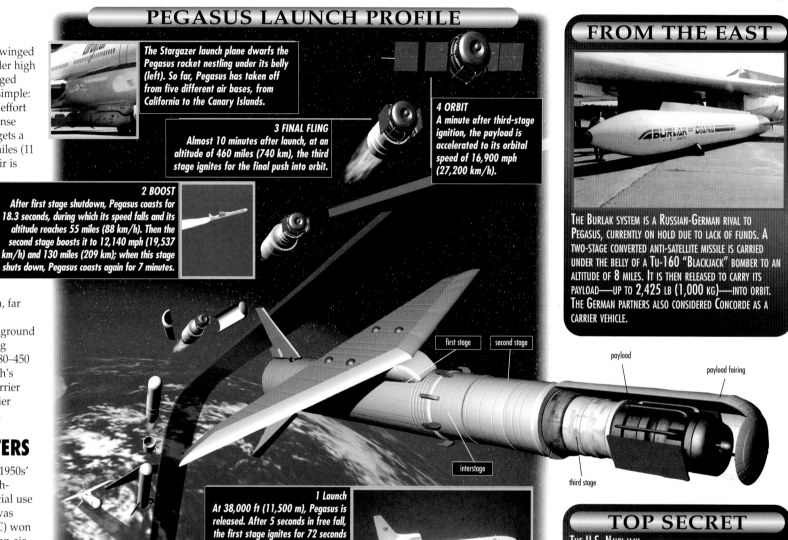

PEGASUS LAUNCH PROFILE

The Stargazer launch plane dwarfs the Pegasus rocket nestling under its belly (left). So far, Pegasus has taken off from five different air bases, from California to the Canary Islands.

3 FINAL FLING
Almost 10 minutes after launch, at an altitude of 460 miles (740 km), the third stage ignites for the final push into orbit.

4 ORBIT
A minute after third-stage ignition, the payload is accelerated to its orbital speed of 16,900 mph (27,200 km/h).

2 BOOST
After first stage shutdown, Pegasus coasts for 18.3 seconds, during which its speed falls and its altitude reaches 55 miles (88 km/h). Then the second stage boosts it to 12,140 mph (19,537 km/h) and 130 miles (209 km); when this stage shuts down, Pegasus coasts again for 7 minutes.

1 Launch
At 38,000 ft (11,500 m), Pegasus is released. After 5 seconds in free fall, the first stage ignites for 72 seconds to boost the craft from 525 mph (844 km/h) to 5,600 mph (9,000 km/h) and an altitude of 39 miles (63 km). The first-stage wings create lift to help it on its way into orbit.

first stage

second stage

interstage

third stage

payload

payload fairing

DELTA LAUNCHERS

O n August 12, 1960, a 100-ft (30-m) diameter aluminized balloon called Echo inflated in orbit around the Earth. It was the world's first real communication satellite—a major breakthrough for the U.S. And it had been lifted into space by a Delta rocket. Delta was the American answer to the Soviet Union's R-7 booster, which secured Soviet space domination during the 1950s and early 1960s. Delta became the mainstay of U.S. crewless space exploration and is still in use today, after almost 40 years of reliable service.

WHAT IF...

...THE DELTA LAUNCHER SERIES CONTINUES?

T he rockets of the Delta series are already the most successful launchers the U.S. has ever built. They were the workhorses of the satellite communications revolution in the 1980s, they have lofted most of the world's weather satellites and they have boosted space probes around the solar system. In upgraded form, Delta launchers are certain to be blasting off well into the next century.

The Delta IV, which first flew in 2002, can lift a 50,000-lb (22,700-kg) payload to low Earth orbit. It was removed from the commercial satellite launch marketplace in 2003 due to a downturn in the industry. The Delta IV family, with five configurations possible depending upon mission requirements, continues to launch US Government missions.

Delta IV Heavy, a heavy-lift variant with a new five-metre high upper stage, made its first flight in December 2004. For very large payloads three Delta IV core vehicles can be combined in a side-by-side arrangement. Delta IV Heavy is part of the Evolved Expendable Launch Vehicle programme and probably represents the heaviest launch capability that will be necessary for many years. Subsequent improvements are more likely

Boeing's Delta IV made its first flights in 2002, with three times the payload capacity of its predecessor. Even if reusable vehicles are developed, there will still be a big market for reliable one-shot launchers.

to be in efficiency than in brute lifting capacity, though. Few commercial payloads are likely to exceed 25 tons (22.6 tonnes)—at least in the foreseeable future.

Governments might have different requirements. In 1987, at the height of President Ronald Reagan's "Star Wars" Strategic Defense Initiative, planners needed all the lifting power they could get. To haul laser battle stations into orbit, rocket engineers designed a super-Delta. Aptly known as the "Barbarian," the monster rocket added three Space Shuttle boosters to a cluster of six normal Delta boosters around a standard Delta core. The combination could have launched a payload of 100,000 lb (45,500 kg). But the super-rocket would have cost half a billion dollars for a single launch, and when "Star Wars" funding was cut back, the Barbarian was among the first projects to be axed.

DELTA SERVICE RECORD

APRIL 1959 NASA COMMISSIONS THE DOUGLAS AIRCRAFT COMPANY TO BUILD 12 LAUNCH VEHICLES TO PUT SATELLITES IN ORBIT. THE FIRST DELTA IS READY 18 MONTHS LATER.	**1974** DELTA 2000 IS INTRODUCED, FOLLOWED IN **1979** BY DELTA 3000.
AUGUST 12, 1960 AFTER AN INITIAL FAILURE ON MAY 13, 1960, DELTA LAUNCHES ITS FIRST SATELLITE. THE REST OF THE FIRST BATCH OF 12 DELTAS ARE EQUALLY SUCCESSFUL.	**1984** DELTA PRODUCTION SHUTS DOWN DUE TO COMPETITION FROM THE SPACE SHUTTLE.
	1986 AFTER CHALLENGER DISASTER, NASA REVERTS TO USING EXPENDABLE LAUNCH VEHICLES. MCDONNELL-DOUGLAS UNVEILS DELTA II.
1962 DOUGLAS BEGINS A SERIES OF MODIFICATIONS THAT WILL INCREASE DELTA'S PAYLOAD CAPACITY TENFOLD WITHIN NINE YEARS.	**1997** BOEING BUYS OUT MCDD AND DEVELOPS DELTA III, WITH MORE THAN TWICE THE PAYLOAD CAPACITY OF ITS PREDECESSOR. WORK STARTS ON DELTA IV, WITH A PAYLOAD THREE TIMES GREATER STILL. LAUNCHPADS IN CALIFORNIA AND FLORIDA ARE GIVEN TO BOEING BY NASA AND THE U.S. AIR FORCE: DELTA REMAINS A WORKHORSE OF THE GOVERNMENT LAUNCH PROGRAMME.
1972 MCDONNELL-DOUGLAS (THE RESULT OF A 1967 MERGER) PRODUCES THE EVEN MORE POWERFUL DELTA-1000 SERIES.	

TRIED AND TRUE

When the Soviet Union launched Sputnik 1 into orbit in 1957, the U.S. was dismayed. The world's first artificial satellite had been launched by the R-7 rocket, the world's first intercontinental ballistic missile (ICBM). That alone was more than enough to worry U.S. defense planners. But the mighty R-7 also gave a huge advantage to the Soviet space program. All the U.S. had to offer, by comparison, was the Thor intermediate-range ballistic missile (IRBM), manufactured by the Douglas Aircraft Company. Far short of the R-7's power, Thor had a range of just 1,500 miles (2,400 km)—barely enough to fly from England to Moscow.

Dollars were pumped into Thor's improvement. By 1960 it had gone through four modifications to produce a new three-stage booster named Thor-Delta. Problems with the rocket's attitude control system made the first launch a failure. But the second liftoff, on August 12, 1960, successfully carried Echo, a rudimentary communications satellite, into orbit. It was a long way from equaling the Soviet achievements, but Echo was a genuine American first and a good start to the program. The Thor name was soon dropped from the commercial version, and the rocket family became known simply as Delta.

Between 1960 and 1962 Delta's reliability was proved by the launch of 14 satellites. Among these were the OSO-1 (Orbiting Solar Observatory), which provided the first

observations of solar gamma rays, and Telstar, the satellite that successfully transmitted the first transatlantic television signals. Initially, Delta could do no more than put a 100-pound payload into geostationary transfer orbit (GTO)—an orbit from which the payload's own thrusters can achieve a geostationary orbit. After 1962, though, successive redesigns steadily increased its thrust and lifting power. By the mid-1960s, Delta had become the primary workhorse rocket for the U.S.

Delta worked more or less flawlessly throughout the 1960s. By the 1970s, the booster's lifting capacity had been increased tenfold, largely due to strap-on booster rockets that were added to its lower stages.

In the 1980s the development of the reusable Space Shuttle threatened Delta's future. But a U.S. Air Force order for seven medium-power

launch vehicles in 1987 gave it a new lease on life. Delta II was introduced in an economic climate in which the world's communication companies were competing with governments for launch power. Delta's reliability made it a solid commercial success and ensured its continued existence. The current Delta III has a payload of 8,400 lb (3,800 kg) to GTO. The Delta IV family can lift up to 27,400 lb (12,500 kg) depending upon the configuration in use.

SPY CATCHER

In August 1960, Delta's predecessor, Thor, launched the first U.S. spy satellite. Corona, as it was called, took photos of the Soviet Union, collecting exposed films in a heat-resistant bucket at its nose. The bucket was then jettisoned into the atmosphere, to be caught by a U.S. aircraft over the Pacific Ocean. Amazingly, this bizarre system (mapped out in the illustration) worked. Corona's bucket films provided the first-ever photos of the Soviet Union's Plesetsk rocket base.

THE DELTA III LAUNCH VEHICLE

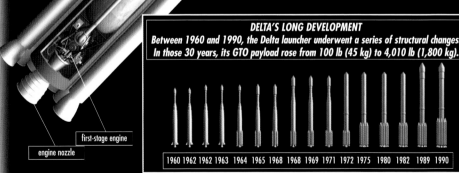

- composite payload fairing
- payload envelope
- payload fairing separation system
- intertank truss
- interstage
- nozzle
- liquid hydrogen tank
- liquid oxygen tank
- second-stage engine
- first-stage fuel tank (kerosene)
- center body
- liquid oxygen tank
- stretched graphite epoxy strap-on boosters
- first-stage engine
- engine nozzle

Delta III provides twice the payload capacity of its predecessor, Delta II. Other major improvements to emerge in the Delta III include bigger solid rocket boosters, giving 25% more thrust, and a new second stage single engine that uses liquid hydrogen fuel.

DELTA'S LONG DEVELOPMENT
Between 1960 and 1990, the Delta launcher underwent a series of structural changes. In those 30 years, its GTO payload rose from 100 lb (45 kg) to 4,010 lb (1,800 kg).

1960 1962 1962 1963 1964 1965 1968 1968 1969 1971 1972 1975 1980 1982 1989 1990

1 FIRST STAGE
The first stage of a Delta II rocket is transported to Launch Complex 17A at Cape Canaveral, Florida.

2 ON PAD
The first stage is lifted into launch position on the pad. This rocket carries a research satellite.

3 BOOSTERS
Solid rocket boosters are strapped to the main vehicle as it stands on the launch platform.

4 SECOND STAGE
The second stage—its engine powered by liquid hydrogen and liquid oxygen—is made ready.

5 PAYLOAD
Once the second stage is in place, the satellite payload is lifted into position in the rocket nosecone.

6 LIFTOFF
Boosters and first stage roaring with flame, the Delta II and its cargo lifts off the pad.

SPACESHIPONE

O n 17 December, 2003, exactly 100 years after the first powered flight by the Wright Brothers, Brian Binnie became the first person to fly supersonically in a craft developed by a small company without government funding, and in doing so reach the edges of space. The small single-engined SpaceShipOne is immediately recognisable as a design from the stable of prolific designer Burt Rutan's Scaled Composites Inc, as is its carrier plane or mothership, known as White Knight. Together they form "Tier One," the first step towards commercial space tourism.

WHO ELSE COMPETED FOR...

...THE ANSARI X-PRIZE?

L aunched in 1996, the X-Prize trophy and $10 million cash was to be awarded to the first team that: privately financed, built and launched a spaceship able to carry three people to 62.5 miles (100 km) altitude; returned it safely to Earth and repeated the launch with the same ship within two weeks. The FAI define the upper extent of Earth's atmosphere (and thus the boundary of space) to be 328,000 ft (100,000 m).

In total 26 teams from seven nations registered to take part in the X-Prize race. A handful reached the flight test stage.

THE ROTON ROCKET

Built by the Rotary Rocket Company for about $5 million, the Roton Rocket Atmospheric Test Vehicle (ATV) was a hybrid rocket/rotorcraft proof-of-concept vehicle for a SSTO (Single Stage To Orbit) spacecraft. Scaled Composites helped fabricate the ATV, which successfully flew in the low-speed, low-altitude regime in 1999, piloted by Brian Binnie. Development halted and the ATV was retired to a museum.

THE CANADIAN ARROW

Modelled after the V-2, but with a reuseable airframe and engine, the Canadian Arrow was intended to splash down in a lake at the end of its sub-orbital flight.

A head-on view of SpaceShipOne emphasizes its highly unusual and unmistakeable design.

RUBICON 1

Development of The Space Transportation Corporation's Rubicon 1 rocket was set back when it exploded and fell into the sea in August 2003, scattering parts of mannequin 'test pilot' Stevie Austin over a wide area. A Rubicon 2 craft was constructed but not launched.

ARMADILLO AEROSPACE

Armadillo's X-Prize contender prototype ran out of fuel after only 600 feet (182 m) of flight and crashed. The company continues to work on the X-Prize Cup.

THE DA VINCI PROJECT

Another Canadian project, da Vinci's Wild Fire MK VI was designed to be carried to 70,000 feet (21,340m) beneath a large balloon before the rocket engines are fired. The da Vinci team are also continuing work towards the X-Prize Cup.

SPACESHIPONE SPECIFICATIONS

CREW:	1	**MAX ALTITUDE:**	367,442 FT (111,996 M)
POWERPLANT:	ONE 16,535 LB THRUST (74 KN) SPACEDEV HYBRID SOLID ROCKET ENGINE	**SPAN:**	16 FT 4 IN (4.97 M)
		LENGTH:	16 FT 5 IN (5.00 M)
		HEIGHT:	UNKNOWN
MAX SPEED:	MACH 3.09 (2,185 MPH/3,518 KM/H)	**WEIGHT:**	LOADED 7,937 LB (3,600 KG)

RIDING ON A WHITE KNIGHT

Although there has been commercial involvement in the US space program from the beginning, the construction of spacecraft has been the domain of large aerospace companies, funded by NASA or the military. The X-Prize was created as a means of fostering innovation by independent private firms and creating the market for commercial spaceflight.

On 30 September 2004, White Knight launched SpaceShipOne on the first qualifying flight for the X-Prize. It was not without drama for the pilot, Mike Melvill and the tens of thousands watching from the Mojave Airport. At the top of its climb, the spacecraft oscillated and began a series of rolls at nearly Mach 3, causing Melvill to cut the engine 11 seconds early and reach a lower than planned altitude. After landing he said: "Did I plan the roll? I'd like to say I did but I didn't. You're extremely busy at that point. Probably I stepped on something too quickly and caused the roll but it's nice to do a roll at the top of the climb."

SPACESHIPONE

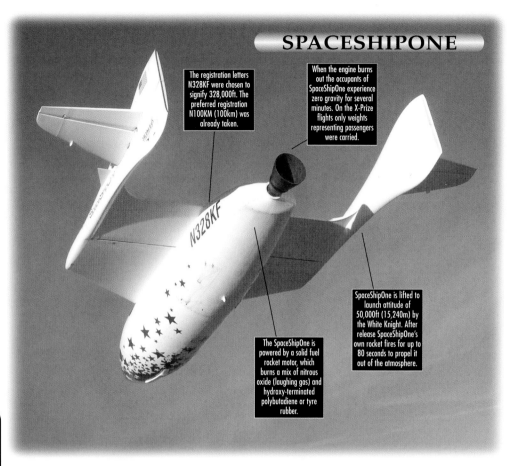

The registration letters N328KF were chosen to signify 328,000ft. The preferred registration N100KM (100km) was already taken.

When the engine burns out the occupants of SpaceShipOne experience zero gravity for several minutes. On the X-Prize flights only weights representing passengers were carried.

SpaceShipOne is lifted to launch altitude of 50,000ft (15,240m) by the White Knight. After release SpaceShipOne's own rocket fires for up to 80 seconds to propel it out of the atmosphere.

The SpaceShipOne is powered by a solid fuel rocket motor, which burns a mix of nitrous oxide (laughing gas) and hydroxy-terminated polybutadiene or tyre rubber.

SpaceShipOne is carried to altitude aboard its mothership, known as White Knight, also designed by Rutan.

On 4 October, well within the stipulated two-week window, Brian Binnie took SpaceShipOne to 368,000ft (112,000 m) in a flawless flight, capturing the $10 million prize. Afterwards he said: "It's a fantastic view; it's a fantastic feeling. There is a freedom there and a sense of wonder that—I tell you what—you all need to experience."

The awarding of the X-Prize was by no means the end of the story. Together with Richard Branson's new company Virgin Galactic, Scaled Composites plans to develop a seven-seat "SpaceShipTwo" to offer commercial flights into space. The X-Prize Cup has been established as the focus of an annual event in New Mexico to encourage the runners up in the original X-Prize to continue developing their craft.

THE PILOTS

MICHAEL W. MELVILL

Born in South Africa in 1941 and raised in England, civilian test pilot Mike Melvill has worked for Burt Rutan for nearly 30 years and has made the first flights in 10 of Rutan's designs. In all he has flown 140 fixed- and rotary-wing aircraft types, accumulating over 7,050 flight hours. He holds nine US and world speed and altitude records, all achieved in Rutan aircraft.

BRIAN BINNIE

Scots-born Brian Binnie flew strike aircraft for most of his 20-year US Navy career, and has over 4,600 hours experience in 59 different types of aircraft. A graduate of the US Navy's Test Pilot School, he has flown evaluations on many new systems and weapons for the A-6, A-7 and F/A-18 aircraft.

MISSION DIARY

MAY 1996
X-Prize offered

APRIL 18, 2003:
Scaled Composites unveiled the existence of their commercial manned space programme, which had been underway at its Mojave facility for two years.

MAY 20, 2003:
First (unmanned) captive flight of SpaceShipOne, duration 1 hour, 48 minutes

JULY 29, 2003:
First release and glide flight. Pilot Mike Melvill, total duration 2 hours, 6 minutes
17 December, 2003: First powered launch, reaching Mach 1.2 and 67,900 ft (20,700 m). Pilot Brian Binnie, duration 18 minutes, 10 seconds

JUNE 21, 2004:
First flight over 62 miles (100 km). Pilot Mike Melvill. Top speed Mach 2.9, duration 24 minutes, 5 seconds

SEPTEMBER 29, 2004:
Mike Melvill makes the first qualifying flight for the X-Prize, reaching 337,000 ft (102,900 m) and a maximum

speed of Mach 2.92 during a flight lasting 24 minutes 11 seconds

OCTOBER 4, 2004: The Tier One team wins the Ansari X-Prize by twice exceeding 100 km in two weeks. Brian Binnie reached 69.5 miles (112 km) at Mach 3.09 on a 23 minute and 56 second flight

SATELLITES

On 5 October 1957, to U.S. and world surprise, the Soviet Union launched Sputnik I ("fellow traveler"), the first man-made object into orbit. Sputnik I was nothing more than a bleeping radio transmitter, but many practical applications have since been found for artificial satellites. Today well over 2,000 are in Earth orbit. Satellites relay telephone calls, data, and television programmes, observe changes in the environment, and provide accurate location data to GPS receivers, which have a wide range of civil and military applications. Purely military uses include monitoring ballistic missile launches, listening to enemy communications and electronic emissions, as well as imaging sites of military interest. The type of orbit is important. A geosynchronous orbit (in which the satellite remains over one point on Earth) allows communications to be relayed over most of the planet. Spy satellites and most other Earth-observing craft need to make moving orbits, passing over their targets for a short period each day. The modern world would barely function without satellites, and new launches will need to be made far into the foreseeable future.

Three Shuttle crewmembers capture the 4.5 ton (4 tonne) INTELSAT VI communications satellite by hand, prior to installing a new engine to reposition the craft in the correct orbit. Satellite deployment and maintenance is a regular part of Shuttle missions.

LAUNCHING SATELLITES

A satellite's journey into space usually involves several phases, each with its own hazards. First, the launching rocket places its payload in low Earth orbit. Then a rocket burn sends the satellite out along an elliptical transfer orbit. Finally, another rocket motor, strapped to the satellite or built into it, puts the craft permanently into its planned orbit. Such maneuvers are all routine for rocket engineers. But a single failure at any point can turn an expensive piece of technology into a piece of space junk.

WHAT IF...

...WE COULD BRING A SATELLITE BACK TO EARTH?

A satellite cannot remain in orbit indefinitely. In geostationary orbit, a satellite is too high to be affected by the Earth's atmosphere. But gentle gravitational tugs from the Moon and the Sun inevitably nudge the spacecraft from its orbital position. Sooner or later, it will run out of the propellant that allows it to adjust its altitude. Before that happens, ground engineers use the last drops of fuel to shift the satellite to a more distant graveyard orbit, where it will present no hazard to other, still-operational spacecraft.

No satellite in geostationary orbit has ever been retrieved for examination. But inactive satellites in low orbit come down to Earth after a few months because of drag from Earth's ultra-thin outer atmosphere. Many of these satellites will have failed due to equipment problems that are caused by the harsh environment of space. On reentry, the satellites burn up in the atmosphere. If the craft are large enough for substantial chunks to survive reentry intact, engineers try to ensure that the dying satellite falls in remote areas of the oceans.

Crewed satellites must obviously land back on Earth in one piece. And scientists are exploring ways to retrieve other, uncrewed, geostationary satellites intact. One promising method uses inflatable heatproof bags that act as a combination balloon and parachute to protect the satellite from the fiery heat of reentry into

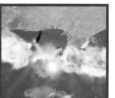

A 21st-century geostationary satellite returns home behind an improvised heat shield. The first satellite retrieved intact from geostationary orbit could become an interesting museum piece.

the atmosphere and to slow the craft down so it does not hit the ground at high speed.

The system was tested in February 2000 on the new Russian Fregat rocket stage and on a European cargo container, the IRDT, onboard Fregat. In the trial, Fregat and the IRDT were spotted on radar by the recovery team. The IRDT was successfully recovered, dented but otherwise intact. But Fregat itself was lost in bad weather that blanketed the landing zone. Project engineers are confident that the craft survived, although it has not yet been found. With luck, such rocket stages might be reused in the future—if they are not too badly damaged.

An orbiting crewed spacecraft usually returns to Earth without waiting for natural orbit decay. To achieve reentry, the craft fires small rocket engines that lower its speed below orbital velocity and send it into a long, arcing fall into our planet's atmosphere.

On reentry, temperatures outside the craft reach 2,000–3,500°F (1000–2000°C). But an expendable or reusable heat shield, such as the Space Shuttle's tiles, protects the crew and maintains room temperature inside. A parachute or runway landing completes the mission.

ORBITAL MILESTONES

First satellite launch into orbit	October 4, 1957 (Sputnik-1)
First US satellite launch into orbit	February 1, 1958 (Explorer-1)
First launch into polar orbit	February 28, 1959 (Discoverer-1)
First launch of several satellites at once	June 22, 1960 (Transit-2A, Sunray)
First landing of a satellite from orbit	August 18, 1960 (Discoverer-14)
First commercial communications satellite launch	July 10, 1962 (Telstar-1)
First launch into geostationary orbit	August 19, 1964 (Syncom-3)
First Shuttle retrieval of a satellite	November 16, 1984 (Palapa B-2 & Westar-6)

KICK START

The first part of a satellite's journey to its working orbit usually begins with a rocket launch that leaves the spacecraft at least 80 miles above the Earth and moving at around 17,000 miles per hour (27,500 km/h)—orbital velocity. At this height, no orbit is stable: The Earth's atmosphere is still thick enough for its drag to bring the satellite down within a few days or weeks. Most spacecraft are designed to work from a higher altitude, from a few hundred miles for most observational satellites to the 22,700-mile (36,500 km) geostationary orbit that is now the home of much communications equipment.

To get the satellite to its planned orbit, at least two more rocket burns are necessary. Both must be made in strict accordance with the laws of orbital mechanics. The first takes place at the satellite's perigee—the point where it is closest to the Earth's surface and moving at its fastest. The extra speed gained from the rocket burn sends the craft on a transfer orbit that joins its original

ON HIRE

THE SPACE SHUTTLE (RIGHT) USED TO LAUNCH COMMERCIAL SATELLITES, RELEASING THEM FROM ITS CARGO BAY WITH MOTORS TO KICK THEM TO THEIR FINAL ORBIT. BUT COMMERCIAL LAUNCHES WERE STOPPED BY THE U.S. GOVERNMENT AFTER THE 1986 CHALLENGER DISASTER. GOVERNMENT SATELLITES ARE STILL LAUNCHED FROM THE SHUTTLE, BUT ONLY A FRACTION OF THE FULL LAUNCH COST— WHICH CAN BE UP TO $500 MILLION—IS CHARGED.

BIG BUSINESS

THE COMMERCIAL SATELLITE INDUSTRY STARTED IN 1962 WITH A LINKUP BETWEEN AT&T AND NASA TO LAUNCH THE TELSTAR-1 COMMUNICATIONS SATELLITE. THE INTERNATIONAL INTELSAT ORGANIZATION WAS THE FIRST BIG BUYER OF SATELLITES. TODAY, MANY COUNTRIES PURCHASE AND LAUNCH SATELLITES. SATELLITE-USING COMPANIES SPEND ABOUT $12 BILLION A YEAR BUYING SATELLITES, AND A SINGLE INTELSAT SATELLITE MAY GENERATE $1 BILLION OF REVENUE OVER ITS LIFETIME.

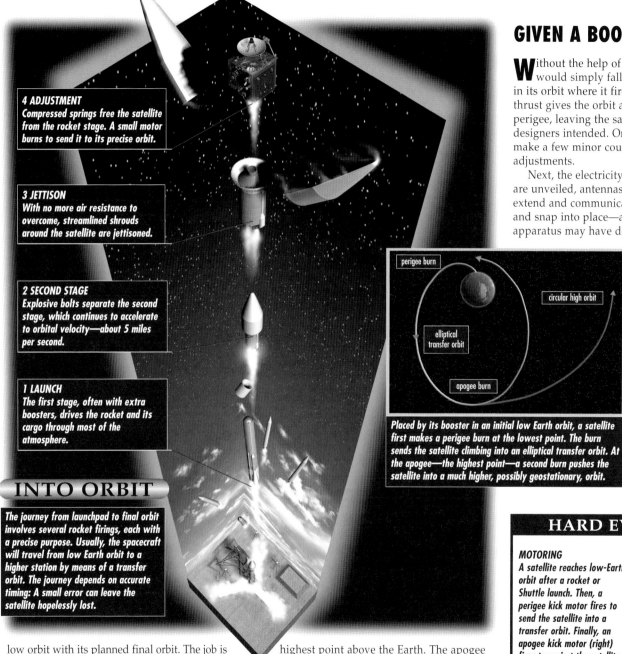

4 ADJUSTMENT
Compressed springs free the satellite from the rocket stage. A small motor burns to send it to its precise orbit.

3 JETTISON
With no more air resistance to overcome, streamlined shrouds around the satellite are jettisoned.

2 SECOND STAGE
Explosive bolts separate the second stage, which continues to accelerate to orbital velocity—about 5 miles per second.

1 LAUNCH
The first stage, often with extra boosters, drives the rocket and its cargo through most of the atmosphere.

INTO ORBIT

The journey from launchpad to final orbit involves several rocket firings, each with a precise purpose. Usually, the spacecraft will travel from low Earth orbit to a higher station by means of a transfer orbit. The journey depends on accurate timing: A small error can leave the satellite hopelessly lost.

perigee burn

circular high orbit

elliptical transfer orbit

apogee burn

Placed by its booster in an initial low Earth orbit, a satellite first makes a perigee burn at the lowest point. The burn sends the satellite climbing into an elliptical transfer orbit. At the apogee—the highest point—a second burn pushes the satellite into a much higher, possibly geostationary, orbit.

GIVEN A BOOST

Without the help of the AKM, the satellite would simply fall back to the low point in its orbit where it fired the PKM. The extra thrust gives the orbit a new and much higher perigee, leaving the satellite where its designers intended. On-board thrusters may make a few minor course and attitude adjustments.

Next, the electricity-producing solar panels are unveiled, antennas unfold, instruments extend and communications dishes open up and snap into place—although some of this apparatus may have deployed earlier during the transfer orbit to enable mission control to talk to the satellite or to run the satellite's electrical systems. With the satellite operational in its final orbit, the launch vehicle is destroyed—unless the satellite has been carried into space by the reusable Shuttle, which returns to Earth on completion of a mission.

HARD EVIDENCE

MOTORING
A satellite reaches low-Earth orbit after a rocket or Shuttle launch. Then, a perigee kick motor fires to send the satellite into a transfer orbit. Finally, an apogee kick motor (right) fires to project the satellite into its final orbit. Some perigee and apogee rocket firings are achieved by the upper stage of the launch rocket—for example, the Russian Proton's Block DM or the U.S. Atlas rocket's Centaur.

low orbit with its planned final orbit. The job is usually performed by the perigee kick motor (PKM), which may be part of the launching rocket or of the satellite itself.

The next burn takes place when the satellite reaches the apogee of its transfer orbit—its highest point above the Earth. The apogee kick motor (AKM), usually mounted on board the satellite, increases the spacecraft's speed once more, and converts the long ellipse of the transfer orbit into the near-perfect circle usually required for its permanent, final orbit.

UNMANNED SPUTNIKS

The Soviet Union amazed the world in 1957 when it launched the world's first artificial satellite, Sputnik 1. But this was just the first in a series of unmanned missions that ranged from scientific research to test flights for crewed space capsules and Venus-probe rocket stages. Together with three satellites that launched dogs into space, the success or failure of every mission was hidden behind the word "sputnik"—the name given to them by the Soviets to conceal their true purpose from the rest of the world.

WHAT IF...

...A SATELLITE'S TRUE PURPOSE WERE KEPT SECRET TODAY?

The world has changed since the days when the obsessively secret Soviet Union concealed the true objectives of its Sputnik missions simply by giving them all the same name. The chances of concealing a satellite's true purpose today are slim.

To begin with, the Earth's surface is peppered with military radar installations and telescopes. For example, the U.S. Air Force's Ground-based Electro-Optical Deep Space Surveillance System (GEODSS) network has giant cameras that take clear photographs of satellites passing overhead. Military analysts at the Department of Defense are then able to make an intelligent guess at the satellite's purpose from its shape, its orbit and the type of radio signals it transmits.

Even the liftoff couldn't be kept a secret. Networks of early-warning stations on the ground in both Europe and North America still monitor launch activity, keeping an eye out for missile attacks. Meanwhile, spacecraft such as the U.S. DSP (Defense Support Program) satellite watch from above for the launch of ballistic missiles. These intelligence satellites are equipped to detect any type of space launch, and can actually identify the rocket by analyzing its heat emissions. Knowing what type of launcher is being used gives an indication of how heavy its satellite payload might be.

The Space Shuttle Atlantis releases the DSP (Defense Support Program) satellite in December 1991. The DSP provides the space-based element of the U.S. defense network.

Information about orbiting satellites can also be obtained by amateur astronomers. Even a medium-size amateur telescope can be fitted with a charge-coupled device (CCD) which, when linked to a laptop computer running satellite-tracking software, enables enthusiastic stargazers to photograph any satellite in Earth orbit.

Yet despite the likelihood of detection, it is still conceivable that a satellite could appear to perform one function while actually hiding a more sinister purpose. A military satellite, for example, could easily be disguised as a communications satellite, designed to relay cellular phone calls, then be used to eavesdrop on confidential government telephone conversations instead.

The U.S. Space Command (SpaceCom) in Colorado Springs, Colorado, catalogs every satellite launch, but even SpaceCom occasionally misses one. Those that do slip through the net tend to be spacecraft whose launch vehicles fail: Within an hour or so of liftoff, these sorry satellites generally end up as charred debris at the bottom of the ocean—even so, the possibility remains.

UNMANNED LAUNCHES

Nickname	Real Name	Launch Date	Launch Vehicle	Reentry Date	Purpose
Sputnik 1	Object PS-1	October 4, 1957	8K71 (A-class)	January 4, 1958	Publicity
Sputnik 3	Object D-1	May 15, 1958	8A91 (A-class)	April 6, 1960	Scientific research
Sputnik 4	Object KS-1	May 15, 1960	8K72 (A-1 class)	October 15, 1965	Vostok capsule test
Sputnik 7	1VA No.1	February 4, 1961	8K78 (A-2-e class)	February 26, 1961	Venus probe launch
Sputnik 8	1VA No.2	February 12, 1961	8K78 (A-2-e class)	February 25, 1961	Venus probe launch

FIRST SHOTS

Sputnik 1, the world's first artificial satellite, was one of the smallest and simplest spacecraft ever built. The 2-ft (60-cm) -wide, 184-lb (83-kg) aluminum alloy ball contained batteries, radio transmitters and four long antennas, and transmitted a "beep...beep ...beep" signal that could be received by amateur radio enthusiasts all over the world. But this simplicity of design was the secret to its success.

Sputnik 2, along with Sputniks 5 and 6, carried dogs into orbit to test the effects of space travel on animals. The next Sputnik without a living passenger was the 2,900-lb (1,315-kg) Sputnik 3. Larger and more sophisticated than Sputnik 1, it was originally intended to be the first Soviet satellite to reach space. Problems with its construction delayed the launch until May 1958—more than seven months after the first Sputnik.

A dozen instruments were fitted to Sputnik 3 to study the Earth's upper atmosphere, magnetic field and radiation belt, as well as any cosmic rays and micro-meteoroids that might be found there. Small solar panels were built around its cone-shaped main body to provide power. Sputnik 3 succeeded in transmitting data back to Earth until April 6, 1960, when its orbit decayed, causing it to burn up in the atmosphere.

The Soviet Union greeted the 1960s with the launch of Sputnik 4. This mission was the first test flight in space of the Vostok capsule—the spacecraft later used by Yuri Gagarin to become the first human in orbit. The engineers behind the new capsule called it Object KS-1

SPY IN THE SKY

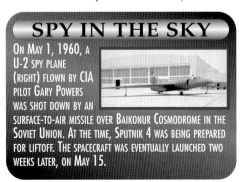

ON MAY 1, 1960, A U-2 SPY PLANE (RIGHT) FLOWN BY CIA PILOT GARY POWERS WAS SHOT DOWN BY AN SURFACE-TO-AIR MISSILE OVER BAIKONUR COSMODROME IN THE SOVIET UNION. AT THE TIME, SPUTNIK 4 WAS BEING PREPARED FOR LIFTOFF. THE SPACECRAFT WAS EVENTUALLY LAUNCHED TWO WEEKS LATER, ON MAY 15.

or "Spaceship-1." This first Vostok was a stripped-down, 7-ft 5-in (2.2-m)-wide test capsule; it had no heat shield, no parachutes and no ejection seat, and was intended to burn up in the atmosphere. On the outside of the ball-like capsule were a pair of small solar panels and some radio antennas. Strapped to the back of the Vostok was a service section containing thrusters and propellant tanks, more antennas, and heat control systems. The service section also housed Sun and Earth sensors for orientation of the capsule, and the main rocket engine used for reentry—but things went wrong when the time for reentry arrived. The capsule was facing the wrong direction when the rocket fired. Instead of descending, it shot into a higher orbit and didn't come down for five years.

Sputnik 7 and 8 were not actually satellites, but the upper stages of rocket launchers designed to send Venera space probes to Venus. Built by Sergei Korolev's Design Bureau OKB-1, these "Block L" rocket stages used a sophisticated motor that burned kerosene and liquid oxygen propellants.

Block L was the first rocket engine designed to be fired while in Earth orbit. In the weightlessness of space, propellants tend to float around inside their tanks, which makes it difficult to pump them into the engine. On Sputniks 7 and 8, small rockets

were fired to force the propellants toward the bottom of their tanks and into the engine pumps. Due to a technical failure, Sputnik 7 failed to release its payload. But on February

12, 1961, Sputnik 8 did succeed in its mission—the upper rocket stage engine ignited in low orbit to propel the Venera 1 probe toward its destination.

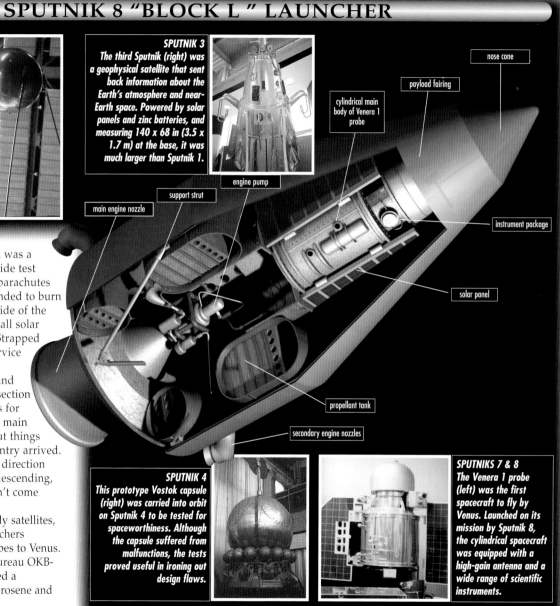

SPUTNIK 8 "BLOCK L" LAUNCHER

SPUTNIK 1
Launched from Baikonur, Soviet Union, on October 4, 1957, Sputnik 1 (replica shown right) was the first human-made object to go into orbit around the Earth. The shiny sphere was carried into space by a modified R-7 booster. The satellite's orbit decayed after three months in space, and it burned up in the atmosphere.

SPUTNIK 3
The third Sputnik (right) was a geophysical satellite that sent back information about the Earth's atmosphere and near-Earth space. Powered by solar panels and zinc batteries, and measuring 140 x 68 in (3.5 x 1.7 m) at the base, it was much larger than Sputnik 1.

nose cone
payload fairing
cylindrical main body of Venera 1 probe
engine pump
support strut
main engine nozzle
instrument package
solar panel
propellant tank
secondary engine nozzles

SPUTNIK 4
This prototype Vostok capsule (right) was carried into orbit on Sputnik 4 to be tested for spaceworthiness. Although the capsule suffered from malfunctions, the tests proved useful in ironing out design flaws.

SPUTNIKS 7 & 8
The Venera 1 probe (left) was the first spacecraft to fly by Venus. Launched on its mission by Sputnik 8, the cylindrical spacecraft was equipped with a high-gain antenna and a wide range of scientific instruments.

TELSTAR 1

One day in July 1962, scientists at a receiving station in France beamed with delight when they heard "The Star-Spangled Banner" and saw a picture of the American flag on a TV screen. They had just received the first-ever transatlantic TV relay, transmitted from the United States by the new Telstar satellite. Although the experimental satellite could only broadcast during a limited period each day, its pioneering transmissions paved the way for a revolution in international communications.

WHAT IF...

...WE DID NOT HAVE ANY COMMUNICATIONS SATELLITES?

Communications and broadcast satellites have become such an integral part of global civilization that we take them for granted. But suppose these great facilitators were absent—that electronic relay stations in space were somehow impossible.

It is not hard to imagine a scenario that could bring this about. In real life, Telstar was almost scuttled by a 1.4-megaton nuclear explosion. The test—Project Starfish, detonated 250 miles (400 km) above Earth the day before Telstar launched—played havoc with radio communications, and it was years before the planet's magnetic field returned to normal. The Soviets quickly matched Starfish with high-altitude explosions of their own.

The Cold War turned even nastier four months later, when the Cuban missile crisis erupted and both sides' nuclear arsenals were put on hair-trigger alert. Fortunately, cool heads prevailed. The United States and the Soviet Union had frightened each other badly enough to realize that their relations had to improve. Soon, treaties banned atmospheric testing, as well as nuclear weapons in space. But there could have been other outcomes, and a full-blown nuclear war—which would have left the world with much more to worry about than the absence of communications satellites—is only one of them.

It is possible to imagine the U.S. and the Soviets engaging in a space bomb race and permanently disrupting the natural radiation belts around the Earth, as well as adding their

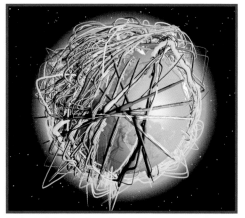

Without TV broadcast satellites, cable companies could expect to do great business. The world would still be linked in a communications net, but costs would be higher and TV far less international.

own radioactive loads to near-Earth space. Sooner or later, the Cold Warriors come to their senses. But by then, the damage has been done.

Fifty years later, in 2012, it is time for some retrospective might-have-beens. At a conference in Maine, near the site of the first and only transatlantic television broadcast, telecommunications experts will consider the legacy of Telstar and ask themselves what might have happened without the Earth's human-made radiation belts. In about 5,000 years, when radiation levels have fallen sufficiently, they may find out. In the meantime, a Communications Futures Conference will be chaired by cable pioneer Red Rurner, CEO of MMN. Items for discussion are: "Cabling and You: Direct Cabling into Every Household," "International Cabling" and "Realtime Cabling: A Cable For Every Occasion." It's a wired-up world.

TELSTAR 1

LAUNCH DATE	JULY 10, 1962	ORBITAL PARAMETERS	586 MILES (943 KM) x 3,499 MILES (5,631 KM)
LAUNCH VEHICLE	DELTA BOOSTER	ORBITAL INCLINATION	44.8°
LAUNCH SITE	CAPE CANAVERAL	FREQUENCIES	UPLINK: 6,390 MHz
WEIGHT	171 LB (77.5 KG)		DOWNLINK: 4,170 MHz

MOVING PICTURES

As early as October 1945, science fiction writer Arthur C. Clarke had discussed the possibility of worldwide television and radio broadcasts using spacecraft as orbiting relays. In Clarke's model, space stations would travel in geostationary orbit, providing total coverage of the Earth for television and radio.

By 1962, the technology required to maintain stations in space was still on the drawing board, but John R. Pierce and his team at AT&T's Bell Laboratories were designing a simpler, less costly kind of relay—the artificial satellite Telstar. Pierce's strategy was ambitious. It called for "a system of 40 satellites in polar orbits, and 15 in equatorial orbits…and about 25 ground stations, so placed as to provide global coverage."

Truly worldwide broadcasts, though, would require worldwide cooperation. The reality of Cold War politics made this impossible, and the goal of the Telstar project was simplified: Put a satellite into low Earth orbit, and use it to transmit television pictures across the Atlantic.

AT&T paid NASA $3 million to launch Telstar 1 on July 10, 1962. The spacecraft was placed into a high, elliptical orbit, where it immediately encountered an unforeseen hazard: The day before launch, the United States had detonated Starfish, a powerful nuclear weapon that exploded high above the atmosphere. Radiation from the blast would take its toll on Telstar in the months to come. But in the hours after launch, the Telstar team stayed glued to their television sets, awaiting the first intercontinental telecast.

INTERNATIONAL

On July 24, 1961, one year before Telstar was launched, President John F. Kennedy sought to bring the new technology of the satellite revolution within the laws of existing U.S. regulatory bodies. But there was a more positive aspect to the policy. Kennedy (above) invited international participation in satellite development. "In the interest of world peace and closer brotherhood among the peoples of the world."

ATLANTIC LINKUP

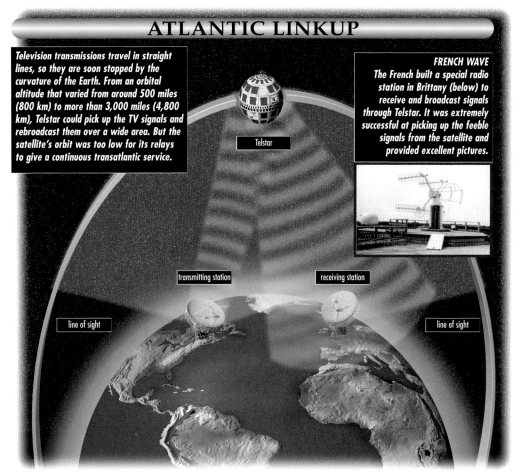

Television transmissions travel in straight lines, so they are soon stopped by the curvature of the Earth. From an orbital altitude that varied from around 500 miles (800 km) to more than 3,000 miles (4,800 km), Telstar could pick up the TV signals and rebroadcast them over a wide area. But the satellite's orbit was too low for its relays to give a continuous transatlantic service.

FRENCH WAVE
The French built a special radio station in Brittany (below) to receive and broadcast signals through Telstar. It was extremely successful at picking up the feeble signals from the satellite and provided excellent pictures.

Telstar

transmitting station   receiving station

line of sight   line of sight

WAVES ACROSS THE SEA

In Andover, Maine, Bell Laboratories flew the American flag in front of its transmitters. In Pleumeur-Bodou in Brittany, France, technicians were poised to broadcast a videotape of an actor, a singer and a guitarist. And at Goonhilly Down in Cornwall, England, the control room was ready to pick up the first images from America.

But all three stations had to wait for Telstar to reach the right place. The satellite's low orbit meant that there was only a limited period during which it was above the horizon—and in the line of sight—for both European and U.S. transceiver stations. Transatlantic TV was only possible for 102 minutes each day.

At the appointed moment, France picked up Old Glory. Technical problems delayed British reception, and it was late in the night before the BBC received pictures of AT&T chairman Frederick Kappel reading a statement. They couldn't hear him because the sound had failed.

Despite technical glitches, Telstar 1 was an outstanding achievement, blazing a trail for the extraordinary growth in global telecommunication. Arthur C. Clarke is the author of many science fiction classics, but his "extraterrestrial relays" have become a reality.

MISSION DIARY: TELSTAR 1

1960 AT&T begins research and development for a satellite communications system: Telstar (right). *1961* J.R. Pierce of Bell Laboratories, an AT&T subsidiary, estimates cost of Telstar project at $500 million. *July 24* Concerned about the prospect of an AT&T monopoly in telecommunications, President Kennedy announces measures to guarantee competition and advocates international partnership in the field. *July 9, 1962* The U.S. military detonates a huge

thermonuclear test weapon, Starfish, at high altitude without advance warning. *July 10, 1962* Telstar 1 launched into elliptical Earth orbit from Cape Canaveral in Florida. *July 11, 1962* Telstar relays TV pictures from a ground station in Andover, Maine, to a receiver on the east coast of France. France sends back pictures of an actor, a singer and a guitarist (above). Britain picks up signal later, at Goonhilly receiving station (far right). *August 31* U.S. government passes Communication Satellite

Act. It sets up a federally funded satellite communications corporation, Comsat. *October* Telstar, already damaged by atmospheric radiation from the U.S. nuclear test in July, is further compromised by a matching Soviet nuclear test. Ground controllers manage to carry out some remote repairs. *December* Telstar finally ceases to operate. *May 7, 1963* Telstar 2 launched. Communications resume.

DIRECT BROADCAST SATELLITES

Satellite technology has turned broadcasting into a global industry. Images beamed from space already reach hundreds of millions of people all over the planet, and from Alaska to Afghanistan, from Siberia to South Carolina, more viewers are signing up every day. New digital compression techniques mean that there is more for them to view. The latest satellites can transmit more than 200 channels simultaneously, and their successors will beam new high-definition TV and videophone links, too.

...SATELLITES TRANSMITTED HIGH-DEFINITION TV?

One of the most exciting aspects of the Direct Broadcast Satellite (DBS) system is its ability to transmit high-definition television (HDTV). The advent of this technology promises big changes in visual communication and the entertainment media. According to some enthusiasts, its impact will be as great as the arrival of television was back in the radio age.

A television picture is composed of lines on the screen, updated many times each second. The more lines and the more updates, the clearer the picture. U.S. broadcasts use 525 lines per frame and 30 frames per second. HDTV will deliver at least 1,125 lines and 60 frames per second—enough to create a picture so clear that a viewer sitting right by the screen will still see a sharp image. HDTV comes with digital stereo sound at a quality that approaches a compact disc player.

To achieve these levels, an HDTV signal has to carry about 10 times as much information as the system it is set to replace. This data burden explains why basic HDTV technology has been around for a while, without ever taking off.

Now, with the arrival of the Direct Broadcast Satellite, there is a cost-effective way to send the huge amounts of data required to make HDTV work. The high-frequency, wide bandwidth of DBS can handle far more information than regular ground transmissions. And the old

High-definition television means more detail, better color and better sound. But only Direct Broadcast Satellites have the power to bring it cheaply to millions of homes throughout the world.

analog TV signals have now been replaced by digital versions—which means that new, smart compression techniques can squeeze more data into the same space. Compression technology distinguishes the parts of a video picture that move from those that stay still. It wastes no space updating an unchanging background, for example.

But HDTV promises more than just improved home entertainment. The movie industry will be able to use it to create high-quality special effects. It will be invaluable in education and in the medical profession. And the military is already harnessing the technology for use in spy satellites.

HDTV is only the most obvious part of a communications revolution that is based on our new ability to transmit huge quantities of data very quickly. The forthcoming National Information Infrastructure (NII) will bring together the Internet, television and multimedia communication in a single service accessible to all. But it won't improve everything. Even in high definition and in stereo, junk TV will still be junk TV.

WORLD SATELLITE LOCATOR

NAME	ORBIT*	LAUNCH DATE	COVERAGE
AMOS 1	4.0°E	MAY 16, 1966	MIDDLE EAST AND EUROPE
APSTAR 1A	133.8°E	JULY 3, 1996	CHINA, JAPAN AND SOUTHEAST ASIA
ASTRA 1E	19.2°E	OCTOBER 19, 1995	EUROPE
BRASILSAT A2	92.0°W	FEBRUARY 8, 1985	BRAZIL
DFS KOPERNIKUS 2	28.5°E	JULY 24, 1990	GERMANY AND SURROUNDING NATIONS
EUTELSAT II-F1	13.0°E	AUGUST 30, 1990	EUROPE
GALAXY 1R	133.0°W	FEBRUARY 9, 1994	U.S. AND PUERTO RICO
INSAT-1D	83.0°E	JUNE 7, 1990	INDIA

*ALL ORBITS ARE ON THE EQUATOR (0° LATITUDE), SO ORBITAL POSITIONS ARE IN DEGREES LONGITUDE (E OR W) ONLY.

TV FROM ORBIT

Television signals have been beamed to Earth from satellites since 1962, when a satellite called Telstar began the era of worldwide TV. But until the 1990s, satellite signals were weak. To receive them, you needed a big dish antenna, up to 12 ft (3.5 m) across. Often, such equipment was owned by cable TV broadcasters, who used it to distribute programs nationwide, and most of the dishes had motorized controls that allowed them to move easily from one satellite to another.

Now, the giant dishes are going the way of the dinosaurs. A new generation of Direct Broadcast Satellites (DBS) have been placed in orbit. These offer easily accessible direct-to-home (DTH) digital broadcasts, which can be received by dishes as small as 15 inches (38 cm) across. The new satellites send out tightly focused, high-power digital signals. And they are capable of transmitting up to 200 TV channels at once—all with high-quality stereo sound.

A DBS receives the programs transmitted from TV companies' ground stations. Onboard devices called transponders relay the signals back to Earth at a different frequency. Dish antennae focus the signals into a narrow, powerful beam, just as reflectors focus a flashlight bulb.

HARD EVIDENCE

GEOSTATIONARY ORBIT
A satellite's speed depends on its height above the Earth. At Space Shuttle altitudes of around 300 miles (500 km), it will complete one orbit every 95 minutes or so. At lunar distance, it will orbit the earth once a month—just as the Moon does. And at what is known as geosynchronous altitude, 22,240 miles (35,800 km) above the planet, it will complete an orbit in exactly 24 hours. For a geosynchronous orbit to be

geostationary, the satellite must be in the plane of the Earth's equator. In that case, as seen from Earth, it will always be in the same position in the sky. Such orbits are essential for direct satellite broadcasts, since home users have a fixed target for their dish antennae.

24-hour orbital period

22,240 miles (35,800 km)

24-hour rotation

CONTINENTAL FOOTPRINT

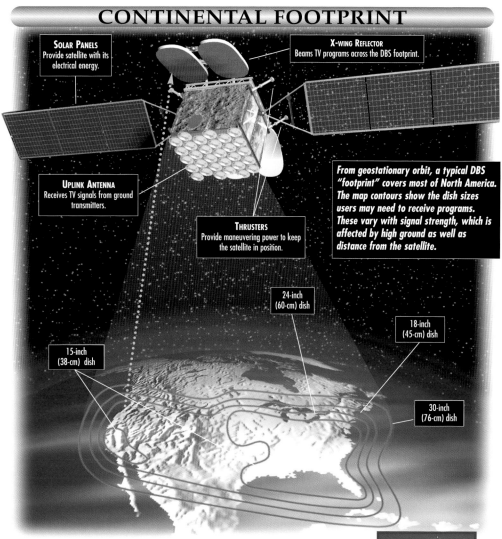

SOLAR PANELS
Provide satellite with its electrical energy.

X-WING REFLECTOR
Beams TV programs across the DBS footprint.

UPLINK ANTENNA
Receives TV signals from ground transmitters.

THRUSTERS
Provide maneuvering power to keep the satellite in position.

From geostationary orbit, a typical DBS "footprint" covers most of North America. The map contours show the dish sizes users may need to receive programs. These vary with signal strength, which is affected by high ground as well as distance from the satellite.

24-inch (60-cm) dish

18-inch (45-cm) dish

15-inch (38-cm) dish

30-inch (76-cm) dish

A VISION OF THE FUTURE

IN 1945, WRITER ARTHUR C. CLARKE (LATER FAMED FOR HIS NOVEL *2001: A SPACE ODYSSEY*) HAD A VERY BRIGHT IDEA. IN THE FEBRUARY ISSUE OF THE MAGAZINE WIRELESS WORLD, HE WROTE: "AN ARTIFICIAL SATELLITE AT THE CORRECT DISTANCE FROM THE EARTH WOULD MAKE ONE REVOLUTION EVERY 24 HOURS; I.E. IT WOULD REMAIN STATIONARY ABOVE THE SAME SPOT AND WOULD BE WITHIN OPTICAL RANGE OF NEARLY HALF THE EARTH'S SURFACE." CLARKE'S VISION OF SATELLITES IN SPACE

HAS BECOME A REALITY. AND ALTHOUGH HE JOKINGLY COMPLAINS THAT HE NEVER RECEIVED ROYALTIES FROM HIS SUGGESTION, GEOSTATIONARY ORBIT HAS BEEN NAMED "CLARKE ORBIT" IN HIS HONOR.

STANDING STILL IN SPACE

A typical DBS satellite has 32 transponders working at any one time—plus a few spares, since there is no easy way to repair a satellite once it has been set in orbit. The beam from each transponder is carefully aimed to cover a particular area of the Earth far below. Together, the transponders create what satellite controllers call a "footprint"—an area where signals are reliably received.

A geostationary orbit takes precisely one day, so the satellite remains permanently above the same spot on the Earth. The receiving dishes down below need only be aimed once and then locked into place.

Because of the effects of lunar and solar gravity, even a perfectly placed geostationary satellite wavers a little off course. A wobble of 100 miles (160 km) or so will not affect reception on Earth, but from time to time ground controllers must maneuver their satellites back into their proper positions. Such adjustments use up only tiny quantities of onboard fuel. Sooner or later, though, the fuel is expended—one reason why satellite lifetimes are limited to around 15 years, despite the reliability of DBS components.

Since all geostationary satellites must orbit in the same plane, there is a limited number of orbital "slots" whose allocation is decided by international agreement. Each slot takes up about 2° of the 360° orbit—around 900 miles (1500 km). Several satellites may be in each slot, spaced about 100 miles (160 km) apart. When transponders on one satellite fail, another satellite can sometimes take over its faulty neighbor's workload so smoothly that television viewers are unaware that anything has gone wrong.

HARD EVIDENCE

THE SATELLITE REVOLUTION
Satellite television in the U.S. began in March 1978, when the Public Broadcasting Service (PBS) first used satellites to distribute programs to its local stations. By the early 1980s, direct-to-home (DTH) satellite receivers were commonplace, especially in rural areas where regular broadcast quality was poor and cable nonexistent. Satellite TV rapidly went global, with North America, Europe and Japan leading the way and virtually the whole world following as fast as it could. DBS satellites themselves have grown in size and complexity: The Galaxy model, built by the Hughes Corporation, weighs 3,800 pounds (1,723 kg) and can transmit more than 200 separate TV channels.

EARLY WARNING SATELLITES

T he chilling words "Missile launch alert!" quicken the hearts of military personnel around the world. They mean that some form of attack could be under way—and if it is, every second of early warning is invaluable. Missile early warning is the mission of the United States military's Defense Support Program (DSP). Providing worldwide coverage, the DSP uses a network of at least five satellites in geosynchronous orbit to monitor the Earth for space and missile launches.

WHAT IF...

...DSP SATELLITES HAD TO BE REPLACED?

D espite operating far beyond their planned service lives, all DSP satellites eventually wear out and have to be retired. But in the future, there will be no point in just replacing a worn-out DSP with another one of the same or a similar design, as has happened until now. The nature of the threat from ballistic missiles has changed. Concerns about a massive strategic ballistic missile attack are low today, but there is great concern over tactical missile attacks on deployed military troops. Because of this, DSP number 23 will be the final satellite in the constellation. Eventually, a new multilayered constellation of satellites will take over the early warning mission. This new system is called the Space Based Infrared System (SBIRS).

SBIRS will consist of two groups of satellites in at least three different orbital planes. SBIRS Low group will have approximately 24 satellites in low Earth orbit, and SBIRS High will have four in geosynchronous orbit and two in highly elliptical orbits. This mix of satellites and orbital planes will provide more accurate and timely data from multiple viewpoints. It will also provide launch locations as well as launch warnings, allowing for quick counterstrikes. Additionally, the satellites will be able to calculate a missile's probable target and pass that information to defensive systems.

The low portion of the SBIRS network is called the Space and Missile Tracking System

The low segment of the SBIRS system will consist of about 24 of these Brilliant Eyes satellites. Working together, they will be able to track a missile and its payload all the way from launch to target.

(SMTS), nicknamed "Brilliant Eyes." Brilliant Eyes satellites will work in pairs to give a stereoscopic view of a launch vehicle and payload. For the first time ever, early warning satellites will be able to track a missile even after its booster engines have stopped firing, and follow the missile from launch to burnout, coast and reentry. The satellites will also have the capability to "talk" to each other, passing tracking data from one to the next to ensure a target is kept under constant surveillance.

Improvements like this do not come easy or cheap. Work on the SBIRS program began in 1996, but the first operational satellite is not expected to launch for several years. Total cost for the program is estimated at $22 billion dollars. This price tag may look fairly hefty at first, but every military commander in the world will tell you that early and accurate knowledge of a missile attack is worth every dollar.

CURRENT DSP SATELLITES

PRIMARY MISSION	STRATEGIC AND TACTICAL MISSILE LAUNCH DETECTION	WEIGHT	5,250 LB (2,381 KG)
ORBITAL ALTITUDE	22,233 MILES (35,780 KM)	POWER GENERATION	SOLAR ARRAYS GENERATING 1,485 WATTS
HEIGHT	32.8 FT (10 M) ON ORBIT,		(SATELLITE USES 1,274 WATTS)
	28 FT (8.5 M) AT LAUNCH	FIRST DEPLOYED	NOVEMBER 6, 1970
DIAMETER	22 FT (6.7 M) ON ORBIT,	UNIT COST	APPROXIMATELY $250 MILLION EACH
	13.7 FT (4.1 M) AT LAUNCH	SATELLITES IN PROGRAM	23

TIMELY WARNINGS

The Defense Support Program (DSP) is a survivable and reliable satellite-based system that uses infrared sensors to detect missile and space launches and nuclear detonations. The program began in 1966, but the first launch of a satellite didn't occur until November 6, 1970. Since then, DSP satellites have fed a constant stream of warning data, 24 hours a day, 365 days a year, to the Missile Warning Center at Cheyenne Mountain, Colorado, via ground stations as far away as Australia and Germany.

As the DSP program progressed, the satellite went through five major design changes, each one increasing its capabilities. Phase 1 satellites had 2,048 sensor elements—optimized to detect a single infrared wavelength—and a planned 3-year life span. By the fifth design, DSPs had 6,000 sensors on two wavelengths and 7- to 9-year lifetimes. They were also hardened against laser jamming and nuclear attack, and they were capable of detecting any imminent physical threat from an attack satellite and maneuvering themselves out of harm's way.

The DSP proved its worth during Operation Desert Storm, in the Gulf War of 1991. Originally designed to detect strategic threats against the continental United States, the DSP satellites turned their attention to the Persian Gulf and detected Iraqi ballistic missiles being launched against the U.S.-led coalition forces.

MIDAS TOUCH

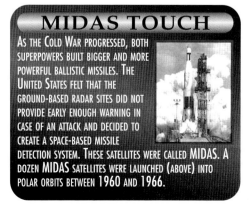

AS THE COLD WAR PROGRESSED, BOTH SUPERPOWERS BUILT BIGGER AND MORE POWERFUL BALLISTIC MISSILES. THE UNITED STATES FELT THAT THE GROUND-BASED RADAR SITES DID NOT PROVIDE EARLY ENOUGH WARNING IN CASE OF AN ATTACK AND DECIDED TO CREATE A SPACE-BASED MISSILE DETECTION SYSTEM. THESE SATELLITES WERE CALLED MIDAS. A DOZEN MIDAS SATELLITES WERE LAUNCHED (ABOVE) INTO POLAR ORBITS BETWEEN 1960 AND 1966.

WATCHING FROM ABOVE

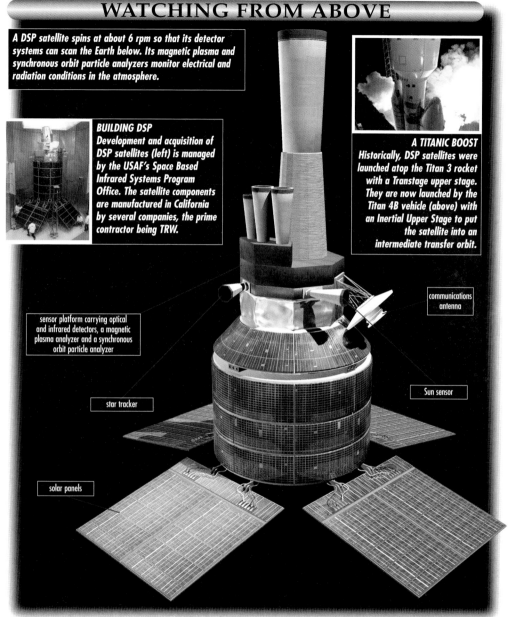

A DSP satellite spins at about 6 rpm so that its detector systems can scan the Earth below. Its magnetic plasma and synchronous orbit particle analyzers monitor electrical and radiation conditions in the atmosphere.

BUILDING DSP
Development and acquisition of DSP satellites (left) is managed by the USAF's Space Based Infrared Systems Program Office. The satellite components are manufactured in California by several companies, the prime contractor being TRW.

A TITANIC BOOST
Historically, DSP satellites were launched atop the Titan 3 rocket with a Transtage upper stage. They are now launched by the Titan 4B vehicle (above) with an Inertial Upper Stage to put the satellite into an intermediate transfer orbit.

sensor platform carrying optical and infrared detectors, a magnetic plasma analyzer and a synchronous orbit particle analyzer

communications antenna

Sun sensor

star tracker

solar panels

SCUDBUSTERS

The DSP's superb operational performance during Operation Desert Storm demonstrated that the system, although designed to detect intercontinental missile launches, could also provide significant early warning against comparatively small tactical missiles. Every Scud missile launch made during the Gulf War was detected within

HARD EVIDENCE

EYE IN THE SKY
the current DSP satellites use sensor arrays of 6,000 lead sulfide and mercury cadmium telluride elements to detect infrared radiation. These sensors detect and track the exhaust heat generated by ballistic missiles (right) and are sensitive enough to detect military jet aircraft operating

on afterburners. Additionally, DSPs carry optical sensors that can detect nuclear detonations and large meteoroids entering the atmosphere. The main sensor barrel of a DSP satellite is tilted 7.5° to the side and the vehicle spins as it travels around its orbit, so the sensors sweep out a cone-like pattern to cover a wide area.

SHUTTLE

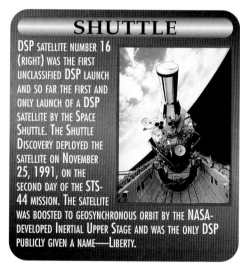

DSP SATELLITE NUMBER 16 (RIGHT) WAS THE FIRST UNCLASSIFIED DSP LAUNCH AND SO FAR THE FIRST AND ONLY LAUNCH OF A DSP SATELLITE BY THE SPACE SHUTTLE. THE SHUTTLE DISCOVERY DEPLOYED THE SATELLITE ON NOVEMBER 25, 1991, ON THE SECOND DAY OF THE STS-44 MISSION. THE SATELLITE WAS BOOSTED TO GEOSYNCHRONOUS ORBIT BY THE NASA-DEVELOPED INERTIAL UPPER STAGE AND WAS THE ONLY DSP PUBLICLY GIVEN A NAME—LIBERTY.

seconds of liftoff. Warnings were relayed to civilian populations and coalition forces, including Patriot missile defense batteries, in Saudi Arabia and Israel.

Without a doubt, the Defense Support Program was also a key factor in America's winning the Cold War. Because of the DSP's global satellite coverage, no missile launch, no matter where in the world it came from, could go unnoticed. No aggressive forces could launch an attack without being detected—and they knew it. The program continues to be one of the U.S. military's most successful programs ever.

SPY SATELLITES

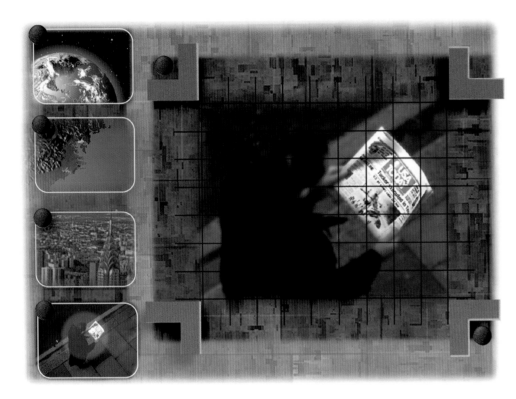

Scud missiles in Iraq, troop movements in Bosnia and the drug barons of Colombia are among the many targets being spied on by military satellites of the U.S., Russia and several other countries. These photoreconnaissance satellites usually fly at an altitude of 100 miles (160 km) or more above the ground and at speeds of over 17,000 miles per hour (27,000 km/h), but they produce highly detailed images that can reveal even such relatively small objects as individual people—and their abilities are constantly being improved.

WHAT IF...

...EVERY ENEMY POSITION COULD BE SEEN FROM SPACE?

As a matter of routine, spy satellites are deployed by both the U.S. and Russia to monitor each other's missile and troop movements and military exercises. Both countries also launch new satellites, or reposition existing ones, to check on other countries' troop deployments at times of crisis such as the Gulf War.

One aim of the developers of spy satellite technology is to produce systems that transmit very detailed, real-time images of the battlefield directly to the troops on the ground. These systems will enable spy satellites to zoom in on details of the enemy's activities, day or night and in any weather, using a combination of optical, infrared and radar imaging technologies.

Armed with this information, battlefield commanders will be better able to defend themselves from enemy attacks. They will also be able to direct their forces more effectively against the opposition's troop and armor formations, and key facilities such as mobile missile launchers and communications bases.

One of the problems faced by the designers of spy satellite systems is that of producing the ever-more-detailed images that the military need. In this respect, the satellite designers have the same problem as astronomers using space telescopes—to

Spy satellite technology may soon be able to provide battlefield commanders with real-time images of the enemy's positions and movements.

see objects with more detail at optical or infrared wavelengths, they need more powerful optics in their satellites.

One way to do this might be to use much larger systems of lenses and telescope mirrors, but this would make the satellites too big, heavy and expensive to be practical. An alternative, already being pioneered by astronomers and probably by the military as well, uses a technique called interferometry to squeeze huge amounts of information from the tiny differences between images of the same object taken from slightly different points in space or time. Instead of using one large telescope or camera, the signals from a number of small ones are combined to produce an image as detailed as one from a very large single instrument.

Another problem for spy satellites is the blurring of images caused when light gets bent as it travels through the Earth's atmosphere. This can be reduced by using adaptive optics—so-called "rubber mirrors" that constantly change shape to compensate for the momentary atmospheric distortion.

When techniques such as interferometry and adaptive optics have been perfected, the monitoring of individual people from space could become a reality.

US SPACE SPIES

	ADVANCED KH-11/IMPROVED CRYSTAL	LACROSSE/VEGA
LAUNCH VEHICLE	SPACE SHUTTLE ATLANTIS, 1990	SPACE SHUTTLE COLUMBIA, 1988
BUILDER	LOCKHEED MARTIN	LOCKHEED MARTIN
SOLAR PANEL WINGSPAN	115 FT (35 M)	150 FT (45 M)
DIAMETER	14 FT (4.2 M)	14 FT (4.2 M)
WEIGHT	18 TONS (16 TONNES)	16 TONS (14.5 TONNES)
MISSION	HIGH-RESOLUTION DIGITAL IMAGES OF SPECIALLY IDENTIFIED TARGETS TRANSMITTED VIA DATA SATELLITES FOR IMMEDIATE USE AT THE NATIONAL RECONNAISSANCE OFFICE.	DAY AND NIGHT COVERAGE OF TROOP AND ARMOR MOVEMENTS, AND RADAR IMAGES TO PROVIDE TARGET IDENTIFICATION FOR KH-11 (CRYSTAL) SATELLITES.

SPIES IN THE SKY

Spying from above is nothing new—balloons were used to spy on enemy positions during many 19th-century wars, and aircraft have been used for reconnaissance since as early as World War I. The coming of space technology and the ability to get satellites into orbit—out of range of missiles—made the spy satellite a logical development in reconnaissance. The ideal orbit is one that takes the satellite over the poles of the Earth at a relatively low altitude of about 200 miles (320 km). At that height, 17 orbits—covering the whole of the Earth— can be flown in a single day.

The U.S. flew its first spy satellites under the name Discoverer in 1959. These top-secret craft, code-named Corona by the CIA, first became operational in 1960 and took pictures on film. The film was returned to Earth in a reentry capsule. The Soviet Union launched its first spy satellite in 1962. Code-named Zenit, this also returned a recoverable film capsule.

Hundreds of spy satellites have been launched since the early 1960s. In addition to the U.S. and Russian programs, spy satellite missions launched by other countries include those of China and Israel, and a joint venture by France, Spain and Italy.

KH-11
The first of the U.S. KH-11 (Crystal) series of spy satellites was launched from the Space Shuttle in 1990. Since then, others have been launched on Titan 4 rockets. The satellites in this series are equipped with digital multispectral imaging systems, as well as electronic intelligence sensors.

SHUTTLE SPY
A SPACE SHUTTLE MISSION LAUNCHED IN NOVEMBER 1991 DEPLOYED A MISSILE EARLY WARNING SATELLITE AND ALSO CARRIED OUT EXPERIMENTS CALLED "TERRA SCOUT" AND "MILITARY MAN IN SPACE." THESE INVOLVED THE FIRST AMERICAN SPACE SPY, THOMAS HENNAN (RIGHT), WHO TESTED THE EFFECTIVENESS OF HUMAN RECONNAISSANCE FROM SPACE USING HIGH-RESOLUTION SENSORS AND CAMERAS.

DIGITAL SYSTEMS

The smallest ground-based object that could be seen in the early satellite images was the size of a house. Today, the technology is improving so much that spy satellites could soon be reading the headlines on a newspaper.

Film capsules are being replaced by high-resolution, multispectral digital systems that take both visible-light and infrared pictures, and transmit these images back to ground stations. Infrared photography produces images of heat patterns, so it can be used by night as well as day, and radar reconnaissance satellites can record images day or night through even the thickest cloud cover.

The digital images from U.S. satellites are transmitted to the National Reconnaissance Office in Washington, D.C., via military communications satellites and NASA's Tracking and Data Relay Satellites.

HIGH RES
Russia operates Zenit, Yantar and Kometa satellites that return high-resolution images on film in recoverable capsules. Since the collapse of the Soviet Union, some of these images have been sold commercially. They show that the cameras could see details as small as 3 ft (1 m) across.

KOSMOS/ZENIT
The Soviet Union attempted to conceal the launches of its Zenit spy satellites by making them part of its Kosmos program of scientific satellites. This Zenit satellite carries a number of flask-like recoverable film capsules that are periodically dropped back to Earth. Newer versions transmit digital images back to ground stations.

MISSILE GAP
ONE OF THE FIRST SPY SATELLITE IMAGES, TAKEN BY DISCOVERER 14 IN 1960 AS PART OF THE CORONA PROGRAM, SHOWED A MILITARY AIRFIELD IN NORTHEAST RUSSIA. THE PROGRAM SHOWED THAT, DESPITE THE CLAIMS OF SOVIET PREMIER NIKITA KRUSCHEV, THE SOVIET UNION DID NOT HAVE MORE INTERCONTINENTAL BALLISTIC MISSILES DEPLOYED THAN THE U.S. DID. CORONA IMAGES BROUGHT THE ESTIMATE OF THE NUMBER OF MISSILES THEN IN THE SOVIET UNION'S NUCLEAR ARMORY DOWN FROM HUNDREDS TO TENS.

EUROPEAN SPACE SPY
Europe's first military spy satellite, Helios 1, was launched in 1995 aboard an Ariane rocket. Helios was based on the successful Spot commercial remote-sensing satellites, one of which took this infrared image of Baghdad, the capital of Iraq.

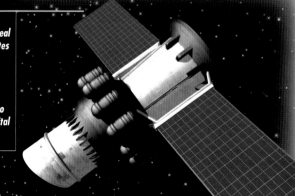

SCIENTIFIC SATELLITES

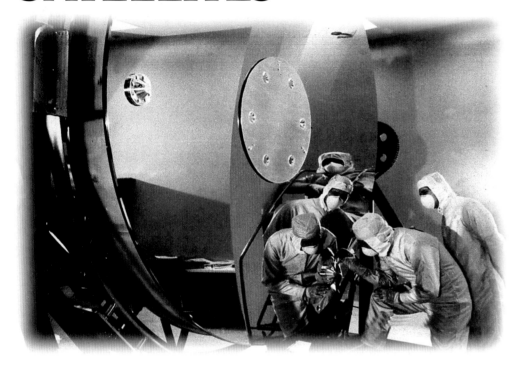

Every minute of every day, scientific satellites look both inward and outward, studying everything from deforestation and the hole in the Earth's ozone layer to exploding stars and supermassive black holes at the hearts of galaxies. Since the dawn of the Space Age, the world's space agencies have launched hundreds of scientific satellites into Earth orbit and sent scores more to study the other bodies of the solar system. Future probes will look deeper into our own planet—and into the farthest reaches of time and space.

WHAT IF...

...WE CONTINUE TO IMPROVE SATELLITE TECHNOLOGY?

Astronomers have discovered several dozen planets orbiting stars resembling our own Sun. But all of these planets are giants, similar to Saturn and Jupiter in our own solar system, so they are unlikely homes for life. Ground-based telescopes cannot detect Earth-size planets around these stars—nor can the orbiting Hubble Space Telescope. That job may be left for Terrestrial Planet Finder (TPF), one of the most ambitious scientific satellites scheduled for the next decade. Its instruments would not only detect Earth-size planets, but measure the contents of their atmosphere for signs of life and even photograph their surfaces.

Other future scientific satellites will produce similarly spectacular results. They will see deeper into space and time with greater clarity than today's satellites. They will also give us a better understanding of humankind's impact on Earth's environment and seek out microscopic life on the planets and moons of our solar system.

Some of these new probes will follow NASA's philosophy of "faster, better, cheaper." For example, the Small Explorer Program (SMEX) will launch a series of small, inexpensive orbiters to address fundamental questions in astronomy and astrophysics. These craft will weigh only a few hundred pounds, but thanks to advances in microelectronics and other technologies, they will produce better scientific results than older satellites

NASA's Small Explorer Program (SMEX) provides frequent flight opportunities for highly focused, inexpensive science missions. Each SMEX satellite (right) costs just $50 million to build and launch.

that were much larger and more expensive.

But for some missions, size still matters. TPF, for example, will deploy several large telescope mirrors to image distant star systems. The Next-Generation Space Telescope's mirror will be substantially larger than that of the Hubble Space Telescope, allowing it to see fainter and more distant objects.

Earth resources satellites will proliferate, too. Some will be small and focus on a single mission, such as measuring ozone depletion, while others will tackle multiple missions, allowing them to view Earth as an overall, integrated system. Such satellites may resolve the question of human responsibility for global warming, and help us plan massive new agricultural enterprises for the developing world.

The technology developed for satellite missions also has applications on Earth. For example, by applying the computerized image enhancement technology developed to read Earth resources satellite photographs, experts are now able to provide "maps" of the human body. Such advances may eventually help solve many of the questions facing science today.

SATELLITE MISSIONS

Satellite	Mission	Satellite	Mission
Landsat 7	Monitor crop resources, minerals and forests	Chandra X-Ray Observatory	Study hot, violent objects and events
TOPEX/ Poseidon	Monitor global ocean circulation and sea levels	Extreme Ultraviolet Explorer	Study white dwarfs and other objects
SoHO	Study the structure and physics of the Sun	Mars Global Surveyor	Map the surface of Mars and study Martian climate
Imager	Produce 3-D images of Earth's magnetosphere	NEAR Shoemaker	Map the surface and chemical content of asteroid Eros
Hubble Space Telescope	Optical ultraviolet, infrared observations of the universe	Galileo	Study Jupiter and its moons
		Cassini	Rendezvous with Saturn

LAB PARTNERS

May 9, 2000, was a busy day around the solar system. From Earth orbit, Landsat 7 surveyed an out-of-control wildfire that was advancing on Los Alamos, New Mexico. Landsat 7's optical and infrared images helped firefighters assess the size and movement of the blaze and pinpoint hot spots. At the same time, the Tropical Rainfall Measuring Mission was peering through the clouds to measure the temperature of the ocean surface. Such measurements can help scientists understand and predict climate changes, such as the El Niño and La Niña events. Far from Earth, the NEAR Shoemaker spacecraft was orbiting less than 30 miles (50 km) above the surface of 433 Eros, a large asteroid. And a half-billion miles away, Galileo was preparing for another encounter with Ganymede, the largest moon of Jupiter.

These and many other scientific satellites are extending our knowledge of Earth and of humanity's impact on our planet, as well as the interaction between the Sun, the Earth and the cosmos—from the Moon and planets to the edge of the visible universe.

Satellites have conducted scientific missions since the very beginning of the Space Age. Explorer 1, the first American satellite, discovered the radiation belts that encircle Earth in 1958. Since then, the world's space

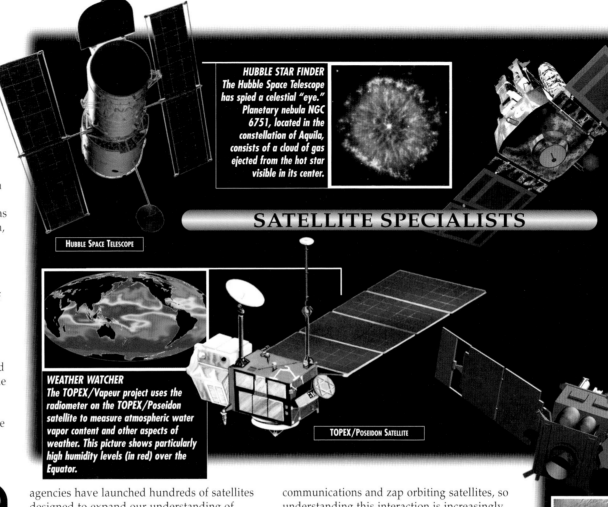

SATELLITE SPECIALISTS

HUBBLE STAR FINDER
The Hubble Space Telescope has spied a celestial "eye." Planetary nebula NGC 6751, located in the constellation of Aquila, consists of a cloud of gas ejected from the hot star visible in its center.

HUBBLE SPACE TELESCOPE

WEATHER WATCHER
The TOPEX/Vapeur project uses the radiometer on the TOPEX/Poseidon satellite to measure atmospheric water vapor content and other aspects of weather. This picture shows particularly high humidity levels (in red) over the Equator.

TOPEX/POSEIDON SATELLITE

SOLAR AND HELIOSPHERIC OBSERVATORY (SoHO)

SOLAR OBSERVER
Using an extreme ultraviolet imaging telescope (EIT), SoHO images the Sun's atmosphere at different wavelengths in order to show stellar material at different temperatures. This image shows stellar material at 1.8 million °F (1 million °C).

MARS GLOBAL SURVEYOR

MARTIAN SURVEYOR
Mars Global Surveyor took this photo of the Apollinaris Patera volcano in March 1999. This ancient volcano is thought to be as much as three miles high. The crater is about 50 miles (80 km) across.

TAKE TWO

AFTER THE SUCCESS OF SPUTNIK 1, ENGINEERS AND LAUNCH CREW WERE CALLED BACK FROM VACATION TO ASSEMBLE SPUTNIK 2 IN JUST ONE MONTH. LAUNCHED ON NOVEMBER 3, 1957, SPUTNIK 2 WAS THE FIRST TRUE SCIENTIFIC SATELLITE. ITS SENSORS WERE DESIGNED TO DETECT ULTRAVIOLET AND X-RAY ENERGY FROM THE SUN. SENSORS ALSO STUDIED THE EFFECTS OF SPACE ON THE PHYSICAL PROCESSES OF LAIKA THE DOG, SPUTNIK 2'S PASSENGER, WHO UNFORTUNATELY PERISHED IN ORBIT.

agencies have launched hundreds of satellites designed to expand our understanding of physics, astronomy, meteorology, oceanography and many other fields. Earth resources satellites locate mineral deposits, monitor deforestation and measure ozone depletion and global warming. City planners use satellite data to help draft new zoning laws, and satellite images have been used as evidence in lawsuits against industrial polluters.

Several craft are studying the interaction of the Sun's magnetic field and the solar wind—a steady flow of charged particles from the Sun's outer atmosphere—with Earth's magnetic field. Outbursts from the Sun can knock out power grids, disrupt radio

communications and zap orbiting satellites, so understanding this interaction is increasingly important for our technological society.

Beyond Earth, probes have scanned the Moon and every planet except Pluto, plus several asteroids and comets. They have discovered ancient river valleys on Mars, possible oceans beneath the icy crust of Jupiter's moon Europa and giant volcanoes on Io, another Jovian moon. Observatories in Earth orbit have discovered the "seeds" from which stars and galaxies grew, imaged powerful jets of hot gas squirting away from black holes, and found evidence of a "dark energy" that may be forcing the universe to expand faster and faster.

These and other satellite missions continue

to push back the boundaries of the known universe, expand our knowledge of our own world, and increase our understanding of our place in the cosmos.

WEATHER SATELLITES

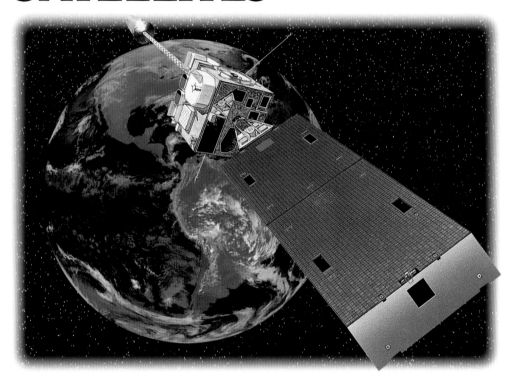

F or as long as people have looked at clouds in the sky, there have been people who have tried to predict what tomorrow's weather will be. Forecasting the weather is a tricky business, but accurate information about it is crucial to modern life—advance knowledge of dangerous winds, rain or snow can help save property crops, and even lives. Fortunately, space science has provided meteorologists with a reliable and accurate tool for predicting the weather: the weather satellite.

WHAT IF...

...YOU HAD YOUR OWN WEATHER SATELLITE DATA?

Most people would describe the weather around them as unpredictable. But what if you could forecast your own weather—and be certain of getting it right? Space engineers are working on long-range plans that could make personalized weather forecasting a reality.

The data from weather satellites is fed into a worldwide computer network that is used to track and predict weather on a global scale. This data is also used regionally to track weather systems as they move across countries, states, and cities. That is as detailed as today's weather satellites can get—but not the satellites of tomorrow. Satellite designs now on the drawing boards of engineers have capabilities that the forecasters of yesteryear could not even dream of.

Engineering visionaries are designing constellations of low-flying satellites that will be digitally linked together like networks of communications satellites. They will be able to cover every square inch of the Earth's surface with super-high-resolution sensors, 24 hours a day. These sensors will let people on the ground zoom in on any location on Earth—to within a few dozen feet—and see the weather systems around it. Multi-spectral infrared scanners will peel open layers of clouds, Doppler radar will tell in which direction and how fast winds are

In the not-too-distant future, highly localized weather satellite information could be available via hand-held receivers or wireless applications protocol (WAP) cellphones.

blowing, and water vapor sensors will give air humidity levels. Combining this precise data with global positioning systems, and making it possible for the public to access it, perhaps via the Internet, will let anyone predict the weather anywhere in the world.

One idea is to combine low-flying cellular telecommunication satellites with wide-area-scanning weather satellites, giving users anywhere the ability to link to the latest imagery from space. Another idea is to electronically link geosynchronous weather, communications and global positioning satellites so users can access all three data feeds simultaneously.

No matter which method the engineers choose, one thing is certain: High-resolution, real-time weather data for anywhere in the world will someday be available to anyone who wants it. And data this detailed would not just be convenient, it could be life-saving. When people can precisely pinpoint dangerous weather systems such as hurricanes or tornadoes, they will be better able to prepare themselves and get out of harm's way, saving thousands of lives every year and minimizing property damage.

WEATHER SATELLITES

AUGUST 1959	LAUNCH OF EXPLORER 6; FIRST PICTURES OF CLOUDS FROM SPACE	NIMBUS (U.S.) 1964–78
APRIL 1, 1960	LAUNCH OF TIROS 1, WORLD'S FIRST WEATHER SATELLITE	ESSA (U.S.) 1966–69
		METEOR (U.S.S.R./RUSSIA) 1969–PRESENT
AUGUST 28, 1964	LAUNCH OF NIMBUS 1, FIRST POLAR WEATHER SATELLITE	NOAA (U.S.) 1970–PRESENT
		GOES (U.S.) 1975–PRESENT
		METEOSAT (EUROPE) 1977–PRESENT
		GMS (JAPAN) 1977–PRESENT
MAJOR WEATHER SATELLITE SYSTEMS		BHASKARA (INDIA) 1979–81
TIROS (U.S.) 1960–65		INSAT (INDIA) 1982–PRESENT
COSMOS (U.S.S.R.) 1962–68		FENG YUN (CHINA) 1988–PRESENT

EYE IN THE SKY

The introduction of weather satellites, combined with the use of other advanced equipment such as supercomputers, greatly improved the accuracy of weather forecasts. At any one time, there are no fewer than a dozen different operational weather satellites from half a dozen countries orbiting the Earth. The images and other data from these satellites provide weather forecasters with details of storm systems, weather fronts, cloud formations, winds, rainfall, fog, ice and snow.

The United States was the first country to test the idea of beaming pictures of clouds back from an orbiting satellite. In August 1959, Explorer 6 radioed the first experimental photos of cloud cover from space. Eight months later, the world's first operational weather satellite, the Television and Infrared Observation Satellite (TIROS), went into orbit. Altogether, 10 TIROS-class satellites were launched in the early 1960s, broadcasting almost 650,000 images.

Other countries around the world soon began launching meteorological satellites, with the Soviet Union launching its first in 1962. The European Space Agency's first came in 1977, and China launched its first in 1988. More than 200 different weather satellites have now been launched into Earth orbit. These include over 60 sent aloft by the United States. Today, more than 120 countries around the world receive most or all of their weather pictures from U.S. satellites.

A NEW ERA

THE AGE OF SPACE-BASED WEATHER FORECASTING BEGAN ON APRIL 1, 1960 WITH THE LAUNCH OF THE TELEVISION AND INFRARED OBSERVATION SATELLITE (TIROS). HARRY WEXLER, DIRECTOR OF RESEARCH FOR THE U.S. GOVERNMENT WEATHER BUREAU, HAD SUGGESTED THAT CAMERAS IN SPACE WOULD BE ABLE TO SEE THE EARTH'S CLOUD PATTERNS, AND TIROS PROVED HIM RIGHT. THOUGH THE 270-POUND (122-KG) SATELLITE OPERATED FOR ONLY 89 DAYS, IT RADIOED BACK 22,952 PHOTOGRAPHS OF CLOUD COVER FROM ITS 450-MILE (725-KM) -HIGH ORBIT, FOREVER CHANGING THE SCIENCE OF WEATHER FORECASTING.

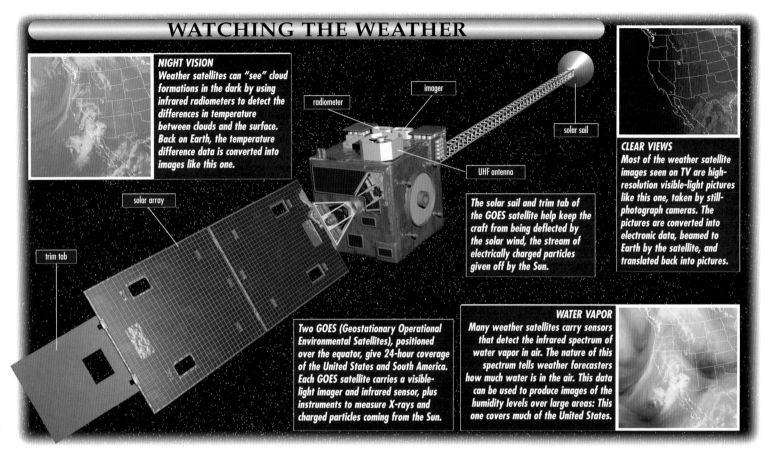

WATCHING THE WEATHER

NIGHT VISION
Weather satellites can "see" cloud formations in the dark by using infrared radiometers to detect the differences in temperature between clouds and the surface. Back on Earth, the temperature difference data is converted into images like this one.

radiometer

imager

UHF antenna

solar sail

solar array

trim tab

CLEAR VIEWS
Most of the weather satellite images seen on TV are high-resolution visible-light pictures like this one, taken by still-photograph cameras. The pictures are converted into electronic data, beamed to Earth by the satellite, and translated back into pictures.

The solar sail and trim tab of the GOES satellite help keep the craft from being deflected by the solar wind, the stream of electrically charged particles given off by the Sun.

Two GOES (Geostationary Operational Environmental Satellites), positioned over the equator, give 24-hour coverage of the United States and South America. Each GOES satellite carries a visible-light imager and infrared sensor, plus instruments to measure X-rays and charged particles coming from the Sun.

WATER VAPOR
Many weather satellites carry sensors that detect the infrared spectrum of water vapor in air. The nature of this spectrum tells weather forecasters how much water is in the air. This data can be used to produce images of the humidity levels over large areas: This one covers much of the United States.

VERSATILITY

As weather satellite designers and engineers gained more experience of the new technology, the designs and functions of weather satellites became increasingly sophisticated. In addition to sending back photographs of clouds and weather systems, weather satellites began recording atmospheric and ocean temperatures at various altitudes and depths, and estimating rainfall.

Today's meteorological satellites can pinpoint weather formations with a high degree of accuracy. They can photograph clouds at night using infrared cameras, and take pictures in several different frequencies of light at the same time. And to peer inside and through the clouds, some weather satellites use instruments that detect the microwave energy given off by the Earth and the atmosphere. These instruments can scan cloud formations to detect hailstorms and tornadoes forming, and can give as much as 25 minutes warning before a tornado strikes. In addition, they provide data about surface wind speeds over the oceans, ground moisture, rainfall, sea ice packs and snow cover.

Weather satellites also produce other useful information about the Earth—for instance, by mapping the world's ocean currents, or by capturing infrared images of the Earth's surface that scientists can use to study land usage and assess crop health. Some of them can even listen for distress signals from ships at sea and downed aircraft.

HARD EVIDENCE

ORBITS
Weather satellites are launched into either polar orbits or geosynchronous orbits. Those in polar orbits usually fly at altitudes of about 450–900 miles (725–1,500 km), crossing over the north and south polar regions in each orbit. They take very detailed close-up pictures as the Earth rotates below, but only cross over any one spot on the Earth once per day. Geosynchronous satellites orbit at 22,500 miles (36,200 km) and stay in the same position over the Earth all the time. This allows them to take continuous pictures of the area below them and track the motions of weather fronts and storm systems.

INFRARED SATELLITES

Throughout the universe, objects too cool to produce much visible light glow brightly at infrared wavelengths. They can be detected by infrared cameras, but the Earth's atmosphere absorbs most incoming infrared energy, so the best way to study them is from space. Astronomers have launched two major infrared observatories, plus an infrared camera aboard the Hubble Space Telescope. In the future, even better infrared space telescopes will provide detailed views of some of the coolest objects in the universe.

WHAT IF...

...WE HAD MORE POWERFUL INFRARED SATELLITES?

The next generation of orbiting infrared observatories will look deeper into the past than earlier satellites, providing new insights into the formation of the first galaxies. They also will see infrared objects more clearly, giving astronomers a more detailed picture of how stars and planets are born. The new observatories will achieve these results with bigger telescopes and more sensitive electronic detectors. With these new satellites, the number of known infrared objects in the universe will climb into the millions. The first of these observatories, the Spitzer Space telescope, was launched in August 2003. It was named in honour of astrophysicist Layman Spitzer, who first proposed putting telescopes in space.

Spitzer's primary mirror, which can gather and focus infrared energy from astronomical objects, spans about 33 inches (84 cm), compared with just 22 inches (56 cm) for the IRAS mirror. A larger mirror gathers more energy, allowing it to detect fainter objects, which are usually farther away than bright ones. Spitzer has enough helium coolant aboard to operate for about five years.

The follow-up to the Hubble Space Telescope and Spitzer is already underway. Originally named the Next-Generation Space Telescope (NGST), the name was changed to James Webb Space Telescope in honour of the NASA administrator who led the Apollo moon missions.

The James Webb Telescope will be able to detect massive planets in other star systems just from their infrared energy. NGST will be the giant of the infrared satellites, with a primary mirror 13 to 26 feet (4–8 m) in diameter—up to three times the size of the Hubble Space Telescope's mirror. Launch is projected for August 2011.

Japan's ASTRO-F infrared satellite is due for launch in late 2005. It will carry a 27-inch (69-cm) infrared telescope and enough coolant to operate for 500 days. Even after the coolant runs out, ASTRO-F will continue to function, but will only be able to detect wavelengths in the near infrared.

Europe's Herschel Space Observatory, originally titled Far-Infrared and Sub-millimetre Telescope (FIRST), is due for launch in 2007. It will carry a monstrous 138-inch (3.5 m) primary mirror. Infrared satellites must maintain a constant temperature to produce their sharpest images, but in Earth orbit they get alternately cold and hot as they pass through Earth's shadow then back into sunlight. So to improve their views of the heavens, Spitzer and Webb must part company with planet Earth. Spitzer trails behind Earth as our planet orbits the Sun, gradually moving farther and farther away. Webb may follow an elliptical orbit around the Sun, or it may orbit a Lagrangian point about 930,000 miles (1.5m km) from Earth, where the gravitational pulls of Earth and Sun are evenly balanced.

INFRARED SATELLITES

Satellite	Telescope	Launched	Ceased Operation
IRAS	22 inches (56 cm)	January 1983	November 1983
ISO	24 inches (61 cm)	November 1995	May 1998
HST*	92 inches (233 cm)	April 1990	Still Working
SIRTF	33 inches (84 cm)	August 2003	Still Working
ASTRO-F	28 inches (71 cm)	Late 2005**	
Herschel	138 inches (3.5 m)	2007**	
James Webb	160–320 inches (4–8 m)	2011**	

*The Hubble Space Telescope is primarily an optical telescope, with one large infrared camera
**Planned launch date

DISTANT PROSPECTS

When the Infrared Astronomy Satellite (IRAS) was launched in early 1983, astronomers had already charted about 6,000 objects that emit most of their energy at infrared wavelengths. By the end of the year, IRAS had identified almost 350,000 sources of infrared energy—a leap in knowledge unparalleled since Galileo turned his first optical telescope toward the heavens almost four centuries earlier.

Infrared light is a form of electromagnetic energy, like radio waves and visible light. Its wavelengths are longer than those of visible light, so it is invisible to human eyes. Instead, we feel infrared as heat. As a rule, though, objects that emit the most infrared are actually cooler than those that emit more visible light.

The catalog of known infrared objects includes cocoons of dust grains that surround newly forming stars; brown dwarfs, which are too small to become stars but too big to be considered planets; and shells of gas and dust expelled by dying stars. In addition, infrared energy can penetrate the clouds of cold dust that cloak galaxies. Infrared telescopes allow astronomers to see into the heart of our own Milky Way and other galaxies.

Water vapor in our atmosphere absorbs infrared energy, so astronomers (or their instruments) must climb above the atmosphere to get a clear view of the infrared sky. To do this, they build telescopes on mountain peaks, carry them on airplanes, float them on research balloons—or launch them into space.

HARD EVIDENCE

PENETRATING THE FOG
The center of our Milky Way galaxy is packed with stars, gas and dust clouds, and possibly a black hole a million times the mass of our Sun. This region is invisible to our eyes because it is hidden behind thick clouds of dust. But because infrared energy emitted by objects within the region penetrates the dust, infrared satellites can get a good view of them. These satellites can also see into the cores of other galaxies that are obscured by their own dust clouds. These two pictures show the Milky Way as seen in visible light (top left), and the same view as seen by an infrared satellite (bottom left).

IRAS – FIRST OF ITS KIND

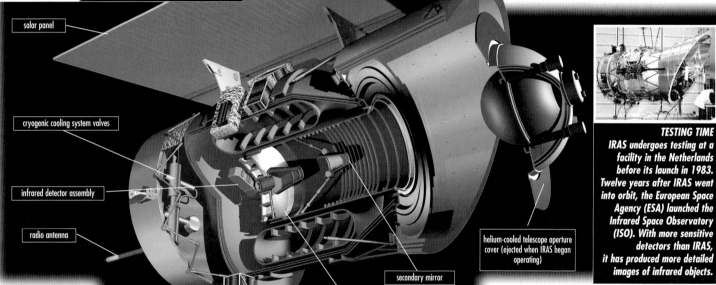

COCOONED DUST
Newly forming stars are embedded in thick "cocoons" of dust grains, which hide them from view (top left). Energy from the young stars warms the dust, causing it to produce infrared energy. Infrared observations (center and bottom left) can thus provide the best information on the process of starbirth.

IRAS, a joint American-British-Dutch project, was the first satellite to conduct a whole-sky survey at infrared wavelengths. It carried an infrared telescope weighing 1,785 pounds (800 kg).

solar panel

cryogenic cooling system valves

infrared detector assembly

radio antenna

helium-cooled telescope aperture cover (ejected when IRAS began operating)

secondary mirror

22.4-inch primary mirror

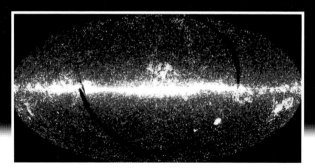

INFRARED SKY
This all-sky map shows all the infrared point sources discovered by IRAS. In total, IRAS detected about 350,000 of these, increasing the number of known sources six-fold.

HOT OBJECTS

MANY ASTRONOMICAL OBJECTS, INCLUDING PLANETS AND COOL STARS, PRODUCE NO VISIBLE LIGHT—BUT THEY USUALLY EMIT INFRARED ENERGY. THE PALE BLOB (BOTTOM LEFT IN THIS IMAGE) CAPTURED BY THE HUBBLE SPACE TELESCOPE'S INFRARED CAMERA, WAS INITIALLY THOUGHT TO BE A PLANET ORBITING A BINARY STAR, BUT IT IS MORE LIKELY TO BE A RED DWARF OR BROWN DWARF STAR.

TESTING TIME
IRAS undergoes testing at a facility in the Netherlands before its launch in 1983. Twelve years after IRAS went into orbit, the European Space Agency (ESA) launched the Infrared Space Observatory (ISO). With more sensitive detectors than IRAS, it has produced more detailed images of infrared objects.

COOL VIEW

Although infrared is a form of heat energy, most infrared objects are relatively cool (compared with stars), with temperatures of just a few hundred degrees. Because of this, their infrared signals can be faint, and a warm telescope emits so much infrared energy that it drowns out the weaker signals from celestial sources. To cool an infrared telescope's detectors, designers encase them in bottles of liquid helium, at a temperature just a few degrees above absolute zero. At such a super-cold temperature, the IRAS instruments could detect the warmth from a 20-watt lightbulb on Pluto.

X-RAY SATELLITES

In 1962, as the U.S. was hitting its stride in the space race, NASA sent up a small suborbital rocket to detect X-rays in space. Once above the atmosphere, which absorbs X-rays, the detector instantly recorded an abundance of these high-energy, short-wavelength rays, and astronomers realized they had found a new "window" on the universe. Since then, dozens of satellites have analyzed X-rays from hundreds of stars and galaxies—and revealed an X-ray "zoo" that includes some of the weirdest objects in the universe.

WHAT IF...

...X-RAY SATELLITES COULD REVEAL MORE?

A At the beginning of the 21st century, there are about a half-dozen functioning X-ray satellites in orbit around the Earth. These include NASA's Rossi XTE and Chandra; the Italian/Dutch BeppoSAX; and the European Space Agency's XMM.

The largest and most powerful of these are Chandra and the XMM. Chandra—the X-ray equivalent of the Hubble Space telescope—was launched in July 1999 and will study all X-ray sources in extraordinary detail. It has 100 times the sensitivity of the first major orbiting X-ray telescope, the Einstein Observatory, which was itself 1,000 times more sensitive than previous instruments.

It will analyze the coronas—the source of stellar X-rays—in all our 100,000 neighboring stars. It will also study X-ray binaries, those systems in which neutron stars and (we assume) black holes circle and consume their vast gaseous companions. It will even be able to examine such systems way beyond our own galaxy, out to the Virgo cluster of galaxies 60 million light-years away.

XMM (X-Ray Multi-Mirror) is the largest science satellite ever built in Europe, and has a more powerful mirror system than Chandra. Like Chandra, it was launched in 1999 and will study the X-ray sources in the distant cosmos for up to 10 years, and perhaps even longer.

The European Space Agency's proposed successor to XMM is XEUS (X-Ray Early

Astronomers hope that data from the latest X-ray telescopes—Chandra and XMM (above)—will help them unravel cosmic mysteries ranging from the nature of black holes to the origin of the universe itself.

Universe Spectroscopy), an X-ray telescope split into two. XEUS will be two separate spacecraft, one carrying the telescope mirrors and the other the telescope's detectors. The mirrors and detectors will orbit 165 feet (50 m) apart, creating a very powerful telescope about 200 times more sensitive than XMM.

NASA plans to follow CHANDRA with Constellation-X, a group of four satellites which, working together, can function as if they were a single extremely powerful X-ray telescope. Constellation-X will study, among other things, Black Holes, galaxy formation and the nature of dark matter. It is possible that Constellation-X and XEUS, which use similar principles, may be combined into a single project.

India's ASTROSAT, intended to launch in 2007, will carry x-ray instruments designed to detect a broad range of wavelengths. ASTROSAT is India's first multi-wavelength astronomy satellite.

X-RAY SATELLITE LAUNCHES

1962	AEROBEE SOUNDING ROCKET DETECTS FIRST COSMIC SOURCE OF X-RAYS	1991	YOHKOH ("SUNBEAM") X-RAY SOLAR OBSERVATION SATELLITE
1970	FIRST X-RAY SATELLITE, UHURU	1993	ARRAY OF LOW-ENERGY X-RAY IMAGING SENSORS (ALEXIS)
1974	ARIEL 5 X-RAY SATELLITE		
1978	EINSTEIN OBSERVATORY (HEAO-2) LAUNCHED, WITH X-RAY TELESCOPE	1995	ROSSI X-RAY TIMING EXPLORER (RXTE)
		1996	SATELLITE FOR X-RAY ASTRONOMY (BEPPOSAX)
1983	EUROPEAN X-RAY OBSERVATORY SATELLITE (EXOSAT)	1999	CHANDRA AND XMM
1990	RÖNTGEN SATELLITE (ROSAT)	2005	ASTRO-E2

X-RAY HUNTERS

By 1970, rocket-borne experiments had identified over 30 X-ray sources in our own galaxy, and several more beyond it. But the exact nature of these sources was hard to pin down, because scientists could not actually see them.

A new window on the universe opened with the first satellite devoted to X-ray astronomy. This was one of the Explorer series of satellites, Explorer 42, originally called Small Astronomical Satellite 1 (SAS-1) but later renamed Uhuru. Launched in 1970, it carried a simple X-ray detector and could measure the strength of X-rays and pinpoint the direction from which they were coming. Uhuru spent three years scanning the sky and recorded 160 X-ray sources. Its data revealed that most of the brightest sources were clustered along the plane of the galaxy, toward its center, and intriguingly, many X-ray sources were variable. One, named Centaurus X-3, pulsed on and off every 4.84 seconds on a 2.1-day cycle. What could the mysterious object be? What made the X-rays? And why the rapid pulse?

Gradually, a theory emerged. A pulsing X-ray source could only be a neutron star—a small but super-dense object—that was sucking in gas from a neighboring star. This gas was being heated to millions of degrees and releasing vast amounts of X-ray energy. The rotation of the neutron star caused the pulsation of the X-rays. To study these extraordinary objects, astronomers needed details that Uhuru could not provide.

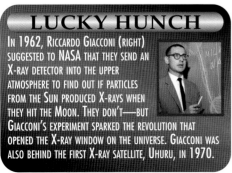

LUCKY HUNCH

IN 1962, RICCARDO GIACCONI (RIGHT) SUGGESTED TO NASA THAT THEY SEND AN X-RAY DETECTOR INTO THE UPPER ATMOSPHERE TO FIND OUT IF PARTICLES FROM THE SUN PRODUCED X-RAYS WHEN THEY HIT THE MOON. THEY DON'T—BUT GIACCONI'S EXPERIMENT SPARKED THE REVOLUTION THAT OPENED THE X-RAY WINDOW ON THE UNIVERSE. GIACCONI WAS ALSO BEHIND THE FIRST X-RAY SATELLITE, UHURU, IN 1970.

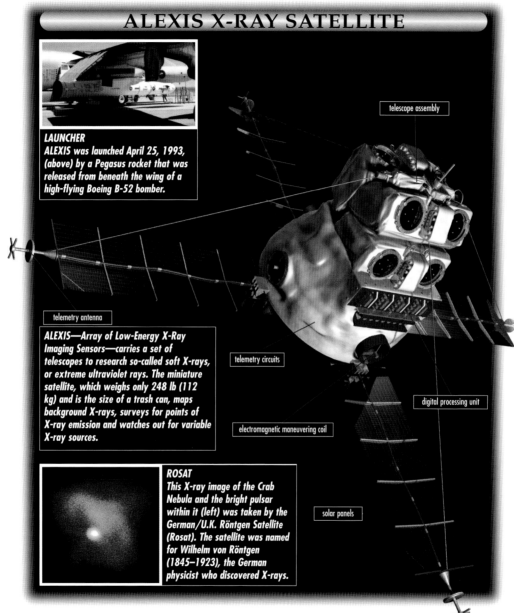

ALEXIS X-RAY SATELLITE

LAUNCHER
ALEXIS was launched April 25, 1993, (above) by a Pegasus rocket that was released from beneath the wing of a high-flying Boeing B-52 bomber.

telescope assembly

telemetry antenna

ALEXIS—Array of Low-Energy X-Ray Imaging Sensors—carries a set of telescopes to research so-called soft X-rays, or extreme ultraviolet rays. The miniature satellite, which weighs only 248 lb (112 kg) and is the size of a trash can, maps background X-rays, surveys for points of X-ray emission and watches out for variable X-ray sources.

telemetry circuits

electromagnetic maneuvering coil

digital processing unit

solar panels

ROSAT
This X-ray image of the Crab Nebula and the bright pulsar within it (left) was taken by the German/U.K. Röntgen Satellite (Rosat). The satellite was named for Wilhelm von Röntgen (1845–1923), the German physicist who discovered X-rays.

X-RAY IMAGING

The success of Uhuru encouraged the U.S., the U.S.S.R. and several European countries to launch many more satellites carrying X-ray detectors. But what astronomers really wanted was a telescope that could "see" X-rays, using carefully shaped mirrors placed almost parallel to the incoming rays to focus them onto detector instruments. This instrument would produce X-ray images instead of the visible-light images produced by ordinary telescopes.

COSMIC RAYS

THE FIRST SATELLITE TO DETECT COSMIC X-RAYS, RATHER THAN X-RAYS FROM THE SUN, WAS ACTUALLY A SOLAR OBSERVATION SATELLITE (RIGHT). THE THIRD ORBITING SOLAR OBSERVATORY (OSO-3) WAS LAUNCHED INTO EARTH ORBIT ON MARCH 8, 1967, AND CARRIED INSTRUMENTS FOR DETECTING GAMMA RAYS AND X-RAYS. IT TRAVELED IN A NEARLY CIRCULAR ORBIT ABOUT 340 MILES (550 KM) ABOVE THE EARTH, AND MADE ITS FINAL DATA TRANSMISSION ON NOVEMBER 10, 1982.

A simple imaging X-ray telescope had been launched on a small rocket in 1965, and it produced crude images of hot spots in the Sun's upper atmosphere. Much better X-ray images of the Sun were produced by the first large focusing X-ray telescope, the Apollo Telescope Mount (ATM) that was flown aboard the Skylab orbiting laboratory in the early 1970s.

The experience gained during the design, construction and use of the ATM was put to good use in the development of the first large mirror-based X-ray telescope. This satellite-based telescope, NASA's High Energy Astrophysical Observatory 2 (HEAO-2), was launched in 1978, and because that year was the centennial of physicist Albert Einstein's birth, HEAO-2 was renamed the Einstein Observatory. It offered a 1,000-fold increase in sensitivity compared with earlier instruments, and by 1979, its 7,000 images had revealed thousands of new X-ray sources, in our own galaxy and in others.

The Einstein Observatory vastly extended the new field of study and discovered many X-ray sources, some of which can also be seen with optical and radio telescopes. More than two decades later, X-ray satellites, including the Chandra and XMM orbiting X-ray telescopes, continue to add to our understanding of the galaxy and the history of its stars.

REMOTE SENSING

Today, dozens of remote sensing satellites scrutinize the entirety of the Earth's surface. With instruments and cameras that see further into the spectrum than the human eye, they chart previously hidden geological features, record the shifting pattern of global land use and uncover hidden archeological sites. They also play a pivotal role in response to natural disasters and pollution control. Remote sensing provides a health check for the entire biosphere, and gives us new insights into climatic change.

WHAT IF...

...WE COULD LINK ALL REMOTE SENSING SATELLITES?

Space-based remote sensing systems surround the Earth in a variety of paths, from near-Earth polar orbits all the way out to geostationary orbits more than 22,000 miles away. These satellites were never intended to work together. Individual spacecraft survey different segments of the Earth in isolation, from the high stratosphere to the biosphere to the surface of the continents and oceans.

But as we learn more about the interconnected nature of our planet, this piecemeal coverage has become unsatisfactory. Evidence from a variety of sources—including space observation—suggests that Earth is due for a series of rapid environmental changes, most of them unpleasant. They include global warming, rises in sea level, deforestation, desertification, ozone depletion, and acid rain. Mass extinctions are a real possibility, too. Since the Earth's climate is a complex interconnected system, if we want to accurately predict its future behavior, we need increased knowledge of how the different parts come together.

In 1989, then-President George Bush proposed that climate change questions could be answered with a fleet of linked remote sensing satellites that would monitor the whole Earth just as intensive care equipment checks the health of a hospital patient. Observations carried out in this manner over decades would provide a reliable insight into global trends. This multi-billion-dollar Earth Observing System (EOS) was placed at the heart of the U.S. Global Change Research Program.

But during the next decade, EOS had a hard time getting off the ground. Skeptical politicians sliced its budget from $18 billion to just $7.25 billion and it went through numerous redesigns. In 1999, the first satellites in the slimmed-down EOS plan finally launched, including Landsat-7 (latest in the 28-year series of Landsat spacecraft), Terra (intended to study soils, land use and the role played by volcanic gases in climate change) and Quikscat (to investigate ocean wind patterns).

The network of EOS satellites continues to grow. The most recent launches are AURA, which studies both the chemistry and the dynamics of Earth's atmosphere, and ICESat, which collects data on land topography as well as cloud and aerosol heights. The next launch should be Glory, designed to investigate the effect of carbon soot and aerosols on climate change.

Just as the first weather satellites revolutionised our view of the climate, EOS will open the door to as new era of climatic understanding. If the results are as dramatic as scientists predict, then EOS could become a permanent fixture, and future generations will look back to the first decade of the 21st century as the time when the Earth grew an electronic nervous system.

REMOTE SENSING SATELLITES

	LANDSAT 5	LANDSAT 7	SPOT 3	RADARSAT	TERRA
OPERATOR	U.S.	U.S.	FRANCE	CANADA	U.S.
LAUNCH SITE	VANDENBURG	VANDENBURG	KOUROU	VANDENBURG	VANDENBURG
LAUNCH VEHICLE	DELTA	DELTA	ARIANE 4	DELTA	ATLAS 2AS
ORBITAL INCLINATION	98.2°	98.2°	98.7°	98.6°	98.2°
MASS	4,266 LB (1,935 KG)	4,332 LB (1,964 KG)	4,195 LB (1,900 KG)	5,969 LB (2,700 KG)	10,679 LB (4,843 KG)
INSTRUMENTS	MULTISPECTRAL SCANNER AND THEMATIC MAPPER	ENHANCED THEMATIC MAPPER	2 HIGH-RESOLUTION CAMERAS	SYNTHETIC APERTURE RADAR	MULTISPECTRA IMAGERS, RADIATION AND GAS DETECTORS

SCENES FROM SPACE

Remote sensing from space detects features invisible to ground-level observers, and has transformed our knowledge of the Earth. Mining companies use remote sensing imagery to guide excavations. Farmers can accurately estimate seasonal crop yields. Before-and-after pictures of communities hit by floods or hurricanes help disaster relief efforts. Authorities identify tankers illegally dumping fuel at sea. And fast-food companies assess suburb growth to locate new restaurants.

Satellite-mounted cameras permit the simultaneous observation of large areas of the Earth's surface. Along with this wide perspective, they also allow a deeper view— into electromagnetic (EM) wavelengths beyond the range of human vision. EM energy from the sun is variously scattered, reflected or absorbed by terrestrial materials. Different materials reflect solar energy in different ways, some of which our eyes detect as the various colors of visible light. But light comprises only a tiny fraction of the entire EM spectrum. For instance, vegetation reflects with more intensity in infrared than it does in visible light, and its exact reflectivity provides a guide to its state of health.

Besides reflected infrared radiation, everything with a temperature of above absolute zero gives out thermal infrared radiation. By measuring this radiation, researchers can learn a material's physical characteristics and temperature. Remote sensing cameras typically image several different visible and infrared EM bands at once. These correspond to the reflected or emitted energy of specific materials—loose sand, say, or cultivated soil, vegetation, water or different mineral types.

As well as these "passive" methods, satellites can also use "active" sensing. Satellite-mounted Synthetic Aperture Radar (SAR), for example, beams radio waves to the ground and records their reflection. SAR can penetrate cloud cover, vegetation or even surface soil.

Remote sensing spacecraft were developed from early weather satellites, which had simple onboard infrared cameras to image cloud formations by night. The first dedicated remote sensing satellite was Landsat-1, equipped with a multispectral camera that transmitted digital data, launched in 1972. There have been Landsats in orbit ever since. Landsat-7 was launched by NASA in April 1999. Its predecessor, Landsat-5, is still in service, although it is now operated by the commercial company Space Imaging EOSAT. The satellites photograph almost the whole Earth every 16 days.

In turn, the Landsats have inspired various foreign equivalents, such as the French SPOT series, the Russian Meteor-Priroda, and counterparts from Europe, Canada, China and Japan. And along with the National Oceanographic and Atmospheric Administration, NASA operates remote sensing satellites to monitor the Earth's oceans and poles.

WHEAT FUTURES

IN A 1970s REMOTE SENSING EXPERIMENT, NASA AND THE DEPARTMENT OF AGRICULTURE USED DATA FROM A LANDSAT SPACECRAFT (RIGHT) TO ACCURATELY FORECAST THE FUTURE WHEAT PRODUCTION OF THE ENTIRE WORLD. TOTAL WHEAT PRODUCTION POTENTIAL WAS CALCULATED FROM LANDSAT IMAGES OF MUCH OF THE EARTH. THEN LOCAL WEATHER FORECASTS WERE USED TO FINE-TUNE ESTIMATES OF THE LIKELY YIELD IN EACH REGION.

HIGH LIGHT

THE POTENTIAL OF REMOTE SENSING WAS DEMONSTRATED EARLY IN THE SPACE PROGRAM WITH OBSERVATIONS BY MERCURY ASTRONAUT GORDON COOPER IN 1963. DURING VISUAL OBSERVATIONS FROM THE TINY PORTHOLE OF HIS FAITH 7 SPACECRAFT, HE WAS ABLE TO SEE A TEST 44,000-WATT XENON LAMP BEAMED AT HIM FROM SOUTH AFRICA, AS WELL AS CITIES, OIL REFINERIES AND EVEN SMOKE FROM INDIVIDUAL HOUSES AROUND THE PLANET.

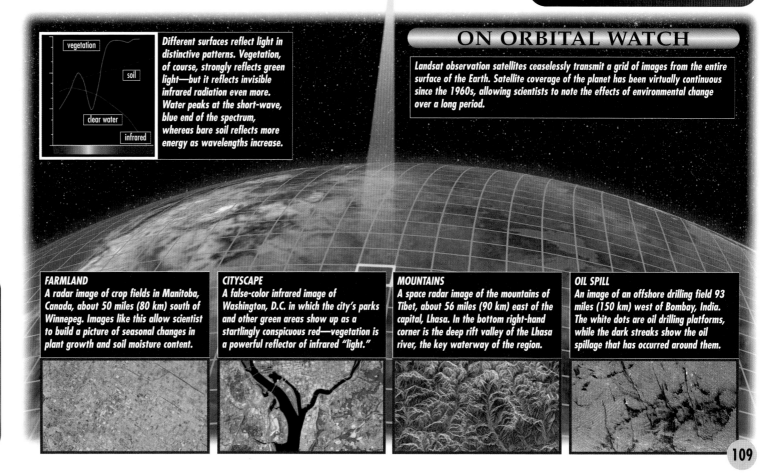

Different surfaces reflect light in distinctive patterns. Vegetation, of course, strongly reflects green light—but it reflects invisible infrared radiation even more. Water peaks at the short-wave, blue end of the spectrum, whereas bare soil reflects more energy as wavelengths increase.

vegetation | soil | clear water | infrared

ON ORBITAL WATCH

Landsat observation satellites ceaselessly transmit a grid of images from the entire surface of the Earth. Satellite coverage of the planet has been virtually continuous since the 1960s, allowing scientists to note the effects of environmental change over a long period.

FARMLAND
A radar image of crop fields in Manitoba, Canada, about 50 miles (80 km) south of Winnepeg. Images like this allow scientist to build a picture of seasonal changes in plant growth and soil moisture content.

CITYSCAPE
A false-color infrared image of Washington, D.C. in which the city's parks and other green areas show up as a startlingly conspicuous red—vegetation is a powerful reflector of infrared "light."

MOUNTAINS
A space radar image of the mountains of Tibet, about 56 miles (90 km) east of the capital, Lhasa. In the bottom right-hand corner is the deep rift valley of the Lhasa river, the key waterway of the region.

OIL SPILL
An image of an offshore drilling field 93 miles (150 km) west of Bombay, India. The white dots are oil drilling platforms, while the dark streaks show the oil spillage that has occurred around them.

SATELLITES AND MOBILE PHONES

T he cellular telephone system has freed phone users from the restrictions of hard-wired landlines and given them the ability to make calls on the move, but it has its limitations. To get a connection, you have to be within range of a base station. If your phone isn't compatible with the systems in other countries, you won't be able to use it abroad. These restrictions would vanish if the new generation of mobile phones could link directly to satellites. However, this poses some impressive technical challenges which have yet to be overcome.

WHAT IF...

...IRIDIUM HAD BEEN A SUCCESS?

V arious attempts have been made to set up a global satellite-phone network. One of the most comprehensive was the Iridium network, so named for the atomic number if the element Iridium = 77. There were originally to be 77 satellites in the Iridium network, though this number was reduced to 66 without provoking a name change to Dysprosium—the element that has 66 electrons per atom.

Iridium stands as a case study in what might have been achieved, and what problems remain to be solved. As with existing telephone networks, the services provided by the satellite networks were to include not only voice communications but also other facilities including fax and computer data transmission. The main technical problem was that data transfer speeds were slow compared with dedicated data links such as ISDN (Integrated Services Digital Network) fiber-optic lines. Satellite phone systems need much higher rates of data flow if they are to carry large streams of financial information or to offer high-quality Internet services.

Several high-speed systems—along with phones smart enough to use the communications power that they make available—were developed by various companies. All of them offered broadcast-quality voice links as well as data communications. The companies developing such systems included Celestri, Astrolink and the Microsoft-backed Teledesic. The Teledesic system was to have 288 satellites that could relay data at very high rates.

Because satellites provide global coverage, it would be possible for each user to have a personal telephone number for life, no matter how many times they might change address, telephone or telephone company.

AIRBORNE ALTERNATIVE

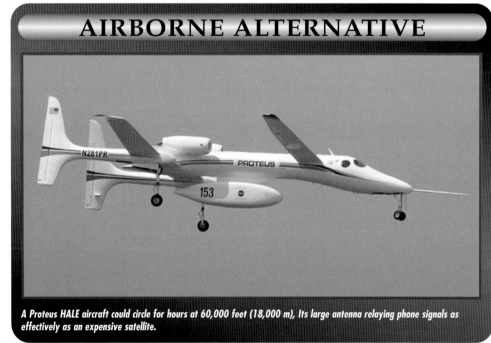

A Proteus HALE aircraft could circle for hours at 60,000 feet (18,000 m), Its large antenna relaying phone signals as effectively as an expensive satellite.

GOING GLOBAL

Iridium was to be the first of the new systems to become operational. Originally intended to use 77 satellites, this was amended to 66, 11 in each of six separate orbits. The satellites would travel at a height of 485 miles (780 km) and take less than two hours to orbit the Earth. An Iridium satellite's signal coverage would cover an area on the ground about the size of the eastern U.S. This area is divided up into 48 overlapping zones. A separate spotbeam of signals would serve each zone, nearly 100 miles (160 km) across.

Calls from Iridium phones would be routed by the satellite. A call from one Iridium phone to another could be sent directly if both phones are covered by the same satellite. If one was farther away, the call would be cross-linked to another Iridium satellite that can "see" the other phone. If the call is to a non-Iridium phone, the message would be routed to an Iridium ground station, either directly or via a number of other Iridium satellites. From the ground station it would enter the ordinary telephone system and proceed to its destination.

Calls to Iridium phones from other networks would be routed in the opposite direction. During the course of a conversation, the satellite handling the call would of course move across the sky and perhaps go "out of sight" of either person talking. If that happened, the system was set up to automatically transfer the call signal to another Iridium satellite in the same orbit or in a neighboring one. The other systems were to work in a broadly similar way.

SATELLITE CONSTELLATIONS

The Ellipso system was to deploy its satellites in two "sub-constellations." One, Ellipso-Borealis, covered the north of the Earth and the other, Ellipso-Concordia, covered the tropical and southern latitudes. Ellipso-Borealis was to have five satellites in each of two orbits; Ellipso-Concordia

would have had six satellites in one orbit and four in the other.

Creating a useful hand-held satellite receiver proved to be an immense technical challenge. Attaché-case sized units are possible, such as the Inmarsat unit, but mobile phone users have become accustomed to small, slim handsets and are not inclined to buy a large and bulky unit that does not offer many advantages over the existing ground-based cellphone system.

Thus Iridium and other satellite phone networks never quite came to be. It is likely that the concept will be revisited at some point in the future, however. In the meantime other technical innovations are entering the market. Landlines and satellites also could be joined by yet another new communications link: High Altitude Long Endurance (HALE).

HALE uses large airborne antenna to relay the signals of phones and ground stations. These antenna will be carried on HALE aircraft—airplanes or balloons—that will fly high over densely populated areas. Each aircraft will stay aloft for 18 hours or more and may operate under remote control. Relief aircraft will take over at refueling time, so HALE will provide continuous service. HALE is not a space-based system, but it is based on the same concepts.

IRIDIUM AND ELLIPSO

IRIDIUM
The Iridium "constellation" would have consisted of a total of 66 satellites, with 11 satellites flying at a height of 485 miles (780 km) in one of six separate orbits.

ELLIPSO SATELLITE
The Ellipso satellites would have each have 61 spotbeams and an operational life of at least five years.

IRIDIUM SATELLITE
The Iridium satellites were developed by Motorola and were to be launched in groups by U.S., Russian and Chinese rockets.

CROSS-LINKED
The Iridium satellites would achieve blanket coverage through their ability to communicate with one another.

ELLIPSO
The Ellipso system was to deploy its satellites in two "sub-constellations." One, Ellipso-Borealis, would cover the north of the Earth and the other, Ellipso-Concordia, would cover the tropical and southern latitudes. Ellipso-Borealis would have had five satellites in each of two orbits; Ellipso-Concordia would have had six satellites in one orbit and four in the other.

ELLIPTICAL ORBITS
Ellipso satellites would have had highly elliptical orbits that allowed them to spend a proportionately longer time over the more densely populated areas of the Earth.

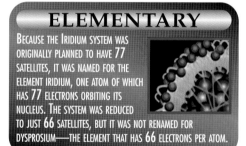

ELEMENTARY

BECAUSE THE IRIDIUM SYSTEM WAS ORIGINALLY PLANNED TO HAVE 77 SATELLITES, IT WAS NAMED FOR THE ELEMENT IRIDIUM, ONE ATOM OF WHICH HAS 77 ELECTRONS ORBITING ITS NUCLEUS. THE SYSTEM WAS REDUCED TO JUST 66 SATELLITES, BUT IT WAS NOT RENAMED FOR DYSPROSIUM—THE ELEMENT THAT HAS 66 ELECTRONS PER ATOM.

COMPETITION

IF THE TECHNOLOGICAL HURDLES CAN BE OVERCOME, SATELLITE CELLULAR PHONES WILL PROVIDE STRONG COMPETITION FOR EXISTING GLOBAL SATELLITE PHONES, SUCH AS THE ATTACHÉ-CASE-SIZE INMARSAT UNITS (RIGHT).

HUBBLE TELESCOPE

I n 1990, the Hubble Space Telescope (HST) was carried into Earth orbit by the Space Shuttle Discovery. Its 15-year mission was to take a closer look at our solar system, the Milky Way and other galaxies and to gaze back in time into the farthest reaches of the universe. Although there are larger telescopes based on Earth, the HST has the benefit of being above the atmosphere, enabling it to give us a much clearer view of the heavens—the pictures it has sent back have been truly stunning.

WHAT IF...

...THE HUBBLE'S MISSION IS FURTHER EXTENDED?

The Hubble Space Telescope (HST), launched in 1990, is a joint venture between ESA and NASA. It was designed to carry a range of instruments – four in axial bays and one in a radial bay. However, this capability was reduced when one of the bays had to be used for corrective optics to counteract a spherical aberration in the main mirror. Currently the Hubble carries an advanced digital camera, an infrared camera and a spectroscope in addition to the wide-field camera which has taken most of the Hubble's most famous images.

The HST's future has hung in the balance for several years. In January 2004 NASA announced that no further manned servicing missions to the Hubble would be mounted. Plans for a robotic servicing mission came to naught, and it seemed that the HST's days were numbered. However, the HST remains a highly useful instrument and the battle to save it continues. By shutting down one of Hubble's three gyros and replacing its data with information derived from the station's fine guidance sensors, the HST's life span has been extended to around 2008.

Meanwhile, NASA has stated that servicing missions to Hubble will be considered after two 'return-to-flight' missions by the Space Shuttle. It seems that if the Shuttle can be returned to full service and funding can be found, Hubble may continue to operate for many years to come.

One possible obstacle is cost. Budgets are tight at NASA. The European Space Agency may be asked to contribute to the costs of extending Hubble's lifespan.

...THE HUBBLE WAS REPLACED?

Scientists and engineers are already hard at work on a replacement for the Hubble Space Telescope, known as the James Webb Space Telescope (JWST). The idea is to site JWST at the point where the pull of gravity from the Earth and the Sun cancel each other out—some 1 million miles (1.6m km) out in space toward the Sun.

The JWST will be three times bigger, yet four times lighter, than Hubble. It will be cheaper, too, with a planned budget of just $500 million. The design features an ultra-lightweight mirror 20 feet (6.5 m) across. The mirror will be a beryllium composite rather than glass.

The new telescope will be launched using a conventional uncrewed rocket, with the mirror folded up inside the nose-cone (the way the mirror unfolds is the major difference between the three designs). One advantage of putting the telescope far out in space is that it will not heat up and cool down as it goes in and out of the Earth's shadow, as the HST does. The drawback is that it cannot be repaired if anything goes wrong. With luck, the NGST will be able to see galaxies forming at the very edge of the universe, and to look for life on planets orbiting other stars.

HUBBLE SPACE TELESCOPE

MASS	11.4 TONS (10.3 TONNES)	ANGULAR RESOLUTION	0.1 ARC-SEC
LENGTH	43 FT (13 M)	POINTING ACCURACY	0.007 ARC-SEC FOR 24 HR
DIAMETER	14 FT (4.2 M)	MAGNITUDE RANGE	5 TO 29
PRIMARY MIRROR	8 FT (2.4 M)	ORBIT	380 MILES (611 KM)
SECONDARY MIRROR	1 FT (30 CM)	ORBITAL PERIOD	94 MIN
WAVELENGTH RANGE	110 NANOMETERS (UV) TO 1 MM (IR)	PLANNED LIFETIME	15 YR

WINDOW ON THE UNIVERSE

No matter how big you build an Earth-based telescope, it has one great drawback: the atmosphere absorbs and deflects incoming light, thereby degrading the view. The solution—putting a telescope into orbit—only became feasible with the development of the Space Shuttle, due to the size and sensitivity of the equipment involved. Even so, plans for a Large Space Telescope or LST (named Hubble in 1983) were already well under way by the time of the launch of the first shuttle, Columbia, in 1981.

GROUND CONTROL

The Hubble is roughly the size of a railroad car and is built to fit inside the Shuttle's cargo bay. It occupies a low Earth orbit, circling at an altitude of just 320 miles (515 km), and is controlled from the ground. Attached to its "eyepiece" are two cameras—one that looks in great detail at a small area of space, and one that focuses on much larger areas or objects. Other instruments analyze the infrared waveband and characteristics of light.

Data from Hubble is relayed to White Sands, New Mexico, then on to mission control at the Goddard Space Flight Center near Washington, D.C. Another link forwards the data to the Space Telescope Science Institute in Baltimore. The HST costs an estimated $8 a second to use. Excluding the cost of shuttle launches, $4 billion will be spent on the Hubble in its lifetime.

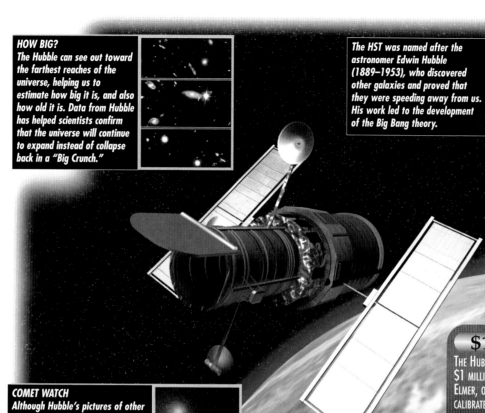

HOW BIG?
The Hubble can see out toward the farthest reaches of the universe, helping us to estimate how big it is, and also how old it is. Data from Hubble has helped scientists confirm that the universe will continue to expand instead of collapse back in a "Big Crunch."

The HST was named after the astronomer Edwin Hubble (1889–1953), who discovered other galaxies and proved that they were speeding away from us. His work led to the development of the Big Bang theory.

COMET WATCH
Although Hubble's pictures of other planets in the solar system cannot compare with those taken by fly-by space probes, it does allow for regular and long-term observation. A prime example of the Hubble's value came when it showed the comet Shoemaker-Levy 9 crashing into Jupiter (right).

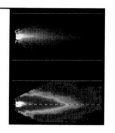

LIFE SEARCH
The Hubble allows us to search for planets orbiting other stars, which no ground-based telescope can see. Where there are planets, there may also be life. Here, Hubble shows a new solar system beginning to form around the star Beta Pictoris.

STAR DEATH
The Hubble has brought us this spectacular picture of the Hourglass Nebula. The nebula is made up of the remnants of a dead star, puffed into space at the end of the star's life. By analyzing images like these, scientists gain new insights into the life cycle and evolution of stars.

$1.6 BILLION BLUNDER

The Hubble's main mirror was ground from a $1 million blank by the U.S. company Perkin Elmer, one of whose instruments was calibrated wrongly. The mistake should have been picked up by a pre-launch check, but Perkin Elmer was under pressure to deliver on time and on budget, so it was not spotted until the Hubble was in orbit. Dubbed "the $1.6 billion blunder" by the media, the fault was eventually corrected during a spacewalk by replacing one of the Hubble's instruments with a refocusing unit. The repair and servicing mission cost $800 million.

MISSION DIARY: HUBBLE SPACE TELESCOPE

1977 Hubble Space Telescope project approved. Budget estimated at $450 million—85% to come from NASA and the rest from the European Space Agency. Launch date is set for 1983.
1979 Construction of Hubble begins. Launch is rescheduled for 1986.
1986 Launch of Hubble is postponed by the Challenger

Shuttle disaster. By now costs have soared to $1.6 billion—three times the original budget.
April 24, 1990 Hubble is finally launched into space in the cargo bay of Shuttle Discovery.
April 24, 1990 Hubble is deployed in low Earth orbit.
May 20, 1990 The space telescope's "first light"—

Hubble is turned toward the star cluster NGC 3532, but the image is out of focus, rendering the HST little better than Earth-based telescopes.
December 5–9, 1993 Astronauts from the Shuttle Endeavour make the longest U.S. spacewalk to date: 29 hr, 40 min.
December 18, 1993 Repairs proclaimed a success.

February 19, 1997 A second service mission adds two new instruments and effects running repairs.
1999 Service mission 3A installs new computer and replaces RSU (Rate Sensing Units)
2002 Service Mission 3B installs Advanced Camera (ACS) and replaces solar arrays
2004 Fourth scheduled service mission is cancelled
2005 Hubble placed on 2-gyro operating mode, extending lifetime
2008 Projected end of service life

INSIDE HUBBLE

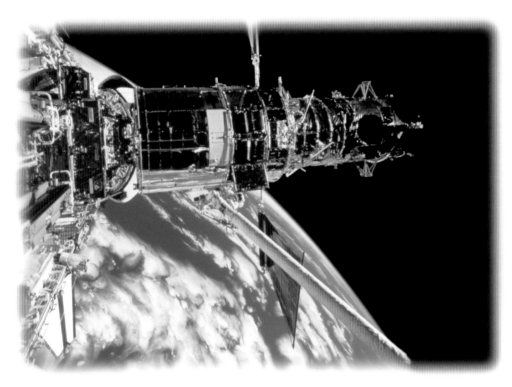

The Hubble Space Telescope (HST) is our single clearest window on the universe, but it is neither the largest nor the farthest-seeing telescope ever built. What makes the $1.6 billion dollar telescope so effective is its position—in orbit 330 miles (530 km) above the Earth. The images seen by terrestrial telescopes are smeared by atmospheric turbulence, so that fine detail is lost. But up in the vacuum of space, the HST's 94½-inch (2.4-m) primary mirror is capable of resolving an image 10 times better than even the best ground-based telescope.

WHAT IF...

...THE HUBBLE TELESCOPE IS RETIRED?

Originally designed for a 15-year lifespan, the Hubble Space Telescope was due for retirement in 2005 but at present looks likely to continue at least until 2008. By that time the National Aeronautical and Space Agency (NASA) will have invested more than $5 billion in the orbiting spacecraft. Plans are already on the drawing board for a more advanced, second generation Space Telescope, but what will happen to its ageing predecessor?

NASA is currently considering two plans that will determine the eventual fate of the HST. The first option is to simply let the Hubble burn up in the atmosphere. Drag from the Earth's atmosphere is gradually causing the HST's orbit to decay anyway. To prevent it from falling back to Earth, the telescope's altitude has to be boosted regularly during Space Shuttle service missions, which have now been curtailed. To prevent an uncontrolled and potentially dangerous de-orbit, a final mission would be necessary in order to attach a rocket thruster to the telescope. The rocket would guide the HST's atmospheric re-entry and prevent pieces of the 13-ton (11.7 tonne) telescope crashing down on populated areas.

The proposed next generation Hubble Space Telescope may be able to look even farther into the cosmos. But if budgets permit, its predecessor could still be operating well into the 21st century.

The second, costlier, option is to return the Hubble back to Earth in the cargo bay of a Space Shuttle. It would be an invaluable resource for aerospace engineers—they could examine the HST's hull and mirrors to learn the effects of a decade and a half of continuous spaceflight. Yet there is no technological reason why the Hubble Space Telescope should be shut down at all. It all comes down to budget—the Hubble would cost $200 million a year to operate and service if Shuttle missions to service and boost it were resumed.

Members of the Hubble team believe this cost can be greatly reduced. By attaching a booster rocket to the telescope, the HST could be lifted to a much higher orbit, freeing it from the effects of atmospheric drag and eliminating the need for regular altitude-boosting missions. And the higher the Hubble is, the farther it can see into space. The HST may be getting older, but is could still serve a useful purpose for many years to come.

HST SPECIFICATIONS

INSTRUMENT	FIELD OF VIEW (ARC-SECONDS)	PROJECTED PIXEL SPACING ON SKY (ARC-SECONDS)	WAVELENGTH RANGE (ANGSTROM UNITS)	MAGNITUDE LIMIT
WF/PC	154 x 154	0.10	1,200–11,000	28.0
	35 x 35	0.0455	1,200–11,000	27.7
FOC	14 x 14	0.014	1,150–6,500	26.2
NICMOS	11 x 11	0.043	8,000–19,000	24.5
	19 x 19	0.075	8,000–25,000	25.0
	51 x 51	0.20	8,000–25,000	25.0
STIS	51 x 51	0.05	2,500–11,000	28.5
	25 x 25	0.024	1,650–3,100	26.5
	25 x 25	0.024	1,150–1,700	24.0

FAR SIGHTED

The Hubble Space Telescope is essentially a telescope like any other, albeit one with 400,000 different parts and 26,000 miles (40,000 km) of electrical wiring, all designed to function in the unforgiving environment of outer space. The HST is an aluminum cylinder, 43½ ft (13.25 m) long, fitted with a 94½-inch (2.4-m) concave mirror at one end. This primary mirror reflects light back to a smaller, secondary mirror, which measures 12½ inches (31.7 cm) in diameter. The secondary mirror redirects the reflected rays through a hole in the larger mirror into a rear bay, where separate cameras and instruments record and analyze the light. These currently include the Faint Object Camera (FOC), which is used to see extremely distant or dimly lit objects, and the Near Infrared Camera and Multi-Object Spectrometer (NICMOS). The latter is used to examine cool celestial objects and clouds that radiate infrared instead of visible light. Another piece of essential hardware is the Space Telescope Imaging Spectrograph (STIS)—this covers the ultraviolet, visible and near-infrared wavelengths and can simultaneously divide light into its component colors at 500 separate points in a single image.

HARD EVIDENCE

ABOVE THE ATMOSPHERE
Air turbulence reduces the clarity of images received by ground-based telescopes. This random movement of air currents spreads a fuzzy patch of light around the center of stars, making them appear to twinkle. Astronomers favor

Gaseous Pillars · M16   HST · WFPC2
PRC95-44a · ST ScI OPO · November 2, 1995

observatories on mountain tops, where the skies are less polluted by artificial light and the air is thinner. High above the atmosphere, the Hubble is capable of resolving an image in space 10 times more clearly than the best terrestrial telescope. Although the Space Telescope has made few fundamental new discoveries, it has given astronomers a more detailed view of objects that were already well-known, such as the famous gas pillars of the M16 Eagle Nebula (above).

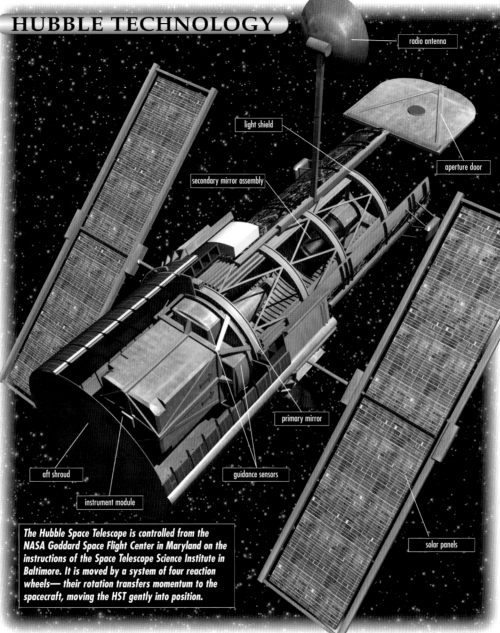

HUBBLE TECHNOLOGY

radio antenna

light shield

aperture door

secondary mirror assembly

primary mirror

aft shroud

guidance sensors

instrument module

solar panels

The Hubble Space Telescope is controlled from the NASA Goddard Space Flight Center in Maryland on the instructions of the Space Telescope Science Institute in Baltimore. It is moved by a system of four reaction wheels— their rotation transfers momentum to the spacecraft, moving the HST gently into position.

SPACE SENSORS

STIS can provide an instant chemical "fingerprint" of a planet's atmosphere, a dust cloud or numerous stars within a galaxy.

It gives information about the target's temperature, chemical composition and motion. To steer the Hubble toward its astronomical targets, the instrument bay also

HITCHES

WHEN THE FIRST PICTURES CAME BACK FROM THE NEWLY-LAUNCHED HUBBLE IN MAY 1990, ASTRONOMERS NOTICED A DISTURBING HALO AROUND THE IMAGES THEY RECEIVED. IT TURNED OUT THAT THERE WAS AN OPTICAL FLAW IN THE HST's PRIMARY MIRROR. IN 1993, A SHUTTLE SERVICING MISSION CORRECTED THE SPACE TELESCOPE'S MYOPIA. A SYSTEM OF CORRECTIVE MIRRORS KNOWN AS THE CORRECTIVE OPTICS SPACE TELESCOPE AXIAL REPLACEMENT (COSTAR) WAS FITTED TO REFOCUS THE LIGHT RECEIVED BY THE TELESCOPE. A REPLACEMENT WIDE FIELD/PLANETARY CAMERA (WF/PC) WAS ALSO INSTALLED.

contains three Fine Guidance Sensors (FGS). A star catalog listing 15 million potential guide stars is used by the FGS as a reference. Another camera, known as the Wide Field/Planetary Camera (WF/PC), occupies a separate bay. It gathers light with a mirror mounted at 45°, which intercepts part of the secondary mirror's beam. The WF/PC has taken many of the Hubble's most famous pictures, such as the famous image of the M16 Eagle Nebula. Like the Space Telescope's other instruments the WF/PC doesn't use photographic film—instead, it uses Charge Coupled Devices (CCDs) arranged in a distinctive "L" shape. CCDs are light-sensitive computer chips that are also found in consumer digital and video cameras. One of the WF/PC's four CCDs can be switched into Planetary Camera mode for a detailed view of a narrower area, such as the observation of a dust cloud on Mars. The CCDs are a hundred million times more sensitive than the human eye, but very vulnerable to damage—the HST cannot be pointed too close to the Sun or they will be burned out. As an extra safeguard, the Hubble has an aperture door. Should the spacecraft ever go out of control, it will shut automatically.

EARTH OBSERVING-1

N ASA's latest Earth-watching spacecraft can fly itself. Earth Observing-1 (EO-1) is a testbed for the next generation of satellites. Onboard artificial intelligence software will carry out complex maneuvers without help from mission control. And EO-1's experimental camera should match the performance of existing Landsat satellites—despite being a fraction of their size. Another instrument will split light from the Earth's atmosphere into dozens of spectral bands, providing unprecedented detail on target areas.

WHAT IF...

...EO-1 TECHNOLOGIES ARE ADAPTED?

E O-1 is not as much a science mission as a technology demonstration. It will not continually gather data from the Earth's surface like Landsat-7. In fact, for 90% of its time in orbit, the EO-1 is in stand-by mode only. The images it does take will not be used for science but for calibrating its instruments.

The success of EO-1 will ultimately be judged by its influence on future Earth-observation missions. Versions of its payload will end up being used on future Landsats, while elements of EO-1's innovative spacecraft design should find broader applications still. EO-1's onboard computer, the Wideband Advanced Recorder Processor (WARP), is the highest-rate solid state recorder that NASA has ever flown. With it, land imaging data can be collected, compressed and processed for transmission back to Earth. Data can be returned to Earth with a revolutionary flat transmitter known as the X-band Phased Array Antenna (XPAA). This is a flat grid of many radiating elements whose combined signals produce a kind of virtual antenna. The XPAA cuts spacecraft weight and the required size of ground stations. And EO-1's lightweight Pulse Plasma Thruster will enable future missions to carry out extremely delicate positioning corrections. Even EO-1's solar panels are deployed in a new way. Instead of swinging open on hinges—which risks damage to delicate spacecraft components—they will be deployed using shaped memory

This 3-D Landsat perspective view looks south along the southeast coast of the North Island of New Zealand, west of the capital city of Wellington. The dark green areas are thick pine forests.

alloy. This is a metal mix that is easily deformed while its temperature is low, but when an electric current heats it up, it returns to its original shape.

If the technologies demonstrated are adopted for use in other spacecraft, EO-1 will be the model for a whole new generation of Earth-watching satellites. Success of EO-1's artificial intelligence software will slash the cost of day-to-day satellite operations. Fewer operatives on the ground will be needed to babysit spacecraft that can basically fly themselves. And EO-1's Enhanced Formation Flying (EFF) software also makes possible the simultaneous operation of fleets of satellites. Known as constellations, such formation-flying satellites have many advantages. Several inexpensive satellites could take the place of one larger spacecraft to form what is known as a virtual platform. This lowers total mission risk, since overall coverage would continue even if a single satellite is lost. It also adds flexibility to future mission plans, for instance by allowing the size of the area surveyed to be altered at will or allowing additional satellites to be added to a constellation.

EARTH OBSERVING-1

VOLUME	APPROXIMATELY 4 × 2.4 FT (1.2 × 0.75 M)	ORBITAL INCLINATION	98.2°, SUN-SYNCHRONOUS ORBIT
MASS	1,164 LB (528 KG)		
DATA STORAGE	1.8 GIGABITS OF TELEMETRY AND COMMAND STORAGE, 40+ GIGABITS OF SCIENCE DATA WITH WIDEBAND ADVANCED RECORDER PROCESSOR	ALTITUDE	438 MILES (705 KM)
		HYDRAZINE PROPULSION SYSTEM	FOUR 0.22 LB (1-NEWTON) THRUSTERS WITH 49 LB (22 KG) OF PROPELLANT

LITTLE BROTHER

For 28 years, the Landsat series of spacecraft have gathered images of the Earth's surface for use by scientists and industry. Because the satellites have all used the same type of multispectral camera, their legacy has been a unique long-term record of the changing face of our planet. Politicians have charged NASA with responsibility for continuing the series past the anticipated 6-year lifespan of Landsat-7, which was launched in 1999. NASA engineers believe that technological improvements mean future Landsats can be smaller and cheaper—and at the same time have much more advanced capabilities.

NASA's New Millennium Program—tasked with developing 21st-century spacecraft designs—has put together Earth Observing-1 (EO-1) to test this hypothesis. It was launched in November 2000 by a Delta II rocket from Vandenburg Air Force Base into the same polar orbit as Landsat-7. Wherever Landsat-7 flies, it will be followed a minute later by EO-1. The Terra satellite follows a minute behind EO-1, also in the same orbit.

Landsat's companion brought some powerful new technology to Earth observation. The main instrument aboard EO-1 is the Advanced Land Imager (ALI), a multispectral camera just ⅕th the mass, power consumption and volume of the Landsat-7 version, the Enhanced Thematic Mapper Plus. Since both satellites were flying over the same area close together, images from EO-1 could be compared against Landsat images of the same region.

LOOKING AWAY

EO-1'S BASIC SPACECRAFT BUS DESIGN WAS REUSED FOR NASA'S MICROWAVE ANISOTROPY PROBE (MAP) LAUNCHED IN 2001. INSTEAD OF FACING EARTHWARD, THE MAP SPACECRAFT LOOKED OUTWARD INTO THE COSMOS, MAPPING TINY TEMPERATURE FLUCTUATIONS THOUGHT TO BE ECHOES OF THE BIG BANG.

IMPROVING SATELLITE IMAGERY

The EO-1 is designed to test breakthrough technologies in lightweight materials, high performance integrated detector arrays and precision spectrometers, as well as advances in communications, power, propulsion and data storage.

LAUNCH VEHICLE
The first EO-1 was launched by a Delta II rocket like this one from Vandenburg Air Force Base. However, with a small redesign, the EO-1 can be made compatible with a wide range of launch vehicles.

deployable light-weight flexible solar array

Advanced Land Imager | Hyperion

ADVANCED IMAGER
The ALI uses wide-angle optics and a multispectral spectrometer to filter light reflected from the ground into specific spectral bands.

UNDER CONSTRUCTION
The EO-1 was constructed by Swales Aerospace and Litton Industries and managed by Goddard Space Flight Center.

SEEING FARTHER AND BETTER

The ALI has ground resolution down to 33 feet (10 m) in black and white and 99 feet (30 m) in other spectral bands, matching seven of the eight visible and infrared bands covered by Landsat-7. The camera is basically a reflecting telescope built out of light silicon carbide, with a 4-chip array placed at the focal plane.

HARD EVIDENCE

WORKING TOGETHER
Enhanced Formation Flying (EFF) software allows multiple spacecraft to detect errors and agree together on the appropriate maneuvers to maintain correct position and orientation in orbit.
With this technology, many small, inexpensive spacecraft can fly in formation and gather data together. After deployment, EO-1 adopted a 437-mile (700-km) circular Sun-synchronous orbit at a 98.7° inclination. This matches within one minute the Landsat-7 orbit.

There are two other science payloads on the EO-1. The Atmospheric Corrector (AC) detects what amount of the light reflected from the Earth's surface is absorbed by water vapor or aerosols during its passage through the atmosphere. This will help increase the accuracy of future remote sensing missions. And the Hyperion is a hyperspectral instrument with extreme sensitivity to color. Originally developed for the ill-fated Lewis spacecraft, which was lost after launch, the Hyperion can resolve light from surface features into 220 separate bands. This is enough detail to identify individual species of trees in a forest.

Beyond EO-1's test payload, the entire spacecraft design is experimental. Its radiator is made of light-weight carbon composite, while its Light Weight Solar Array (LWSA) has twice the power efficiency of existing arrays. Most strikingly, EO-1 is one of the most autonomous spacecraft ever flown. An onboard processing capability turns its raw data into usable graphics, a task usually done by Earth-based computers. Artificial intelligence and fuzzy logic software will allow EO-1 to maintain a "frozen orbit" within strict parameters, and it can complete complicated orders from mission control without a long series of uplinked commands. The spacecraft keeps track of its position using a star tracker as well as the Global Positioning System (GPS). In comparison to the crowded ground control centers that support many space missions, EO-1 requires only three supervisors.

SHUTTLES AND STATIONS

Mankind has long dreamed of a permanent habitation in space. Science fiction writers, artists, and filmmakers depicted massive stations serving as communities or staging posts for missions to the stars. The massive wheel-like space stations of sci-fi publications and films such as *2001: a Space Odyssey* have not yet come to pass. The Soviet Soyuz and U.S. Skylab were followed by the long-lived Mir, but all eventually fell to earth when their working lives were over, and today the International Space Station (ISS) is the only permanently manned outpost in space. Of course, a space station needs a means of delivering and returning its residents and their provisions, as well as getting its components into orbit in the first place. Constructing and servicing a station was one rationale behind the Space Transportation System or Space Shuttle, first proposed at the end of the Apollo programme. Critics of NASA say that today the Space Shuttle's only justification is to support the ISS, and the ISS only exists to give the Shuttle a mission. Losses of Shuttles in 1986 and 2003 have added urgency to development of the new Crew Exploration Vehicle, although there is likely to be a gap of several years when the U.S. will not have its own manned launch capability.

The launch of the Space Shuttle Atlantis *on December 2, 1988. At launch the combined rockets develop a thrust of 7.82 million lb (34.8 mN). The Shuttle takes about 8½ minutes to accelerate to a speed of over 17,000 mph (27,300 km/h) and go into orbit.*

SALYUTS 1–5

The early Soviet space stations, launched in the 1970s, went from problems to solutions and tragedy to triumph. What started as a hasty modification of the Almaz military space station—to regain Soviet prestige after the Americans beat them to the Moon—developed into a world-leading mastery of space technology. During the Salyut program, ingenious designers had to struggle with unreliable launch systems and changing priorities, while brave cosmonauts faced the constant threat of death.

WHAT IF...

...THE MILITARY SALYUTS HAD CONTINUED?

The Almaz space stations, the military Salyuts, tested many different technologies for the Soviet armed forces. Primarily, they were observation platforms from which Soviet military personnel could observe the Earth below, and the crews of Salyut 3 and Salyut 5 observed Russian military exercises in experiments designed to test how much they could see. The Almaz crews, talking to mission control through encrypted voice channels, could report in real time, and quickly redirect their massive spy cameras at new targets. This gave them a big advantage over spy satellites, which cannot be repositioned quickly and can take some time to transmit or deliver their intelligence back to Earth.

In 1980, the fourth Almaz station was cancelled. The Soviet Union decided that orbiting observation posts were not worth the effort. But what if they had made the opposite decision?

By 1988, Mir is not the only Russian space station. A similarly sized Almaz-based station, called "Cosmos 1500," flies in a low orbit, supplied by modules launched on Proton rockets and on Buran shuttles boosted into orbit by the giant Energia launcher. A crew of six monitors U.S. air and naval bases, and uses high-power radar to track U.S. carrier groups at sea. The crew also experiments with lidar—laser radar—for detecting submarines underwater. Every few hours, the

In a Soviet mission control room, military specialists monitor the operations of their Almaz space stations and analyze the data and pictures sent back from them.

cosmonauts spend 20 minutes staying perfectly still—their telescope camera is taking photographs of a battle in Afghanistan, and they must keep the station steady so that vibrations do not blur the pictures.

A pair of Topaz nuclear reactors at the end of a long boom supply electricity for the station, especially vital for its power-hungry radar. Solar panels are of little use on this craft because they have too much drag at the low altitudes needed for visual and radar reconnaissance. And on a crewed craft, Topaz failures can be repaired—on uncrewed satellites, a malfunction could lead to a propaganda disaster like the real-life Cosmos 954, whose failure scattered nuclear waste on Canada.

The station is useful for Cold War maneuvers, but the Soviets know it will not last long if the war turns hot. Both sides have anti-satellite weapons (ASATs), and defending a big, pressurized, space station against them is almost impossible. If World War III starts, the crew of Cosmos 1500 have got less than half an hour to get out.

MISSIONS TO SALYUTS

Mission	Date	Result
Soyuz 10	April 1971	Failure. Unable to dock properly
Soyuz 11	June 1971	Failure. Crew killed on re-entry
Soyuz 14	July 1974	Success. Military mission to Salyut 3
Soyuz 15	August 1974	Failure. Nearly rams Salyut 3
Soyuz 17	January 1975	Success. First mission to Salyut 4
Soyuz 18–1	April 1975	Failure. Sub-orbital launch abort
Soyuz 18	May 1975	Success. Crew spends two months on Salyut 4
Soyuz 21	July 1976	Partial success. Crew makes emergency return from Salyut 5
Soyuz 23	October 76	Failure. Docking fails: splashdown return
Soyuz 24	February 77	Success. End of operations on Salyut 5

SPACE OUTPOSTS

In 1969, the Soviets lost the race to the Moon. But they had a functional spacecraft—the Soyuz moonship—and space-station hulls built for the Almaz military program. By combining these two technologies, they could build a space station and beat the Americans that way.

Salyut 1 was an Almaz with Soyuz systems bolted on. Two small sets of solar panels were fitted to provide electrical power, and a Soyuz service module was placed at the rear. The front docking port replaced the Almaz's return capsule and its self-defense gun (a Nudelman 23mm cannon). Then came two examples of the "standard" military Almaz, Salyuts 2 and 3. Salyut 4 was civilian, with improved systems, but Salyut 5 was another Almaz.

With each new Salyut, increasingly advanced technology was tested. Salyuts 3 and 5 were fitted with gyroscopes that controlled the stations' orientation. Similar systems have since been used for Mir and the International Space Station (ISS). Salyut 4 incorporated a computer that automatically kept the station pointed in the correct direction. And in November 1975, Soyuz 20 and Salyut 4 achieved the world's first uncrewed docking, proving the technology used by Progress ferries and the later Soviet stations.

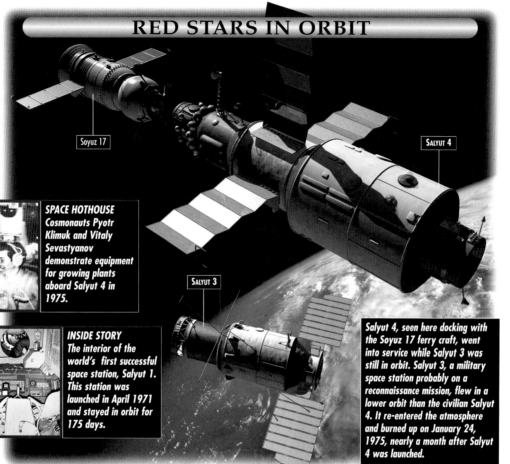

RED STARS IN ORBIT

Soyuz 17

Salyut 4

Salyut 3

SPACE HOTHOUSE
Cosmonauts Pyotr Klimuk and Vitaly Sevastyanov demonstrate equipment for growing plants aboard Salyut 4 in 1975.

INSIDE STORY
The interior of the world's first successful space station, Salyut 1. This station was launched in April 1971 and stayed in orbit for 175 days.

Salyut 4, seen here docking with the Soyuz 17 ferry craft, went into service while Salyut 3 was still in orbit. Salyut 3, a military space station probably on a reconnaissance mission, flew in a lower orbit than the civilian Salyut 4. It re-entered the atmosphere and burned up on January 24, 1975, nearly a month after Salyut 4 was launched.

SCIENCE ON SALYUT

The military origins of Salyut meant that much of the main compartment was taken up by a housing for a spy camera or telescope. Although Moscow was keen to publicize the economic advantages of photographing the Earth, much of the Salyuts' photographic activity was a cover for military missions to evaluate the potential of crewed spy satellites. But some observations were very important: Salyuts 4 and 5 used an infrared spectrometer to measure the water content of the stratosphere, and discovered the first signs of ozone depletion.

Biology was covered by Salyuts 1 and 4, which had miniature greenhouses where cosmonauts tried to grow plants. These stations also carried out many useful astronomical observations, mainly in wavelengths blocked by the atmosphere. Salyut 1's primary instrument was an ultraviolet telescope. Salyut 4's was a solar telescope. A broken sensor had crippled this instrument, but the Soyuz 17 crew learned to steer it by controlling its servo motors by ear.

During the operational lives of the first Salyut stations, the Soviets also made great progress in the medical evaluation of long-duration flights, and by 1977, 13 cosmonauts had each flown in space for more than two weeks at a time.

NO ENTRY

BECAUSE COSMONAUT NIKOLAI RUKAVISHNIKOV FAILED THREE TIMES TO COMPLETE A SALYUT MISSION SUCCESSFULLY, HIS FELLOW COSMONAUTS JOKED THAT SALYUTS WERE FITTED WITH "ANTI-RUKAVISHNIKOV" DEVICES. HIS FIRST ATTEMPT WAS THWARTED BY DOCKING PROBLEMS, THE SECOND BY ENGINE FAILURE. AND THE THIRD TIME HE WAS DUE TO FLY ON A SALYUT MISSION, HE HAD TO DROP OUT BECAUSE HE HAD A COLD.

MISSION DIARY: SALYUTS 1–5

APRIL 19, 1971 SALYUT 1 IS LAUNCHED FROM BAIKONUR COSMODROME ON BOARD A PROTON ROCKET.

APRIL 23, 1971 THE SOYUZ 10 SPACECRAFT TAKES OFF CARRYING THE FIRST CREW (RIGHT) TO VISIT SALYUT 1. THE CREWS ARE UNABLE TO DOCK SUCCESSFULLY WITH THE SPACE STATION AND RETURN TO EARTH ON APRIL 25.

JUNE 7, 1971 THE CREW OF SOYUZ 11—GEORGY DOBROVOLSKY, VIKTOR PATSAYEV AND VLADISLAV VOLKOV — (RIGHT) SUCCESSFULLY DOCK WITH SALYUT 1 AND STAY ABOARD UNTIL JUNE 29. THE MISSION ENDS IN TRAGEDY WHEN ALL THREE

DIE AFTER THEIR CAPSULE CABIN PRESSURE FAILS DURING RE-ENTRY.

JULY 29, 1972 A SOVIET ATTEMPT TO PUT ANOTHER SPACE STATION INTO ORBIT FAILS WHEN IT IS DESTROYED DURING LAUNCH.

APRIL 3, 1973 SALYUT 2 IS LAUNCHED SUCCESSFULLY FROM BAIKONUR, BUT IT BREAKS UP IN ORBIT.

MAY 11, 1973 ANOTHER SALYUT, CODENAMED COSMOS 557, IS LAUNCHED BUT GOES OUT OF CONTROL AND RE-ENTERS THE ATMOSPHERE ON MAY 22.

JUNE 25, 1974 SALYUT 3, A MILITARY SPACE STATION, IS LAUNCHED FROM BAIKONUR.

JULY 3, 1974 SOYUZ 14 IS LAUNCHED, CARRYING COSMONAUTS PAVEL POPOVICH AND YURI ARTYUKHIN TO SALYUT 3. THEY STAY ABOARD THE STATION UNTIL JULY 19.

DECEMBER 26, 1974 SALYUT 4 IS LAUNCHED.

JANUARY 11, 1975 ALEXEI GUBAREV AND GEORGY GRECHKO TAKE OFF ON SOYUZ 17 FOR A FOUR-WEEK VISIT TO SALYUT 4.

JUNE 22, 1976 SALYUT 5, A MILITARY SPACE STATION, IS LAUNCHED SUCCESSFULLY FROM BAIKONUR.

JULY 6, 1976 SOYUZ 21 CARRIES BORIS VOLYNOV AND VITALY ZHOLOBOV TO SALYUT 5.

FEBRUARY 7, 1976 YURI GLAZKOV AND VIKTOR GORBATKO LIFT OFF ON SOYUZ 24, THE FINAL MISSION TO SALYUT 5.

SALYUTS 6 AND 7

The operation of the two space stations Salyuts 6 and 7 from 1977 to 1986 was a vital stage in the development of human spaceflight. The experience gained was crucial to later successful operations on the Mir space station and for future human missions into the solar system. Cosmonauts worked routinely aboard the stations, carrying out a wide range of experiments as well as acting as orbiting repairmen. Salyuts 6 and 7 clearly placed the Soviet Union ahead of the rest in the field of long-duration spaceflight.

INSIDE STORY

SALYUTS 6 AND 7

Salyut 6's last crew of Vladimir Kovalenok and Victor Savinykh departed the station in May 1981 after spending 75 days aboard. The two cosmonauts, who had completed hundreds of experiments covering astronomy, Earth observation, materials processing and biology, returned home 5.5 lb (2.5 kg) lighter due to the effects of living in space. After they returned to Earth, the Soviets announced that there would be no further crewed flights to the station.

Although no more cosmonauts would be launched to Salyut 6, its mission was not yet over. A large module was docked with the station to boost its orbit and to test the joined combination. In July 1982, three months after its successor Salyut 7 reached orbit, Salyut 6 was guided by its earthbound controllers to burn up in the Earth's atmosphere over the Pacific Ocean, where any large pieces of debris would fall harmlessly into the ocean. Salyut 6 had been designed to operate for 22 months but reached the grand old age of 58 months.

Salyut 7's final crew arrived not from the Baikonur cosmodrome but from another Soviet space station—Mir. Leonid Kizim and Vladimir Solovyov brought the empty station back to life, carried out repairs and completed experiments. After seven weeks of work on Salyut 7, they closed down the station in June 1986 and returned to Mir with equipment and experiments from Salyut. The

Salyut 7 starts to break up as it enters the atmosphere. The inhabitants of the town of Capitan Bermudez, Argentina, witnessed a spectacular meteor shower as the fragments fell to Earth.

Soviets considered a future mission to Salyut 7, not to live on the station, but to return it to Earth aboard a Soviet space shuttle for examination and later display. But the mission was not carried out, and Salyut 7 reentered the atmosphere in February 1991. Controllers were hoping to guide the station to a mid-ocean demise, but fluctuations in the Earth's atmosphere caused by a sharp rise in solar activity forced the station to impact in South America. Hundreds of fragments survived the fiery journey through the atmosphere and landed in the Andes Mountains on the Chile-Argentina border. Several large pieces were found, although no injuries from falling space debris were reported.

Salyut 6 and 7 far surpassed their original objectives and set the scene for the more advanced 3rd-generation space station—the Mir complex. Not only did these Soviet orbital outposts prove that humans and spacecraft could operate in space for lengthy periods of time, but they also brought humankind one step closer to the human colonization of space.

SALYUTS 6 AND 7 STATS

	Salyut 6	Salyut 7
Launch date	September 29, 1977	April 19, 1982
Weight	22 tons (19.9 tonnes)	22 tons (19.9 tonnes)
Length/diameter	47 feet by 14 feet (14 x 4 m)	47 feet by 14 feet (14 x 4 m)
Dockings	36	29
Crew visits	16	10
Spacewalks	3	13
Visiting international crew	9	2
Longest crew stay	185 days (Soyuz 35)	237 days (Soyuz T-10)
Reentry	July 29, 1982	February 7, 1991
Time in orbit	1,764 days	3,215 days

THE LONG RUN

When Salyut 6 reached orbit in September 1977, it marked an important step toward the construction of a permanent orbiting space station. The improved Salyut station had two docking ports and could receive two spacecraft at once. This meant that refueling of the propulsion engine and crew changes could both take place in orbit.

Salyut 6 was planned to support long duration missions of between 90 and 180 days. But the Soyuz ferry spacecraft were only designed to stay in space for about 80 days. So the main crew would spend a long mission on the space station and other crews on short missions would arrive in a new Soyuz and depart in the main crew's Soyuz. These secondary crews would often include a non-Soviet guest cosmonaut from socialist countries such as Czechoslovakia, Cuba, Vietnam, East Germany, Mongolia and Poland.

Salyut 6 marked the first use of the uncrewed Progress cargo ferries. These brought supplies for the crew as well as fuel for the station. Progress was also used as a space tug to boost the station's orbit. Before releasing the cargo ship to burn up in the atmosphere the crews would stuff all their refuse, dirty clothes and redundant equipment into the ferry.

Over a 4-year period, five long-stay crews visited the outpost, notching up missions of 96, 140, 175, 185 and 75 days. In addition, 11 short-stay crews visited. The crews carried out hundreds of experiments and occupied Salyut 6 for 676 days, far ahead of the 171 days' occupation of the U.S. Skylab station.

RECORD BREAKER

Salyut 7 entered orbit with several improvements, including larger and stronger docking ports, an improved computer, a refrigerator and even hot and cold running water. The first crew aboard Salyut 7 set a new world record by spending 211 days in orbit, and the third long-stay crew managed 237 days.

Salyut 7 suffered many technical breakdowns during its life and its crews had to repeatedly carry out repairs both inside and outside the

SALYUT 7 IN ORBIT

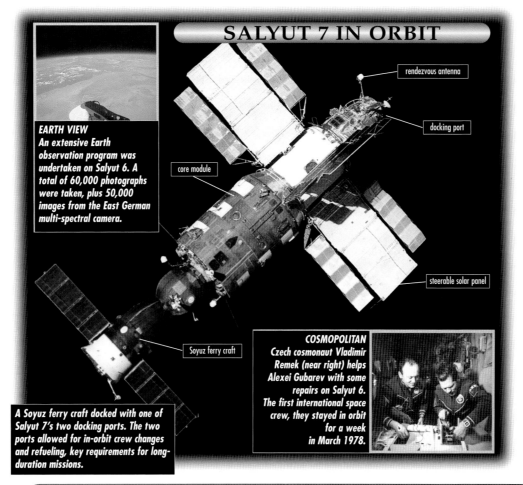

rendezvous antenna

docking port

core module

steerable solar panel

Soyuz ferry craft

EARTH VIEW
An extensive Earth observation program was undertaken on Salyut 6. A total of 60,000 photographs were taken, plus 50,000 images from the East German multi-spectral camera.

A Soyuz ferry craft docked with one of Salyut 7's two docking ports. The two ports allowed for in-orbit crew changes and refueling, key requirements for long-duration missions.

COSMOPOLITAN
Czech cosmonaut Vladimir Remek (near right) helps Alexei Gubarev with some repairs on Salyut 6. The first international space crew, they stayed in orbit for a week in March 1978.

LADIES FIRST

SVETLANA SAVITSKAYA FLEW ABOARD SOYUZ T-12 TO SALYUT 7 AND BECAME NOT ONLY THE FIRST WOMAN TO BE LAUNCHED ON A SECOND SPACEFLIGHT—SHE HAD PREVIOUSLY FLOWN TO SALYUT 7 WITH THE SOYUZ T-7 MISSION—BUT ALSO THE FIRST TO COMPLETE A SPACEWALK. SAVITSKAYA CUT AND SOLDERED SEVERAL METAL PLATES DURING A 3½-HOUR SPACEWALK. THE MISSION'S FLIGHT DIRECTOR IN MOSCOW STATED THAT SVETLANA'S SPACEWALK PROVED THAT "WOMEN CAN DO EVERYTHING."

station. During its time in orbit, two large modules docked with Salyut 7, bringing supplies and providing a larger work area for the crew. These modules were the forerunners of the type later used on the Mir station.

Salyut 7 crews had their fair share of frightening moments. During the Soyuz T-9 flight, the crew began to evacuate the station when they heard a loud crack. On investigation they discovered that a micrometeorite or piece of space debris had hit one of the station's windows leaving a tiny crater.

Although not as successful as Salyut 6, Salyut 7 met its objectives—and established the Soviet Union as the leader in long-duration crewed spaceflight.

MISSION DIARY: SALYUTS 6 AND 7

SEPTEMBER 29, 1977 SALYUT 6 IS LAUNCHED FROM BAIKONUR COSMODROME.
OCTOBER 9, 1977 THE FIRST CREW TO VISIT SALYUT 6, VALERY RYUMIN AND VLADIMIR KOVALENOK (ABOVE, LEFT TO RIGHT), FAIL TO DOCK WITH THE STATION.
JANUARY 23–4, 1978 SALYUT 6 IS REFUELED FOR THE FIRST TIME FROM THE TANKS OF THE DOCKED PROGRESS 1 CARGO SPACECRAFT.
MARCH 2, 1978 SOYUZ 28 FERRIES THE FIRST INTERNATIONAL CREW TO SALYUT 6.
APRIL 9, 1980 COSMONAUTS LEONID POPOV AND VALERY

RYUMIN ARE LAUNCHED ABOARD SOYUZ 35. THEY SPEND A RECORD 185 DAYS IN ORBIT.
MARCH 1981 VICTOR SAVINYKH BECOMES THE 100TH TRAVELER IN SPACE ON A MISSION TO SALYUT 6.
APRIL 19, 1982 SALYUT 7 (ABOVE, BEING ASSEMBLED) IS LAUNCHED.
MAY 13–DECEMBER 10, 1982 SALYUT 7'S FIRST CREW SPEND 211 DAYS ABOARD THE SPACE STATION.
JUNE 24, 1982 FRENCH "SPATIONAUTE" JEAN-LOUP CHRETIEN SPENDS A WEEK ABOARD SALYUT 7.

FEBRUARY 1985 MISSION CONTROLLERS LOSE CONTACT WITH THE UNCREWED SALYUT 7.
JUNE 1985 THE SOYUZ T-13 CREW BRINGS THE TUMBLING, FROZEN SALYUT 7 BACK TO LIFE (RIGHT).
MAY 6, 1986 THE SOYUZ T-15 CREW FLY FROM MIR TO THE UNMANNED SALYUT 7 TO COMPLETE EXPERIMENTS UNFINISHED BY THE SOYUZ T-14 CREW.
FEBRUARY 7, 1991 SALYUT 7 REENTERS EARTH'S ATMOSPHERE OVER SOUTH AMERICA.

SOYUZ 11 DISASTER

In the early hours of June 30, 1971, recovery crews were gathered near the Soviet space complex at Baikonur, on the prairie-like steppes of Kazakhstan, to await the landing of the Soyuz 11 spacecraft. Its crew of three—cosmonauts Dobrovolsky, Patsayev and Volkov—had completed a record-breaking 23 days in orbit in the world's first space station, Salyut 1. A heroes' welcome had been prepared, and the ground crew was overjoyed as the craft made a perfect landing—but all was not as it seemed.

WHAT IF...

...SOYUZ 11 HAD NOT FAILED?

The Soyuz 11 disaster was a tremendous shock to the Soviet space program. It came out of the blue after a string of successes. For a while, both the people and the organization were like accident victims who cannot yet take in what has happened. And the accident eclipsed the fact that the first space station had been established—the latest achievement in a string of Soviet firsts in space.

The Soviet crewed space program was suspended for a year while systems were given a complete overhaul. No further crews were sent to Salyut 1, and on October 11, 1971, it was ordered to destroy itself by re-entering the atmosphere and burning up.

Once the cause of the disaster was known, the Soviets rethought their Soyuz program. They completed the design of a new pressurized space suit—their first to use the closed air regeneration system. They also redesigned the interior of the Soyuz vehicle so that it would carry two cosmonauts wearing pressure suits, instead of three without, and fitted it with repressurization systems.

But some of the impetus and heart had gone out of the Soviet efforts. There was a string of crewless vehicle launch failures and the crewed lunar program sputtered and died in 1974. And, while the Soyuz-Salyut series carried on cautiously gaining experience for longer and longer periods, it was 15 years before the space station Mir was put into orbit.

The first, and so far only, American space station, Skylab, was put into orbit in 1973. It was decommissioned before the end of the year.

The crew of Soyuz 11 died because they were not wearing pressure suits like these. After the disaster, all crewed Soyuz missions carried two cosmonauts wearing pressure suits instead of three without suits.

After the 1975 Apollo-Soyuz Test Project, in which Apollo 18 and Soyuz 19 docked in orbit, America launched no crewed space missions until the first Space Shuttle flight in 1981.

But if Soyuz 11 had not failed and the mission had been seen as the triumph it was—the first step to human colonization of space—public opinion in both the West and the East could have backed the continuation of the crewed space program. Skylab would have been used extensively and the U.S. and the Soviet Union might have built on the spirit of cooperation that had brought about the Apollo-Soyuz Test Project.

The result could have been a full working International Space Station in orbit shortly after the development of the Shuttle. Humanity's first solid base on the highway to space could have been in operation back in the early 1980s.

THE SOYUZ 11 CREW

VLADISLAV NIKOLAYEVICH VOLKOV	GEORGY TIMAFEYEVICH DOBROVOLSKY	VIKTOR PATSAYEV
BORN: NOVEMBER 23, 1935, IN MOSCOW, SOVIET UNION (NOW IN RUSSIA)	BORN: JUNE 1, 1928, IN ODESSA, SOVIET UNION (NOW IN UKRAINE)	BORN: JUNE 19, 1933, IN AKTYUBINSK, SOVIET UNION (NOW IN KAZAKHSTAN)
PREVIOUS SPACEFLIGHT: SOYUZ 7, 1969	PREVIOUS SPACEFLIGHTS: NONE	PREVIOUS SPACEFLIGHTS: NONE

DESCENT TO DEATH

The modern era of crewed space exploration, which focuses on orbiting space stations, began on June 7, 1971. On that day, Georgy Dobrovolsky, Viktor Patsayev and Vladislav Volkov squeezed through the docking port of their Soyuz 11 spacecraft—callsign Yantar ("Amber")—into Salyut 1, the world's first space station. Soviet space scientists, and many others, saw space stations with replaceable crews as the main highway into space. On these orbiting platforms, the craft that would take humanity to explore the solar system and beyond would be built and launched.

The crew's first task was to check all of the systems in the united craft—especially those of Salyut, which had already been in orbit for almost two months. They then settled into a routine of Earth-science observations and medical and biological experiments. They were to have made solar observations, but the large solar telescope was inoperable because its cover had failed to jettison.

They also had exercise equipment—a treadmill and a bungee-string—to help prevent their muscles from wasting away through lack of use. But what looked good on the ground turned out to be unusable in space. Just one 180-pound (82-kg) man throwing himself around proved more than the combined vehicles could take, and the exercises were abandoned.

Then, about three weeks into the mission, the station itself was abandoned. A series of difficulties, and a small electrical fire a week earlier, had persuaded the Soviets to cut the mission short. The crew gathered up their data, transferred back into Soyuz 11 and returned to Earth.

FINAL JOURNEY

At 9:28 on the evening of June 29, Georgy Dobrovolsky undocked the Soyuz craft from Salyut 1. After three orbits of the Earth, Dobrovolsky called Mission Control to tell them that they were beginning their descent. Mission Control radioed back, "Goodbye, Yantar, till we see you soon on Mother Earth." At 1:35 a.m., the craft's retrorockets fired and it

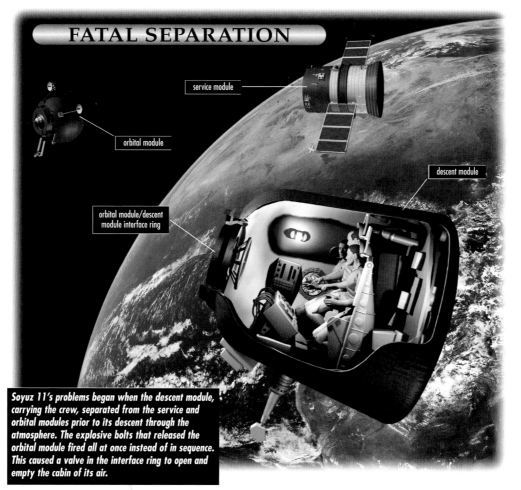

FATAL SEPARATION

service module

orbital module

descent module

orbital module/descent module interface ring

Soyuz 11's problems began when the descent module, carrying the crew, separated from the service and orbital modules prior to its descent through the atmosphere. The explosive bolts that released the orbital module fired all at once instead of in sequence. This caused a valve in the interface ring to open and empty the cabin of its air.

began its descent through the atmosphere. Then its parachutes deployed and it floated gently down to the ground.

But the three cosmonauts never reached the ground alive. Some 723 seconds after the retrorockets had fired, the 12 explosive bolts that released the descent module from the orbital module fired all at once instead of in a controlled sequence. This shook open an air pressure equalization valve, at an altitude of 104 miles (167 km). With appaling speed, the capsule's air hissed into space.

Patsayev unbuckled his safety harness and tried desperately to close or block the valve, but failed. Within about nine or 10 seconds, decompression emptied the crew's lungs— none of them was wearing space suits—and they died within about 30 seconds. A mission that should have been a triumph of Soviet technology had turned into a tragedy.

MISSION DIARY: SOYUZ 11

JUNE 6, 1971, 04:55 A.M. SOYUZ 11 LIFTS OFF FROM THE BAIKONUR SPACE COMPLEX IN KAZAKHSTAN.
JUNE 7 SOYUZ 11 DOCKS WITH THE SALYUT 1 SPACE STATION. THE CREW ENTER SALYUT AND CARRY OUT CHECKS OF ALL SYSTEMS.
JUNE 7 TO JUNE 28 CREW SETTLES INTO ROUTINE OF EXPERIMENTS AND EARTH OBSERVATIONS.
JUNE 17 A SMALL FIRE BREAKS OUT IN SOME ELECTRICAL CABLES. THE CREW PREPARE TO ABANDON THE STATION, BUT THE FIRE IS EXTINGUISHED. THE CREW'S RETURN TO EARTH IS POSTPONED, BUT MISSION CONTROL LATER DECIDES TO END THE PLANNED 30-DAY MISSION EARLY.
JUNE 29, 1971, 9:28 P.M. SOYUZ 11 DISENGAGES FROM

SPACE STATION.
JUNE 30, 1971, 1:35 A.M. RETROROCKETS FIRE, SOYUZ 11 BEGINS ITS DESCENT AND A PARACHUTE SYSTEM DEPLOYS TO BRING IT GENTLY TO THE GROUND. DURING DESCENT, GROUND CONTROL TRIES TO CONTACT COSMONAUTS BUT IS NOT ALARMED WHEN THEY FAIL TO RESPOND.
JUNE 30, 1971 DISBELIEVING RECOVERY TEAM MEMBERS DISCOVER THAT THE COSMONAUTS OF SOYUZ 11 ARE DEAD, AND TRY FRANTICALLY TO RESUSCITATE THEM.

JULY 1, 1971 THOUSANDS OF RUSSIAN MOURNERS FILE PAST THE COFFINS OF THE THREE COSMONAUTS IN RED SQUARE, MOSCOW. SOVIET PRESIDENT NIKOLAI PODGORNY, PRIME MINISTER KOSYGIN AND COMMUNIST PARTY GENERAL SECRETARY LEONID BREZHNEV TAKE TURNS STANDING WATCH AS PART OF THE HONOR GUARD (RIGHT). PRESIDENT NIXON SENDS THE SYMPATHY OF THE AMERICAN PEOPLE TO THE SOVIET UNION, AND AMERICAN ASTRONAUT TOM STAFFORD ATTENDS THE CEREMONY.

SKYLAB

The Apollo Applications Program was started in 1966 to conduct extended lunar operations and long-duration crewed missions in Earth orbit. Using the vast power of the Saturn 5 rocket, the program planned an ambitious series of space stations in orbit. When budget cuts forced mission planners to scale down their ideas, one project survived: The Skylab Orbital Workshop. Launched in 1973, Skylab paved the way for the International Space Station—and still holds the record as the world's largest orbiting spacecraft.

WHAT IF...

...SKYLAB WERE UPDATED?

When Skylab fell to Earth on July 11, 1979, it was not the last of its kind. In the days of Apollo and Saturn, NASA always built backup spacecraft, in case primary vehicles failed and had to be replaced. Skylab was no exception. NASA hoped to fly this second Skylab in the late 1970s and service it with Space Shuttles, but budget problems, and the late development of the Shuttle, prevented this. Eventually, the second Skylab was disassembled, and parts of it are now a permanent exhibit at the Smithsonian Air and Space Museum in Washington, D.C.

A few years ago, some space enthusiasts suggested that the Smithsonian's Skylab could be launched into space aboard a now-defunct Russian Energia booster, to serve as additional living quarters for the new International Space Station (ISS). But it is unlikely that this plan could succeed. After years in a museum, the cost of a refit would probably cost more than building the spacecraft from scratch.

In any case, astronauts and cosmonauts have continued the legacy of Skylab in the missions of other space stations. Whether flying on Skylab, Mir or, ultimately, the ISS, the rules are much the same—and the hardest lessons learned from living and working in Skylab were not technical but psychological.

Skylab's walls, floors and ceilings were cluttered with experiments and control equipment in order to make the best use of the available storage space. But all of the astronauts said they would have preferred a

If the spare Skylab's orbital workshop had ended up in space instead of in the Smithsonian (left), its gold coating would have reflected sunlight and helped control the temperature inside.

more straightforward up-and-down layout, as though they were in a house back on Earth. In weightless conditions, a clear visual sense of direction can help to prevent nausea and disorientation.

With Skylab, NASA learned that windows were the most important part of the station—not for science, but for relaxation. Yet neither the lab nor habitation modules of the U.S. part of the ISS were originally designed with windows. It took the Russians' module design—which was full of windows—and the insistent requests of the U.S. astronauts to eventually change NASA's mind. There is now a single small window in the U.S. lab for—as mission control says—"science."

Thanks to the experiences of the Skylab crew more than a quarter of a century ago, crews on modern space stations may enjoy a more democratic and less stressful life in orbit—something that is often not appreciated from mission control on the ground.

SKYLAB SPECS

OVERALL LENGTH (INCLUDING CSM)	118.5 FT (36 M)	ORBITAL WORKSHOP	
OVERALL WORKING VOLUME	11,700 CU FT (481 M³)	LENGTH	48.1 FT (14.6 M)
POWER OUTPUT	4,000 WATTS AT	DIAMETER	21.6 FT (6.6 M)
	28 VOLTS DC	WEIGHT (WITH SOLAR PANEL)	167,850 LB (76,135 KG)
		WORKING VOLUME	9,550 CU FT (270 M³)
		AMBIENT TEMPERATURE	70°F (21°C)

BASE IN SPACE

The great advantage of Skylab was that it was already half-built even before it was formally approved as a new project in 1969. The idea had occurred to rocket master Wernher von Braun four years earlier. During lunch one day, he casually doodled on a paper napkin how his invention, the Saturn 5 rocket, could be recycled as an Earth-orbiting space station.

Normally, the third stage of the Saturn 5 carried fuel for the Apollo spacecraft's trip out of Earth orbit to the Moon. But if this stage remained in Earth orbit, the propellants and rocket engines for the Moon journey would not be needed, and neither would the fuel to power them. The huge tanks could be filled with air instead, divided into compartments and made into a giant facility for astronauts to live and work in.

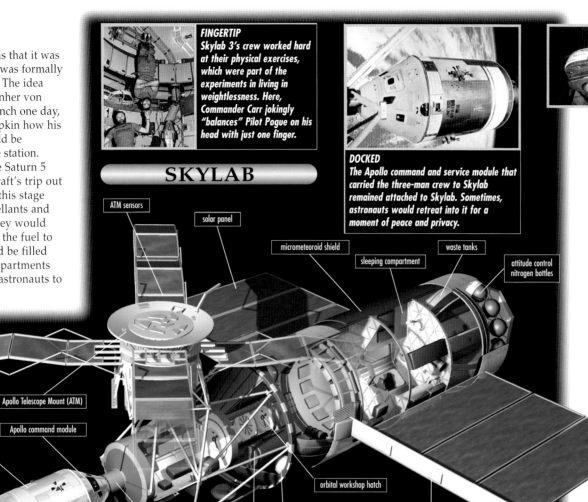

FINGERTIP
Skylab 3's crew worked hard at their physical exercises, which were part of the experiments in living in weightlessness. Here, Commander Carr jokingly "balances" Pilot Pogue on his head with just one finger.

SKYLAB

DOCKED
The Apollo command and service module that carried the three-man crew to Skylab remained attached to Skylab. Sometimes, astronauts would retreat into it for a moment of peace and privacy.

ATM sensors

solar panel

micrometeoroid shield

sleeping compartment

waste tanks

attitude control nitrogen bottles

solar panel for ATM

Apollo Telescope Mount (ATM)

Apollo command module

Apollo service module

orbital workshop hatch

nitrogen tank

solar panel deployment boom

oxygen tank

docking hatch

ATM support struts

propulsion engine nozzle

vernier control motors

Skylab was reconstituted out of the S-IVB stage of a Saturn 5 rocket. It provided a long-term home away from home for U.S. astronauts and also functioned as a large orbital laboratory.

FOOTHOLD
Triangular weights on the Skylab astronauts' shoes gripped the wire grid floor to prevent the crew from floating away.

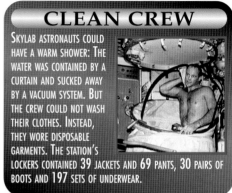

CLEAN CREW

SKYLAB ASTRONAUTS COULD HAVE A WARM SHOWER: THE WATER WAS CONTAINED BY A CURTAIN AND SUCKED AWAY BY A VACUUM SYSTEM. BUT THE CREW COULD NOT WASH THEIR CLOTHES. INSTEAD, THEY WORE DISPOSABLE GARMENTS. THE STATION'S LOCKERS CONTAINED 39 JACKETS AND 69 PANTS, 30 PAIRS OF BOOTS AND 197 SETS OF UNDERWEAR.

Outside, Skylab carried what were then the largest solar panels ever used on a spaceship. On top was the Apollo Telescope Mount (ATM), a solar observatory with an array of instruments including X-ray, infrared and visible light cameras. ATM allowed the Sun's structure and chemistry to be observed in great detail for the first time. Skylab was launched unmanned on May 14, 1973, on the last-ever Saturn 5 booster to fly. In all, three crews visited it over two years, and the space station provided NASA and these nine astronauts with their first valuable experiences of living and working in orbit for long periods.

After six years and 34,981 orbits, Skylab met its end. Increased sunspot activity had expanded the Earth's atmosphere, and this, together with difficulties in maintaining a low-drag attitude, meant that the space station was drawn inexorably Earthward. On July 11, 1979, Skylab crashed home. Though some large pieces of the space station landed in Western Australia, most of the debris—mainly the craft's "skin"—fell harmlessly into the Indian Ocean.

CONVERTED ROCKET

In 1970, when two Apollo Moon landing missions were canceled, the available Saturn 5 was converted into Skylab. Inside, what was once the hydrogen tank was converted into a

two-story space where three astronauts could live and work together. They would breathe a mixture of nitrogen and oxygen, while a thermal and ventilation system provided an ambient temperature of 70°F (21°C). The first story was divided into living areas, with a

ward room, sleeping compartments and a bathroom. Above was the work space, where the crew could "swim" in weightless conditions and carry out experiments. Enough food, water and clothing was stowed on board for all three missions scheduled to visit Skylab.

COLUMBIA'S FIRST FLIGHT

U ntil the launch of the Space Shuttle Columbia in 1981, almost all rockets were used only once—an immensely wasteful practice. Columbia was designed to take off vertically as a rocket and land horizontally as an airplane. Only the main tank was disposable. The new Space Shuttle's orbiter section, with crew cabin, payload bay, wings and engines, was built to survive up to 100 missions over 20 years. Even the solid fuel boosters were reusable. But Columbia's debut flight had some alarming moments.

WHAT HAPPENED...

...TO THE CREW OF COLUMBIA'S FIRST FLIGHT?

T he first flight of the Space Shuttle Columbia was commander John Young's fifth space mission—and it was not to be his last. Over the course of his career, Young clocked up a total of 835 hours in space and became the first person to undertake six spaceflights.

Young became an astronaut in September 1962. On March 23, 1965, he went into space with Gus Grissom on board Gemini 3.

Young went on to command Gemini 10 in 1966, and in 1969 he was Command Module pilot on Apollo 10. In 1972, on his fourth spaceflight, Young, as commander of Apollo 16, revisited the Moon. This time, he explored the surface with fellow astronaut Charlie Duke. They collected 200 lb (90 kg) of rock and drove 16 miles (25 km) in the Lunar Rover.

After Columbia, Young's sixth flight was as commander of Shuttle STS-9 in 1983. It was the first Spacelab mission, and a great deal of scientific research was successfully carried out.

In January 1984, Young became Chief of the Astronaut Office, where he served until May 1987. From May 1987 to February 1996 he worked as Special Assistant to the Director of the Johnson Space Center (JSC) for Engineering, Operations and Safety. In February 1996 he was assigned as Associate Director (Technical) of JSC.

Young's pilot, Robert Crippen, became

John Young and Robert Crippen (above, left and right respectively) went on to command four additional Space Shuttle flights between them. But it is their involvement in Columbia's first flight that has secured their places in history.

an astronaut in September 1969. He was a crew member on the Skylab Medical Experiments Altitude Test—an intensive 56-day simulation for the Skylab mission, which conducted vital medical experiments and tested equipment and procedures—but he didn't actually fly in space until STS-1.

Crippen was commander of STS-7 in 1983—the first Shuttle to carry five people—and STS 41-C in April 1984. He also commanded STS 41-G in October 1984. He was deputy director and later director of the Kennedy Space Center before he left to start a new career in business.

STS-1 STATISTICS

CRAFT	COLUMBIA	TOTAL LIFTOFF WEIGHT	4,457,111 LB (2,021,711 KG)
MISSION	STS-1 (SPACE TRANSPORTATION SYSTEM 01)	LANDING WEIGHT	195,472 LB (88,664 KG)
MISSION DURATION	2 DAYS, 6 HOURS, 20 MINUTES	CREW	COMMANDER: JOHN YOUNG
ORBITS	36		PILOT: ROBERT CRIPPEN

FIRST FLIGHT

Six years had passed since the last U.S. astronauts were launched into space. NASA's vaunted Shuttle was two years late in its development, and had been much more difficult and expensive to prepare than managers had expected. So when commander John Young and pilot Robert Crippen strapped themselves into Columbia's cabin for launch on April 10, 1981, everyone was tense.

Inside the crew cabin, five computers cross-checked results to guarantee accuracy. For safety reasons, at least four of them had to match before Columbia could fly. But 20 minutes before the scheduled liftoff time, the onboard computers could not agree. Young and Crippen had to turn their spacecraft over to the engineers.

Two days later, pilot and commander again climbed into their bulky, uncomfortable ejection seats. These seats would be replaced with seven crew couches, once the Shuttle had proved itself reliable. On a first test flight, though, everyone was glad that the seats were there.

LOOSE TILES

WHEN YOUNG AND CRIPPEN LOOKED THROUGH THE CREW CABIN'S SMALL REAR-FACING WINDOWS, THEY WERE STARTLED TO SEE THAT SIXTEEN HEAT SHIELD TILES HAD FALLEN OFF DURING LAUNCH. LATER INVESTIGATIONS (RIGHT) SHOWED THAT ANOTHER 148 WERE SLIGHTLY DAMAGED. NASA WAS CONCERNED THAT THE SHUTTLE MIGHT BURN UP DURING REENTRY. BUT THE CRUCIAL TILES ON THE BASE OF THE VEHICLE WERE STILL INTACT, AND COLUMBIA CAME HOME SAFELY.

GROUNDED

COLUMBIA WAS NOT THE FIRST SHUTTLE. IN 1977, THE ENTERPRISE WAS CARRIED ALOFT BY A SPECIAL BOEING 747 AND THEN DROPPED FOR GLIDE AND LANDING TESTS. BUT ENTERPRISE NEVER FLEW IN SPACE.

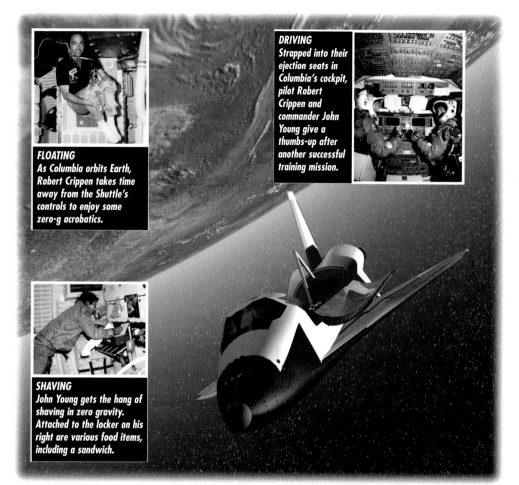

FLOATING
As Columbia orbits Earth, Robert Crippen takes time away from the Shuttle's controls to enjoy some zero-g acrobatics.

SHAVING
John Young gets the hang of shaving in zero gravity. Attached to the locker on his right are various food items, including a sandwich.

DRIVING
Strapped into their ejection seats in Columbia's cockpit, pilot Robert Crippen and commander John Young give a thumbs-up after another successful training mission.

RETURN TO SPACE

At last, just after 7 a.m. on April 12, Columbia got off the launchpad and ascended flawlessly to orbit. America was back in space, and public reaction to the flight was tremendous. More than a million spectators thronged the beaches and fields beyond the Kennedy Space Center to watch the liftoff.

Once in orbit, Young and Crippen benefited from Columbia's large cabin, which was much roomier than previous space capsules. There was even a second deck beneath the cockpit, with food storage, spacesuit racks, an airlock, and a private washroom cubicle.

Two days later, Columbia plunged back into the Earth's atmosphere and glided toward Edwards Air Force Base. It was a critical moment. With no engine power available, the pilots had to land first time around. To NASA's huge relief, they made a perfect touchdown.

For the first time, NASA had a vehicle that could deliver humans and cargo into space, and then be refurbished for further missions. Behind the crew cabin was a payload bay 60 feet (18 m) long and 15 feet (4.5 m) wide. One day soon it would carry pressurized space laboratories, or large satellites and planetary probes ready for release into space.

MISSION DIARY: SHUTTLE MISSION STS-1

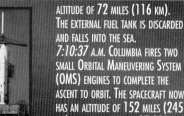

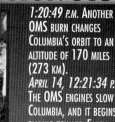

APRIL 12, 1981, 07:00:03 A.M. EST COLUMBIA'S THREE MAIN ENGINES ARE IGNITED, FUELED BY LIQUID HYDROGEN AND LIQUID OXYGEN (RIGHT).
7:00:09 A.M. THE TWIN SOLID ROCKET BOOSTERS FIRE. THE SPACE SHUTTLE LIFTS OFF THE LAUNCH PAD.
7:02:10 A.M. THE SOLID BOOSTERS COMPLETE THEIR BURN. COLUMBIA IS NOW AT A 31-MILE (50-KM) ALTITUDE. THE BOOSTERS FALL AWAY. LATER, THEY ARE RECOVERED FROM THE SEA AND REUSED.
7:08:38 A.M. COLUMBIA'S MAIN ENGINES SHUT DOWN AT AN

ALTITUDE OF 72 MILES (116 KM). THE EXTERNAL FUEL TANK IS DISCARDED AND FALLS INTO THE SEA.
7:10:37 A.M. COLUMBIA FIRES TWO SMALL ORBITAL MANEUVERING SYSTEM (OMS) ENGINES TO COMPLETE THE ASCENT TO ORBIT. THE SPACECRAFT NOW HAS AN ALTITUDE OF 152 MILES (245 KM), AND A VELOCITY OF 17,322 MPH (27,877 KM/H).
7:52 A.M. THE CARGO BAY DOORS ARE OPENED (ABOVE) TO EXPOSE SOLAR PANELS THAT HELP POWER THE SHUTTLE'S EQUIPMENT. YOUNG AND CRIPPEN BEGIN TWO DAYS OF SYSTEMS TESTING.

1:20:49 P.M. ANOTHER OMS BURN CHANGES COLUMBIA'S ORBIT TO AN ALTITUDE OF 170 MILES (273 KM).
APRIL 14, 12:21:34 P.M. THE OMS ENGINES SLOW COLUMBIA, AND IT BEGINS FALLING TOWARD EARTH. AFTER 17 MINUTES, THE SHUTTLE BEGINS TO HEAT UP FROM THE FRICTION OF REENTRY.
1:20:56 P.M. COLUMBIA TOUCHES DOWN ON THE RUNWAY (ABOVE) AT EDWARDS AIR FORCE BASE IN CALIFORNIA.

SPACELAB

Spacelab was born in the early 1970s of a collaboration between two of the world's leading space agencies. NASA had the go-ahead for the Space Shuttle but was denied funding for a space station to go with it. The European Space Agency wanted to send scientists into space but had no means of getting them there. The answer was to build a laboratory module that could be fitted inside the Shuttle's payload bay. First launched in 1983, Spacelab was a great success and heralded the age of space science.

WHAT IF...

...SPACELAB WERE AUTOMATIC?

At first, humans did everything on board Spacelab—including turning switches on and off. But although the idea of sending scientists into space is appealing, mission planners soon realized that it is both safer and more economical to conduct experiments from Earth using robot laboratories.

At first, scientists had to conduct remote-controlled experiments from a NASA control center—in Houston or at the Marshall Space Center in Huntsville, Alabama. Later they could view results and send commands in real time from facilities within their home countries, and the day cannot be far off when scientists will be able to run space experiments from their own labs—or even from home.

Sometimes, though, it is still preferable to have a payload specialist on the spot—especially when things go wrong. The world's scientists needn't abandon hope of going on the ultimate field trip just yet.

...MISSIONS WERE LONGER?

Most Spacelab missions lasted around two weeks. But in the late 1980s, plans were made for tours of up to 28 days in orbit to

By the late 21st century, most day-to-day experimental work will be done by remote-controlled orbiting "science parks" (center), with supplies brought in by crewless vehicles such as NASA's proposed VentureStar shuttle (left). The Space Shuttle (right) may still be employed to transport visiting technicians.

allow long stays in space without relying on the delay-prone International Space Station. The Orbiter Endeavour, under construction at the time, was fitted so that it could be used in this way. NASA anticipated that the biggest problem would be generating power for such long periods. Plans were therefore made to install two cryo-wafers—pallets carrying extra tanks of liquid hydrogen and oxygen for the Shuttle's fuel cells.

Another proposal called SPEDO—Solar-Powered Extended Duration Operations—involved unfurling two large solar arrays in space to produce up to 12 kilowatts of electricity. This would have allowed missions of 40 to 80 days. However, neither plan was carried out due to budget constraints.

SPACELAB STATS

MAX. WEIGHT 32,000 LB (14,500 KG)	CORE MODULE/EXPERIMENT SEGMENT (PRESSURIZED)	DIAMETER 3.3 FT (1 M)
PARTICIPATING ESA NATIONS		INSTRUMENT POINTING SYSTEM
BELGIUM, DENMARK, FRANCE, GERMANY	LENGTH 9 FT (2.7 M)	WEIGHT 2,600 LB (1,180 KG)
IRELAND, ITALY, NETHERLANDS, SPAIN,	DIAMETER 13.5 FT (4 M)	PAYLOAD 6,600 LB (2,993 LB)
SWITZERLAND, U.K.		
BUILDERS	PALLET (EXPERIMENTS ONLY)	IGLOO (INSTRUMENTS ONLY)
ERNO-VFW FOKKER CONSORTIUM	LENGTH 10 FT (3 M)	HEIGHT 7.9 FT (2.4 M)
(PRESSURIZED MODULES), BRITISH	DIAMETER 13.1 FT (4 M)	DIAMETER 3.6 FT (1.1 M)
AEROSPACE (PALLETS), SABCA (IGLOO),	WEIGHT 1,600 LB (725 KG)	WEIGHT 1,400 LB (635 KG)
DORNIER (IPS), MCDONNELL-DOUGLAS	PAYLOAD 3 TONS MAXIMUM	
(TUNNEL) MISSIONS 15 (CREWED	TUNNEL LENGTH 8.7 FT/18.9 FT	
MODULE), 6 (IGLOO/PALLETS)	(2.65/5.7 FT)	

SCIENCE LAB IN ORBIT

NASA's primary objectives after the Apollo Moon landings were to develop a reusable launch vehicle and to have human beings living and working in space for long periods of time. The government agreed to the first objective and gave a green light to the Space Shuttle program. But after Skylab, there was to be no money for the second aim. If NASA wanted a space station, it would have to go somewhere else to find it.

NASA went to Europe. In 1973, it signed an agreement with the European Space Agency (ESA) to develop a scientific research lab that could be carried into space in the Shuttle's payload bay. The result was Spacelab, the first major U.S.-European collaboration in space and Europe's first chance to put humans in orbit.

The Spacelab system is a set of modules that can be used in different combinations, depending on the kind of work to be done. The core segment is a pressurized laboratory, linked to the Orbiter's crew compartment by a 3-ft (1-m) -wide tunnel. The lab is fitted with a workbench and equipment racks, each loaded with up to 645 lb (292 kg) of experiments. Usually, an almost identical experiment segment is bolted on to the core segment to extend the lab space. Each segment has a hole in the roof that can be fitted with either a window for photography or an airlock for exposing experiments to open space. Oxygen, power, heat and communications with Earth are all provided by the Shuttle.

GETAWAY

FOR A FEW THOUSAND DOLLARS, NASA'S GETAWAY SPECIAL (GAS) PROGRAM—INTENDED TO RUN ALONGSIDE SPACELAB—WAS ABLE TO SEND YOUR EXPERIMENT INTO SPACE. IT HAD TO WEIGH NO MORE THAN 200 LB (90 KG) AND OCCUPY A MAXIMUM OF 5 CUBIC FEET (0.14 M³). THE GAS CANISTERS (RIGHT) BEING LOADED INTO THE SHUTTLE'S PAYLOAD BAY FOR A 1998 LAUNCH HOLD EXPERIMENTS FROM UTAH STATE UNIVERSITY AND BROWNARD COMMUNITY COLLEGE IN FLORIDA.

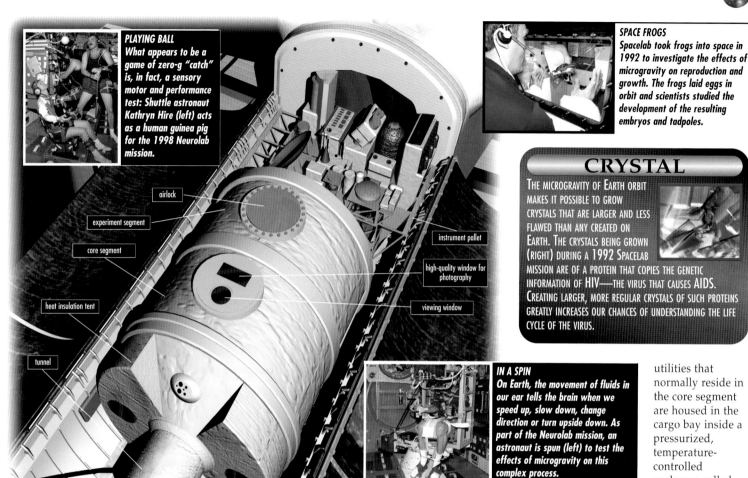

PLAYING BALL
What appears to be a game of zero-g "catch" is, in fact, a sensory motor and performance test: Shuttle astronaut Kathryn Hire (left) acts as a human guinea pig for the 1998 Neurolab mission.

airlock

experiment segment

core segment

heat insulation tent

tunnel

instrument pallet

high-quality window for photography

viewing window

SPACE FROGS
Spacelab took frogs into space in 1992 to investigate the effects of microgravity on reproduction and growth. The frogs laid eggs in orbit and scientists studied the development of the resulting embryos and tadpoles.

CRYSTAL

THE MICROGRAVITY OF EARTH ORBIT MAKES IT POSSIBLE TO GROW CRYSTALS THAT ARE LARGER AND LESS FLAWED THAN ANY CREATED ON EARTH. THE CRYSTALS BEING GROWN (RIGHT) DURING A 1992 SPACELAB MISSION ARE OF A PROTEIN THAT COPIES THE GENETIC INFORMATION OF HIV—THE VIRUS THAT CAUSES AIDS. CREATING LARGER, MORE REGULAR CRYSTALS OF SUCH PROTEINS GREATLY INCREASES OUR CHANCES OF UNDERSTANDING THE LIFE CYCLE OF THE VIRUS.

IN A SPIN
On Earth, the movement of fluids in our ear tells the brain when we speed up, slow down, change direction or turn upside down. As part of the Neurolab mission, an astronaut is spun (left) to test the effects of microgravity on this complex process.

ON TARGET
An instrument pointing system (left) keeps Spacelab's telescopes and other instruments accurately trained on their targets or steers them through precision scans of the night sky. The crew control it directly or leave the work to computers guided by Sun sensors and star trackers.

ON THE OUTSIDE

The other components of the Spacelab system are designed to operate by remote control in open space. They consist of a set of 10 U-shaped structures called pallets that mount scientific equipment in the Orbiter's payload bay. Some Spacelab missions are designated "pallets-only" and do without the pressurized lab segments. On such missions, up to five pallets can be carried, and the experiments are controlled by mission specialists from a console at the back of the Orbiter's flight deck. The control systems and utilities that normally reside in the core segment are housed in the cargo bay inside a pressurized, temperature-controlled enclosure called the igloo.

Spacelab is not an independent space station like Skylab or Mir, and while in orbit it remains firmly clamped in the Shuttle's cargo bay. Yet its low-key missions are important all the same. Along with its partner Shuttles, Spacelab has proved that space technology can be reusable—each module is designed to last 50 missions. Spacelab's greatest legacy, though, is likely to be the experience gained by humans, not with hardware. Each flight takes one to four civilian scientists into orbit, where only career astronauts had gone before. It is a significant step toward the large-scale habitation of space.

FLYING THE SHUTTLE

...THE OLD WORKHORSE SOLDIERS ON?

Although work on VentureStar, the Shuttle's projected replacement, is now underway, there is no reusable vehicle to take over the Shuttle's mission in the near-term if it is retired.

Concerns about the safety of the existing Shuttle, in addition to the fact that it is approaching the end its projected life span, may ground the remaining Shuttles and cause the project to be terminated.

NASA did commit to keeping the Shuttle in service until at least 2008, but the loss of

In years to come, Shuttle-based specialists may still make spacewalks to carry out construction and repair work on the International Space Station and other orbiting structures.

Columbia and faults discovered with others has at the time of writing cast doubt over this possibility. However, 'return-to-flight' missions have now commenced – not without difficulties – and it is probable that the Shuttle will haul itself back into space by the bootstraps if that is what it takes.

The present Shuttle system will, however, need to be upgraded to improve performance and safety if the Shuttle fleet is to have a long-term future. Eventually, the Shuttle's solid rocket boosters will be replaced by more controllable liquid-fuelled ones. Whereas the existing boosters splash down in the sea after being jettisoned and have to be recovered by ships at enormous cost, the new boosters may have the ability to fly back to base under remote control.

Other improvements will be made to the avionics, computers and other systems, including equipment to enable the Shuttle to perform missions longer than the current record of 17 days, 15 hours. As the Shuttle is now operated privately for NASA by the United Space Alliance, it is likely that it will be used increasingly for commercial missions such as satellite deployments, space repairs and salvage. There is even a possibility of passenger flights—for the elite few who could afford the estimated $20 million or more required for training and the flight itself!

Capturing a faulty satellite... Docking with a space station to bring new crews and equipment... Operating a space laboratory... Servicing a space telescope... Such activities are routine for the astronauts involved in the Space Shuttle program. Classed as a partially reusable manned spaceplane, the Shuttle was the most versatile spacecraft ever built. Even though it has completed scores of missions since its maiden flight in 1981, Shuttle launches still get live TV coverage and attract thousands of spectators to the Kennedy Space Center. However, its future is in serious doubt.

SHUTTLE MISSIONS

ENTERPRISE (OV-101) WAS A TEST VEHICLE AND WAS NOT INTENDED FOR SPACE MISSIONS.

CHALLENGER (OV-99) FLEW 10 MISSIONS FROM 1983 TO 1986. SHE WAS LOST APPROXIMATELY 73 SECONDS AFTER TAKEOFF ON MISSION STS-51-L IN 1986.

ATLANTIS (OV-104) HAS FLOWN 26 MISSIONS FROM 1985 TO 2002, AND REMAINS IN SERVICE.

ENDEAVOUR (OV-105) HAS FLOWN 19 MISSIONS FROM 1992 TO 2002. ENDEAVOUR REMAINS IN SERVICE.

COLUMBIA (OV-102) FLEW 28 MISSIONS FROM 1981-2003, INCLUDING STS-1, THE VERY FIRST SHUTTLE MISSION. SHE WAS LOST SHORTLY BEFORE TOUCHDOWN ON MISSION STS-107.

DISCOVERY (OV-103) HAS FLOWN 31 MISSIONS FROM 1984 TO THE PRESENT. DISCOVERY MADE THE FIRST RETURN-TO-FLIGHT MISSION AFTER THE LOSS OF COLOMBIA IN AUGUST 2005.

SPACE WAGON

The Space Shuttle is a general-purpose space truck, whose job is to ferry people and cargo to and from orbit above the Earth. Given the limitations of current technology, it performs its task with remarkable efficiency.

The Shuttle is launched like a conventional rocket, and jettisons its twin reusable solid-fuel boosters two minutes after liftoff. The Orbiter spaceplane then climbs under the power of its own rocket engines, which are fed with liquid propellant from a giant external fuel tank. Six minutes later the Shuttle goes into orbit and sheds its tank, which burns up in the Earth's atmosphere. A short time after this, the crew of astronauts and payload specialists goes to work.

WORKING IN SPACE

The Orbiter's payload bay may contain satellites, repair equipment or pressurized modules for conducting space experiments. Some payloads are deployed by remote control; others are operated directly by the crew who access the payload bay via a tunnel.

Missions last an average of nine days, though some extend to three weeks. This can be tough on the crew, who have to live and

WHAT THE SHUTTLE DOES

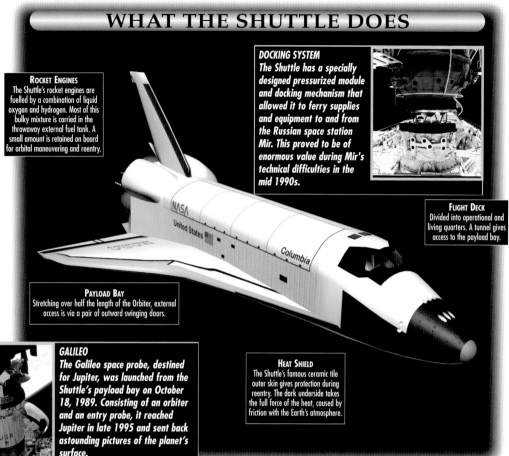

ROCKET ENGINES
The Shuttle's rocket engines are fuelled by a combination of liquid oxygen and hydrogen. Most of this bulky mixture is carried in the throwaway external fuel tank. A small amount is retained on board for orbital maneuvering and reentry.

DOCKING SYSTEM
The Shuttle has a specially designed pressurized module and docking mechanism that allowed it to ferry supplies and equipment to and from the Russian space station Mir. This proved to be of enormous value during Mir's technical difficulties in the mid 1990s.

FLIGHT DECK
Divided into operational and living quarters. A tunnel gives access to the payload bay.

PAYLOAD BAY
Stretching over half the length of the Orbiter, external access is via a pair of outward swinging doors.

GALILEO
The Galileo space probe, destined for Jupiter, was launched from the Shuttle's payload bay on October 18, 1989. Consisting of an orbiter and an entry probe, it reached Jupiter in late 1995 and sent back astounding pictures of the planet's surface.

HEAT SHIELD
The Shuttle's famous ceramic tile outer skin gives protection during reentry. The dark underside takes the full force of the heat, caused by friction with the Earth's atmosphere.

RETRIEVAL
A Japanese satellite is retrieved (left) from orbit by the robot arm of the Shuttle Endeavour on January 13, 1996. The square object in the foreground, another satellite, was launched during the same mission on January 15 and retrieved two days later.

SPACELAB
This was the name given to the Shuttle's original pressurized laboratory module. Spacelab is primarily used by engineers, scientists and astronauts to study the long-term effects of weightlessness on living things.

work in the cramped flight deck or lower mid-deck—which also serves as a galley, sleeping quarters and bathroom. Some missions are divided into shifts to permit 24-hour work days.

On completion of a mission, the Shuttle's engines fire to redirect it back to Earth. After a bumpy ride through the atmosphere, protected by its heat-shield tiles, it glides in to land like an airplane.

SPACESIDE

During the STS-49 Endeavour mission in 1992, a stranded communications satellite, Intelsat 6, was manually captured by three spacewalking astronauts, who fitted it with a rocket engine and redeployed it. The rocket engine was then activated to shoot the satellite into geostationary orbit, from where it is still operating today. NASA considers this "space rescue" to be the best demonstration yet of the versatility of the Space Shuttle, and of the value of being able to deploy humans in space.

MISSION DIARY: SHUTTLE MISSION STS-87

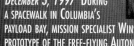

November 19, 1997 The crew of Mission STS-87 makes its way to the launch pad where they board the Space Shuttle Columbia for a 16-day flight. The six members of the crew are Kalpana Chawla, Kevin Kregel (Mission Commander), Takao Doi, Winston Scott, Leonid Kadenyuk and Steven Lindsey.
November 22, 1997 Shuttle pilot Steven Lindsey and mission specialist Kalpana Chawla check on the progress of an experiment carried in Columbia's mid-deck area.

Experiments scheduled to be performed during this mission include the pollination of plants and the processing of materials under weightless conditions.
December 3, 1997 During a spacewalk in Columbia's payload bay, mission specialist Winston Scott releases a prototype of the free-flying Autonomous Extravehicular Activity Robotic Camera Sprint (AERCam Sprint). This spherical, basketball-sized device houses a TV camera and

is intended to be used for remote-controlled inspections of the exterior of the International Space Station.
December 5, 1997 With its drag 'chute deployed to help slow it down, Columbia lands at 7:20:04 a.m. EST on Runway 33 at the Kennedy Space Center. Its mission, the 87th in the Shuttle program, has taken it a total distance of 6.5 million miles (10.5m km) and has lasted for exactly 15 days, 16 hr and 34 min.

SHUTTLE REENTRY

F ew rides can be as hair-raising as the final stage of a Shuttle mission: reentry into Earth's atmosphere from space. As the orbiter heads home, it must transform itself from spacecraft to glider—and lose most of its 17,500-mph (28,000 km/h) orbital velocity. The Shuttle must turn in space, fire braking rockets, and then plunge into the atmosphere for a fiery, 3,000°F (1,650°C) ride until friction slows it to orthodox flying speeds. Then, with engines silenced, it glides toward the welcome sight of the landing strip.

WHAT IF...

...THE ORBITER HAD TO BE DITCHED?

R eturning to Earth on the Shuttle is dangerous enough, but what if something went seriously wrong while it was in orbit or in the first fiery stages of descent to Earth? In space, there would be no escape for the crew. But once in a controlled glide, they would have several options.

At best, they could bail out at about 60,000 ft (18,000 m), but there is no equivalent to the ejection seat of a fighter jet. Instead, the crew would have to exit as if from a burning building, going through a series of intricate operations, with safety margins of seconds. As many as eight astronauts would have to evacuate the craft, encumbered by their 70-lb (32 kg) protection suits.

When the stricken orbiter reached 30,000 ft (9,100 m), it would have slowed to little more than its landing speed—some 230 mph (370 km/h). At 25,000 ft (7,600 m), a crewmember would pull a lever to depressurize the cabin. The crew would then blow out a hatch in the mid-deck of the crew compartment, which would leave them just three minutes to escape, as the orbiter glided automatically from 25,000 ft (7,600 m) to 2,000 ft (600 m). They would then deploy a telescopic escape pole which juts almost 10 feet out of the fuselage, attach their parachute ripcords to a lanyard, and slide down the pole at 12-second intervals. It would take 1.5 minutes for all eight astronauts to exit, emerging below the orbiter's left wing.

Another course would be to steer the

As seen from the Shuttle flight deck, the landing strip at the Kennedy Space Center looks like home. But the pilots only have one chance to put their craft safely on the ground.

orbiter to land at an alternative landing site—if the crew felt that the orbiter could be saved. There are dozens of these dotted around the world, in case the Shuttle has to make an unscheduled descent. In an emergency on a planned reentry, the crew can redirect their craft anywhere within a 1,250-mile (2,000 km) radius to one of many such U.S. sites.

Or the crew could ditch in the ocean. This is the least desirable alternative—because the escape hatch would be underwater and emergency services would have to reach and rescue them within minutes. In NASA's own words, "The probability of the flight crew surviving a ditching is slim."

REENTRY SPECS

ALTITUDE	OPERATION	SPEED	MILES (KM) FROM TOUCHDOWN
557,000 FT (170,000 M)	DEORBIT	17,000 MPH (27,500 KM/H)	5,000 (8,000)
400,000 FT (122,000 M)	ENTRY INTERFACE	17,000 MPH (27,500 KM/H)	3,500 (5,600)
83,000 FT (25,000 M)	TERMINAL AREA	1,700 MPH (2,750 KM/H)	60 (100)
49,000 FT (15,000 M)	SUBSONIC	760 MPH (1,220 KM/H)	25 (40)
10,000 FT (3,000 M)	GLIDE SLOPE	330 MPH (530 KM/H)	8 (12)

HELLO EARTH

Preparation for reentry starts about four hours before the Shuttle reenters the atmosphere. Crews complete their work, life support systems are rechecked, and the star-guidance system is shut down.

With an hour to go, crewmembers strap themselves into their seats. The pilot nudges the craft around so that it faces backward, ready for the engines to fire and start the descent from orbit. The crew enters the correct coordinates into the computer system: height, speed, distance from touchdown. With this information, the computer fires the engines for about 2.5 minutes, cutting the speed by about 500 mph (800 km/h) to some 17,000 mph (27,400 km/h), to start the long, slow fall into the upper atmosphere.

Another piloted maneuver swings the orbiter around again to fly nose-first, descending at precisely the right angle. Accuracy is critical—2° shallower and the orbiter would skip back into space, 2° steeper and it would burn up in the air like a meteor.

From now on, the orbiter can approach touchdown entirely under the control of its computer, although the autopilot can be overridden. Another 25 minutes of downhill coasting brings the orbiter to an altitude of 100 miles (160 km), with some 5,000 miles (8,000 km) to go to touch-down.

Five minutes later, at an altitude of 75 miles (120 km), with some 3,500 miles (5,600 km) to go, sensors pick up the first tenuous traces of atmosphere. The orbiter plows into the air belly-first, its six maneuvering jets ensuring that it presents the right amount of surface area to the air.

HOT BELLY

This first contact with atmosphere—Entry Interface (EI)— creates great friction, and the temperature of the skin of the spacecraft soars. The heat is dissipated by several heat shields, but it is mostly the 23,000 ceramic tiles that protect the underside and leading edges of the main body. Damage to these tiles is what caused the tragic loss of the shuttle

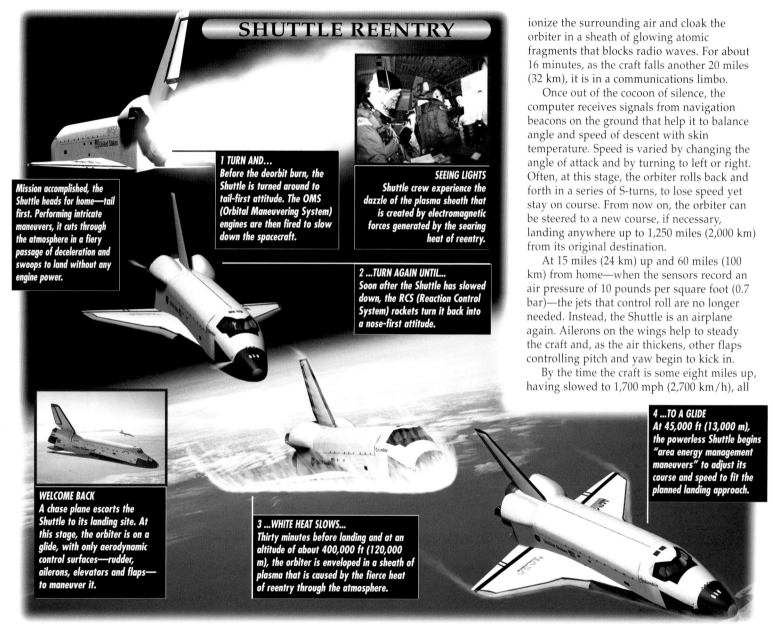

SHUTTLE REENTRY

Mission accomplished, the Shuttle heads for home—tail first. Performing intricate maneuvers, it cuts through the atmosphere in a fiery passage of deceleration and swoops to land without any engine power.

1 TURN AND...
Before the deorbit burn, the Shuttle is turned around to tail-first attitude. The OMS (Orbital Maneuvering System) engines are then fired to slow down the spacecraft.

SEEING LIGHTS
Shuttle crew experience the dazzle of the plasma sheath that is created by electromagnetic forces generated by the searing heat of reentry.

2 ...TURN AGAIN UNTIL...
Soon after the Shuttle has slowed down, the RCS (Reaction Control System) rockets turn it back into a nose-first attitude.

WELCOME BACK
A chase plane escorts the Shuttle to its landing site. At this stage, the orbiter is on a glide, with only aerodynamic control surfaces—rudder, ailerons, elevators and flaps—to maneuver it.

3 ...WHITE HEAT SLOWS...
Thirty minutes before landing and at an altitude of about 400,000 ft (120,000 m), the orbiter is enveloped in a sheath of plasma that is caused by the fierce heat of reentry through the atmosphere.

4 ...TO A GLIDE
At 45,000 ft (13,000 m), the powerless Shuttle begins "area energy management maneuvers" to adjust its course and speed to fit the planned landing approach.

ionize the surrounding air and cloak the orbiter in a sheath of glowing atomic fragments that blocks radio waves. For about 16 minutes, as the craft falls another 20 miles (32 km), it is in a communications limbo.

Once out of the cocoon of silence, the computer receives signals from navigation beacons on the ground that help it to balance angle and speed of descent with skin temperature. Speed is varied by changing the angle of attack and by turning to left or right. Often, at this stage, the orbiter rolls back and forth in a series of S-turns, to lose speed yet stay on course. From now on, the orbiter can be steered to a new course, if necessary, landing anywhere up to 1,250 miles (2,000 km) from its original destination.

At 15 miles (24 km) up and 60 miles (100 km) from home—when the sensors record an air pressure of 10 pounds per square foot (0.7 bar)—the jets that control roll are no longer needed. Instead, the Shuttle is an airplane again. Ailerons on the wings help to steady the craft and, as the air thickens, other flaps controlling pitch and yaw begin to kick in.

By the time the craft is some eight miles up, having slowed to 1,700 mph (2,700 km/h), all

Columbia in 2003, allowing superheated gas to penetrate the shuttle's wing.

There is virtually no chance of survival for a Shuttle whose heat shield fails, so techniques for discovering, locating and repairing damage – and for preventing it in the first place – were a

critical part of the Return to Flight mission planning. During their August 2005 mission, Discovery's crew were able to demonstrate these techniques and conduct repairs to their craft.

About 50 miles (80 km) up, the temperature reaches almost 3,000°F (1,650°C)—enough to

jets are off and the orbiter has become a supersonic glider. Eight miles from touchdown, at 10,000 ft (3,000 m), its speed drops to some 330 mph (530 km/h) and the orbiter enters its final approach as it prepares to make its landing.

SHUTTLE LANDING

O n April 19, 1985, Shuttle Mission 51D Commander Karol Bobko applied the brakes to slow the orbiter Discovery for landing. As its nosewheel touched down, the crew heard a clatter as the brakes locked, followed by a loud bang. Fortunately, the Shuttle rolled safely to a stop. Onlookers told of a puff of smoke from the undercarriage that was later traced to a pair of blown tires—shredded due to the locking brakes. It all goes to show that landing the Shuttle in one piece can be as nerve-racking as launching it into space.

WHAT IF...

...THERE WAS AN EMERGENCY?

O ne of the chief hazards faced by the Shuttle during landing is poor weather—and in this respect, hurricane-prone Florida is hardly the ideal home base. So far, when faced with adverse conditions over the Cape, the Shuttle has always been able to divert to Edwards AFB in California, where the weather is invariably good. But in the event of a major malfunction during the critical reentry phase, the glide-only orbiter would be literally powerless to undertake more elaborate maneuvers; Mission control would have to make a quick decision to land the craft somewhere else on the globe.

In fact, this contingency has always been allowed for by NASA, and there are currently 50 designated airstrips around the world where the Shuttle could land in an emergency. Even so, it would be a hazardous undertaking: As any glider pilot knows, an airplane without engines is largely at the mercy of the air currents it travels through.

The next generation of crewless reusable spacecraft, led by the X-33 and X-38 technology demonstration vehicles, is being designed to take the guesswork out of landing. The Soviet Union paved the way in 1988, with the successful landing of its shuttle Buran after its one and only crewless flight into space. But the Space Shuttle, too, has the capacity to make a

The enigmatic statues on Easter Island in the Pacific Ocean play host to the strangest visitor in their long history. But it is no fantasy: In a landing emergency, the island's airstrip could save the Shuttle from disaster.

fully automated landing. On some missions the crew have remained hands-off until the very last moment, and the Shuttle may make a simulated crewless landing before it is finally retired.

The reality, though, is that shuttle landings will continue to be fraught with hazards until reusable spacecraft can be equipped with powerplants capable of operating within the Earth's atmosphere. Rockets do this, of course, but the weight of fuel needed for a rocket-powered return trip more or less rules out their use. Future missions, too, are likely to demand even more flexibility—for example, to make an emergency crew rescue from a stricken ISS, or to return a hazardous piece of space debris safely to Earth. At the moment the ball is firmly in the court of the aero-engine designers. There will be relief in many quarters when they finally come up with the goods.

FOREIGN LANDING SITES

In addition to the 20 U.S. Shuttle landing sites, there are another 30 dotted around the globe.		Crete	Souda	Morocco	Ben Guerir
		Diego Garcia	Diego Garcia	Polynesia	Hao
		Easter Island	Mataveri	Saudi Arabia	King Khalid
Algeria	Tamanrasset	England	Brize Norton	South Africa	Hoedspruit
Australia	Amberley		Mildenhall	Spain	Gran Canaria
	Darwin		Upper Heyford		Moron
Azores	Lajes	Gambia	Yundum		Zaragoza
Bahamas	Nassau	Germany	Cologne/Bonn	Sweden	Arlanda
Bermuda	Bermuda	Guam	Andersen	Turkey	Esenborg
Cape Verde	Amilcar Cabral	Liberia	Monrovia/Roberts	Zaire	Kinshasa/N. Djile

TOUCH DOWN

It is tempting to think of Shuttle landings as routine, like those made by countless passenger jets around the world each day. The reality is somewhat different. For a start, the Shuttle is not a true "spaceplane" because the rocket engines and attitude thrusters that power it in orbit are of no use on the trip back to Earth. Instead, Shuttle commanders find themselves at the controls of a craft that has more in common with the gliders at their local airport. The difference is that this glider weighs in at 100 tons (90 tonnes).

Early tests of the landing sequence were carried out in 1977 using the orbiter Enterprise, which was launched from the back of a specially modified Boeing 747. The routines drew heavily on experience gained with the X-series rocket planes, and were entirely successful. But the moment of truth came in April 1981, when the orbiter Columbia made the first, historic touchdown from orbit at Edwards Air Force Base, California. A relieved Commander John Young asked the control tower if they wanted the vehicle taken straight to its hangar. The tower replied, "It needs to be dusted off first."

Early landings were all made on the long, dry lakebed runway at Edwards in case unforeseen problems caused the Shuttle to overshoot or veer off course. But in 1984, on the tenth mission, NASA felt confident enough to land the Shuttle on a regular airstrip adjacent to Pad LC39 at Cape Canaveral. Since then, the Shuttle has usually landed at the Cape, only diverting to Edwards when adverse weather conditions demand it.

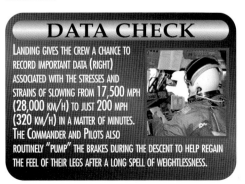

DATA CHECK

LANDING GIVES THE CREW A CHANCE TO RECORD IMPORTANT DATA (RIGHT) ASSOCIATED WITH THE STRESSES AND STRAINS OF SLOWING FROM 17,500 MPH (28,000 KM/H) TO JUST 200 MPH (320 KM/H) IN A MATTER OF MINUTES. THE COMMANDER AND PILOTS ALSO ROUTINELY "PUMP" THE BRAKES DURING THE DESCENT TO HELP REGAIN THE FEEL OF THEIR LEGS AFTER A LONG SPELL OF WEIGHTLESSNESS.

THE LANDING SEQUENCE

1 DIVE
Around 12 miles up, the Shuttle emerges from its 10–15 minute blackout period and goes into a series of S-turns to slow the orbiter.

2 GLIDE
At 13,300 feet (4,050 m), the Commander pulls back on the controls to bring the spacecraft into a flat glide. With no power, the Shuttle is now fully committed to its designated landing site.

3 APPROACH
Adopting a 22° angle of attack, the Shuttle makes its final approach to the runway. Four seconds before touchdown, the wheels drop down under the force of gravity.

4 LANDING
As the nosewheel touches down, the brakes are applied and the main drag chute is deployed to bring the craft to a halt.

CHASED DOWN
During the final moments of landing, the Shuttle is joined by a T-38 chase airplane. The chase pilots repeatedly call speed and altitude readings to the Shuttle's crew to help them land the world's heaviest glider.

DRAG STAR
A few seconds after touchdown, the main drag chute is deployed to help slow the Shuttle from its landing speed of around 215 mph (350 km/h). The procedure became routine after 1992, in an effort to save wear and tear on brakes and tires.

GROUNDED
Once the orbiter has reached wheel lock, the recovery convoy moves in. A 14-foot (4-m) fan is blown over the Shuttle to cool it and disperse any toxic gases before the crewmembers are allowed to exit.

The Shuttle is capable of completing its landing sequence on autopilot, but Shuttle crews generally prefer to switch to manual override for the last crucial seconds.

shallower angle for its final approach, and there is a double sonic boom as it passes through Mach 1. By this time, the automatic pilot will be engaged, though the pilots can override it at any stage—and usually do so in the final moments. After circling the landing area, the Shuttle approaches the runway at a 22° angle.

COMING IN TO LAND

The landing sequence begins as the Shuttle emerges from its 10- to 15-minute "blackout period," after which the crew guides the vehicle through sweeping S-shaped maneuvers that slow it from around 17,500 mph (28,000 km/h) to 1,700 mph (2,730 km/h).

At around 13,300 feet (4,000 m), with 90 seconds to go, the Shuttle eases back to a shallower angle for its final approach.

The main landing gear touches down at 215 mph (350 km/h), followed seconds later by the nosewheel. Finally, the combination of the drag chute and some deftly applied braking brings the orbiter safely to a halt.

CHALLENGER DISASTER

T here was an almost carnival atmosphere at the Kennedy Space Center, Florida, on January 28, 1986, as crowds gathered for the 25th flight of the Space Shuttle. On board the Shuttle Orbiter Challenger was the first "ordinary American citizen," a teacher named Christa McAuliffe, whose cheerful personality had already endeared her to the public. After a long series of delays, at 11:38 EST the Shuttle lifted off the launchpad into a clear, blue sky. Seventy-three seconds later, it was gone.

WHAT IF...

...THE SHUTTLE DISASTERS HADN'T HAPPENED?

T he space shuttle programme has suffered two catastrophes – the loss of Challenger during takeoff in 1986 and of Columbia during re-entry in 2003. These accidents, each of which claimed seven lives, shook confidence in the orbiter fleet and raised doubts over the future of the Shuttle.

Had these disasters not happened, the Space Shuttle would have continued to fly at ever more regular intervals. An increase in civilian passengers would have kept the space program in the public eye, improved NASA's reputation and ensured regular funding.

Yet it now seems that an accident of some sort was inevitable. At the time of the 1986 Challenger disaster an overconfident NASA was planning a pair of flights the following May within just days of each other. One of the missions involved carrying the Galileo Jupiter probe into orbit, together with the Centaur upper stage that was to power the probe on its journey across the solar system. The Centaur rocket was a new design that obtained extra power by storing its liquid oxygen and hydrogen fuel at ultra-low (cryogenic) temperatures. The consequences of even a small leak of these fuels in the Orbiter's payload bay would have been catastrophic. The risks were so great that astronauts dubbed it the "Death Star" mission.

Following the Challenger disaster, a reappraisal of the project concluded that it was unnecessarily dangerous. Use of the Centaur with Galileo was cancelled and was replaced by a less efficient, but far safer, solid-fuel Inertial Upper Stage (IUS).

The inquiry into Challenger also uncovered problems aside from the faulty design of the rocket boosters. Poor communication between the various NASA centers and the space industry was highlighted, and there were concerns about the general standard of safety and hazard analysis on the Shuttle program. NASA was also criticized for its overconfidence, which had resulted in an unrealistic flight rate. All these areas were investigated and addressed to make the Shuttle program safer.

After the loss of Challenger, the Shuttle re-established itself as most reliable means of carrying payloads into orbit. For seventeen years the orbiter fleet was a reliable workhorse. Shuttle missions long ago ceased to be anything out of the ordinary; only those fascinated by spaceflight paid much attention.

Perhaps complacency had descended once again when in 2003 the Shuttle Columbia disintegrated during its landing approach. The orbiter fleet was once again grounded and there were indications that this time it might be permanent. Space programmes such as the construction of the International Space Station and servicing of the Hubble Space telescope have been set back or cancelled. Without a reusable launch vehicle the cost of exploring space is greatly increased. At the time of writing, the Shuttle has a fighting chance.

THE CHALLENGER CREW

BACK ROW, LEFT TO RIGHT
ELLISON SHOJI ONIZUKA (MISSION SPECIALIST), BORN JUNE 14, 1946
SHARON CHRISTA CORRIGAN MCAULIFFE (PAYLOAD SPECIALIST), BORN SEPTEMBER 2, 1948
GREGORY BRUCE JARVIS (PAYLOAD SPECIALIST), BORN AUGUST 24, 1944
JUDITH ARLENE RESNIK (MISSION SPECIALIST), BORN APRIL 5, 1949

FRONT ROW, LEFT TO RIGHT
MICHAEL JOHN SMITH (PILOT), COMMANDER U.S. NAVY, BORN APRIL 30, 1945
FRANCIS RICHARD SCOBEE (COMMANDER), BORN MAY 19, 1939
RONALD ERWIN MCNAIR (MISSION SPECIALIST), BORN OCTOBER 21, 1950

FATAL FLAW

NASA intended 1986 to be a landmark year for the space program. The Space Shuttle was to make 15 flights, and President Reagan was confidently expected to give the go-ahead for the project that would see a U.S. space station in orbit by 1994. It needed to be a good year. Mission 61-C Columbia had been progressively delayed from the year before, public interest in space was on the wane, and rumors were circulating in the media of NASA incompetence. Its reputation as the foremost space agency in the world was beginning to slip.

It seemed that if anything could restore government and public confidence, it would be Shuttle launch 51-L Challenger. NASA saw to it that the inclusion of the first woman civilian astronaut, schoolteacher Christa McAuliffe, received maximum media coverage. The nation warmed to her—and to the idea that ordinary people might soon be going into space.

WAITING TO HAPPEN

Behind the scenes, there was growing concern among NASA technicians about the safety of the Shuttle. The focus of this concern was the flexible sealing system of putty and synthetic-rubber O-rings used on the "field joints" between sections of the reusable solid rocket boosters (SRBs). In-flight damage to the seals had been noticed on previous missions,

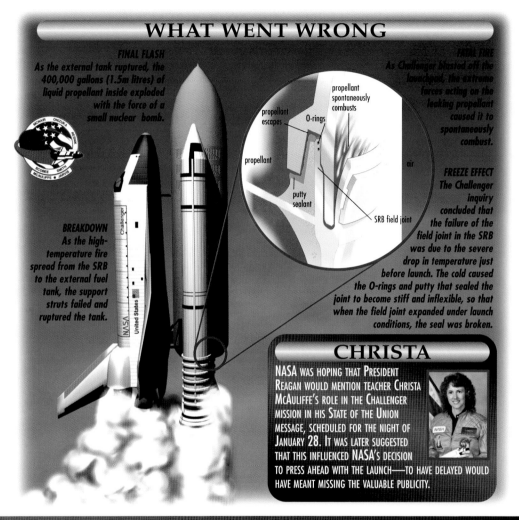

WHAT WENT WRONG

FINAL FLASH
As the external tank ruptured, the 400,000 gallons (1.5m litres) of liquid propellant inside exploded with the force of a small nuclear bomb.

BREAKDOWN
As the high-temperature fire spread from the SRB to the external fuel tank, the support struts failed and ruptured the tank.

FATAL FIRE
As Challenger blasted off the launchpad, the extreme forces acting on the leaking propellant caused it to spontaneously combust.

FREEZE EFFECT
The Challenger inquiry concluded that the failure of the field joint in the SRB was due to the severe drop in temperature just before launch. The cold caused the O-rings and putty that sealed the joint to become stiff and inflexible, so that when the field joint expanded under launch conditions, the seal was broken.

propellant escapes — propellant — putty sealant — O-rings — propellant spontaneously combusts — air — SRB field joint

CHRISTA

NASA WAS HOPING THAT PRESIDENT REAGAN WOULD MENTION TEACHER CHRISTA MCAULIFFE'S ROLE IN THE CHALLENGER MISSION IN HIS STATE OF THE UNION MESSAGE, SCHEDULED FOR THE NIGHT OF JANUARY 28. IT WAS LATER SUGGESTED THAT THIS INFLUENCED NASA'S DECISION TO PRESS AHEAD WITH THE LAUNCH—TO HAVE DELAYED WOULD HAVE MEANT MISSING THE VALUABLE PUBLICITY.

JUDITH RESNIK

RESNIK WAS SELECTED AS AN ASTRONAUT IN 1978, THE FIRST YEAR IN WHICH WOMEN WERE ACCEPTED. SHE WENT ON TO BECOME ONLY THE SECOND AMERICAN WOMAN IN ORBIT DURING THE MAIDEN FLIGHT OF DISCOVERY IN 1984. HER VAST EXPERIENCE WAS SADLY MISSED BY THE U.S. SPACE PROGRAM.

implying a design fault, but no action was taken to solve the problem.

With Challenger already behind schedule after delays in the Columbia mission, the pressure mounted for a swift launch. The crew finally boarded on January 27, only to be told, just 30 minutes from liftoff, that the mission was postponed due to a hatch fault. Meanwhile, engineers at Morton Thiokol, makers of the SRBs, voiced serious concerns about the effects of cold weather on the O-ring seals. A severe cold front was forecast to hit Florida the following day that would see temperatures plummet to 23°F (-5°C). The engineers suggested postponing the launch until the cold snap passed, but their suggestion was overruled by NASA officials and the launch went ahead.

One week later, as the nation mourned its dead, President Reagan assigned a commission to investigate the accident. Their report, published on June 6, 1986, stated that the cause "was the failure of the pressure seal in the aft field joint of the right-hand solid rocket booster...due to a faulty design..."

MISSION DIARY: COUNTDOWN TO DISASTER

T+0.6 SEC A LAUNCH CAMERA PHOTOGRAPHS PUFFS OF SMOKE LOW-DOWN ON THE RIGHT-HAND SOLID ROCKET BOOSTER. THE SMOKE IS THE FIRST SIGN OF THE PROPELLANT THAT HAS BEGUN TO LEAK FROM A FAULTY O-RING SEAL IN THE AFT FIELD JOINT.

T+35 SEC AFTER A SEEMINGLY PERFECT LAUNCH, THE CHALLENGER'S MAIN ENGINES ARE THROTTLED DOWN TO 65% AS THE CRAFT ENTERS THE PERIOD OF MAXIMUM DYNAMIC PRESSURE.

T+37 SEC SEVERE FLUCTUATIONS IN THE FLIGHT PATH BEGIN; THESE CONTINUE FOR ANOTHER 26 SECONDS AND ARE LATER ATTRIBUTED TO "WIND SHEAR"—AN EXPLANATION QUESTIONED BY MANY ANALYSTS.

T+51 SEC ENGINES ARE SET TO FULL THROTTLE.

T+58 SEC FLAMES APPEAR AROUND THE FAULTY FIELD JOINT (CIRCLED, RIGHT). BY THIS TIME THE O-RING SEAL HAS FAILED COMPLETELY.

T+60 SEC THE FLAMES SPREAD RAPIDLY OVER THE STRUTS THAT SECURE THE SRBS AND THE EXTERNAL FUEL TANK (ET).

T+64 SEC THE ET IS BREACHED, BRINGING THE FLAMES INTO CONTACT WITH THE LIQUID HYDROGEN THAT IS LEAKING FROM THE TANK.

T+72 SEC IN A RAPID SEQUENCE OF EVENTS: THE STRUT LINKING THE RIGHT-HAND SRB TO THE ET FRACTURES; THE AFT DOME FALLS AWAY, RIPPING THE UPPER PARTS OF THE TANK APART; THEN THE SEVERED SRB HITS THE ET, CAUSING LIQUID OXYGEN TO MIX WITH THE LIQUID HYDROGEN IN A LETHAL COCKTAIL.

T+73 SEC THE MIXTURE OF LEAKING LIQUID FUELS ERUPTS IN AN EXPLOSION (RIGHT). THE ET DISINTEGRATES AND BOTH SRBS FLY OFF OUT OF CONTROL, LATER TO BE REMOTE-DETONATED BY THE RANGE SAFETY OFFICER. THE ORBITER IS BLOWN CLEAR BUT INSTANTLY BREAKS UP DUE TO AERODYNAMIC FORCES. THE ONBOARD FUEL EXPLODES, THROWING THE CABIN CLEAR OF THE

FIREBALL AND LEAVING THE WRECKAGE TO PLUNGE INTO THE ATLANTIC OCEAN. IT IS BELIEVED THAT SOME OF THE CREW INITIALLY SURVIVED THE EXPLOSION.

BURAN

I n 1974, a Soviet military analyst pointed out to Premier Leonid Brezhnev that the planned U.S. Shuttle might overfly Moscow— "possibly with a dangerous cargo." Brezhnev erupted: "We are not country bumpkins here!" and ordered work to begin at once on a Soviet equivalent. The result was Buran ("Snowstorm"). With no crew, it was carried into orbit by a giant new rocket, Energia—and then landed automatically. But this impressive start led nowhere. The Soviet Union fell apart, and Buran never flew again.

WHAT IF...

...BURAN AND ENERGIA WERE STILL IN USE?

W hen the Soviet Union had its own ambitious space program, Buran's role was clear. As well as countering the supposed threat from the U.S. Shuttle, it would have contributed to the Russian space station, Mir, which went into orbit in 1986. After two modules had been added to Mir, Buran was supposed to have ferried up the third, Kristall, in 1990, providing the space station with additional scientific equipment, retractable solar arrays and a docking node for spacecraft that massed up to 100 tons (90 tonnes). If Buran had accomplished such missions, it would perhaps have graduated to ferrying up parts of the International Space Station (ISS).

In fact, economics and politics combined to make both Buran and Energia redundant. By the time the whole project was canceled in 1993, the U.S. and Russia were partners, not rivals. Buran's role was taken over by the Space Shuttle, which will do all the heavy lifting for the ISS. And because Buran is so similar to the Shuttle in design and purpose, no one believes that it will ever be worth the effort and expense of reviving the project.

The work on Energia was not entirely wasted. Its boosters are used independently as Zenit rockets for the launch of observation and communication satellites. One of them, the 10-ton Tselina-2, was almost certainly monitoring the communications of U.S. and Allied troops during the war in Kosovo in the summer of 1999.

Once the p ride of Soviet cosmonautics, Buran is now grounded in a hangar (above) at Baikonur, Kazakhstan. Its sup er-p owerful carrier, Energia (left), suffered an even more ignominious fate: Many of the launchers were cut up and sold as scrap.

And, in theory, Energia itself could have fulfilled one original aim: to place in orbit elements of the ISS. One Energia flight could match several Shuttles. But the money would have to come mostly from the United States, and it was not in American interests to invest in a rival to the Shuttle. Another suggestion is that Energia could be used to launch a crewed flight to Mars. At present, there are no plans to turn this fantasy into reality. Most experts think Energia will continue to gather dust.

BURAN vs SHUTTLE

BURAN		SHUTTLE	
LENGTH	119 FT (36 M)	LENGTH	122 FT 9 IN (37 M)
WINGSPAN	78 FT (24 M)	WINGSPAN	78 FT 6 IN (24 M)
PAYLOAD BAY	15 x 60 FT (4.5 x 18 M)	PAYLOAD BAY	15 x 60 FT (4.5 x 18 M)
LOW-ORBIT PAYLOAD	33 TONS (30 TONNES)	LOW-ORBIT PAYLOAD	27 TONS (25 TONNES)
COMBINED LAUNCH WEIGHT	2,600 TONS (2,358 TONNES)	COMBINED LAUNCH WEIGHT	2,245 TONS (2,036 TONNES)

SOVIET ORBITER

Five years after Soviet President Leonid Brezhnev responded to the U.S. Shuttle challenge, Buran's first pilots started training. But Buran was not just another product of U.S.-Soviet space rivalry. There were also good scientific reasons for the Soviets to develop their own orbiter. Like the U.S. prototype, Buran avoided the waste of traditional rockets, most parts of which are discarded after launch, leaving astronauts to parachute down in capsules. In orbiters like the Shuttle, astronauts have more workspace and can deliver, reclaim and repair satellites and fly safely back to Earth.

A series of 24 test flights with a prototype Buran began in 1985. By then, rumors were already circulating in the West that the Soviets were working from blueprints of the U.S. Shuttle.

The two craft certainly were superficially similar. Both had a stubby, delta-wing shape. Both were designed to carry approximately 25 to 30 tons (22.5–27 tonnes) of payload into orbit. And both used thousands of ceramic tiles to shield them from the heat of reentry.

But there were differences. The U.S. Shuttle has its own engines, which help the two boosters lift it into orbit. Buran, with only small motors for maneuvering in orbit, rode piggyback on Energia ("Energy"), a huge rocket that could also be used separately for other missions. In addition, Buran could be flown and landed either by two pilots or on autopilot—a useful way to cut costs and risks for some operations.

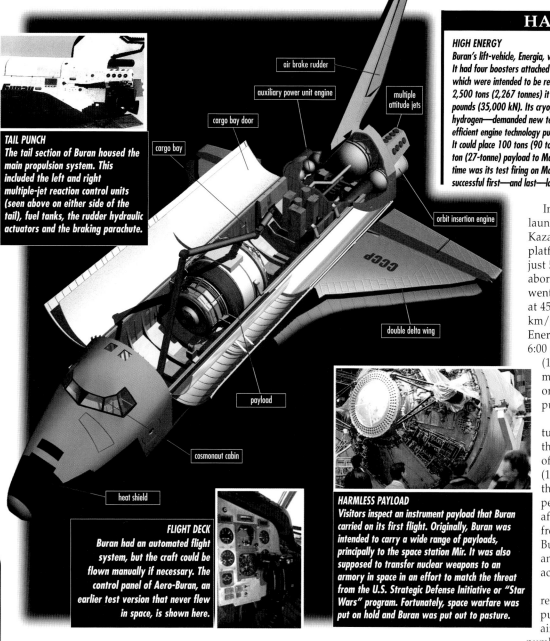

TAIL PUNCH
The tail section of Buran housed the main propulsion system. This included the left and right multiple-jet reaction control units (seen above on either side of the tail), fuel tanks, the rudder hydraulic actuators and the braking parachute.

air brake rudder

auxiliary power unit engine

multiple attitude jets

cargo bay door

cargo bay

orbit insertion engine

double delta wing

CCCP

payload

cosmonaut cabin

heat shield

FLIGHT DECK
Buran had an automated flight system, but the craft could be flown manually if necessary. The control panel of Aero-Buran, an earlier test version that never flew in space, is shown here.

TO CREW OR NOT TO CREW

There was much debate whether Buran's maiden flight would be crewed or uncrewed. Workers on the project, including cosmonauts, sent a letter to the government urging that the new spacecraft be launched with pilots. But a commission set up to investigate the matter ruled in favor of a 2-orbit automated flight.

HARD EVIDENCE

HIGH ENERGY
Buran's lift-vehicle, Energia, was the world's most powerful rocket. It had four boosters attached to a central core of two stages, which were intended to be reusable. With a launch weight of about 2,500 tons (2,267 tonnes) it produced a thrust of 7.8 million pounds (35,000 kN). Its cryogenic fuels—liquid oxygen and liquid hydrogen—demanded new techniques and materials. Its highly efficient engine technology put it years ahead of any other rocket. It could place 100 tons (90 tonnes) into Earth orbit, or hurl a 30-ton (27-tonne) payload to Mars. Yet it flew only twice. The first time was its test firing on May 15, 1987. The second was the successful first—and last—launch of Buran.

In the end, the flight was planned for launch from the Baikonur Cosmodrome, Kazakhstan, on October 29, 1988. But an access platform failed to retract quickly enough. With just 51 seconds to takeoff, the launch was aborted. The second attempt, on November 15, went according to plan. Despite snow flurries at 45 mph (72 km/h)—almost 10 mph (15 km/h) above the launch abort criteria—the Energia rocket blasted off on schedule at 6:00 a.m. Eight minutes later and 100 miles (160 km) up, Buran separated, fired its little maneuvering engines, and settled into orbit. Only then was the launch made public.

After the scheduled two orbits, Buran turned, fired its retrorockets and reentered the atmosphere. Emerging from 20 minutes of radio blackout, caused by the 2,900°F (1,600°C) heat of reentry, Buran coasted through a 40 mph (64 km/h) crosswind to a perfect touchdown three-and-a-half hours after takeoff. It landed just 12 feet (3.5 m) from the center line of Baikonur's runway. Buran's launch, orbital maneuver, deorbit and beautiful landing had all been accomplished without the help of a crew.

This impressive debut was a public relations triumph. Buran received huge publicity when it was displayed at the Paris airshow in June 1989. But its days were numbered. By 1991, the Soviet Union was no more, and its economy was in ruins. Lacking funds and projects, Buran was officially mothballed in 1993. Two other Burans under construction were left unfinished, and in 1995, they were dismantled.

HARMLESS PAYLOAD
Visitors inspect an instrument payload that Buran carried on its first flight. Originally, Buran was intended to carry a wide range of payloads, principally to the space station Mir. It was also supposed to transfer nuclear weapons to an armory in space in an effort to match the threat from the U.S. Strategic Defense Initiative or "Star Wars" program. Fortunately, space warfare was put on hold and Buran was put out to pasture.

MIR

Mir was the first permanent space station, a unique complex designed for expansion. Visiting spacecraft could attach as many as nine modules to it. With many docking compartments to choose from, no one could predict what would be added, or when, or where. In the end, the space station looked like the product of a bizarre engineering experiment in zero gravity. Despite Mir's many problems and near disasters, its relay of crews made it work astonishingly well—and for far longer than expected.

WHAT IF...

...MIR NEVER EXISTED?

Without Mir, we would still be tiptoeing our way into orbit. It gave us proof that spacemen and women—even those once on opposite sides of the Iron Curtain—could live and work together in space for long periods.

Mir also taught space engineers how to improvise. The Kvant 1 module, originally designed for Salyut 7, was "off-the-shelf" hardware. It had to be placed at Mir's "stern" so that it didn't block the docking adapter, where it would have limited further expansion. But, at the rear, it blocked the engines. So, for maneuvering, Mir came to depend on the engines of visiting Progress ferries. Kvant 2 improved living conditions, but it turned Mir into an unstable L-shape. The next unit, Kristall, gave it a stable T-shape. But the computer had to be upgraded to deal with the new, larger shape. This demanded more power: Solar panels were repositioned. So was Mir itself—and to save fuel, crews built a girder and placed a thruster on the end.

All this took years, and cosmonauts became expert mechanics, solving problems no one had even thought of. By the time the station was completed, the base-block, originally designed to last three years, was 10 years old.

Mir was also one long experiment in human space biology, providing a vast range of information on the effects of prolonged weightlessness: Valeri Polyakov's 14 months in

The Space Shuttle docks with the Soviet-built Mir—a landmark in space exploration. Such cooperative efforts should serve well toward Mir's successor, the International Space Station.

space is likely to stand as a record for many years. Mir also showed that it takes a month to settle in space, crews become exhausted if they work anything but a standard 24-hour day, and sleep periods decline to about five hours. Without Mir, we would never have known that it is possible to spend so long in space—and yet recover in a few weeks back on Earth.

Most importantly, the scientific experiments carried out by Mir provided vital data. Freed from Earth's atmosphere and gravity, instruments could monitor stars and galaxies at many different wavelengths. Human eyes, watching weather systems mature or a volcano erupt in 3-D, could often see what instruments could not. Alloys and vaccines made impossible by gravity were manufactured in space. Even food was produced in space: Wheat was grown and harvested, and fish survived in an aquarium.

Questions about the long-term effects of radiation remain, but Mir showed that people can live in space for a year—long enough to plan for followup scientific work on the International Space Station, and long enough for a round trip to Mars.

MIR TIMELINE

1986		**1990**		**JULY 17**	**KRISTALL MOVED TO FINAL POSITION**
FEBRUARY 19	MIR LAUNCHED	JUNE 10–11	KRISTALL ADDED		
MARCH 14	SOYUZ T-15 PROVIDES	**1995**		**1996**	
	FIRST CREW	MAY 27	KRISTALL MOVED TO NEW	APRIL 27	PRIRODA ADDED
1987			DOCK	**1999**	
APRIL 9	KVANT 1 ADDED	JUNE 2	SPEKTR ADDED	AUGUST 27	MIR EVACUATED
1989		JUNE 10	KRISTALL MOVED AGAIN		
DECEMBER 6–8	KVANT 2 ADDED				

COSMIC JALOPY

During its 14-year existence, Mir grew from a single core unit to a complex of six. At its heart was the central module, with a work room, an intermediate room and an adapter: a spherical 7-ft (2.1-m) unit with five docking ports. Cosmonauts lived and worked in the main room, 43 feet (13 m) long and 13.6 feet (4 m) across. It contained the control station, physical fitness machinery, cabins and two tables with special compartments for working in zero gravity.

In April 1987, after two crewed missions ensured that Mir was in good running order, it acquired its first addition—Kvant, an astrophysics research module. Without engines of its own, Kvant had a little "tug" that maneuvered it in orbit to its docking port. Instruments on the 19-foot (5.7-m), 22-ton (20-tonne) module included an X-ray observatory and powerful gyroscopes that would keep the station stable, so that the telescopes could remain fixed on faint objects.

Over the next 2 years and 8 months, Mir-Kvant received 20 visits, several of them international-crewed space missions. Progress craft—uncrewed space ferries—arrived every month or two, to bring new supplies and carry away trash.

The next add-on unit, Kvant 2, arrived in December 1989, turning Mir from a straight-line station into an L-shape. It brought extra power to make Mir more self sufficient: The station could now remove carbon dioxide and recycle water vapor and urine to make oxygen.

BON APPETIT

LIFE ON MIR MEANT CRAMPED QUARTERS, A LACK OF PRIVACY, AN INTENSE WORKLOAD AND POOR HYGIENE. BUT THE FOOD WAS GOOD. HERE, U.S. ASTRONAUT SHANNON LUCID (FAR RIGHT) AND HER COSMONAUT HOSTS CHECK OUT THE NEW SUPPLIES THAT LUCID BROUGHT IN MARCH 1996. IN 1988, FRENCHMAN JEAN-LOUP CHRÉTIEN ARRIVED WITH VEGETABLE SOUP, FISH, HAM, CHEESES AND PÂTÉS, AND THE RUSSIANS GREETED HIM WITH JELLIED SALMON, QUAIL AND CANDIED FRUIT.

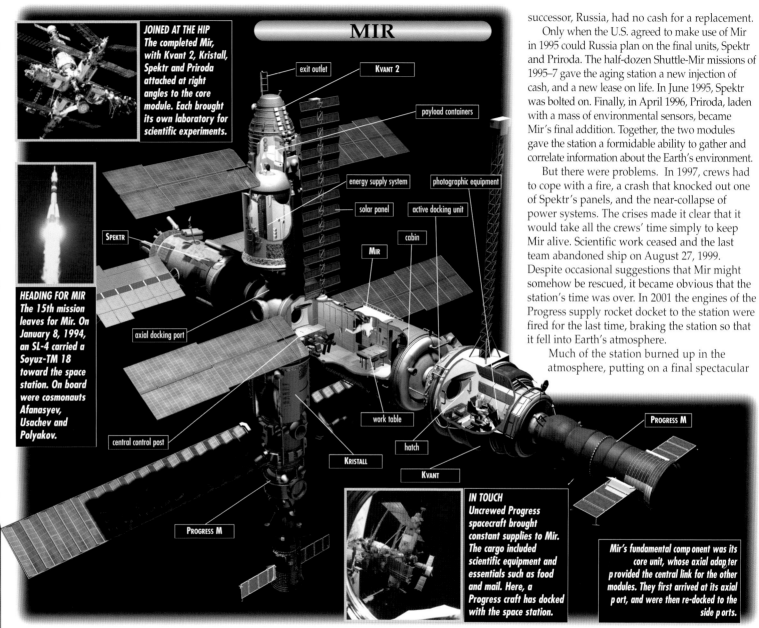

MIR

JOINED AT THE HIP
The completed Mir, with Kvant 2, Kristall, Spektr and Priroda attached at right angles to the core module. Each brought its own laboratory for scientific experiments.

HEADING FOR MIR
The 15th mission leaves for Mir. On January 8, 1994, an SL-4 carried a Soyuz-TM 18 toward the space station. On board were cosmonauts Afanasyev, Usachev and Polyakov.

exit outlet

KVANT 2

payload containers

energy supply system

photographic equipment

solar panel

active docking unit

SPEKTR

cabin

MIR

axial docking port

work table

central control post

hatch

KRISTALL

PROGRESS M

KVANT

PROGRESS M

IN TOUCH
Uncrewed Progress spacecraft brought constant supplies to Mir. The cargo included scientific equipment and essentials such as food and mail. Here, a Progress craft has docked with the space station.

Mir's fundamental component was its core unit, whose axial adapter provided the central link for the other modules. They first arrived at its axial port, and were then re-docked to the side ports.

HOME IMPROVEMENTS

In June 1990, the third module, Kristall, was attached. It was a specialist unit for research into the manufacture of materials for semiconductors. Its retractable and detachable solar panels upgraded the power supply.

But now Mir was reaching the limits of its capacity. Originally, it was to have been replaced around 1992 by Mir 2, but the Soviet Union's successor, Russia, had no cash for a replacement.

Only when the U.S. agreed to make use of Mir in 1995 could Russia plan on the final units, Spektr and Priroda. The half-dozen Shuttle-Mir missions of 1995–7 gave the aging station a new injection of cash, and a new lease on life. In June 1995, Spektr was bolted on. Finally, in April 1996, Priroda, laden with a mass of environmental sensors, became Mir's final addition. Together, the two modules gave the station a formidable ability to gather and correlate information about the Earth's environment.

But there were problems. In 1997, crews had to cope with a fire, a crash that knocked out one of Spektr's panels, and the near-collapse of power systems. The crises made it clear that it would take all the crews' time simply to keep Mir alive. Scientific work ceased and the last team abandoned ship on August 27, 1999. Despite occasional suggestions that Mir might somehow be rescued, it became obvious that the station's time was over. In 2001 the engines of the Progress supply rocket docket to the station were fired for the last time, braking the station so that it fell into Earth's atmosphere.

Much of the station burned up in the atmosphere, putting on a final spectacular show. Parts of the station – possibly as much as 25 tons (22.5 tonnes) – fell harmlessly into the Pacific Ocean in a target zone specifically chosen to eliminate any risk to people on the ground.

FOALE ON MIR

On May 17, 1997, the Shuttle Atlantis prepared to dock with Russia's Mir space station. Aboard Atlantis was Michael Foale, arriving to relieve fellow NASA astronaut Jerry Linenger. Mir's crew had coped with several emergencies in the preceding few months, but now the station was in good repair. The jinx seemed to have lifted. For the first month of Foale's tour of duty, all went smoothly—until a risky docking with a supply ship put the lives of Foale and his crewmates in serious danger.

WHAT IF...

...THE ISS HAS A SERIOUS ACCIDENT?

When Michael Foale and his Russian crewmates found themselves facing imminent disaster aboard the Mir space station, the first step in the emergency drill was to get the Soyuz TM "lifeboat" powered up and ready to leave as quickly as possible.

The Russians have been flying the Soyuz class of crewed spacecraft since the mid-1960s. The TM model can carry up to three people, and was developed specifically for the task of ferrying cosmonauts to and from Mir. It has been in continuous service since 1987. When Foale, Tsibliyev and Lazutkin came close to abandoning ship, the Soyuz rescue craft required just 10 minutes' preparation to get it ready for launch.

But what if the new International Space Station (ISS) suffers a collision? What kind of rescue vehicle will transport an injured astronaut back to Earth for emergency treatment, or evacuate the crew if there is a fire on board the ISS? The answer, for the first few years of the station's life, will be the trusty Soyuz TM. But the size of the ISS crew is projected to rise, and a new, larger vehicle will be needed. The hot favourite for the job was the X-38 spaceplane, the prototype of a new emergency Crew Return Vehicle, or CRV.

Accommodating up to seven passengers, the X-38 was similar in size to the Soyuz TM, but with some distinct differences. Whereas the Soyuz returned its human cargo to Earth by means of the

The first X-38 test vehicle, code-named Vehicle 131, is suspended in a hangar at Edwards Air Force Base, California.

traditional Russian parachute landing, the X-38 was to glide down from orbit like the Space Shuttle. Its crew would be able to choose a suitable site and then maneuver their vehicle toward it with a steerable parachute. The plane was to use skids rather than wheels on touchdown.

The X-38 had a successful first test flight in March 1998, but was cancelled before becoming operational. It is possible that if the Shuttle can return to operations, one might be available to act as a lifeboat or stage a rescue if ISS gets into trouble in the future. However, a dedicated escape or personnel transfer vehicle is a necessity for continued operations of the space station and one will have to be deployed sooner or later. In the meantime, the astronauts and cosmonauts of the International Space Station must put their faith in the tried and tested technology of the Soyuz TM.

DIARY OF DISASTER

FEBRUARY 23	FIRE BREAKS OUT ON BOARD	AUGUST 4	FURTHER FAILURE REPORTED IN OXYGEN GENERATORS
MARCH 7	OXYGEN GENERATOR FAILS		
APRIL 2	MAJOR LEAK IN COOLING SYSTEM	AUGUST 18	MAIN COMPUTER ON MIR SHUTS DOWN DURING DOCKING WITH PROGRESS SUPPLY SHIP; CREW TAKES CONTROL
JUNE 25	COLLISION WITH PROGRESS		
JULY 2	ELECTRONICS FAILURE IN GYROSCOPES		
JULY 13	STATION COMMANDER TSIBLIYEV REPORTS IRREGULAR HEARTBEAT	SEPTEMBER 8	MAIN COMPUTER FAILURE; SHIP-WIDE SYSTEMS TURNED OFF TO CONSERVE POWER
JULY 26	POWER OUTAGES AFTER CREW UNPLUGS WRONG CABLE	SEPTEMBER 16	CREW RETREATS TO SOYUZ LIFEBOAT DURING NEAR COLLISION WITH A U.S. SATELLITE

HIGH DRAMA

By the time he arrived on Mir, Michael Foale already had a great deal of spaceflight experience. Selected by NASA for astronaut training in June 1987, he had flown on three previous Shuttle missions. One of these was STS-63 in February 1995, the first Shuttle-Mir encounter in orbit. At the time, Foale had never thought of jumping ship. But little more than two years later, he was living and working on Mir with his Russian crewmates, station commander Vasily Tsibliyev and cosmonaut Sasha Lazutkin. Foale enjoyed the views of the Earth from the windows, which were larger than he had expected. He was also sleeping better than he had for years, thanks to zero gravity. His scientific work—cultivating plants and examining the behavior of gels in microgravity environments—was going well. Then, on the morning of June 25, 1997, everything changed.

Commander Tsibliyev had been ordered by Russian mission control to attempt a manual docking with a Progress supply ship. Tsibliyev had concerns about the operation—a few weeks earlier, the television link between Progress and Mir had failed. But orders were orders. As the test began, Tsibiliyev was at the helm in the core module, with Lazutkin behind him, checking the positions of Mir and the Progress. Foale was in the Kvant module, where the Progress was due to dock, using a laser to take range readings. The television link was fine, but its pictures revealed that the Progress was coming in much too fast. Lazutkin knew that Mir was in danger. He shouted to Foale: "Go to the Soyuz!"

BUDGET BLUES

RUSSIAN PRESIDENT BORIS YELTSIN LAID THE BLAME FOR THE COLLISION BETWEEN THE PROGRESS SUPPLY SHIP AND MIR SQUARELY ON STATION COMMANDER VASILY TSIBLIYEV. BUT TSIBLIYEV'S JOB WAS MADE DIFFICULT BECAUSE MIR'S AUTOMATED RADAR GUIDANCE SYSTEM WAS SWITCHED OFF. THE FORMER SOVIET REPUBLIC OF UKRAINE MANUFACTURED THE HARDWARE AND WAS ASSERTING ITS NEWLY-WON INDEPENDENCE BY OVERCHARGING RUSSIA FOR THE SYSTEM. SO RUSSIAN MISSION CONTROL—AND TSIBILIYEV—WERE ORDERED TO TRY DOCKING WITHOUT USING RADAR GUIDANCE.

MIR CAN TAKE IT

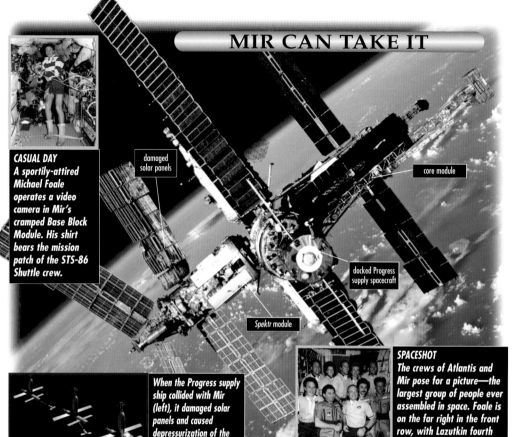

CASUAL DAY
A sportily-attired Michael Foale operates a video camera in Mir's cramped Base Block Module. His shirt bears the mission patch of the STS-86 Shuttle crew.

damaged solar panels

core module

docked Progress supply spacecraft

Spektr module

When the Progress supply ship collided with Mir (left), it damaged solar panels and caused depressurization of the Spektr module.

SPACESHOT
The crews of Atlantis and Mir pose for a picture—the largest group of people ever assembled in space. Foale is on the far right in the front row, with Lazutkin fourth from left and Tsibliyev second from left.

COLLISION IN SPACE

Foale started to float his way to the Soyuz TM, Mir's lifeboat. Before he could get there, he heard and felt a collision. The Progress had overshot Kvant and struck the Spektr module, damaging solar arrays in between. At that moment, Michael Foale wondered if he was about to die. But the hole in Spektr's hull was small: "There was no immediate rush of air out of my lungs, so I thought it must be a slow leak instead," Foale reported later.

With Lazutkin's help, Foale isolated the damaged module. But the collision had sent Mir off balance. The station was tumbling slowly and its solar power arrays were missing vital sunlight. Foale had to find a low-tech means of calculating Mir's rate of spin. He blocked out a star with his thumb and counted how long it took to reappear. The crew then climbed into the Soyuz and fired its thrusters to cancel the spin, putting the solar panels in full sunlight. Mir flickered back to life.

Foale went back to his experiments, though systems failures continued to plague Mir. He finally left the station in October, 1997, riding home aboard Atlantis. But his appetite for spaceflight was still strong—Foale has undertaken six missions into space and holds the US record for time spent in orbit – 374 days.

MISSION DIARY: FOALE'S MISSION ON MIR

MAY 15, 1997 SHUTTLE ATLANTIS LIFTS OFF FROM CAPE CANAVERAL. ITS CREW INCLUDES MICHAEL FOALE AND YELENA KONDAKOVA, VETERAN OF A FIVE-MONTH STAY ON MIR.
MAY 17 ATLANTIS DOCKS WITH MIR IN THE SIXTH LINKUP WITH A NASA SHUTTLE.
MAY 22 JERRY LINENGER, U.S. GUEST COSMONAUT ON MIR SINCE JANUARY, SWAPS PLACES WITH MICHAEL FOALE.
JUNE 13 IN AN INTERVIEW, FOALE (SHOWN ABOVE IN A SHUTTLE FLIGHT SIMULATOR WITH HIS WIFE) REPORTS: "IT'S A PRETTY GOOD LIFE...EVERYTHING'S GOING RATHER WELL."

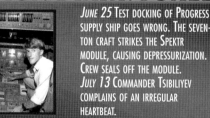

JUNE 25 TEST DOCKING OF PROGRESS SUPPLY SHIP GOES WRONG. THE SEVEN-TON CRAFT STRIKES THE SPEKTR MODULE, CAUSING DEPRESSURIZATION. CREW SEALS OFF THE MODULE.
JULY 13 COMMANDER TSIBILIYEV COMPLAINS OF AN IRREGULAR HEARTBEAT.
AUGUST 4 OXYGEN GENERATOR FAILURE FORCES CREW TO USE OXYGEN "CANDLES."
AUGUST 7 RELIEF COSMONAUTS ANATOLY SOLOVYEV AND PAVEL VINOGRADOV ARRIVE ON MIR.
AUGUST 14 COSMONAUTS TSIBLIYEV AND LAZUTKIN LEAVE MIR AND RETURN TO EARTH.

AUGUST 18 COMPUTER FAILURE DURING PROGRESS DOCKING MANEUVER.
SEPTEMBER 6 FOALE AND SOLOVYEV CONDUCT SIX-HOUR SPACEWALK TO INSPECT EXTERIOR OF SPEKTR MODULE.
SEPTEMBER 27 ATLANTIS DOCKS AT MIR ONCE AGAIN, BRINGING FOALE'S RELIEF ASTRONAUT, DAVID WOLF.
OCTOBER 3 ATLANTIS UNDOCKS FROM MIR AND FLIES AROUND THE STATION TO INSPECT SPEKTR MODULE (LEFT).
OCTOBER 6 FOALE RETURNS TO EARTH AFTER 144 DAYS IN SPACE (ABOVE).

SHUTTLE AND MIR LINKUP

Looking through the overhead windows, astronaut Robert Gibson could see the Mir space station above him. The docking assembly was 10 feet away in the payload bay and out of his line of sight. Using observations from the crew of both craft, along with TV cameras positioned around the Shuttle Atlantis, he closed in and gently nudged the thrusters to make contact. After some anxious moments, history was made. For the first time in 20 years, an American and a Russian spacecraft were linked in Earth orbit.

WHAT IF...

...COOPERATION IN SPACE CONTINUES?

STS-71 came at the start of "phase one" in the construction of the planned new International Space Station (ISS). To NASA's delight, post-Soviet Russia had agreed to help the Americans get their long wished-for space station off the drawing board and into orbit. Soon afterward, 14 other countries signed up for a share in building the station. But there was a lot to be learned before the first modules could be launched.

Phase one was to see U.S. astronauts welcomed aboard Mir in a program of long-term stays, trial run dockings and rendezvous procedures. The practical benefit of all this training to NASA, which lacked Russian experience of long-duration spaceflight, was enormous. But the Americans had to learn fast. The new station (the first sections of which have already been launched) will require 33 Shuttle and 12 Russian missions over five years to deliver and assemble 100 different station components and additional supplies. It will take a staggering 960 hours of daring EVAs to manually bolt together the different pieces—a tall order by any standards. And this is only the beginning.

Once operational, the Russians and Americans will need to ferry new station crew, fresh supplies and scientific experiments to and from Earth as a matter of routine. As far as possible—as on Mir—consumables such as air and water will be

A new era of cooperation: The Shuttle's payload bay camera captures NASA's Unity module as it is lifted into the upright position for mating to the Russian Zarya module during construction of the International Space Station.

recycled, but the logistics involved in maintaining the station are still far in advance of any previous NASA project. Without successful missions such as STS-71, it is inconceivable that the Americans could have been this ambitious.

In one respect the ISS will be very different from a Shuttle-Mir docking—the formal ceremonies and gift exchanges that are familiar to foreign visitors aboard Mir will cease. The station will become a truly international base, where 16 different countries will work alongside each other in space. But as cosmonauts and astronauts spin together around the globe, 200 miles (320 km) above their homelands and completing an entire orbit in just 90 minutes, national differences will probably be the last thing on their minds.

STS-71 MISSION STATS

MISSION	STS-71, 69TH SHUTTLE MISSION; 100TH U.S. HUMAN SPACEFLIGHT LAUNCHED FROM FLORIDA; FIRST SHUTTLE-MIR DOCKING MISSION	CREW COMMANDER	ROBERT L. "HOOT" GIBSON (CAPT. USN), 48, 5TH FLIGHT
		PILOT	CHARLES J. PRECOURT (LT. COLONEL, USAF), 39, 2ND FLIGHT
LAUNCH VEHICLE	SPACE SHUTTLE ATLANTIS OV104 (14TH FLIGHT)/SRB BI-072 /20 ET-70/SSME 2028 (#1), 2034 (#2), 2032 (#3)	MISSION SPECIALIST 1	ELLEN S. BAKER (MD), 42, PAYLOAD COMMANDER, 3RD FLIGHT
		MISSION SPECIALIST 2	GREGORY J. HARBAUGH (CIVILIAN), 34, FLIGHT ENGINEER, 3RD FLIGHT
LAUNCH SITE	LC39A, KENNEDY SPACE CENTER, FLORIDA	MISSION SPECIALIST 3	BONNIE J. DUNBAR (PHD), 46, 4TH FLIGHT

SPACE DOCK

The STS-71 mission launched from the Kennedy Space Center on June 27, 1995, with five astronauts and two cosmonauts—the Mir 19 relief crew—aboard. The flight was to be the first in a planned three-year series of Shuttle-Mir dockings, in preparation for the day when the presence of the International Space Station will make such maneuvers routine. Just as in the historic docking of Apollo and Soyuz in 1975, the Shuttle Atlantis carried a docking mechanism that had been specially constructed for the task. Located in the orbiter's Payload Bay, and linked by a tunnel to the crew cabin, the crew had high hopes that it would prove its space worthiness.

Two days after launch, Atlantis approached Mir. As the two craft closed in, Shuttle commander Gibson carefully guided the orbiter upward from its position below the space station and brought it to a flawless docking with Mir's Kristall Module docking port. Together, the U.S. and the Russians had created the largest spacecraft ever.

RUSSIAN WELCOME

About 90 minutes after the successful linkup, the hatches were opened and Gibson and his Mir 18 counterpart, Vladimir Dezhurov, warmly greeted each other in the docking tunnel. It was the start of five days of

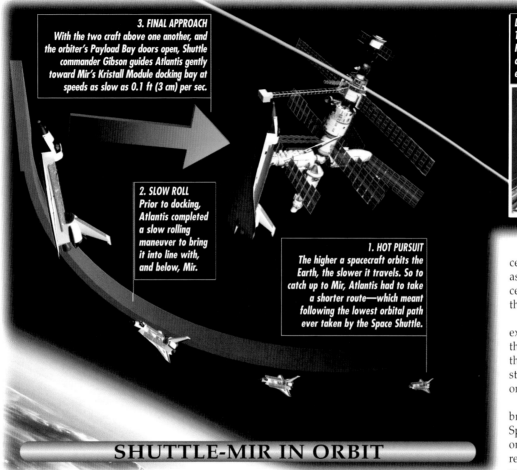

3. FINAL APPROACH
With the two craft above one another, and the orbiter's Payload Bay doors open, Shuttle commander Gibson guides Atlantis gently toward Mir's Kristall Module docking bay at speeds as slow as 0.1 ft (3 cm) per sec.

2. SLOW ROLL
Prior to docking, Atlantis completed a slow rolling maneuver to bring it into line with, and below, Mir.

1. HOT PURSUIT
The higher a spacecraft orbits the Earth, the slower it travels. So to catch up to Mir, Atlantis had to take a shorter route—which meant following the lowest orbital path ever taken by the Space Shuttle.

SHUTTLE-MIR IN ORBIT

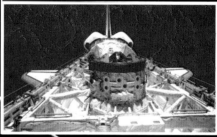

DOCKING TUNNEL
The Orbiter Docking System is installed in the Shuttle's Payload Bay. It is linked by a docking tunnel to the crew cabin airlock and Spacelab airlock positioned at either end of the bay.

celebrations and joint activities for the 10 astronauts and cosmonauts aboard. But ceremonial toasts and gift exchanges aside, there was much hard work to be done.

In the Shuttle's Spacelab module, an extensive program of medical examinations on the Mir 18 crew began. This was the first time that the Americans had had the chance to study the effects of extended weightlessness on the human body since Skylab in the 1970s.

At the end of the 10-day mission, Gibson brought Atlantis to a safe landing at Kennedy Space Center on July 7. It was good to be back on Earth. But for the crew of Mir 18, their return marked the beginning of yet another round of medical tests.

ALIENS LAND!

AFTER THE MIR 18 CREW WERE LAUNCHED FROM KAZAKHSTAN ON MARCH 18, IT WAS REALIZED THAT THE TWO RUSSIAN COSMONAUTS HAD NOT BEEN ISSUED WITH U.S. ENTRY VISAS FOR THEIR SCHEDULED RETURN TO EARTH ONBOARD ATLANTIS. HURRIED COMMUNICATIONS RESULTED IN A VISA WAIVER BEING ISSUED FOR THE FIRST TIME FOR "ALIENS FROM OUTER SPACE."

MISSION DIARY: STS-71

April 26 ATLANTIS SITS ON PAD 39A. PLANNED FOR MAY, THE LAUNCH SLIPS BACK TO THE THIRD WEEK OF JUNE.
June 23 LAUNCH IS SCRUBBED DUE TO BAD WEATHER; RESCHEDULED FOR NEXT DAY, THEN SCRUBBED AGAIN. THE SHUTTLE CREW, BACKUP CREW, AND MIR 19 RELIEF CREW, WAIT ANXIOUSLY.
3:32 P.M. EST June 27 STS-71 FINALLY LAUNCHES FROM PAD 39A IN HOT PURSUIT OF THE ORBITING MIR.
8:00 A.M. June 29 ATLANTIS DOCKS WITH THE KRISTALL MODULE ON MIR WHILE SUSPENDED 216 NAUTICAL MILES (400 KM) ABOVE THE LAKE BAIKAL REGION OF THE RUSSIAN FEDERATION. THE DOCKING SYSTEM WORKS PERFECTLY. ONCE PRESSURE CHECKS HAVE BEEN MADE, THE DOCKING HATCHES ARE

OPENED AND THE U.S. AND RUSSIAN CREWS BEGIN THEIR JOINT MISSION.
June 30 THE CREWS EXCHANGE GIFTS, AND THEN SET TO WORK. HALF A TON OF WATER, 53 LB (24 KG) OF OXYGEN AND 80 LB (36 KG) OF NITROGEN IS TRANSFERRED TO MIR. OTHER WORK INVOLVES SUBJECTING THE MIR 18 CREW TO SEVEN DIFFERENT KINDS OF MEDICAL INVESTIGATION—CARDIOVASCULAR FUNCTIONS, HUMAN METABOLISM, NEUROSCIENCE, HYGIENE, SANITATION AND RADIATION, AND BEHAVIORAL PERFORMANCE AND BIOLOGY.
3:32 P.M. July 3 THE FAREWELL CEREMONY OVER, MIR CLOSES ITS HATCH.
3:48 P.M. July 3 ATLANTIS BOLTS ITS HATCH AND BEGINS

DEPRESSURIZING THE TUNNEL IN PREPARATION FOR UNDOCKING.
7:10 A.M. July 4 ATLANTIS UNDOCKS FROM MIR. THE RUSSIAN SPACECRAFT IS LEFT TEMPORARILY UNOCCUPIED AS THE RECENTLY ARRIVED MIR 19 COSMONAUTS (RIGHT, PICTURED WHILE STILL ON ATLANTIS) UNDOCK THEIR SOYUZ TM TO RECORD THE SHUTTLE'S DEPARTURE.
10:54 A.M. July 7 ATLANTIS GLIDES BACK TO A SAFE LANDING AT THE KENNEDY SPACE CENTER, WITH MIR 18 VISITOR NORMAN E. THAGARD ABOARD.

MIR: COOPERATION AND CRISIS

T he Mir space station set the standard for long-duration spaceflight. Slated for a five-year stay in Earth orbit, Mir was to last 13 years, outliving the Communist regime and playing host to guest cosmonauts from around the world. Western astronauts got their first taste of space station life aboard the former flagship of the Soviet space fleet, and gained much valuable information from it. But they sometimes got more than they bargained for: Technical breakdowns made Mir a dangerous place to be.

WHAT IF...

...WE COULD LEARN LESSONS FROM MIR?

T he idea of a space station, a permanent home in Earth orbit, was discussed in 1883 by Konstantin Tsiolkovsky, the founding father of Russian spaceflight, in a book called Free Space. Tsiolkovsky wrote about the engineering challenges of such an enterprise, but he also wondered about the adaptation of the human body to prolonged periods of weightlessness. A century later, the Russian space station Mir was addressing that very question. The Salyut series of space habitats had given Soviet cosmonauts the chance to experience microgravity, but the ever-increasing duty tours on Mir pushed that experience even further. Medical study revealed that the body takes a full month to adapt, during which time the heart shifts slightly farther up the chest. When cosmonauts came back to Mir for another stay, they found it much easier to adapt—as if the body remembered the adaptation process.

It was discovered during long-duration stays on Mir that crews needed to stick to the familiar solar day, a 24-hour cycle of work and sleep. Early attempts to tailor crew sleep periods to make the best use of time in radio contact were abandoned. The cosmonauts simply grew more and more tired, and were much happier when Mission Control reverted to the solar day.

Long stays in orbit sometimes caused psychological problems. Some Mir residents complained of depression, including NASA

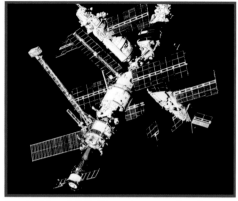

From 1995 through 1998, Mir was the temporary home for seven NASA astronauts. Their research will prove invaluable to the future residents of the International Space Station.

guest John Blaha. Another American, Jerry Linenger, had serious difficulties in his relationships with his Russian crewmates. The differences were cultural, and exacerbated by the stress of maintaining Mir during a period of frequent technical breakdowns. All Mir crew missed their families and friends, of course, as well as fresh food, fresh air and the pleasures of life on Earth.

Keeping a human environment orbiting in the vacuum of space for 13 years was one of the great technological achievements of the 20th century. The lessons learned from the life of Mir are being put to use in the development of the International Space Station (ISS). Russian, American, Japanese, Canadian and European elements will combine to create a working space habitat. But the Russians will be justified in saying, "We did it first."

THE MIR MODULES

Module	Launch Date	Function	Spektr	May 1995	Earth observation laboratory
Core module	February 1986	Station base block			
Kvant 1	March 1987	High energy observatory	Priroda	April 1996	Earth remote sensing
Kvant 2	November 1989	Life-support, EVA airlock			
Kristall	May 1990	Materials research laboratory	Number of NASA guest cosmonauts		7
			Number of non-U.S. guest cosmonauts		26

END OF AN ERA

When the last crew of Mir cosmonauts undocked from the station in August 1999, an eventful chapter in spaceflight history came to a quiet end. During its 13-year working life, Mir had endured a catalog of crises, including a fire, a collision with a supply ship, and numerous technical failures. But it survived—and fulfilled its mission with distinction, serving as a space habitat for a generation of cosmonauts. And it was fitting that Sergei Avdeyev, flight engineer on the last Mir crew, became the most traveled man in history during his last tour on the station, after clocking up more than two years' flight time over three Mir missions.

The Mir core module was launched from Kazakhstan in February 1986. The Soviet Union already had a space station in orbit, Salyut 7, so the timing of the launch took Western observers by surprise. On March 13, a Soyuz craft left Earth to dock with Mir, and the crew made themselves at home. After doing so they visited Salyut 7, to remove some equipment, and returned to Mir. This is the only example to date of crew transfer between two space stations.

By 1991, Mir had acquired three additional modules, providing laboratories for materials research and Earth observation. But on the ground, political geography was now the focus of interest. The Soviet Union had broken up, and Russia became the proud owner of a mighty space program.

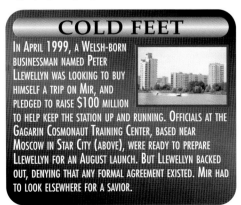

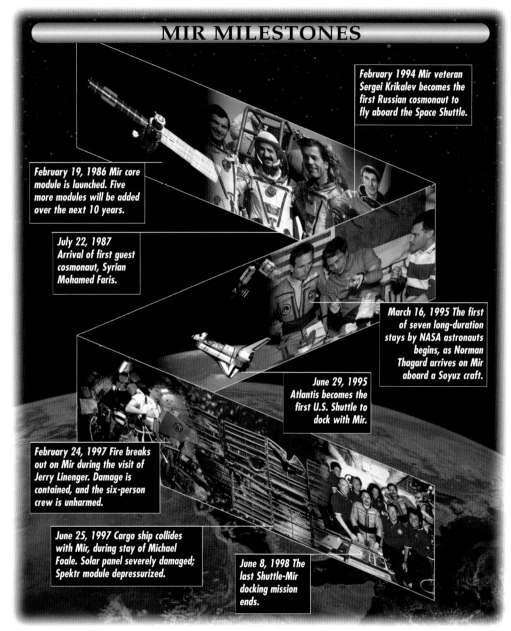

MIR MILESTONES

February 19, 1986 Mir core module is launched. Five more modules will be added over the next 10 years.

July 22, 1987 Arrival of first guest cosmonaut, Syrian Mohamed Faris.

February 1994 Mir veteran Sergei Krikalev becomes the first Russian cosmonaut to fly aboard the Space Shuttle.

March 16, 1995 The first of seven long-duration stays by NASA astronauts begins, as Norman Thagard arrives on Mir aboard a Soyuz craft.

June 29, 1995 Atlantis becomes the first U.S. Shuttle to dock with Mir.

February 24, 1997 Fire breaks out on Mir during the visit of Jerry Linenger. Damage is contained, and the six-person crew is unharmed.

June 25, 1997 Cargo ship collides with Mir, during stay of Michael Foale. Solar panel severely damaged; Spektr module depressurized.

June 8, 1998 The last Shuttle-Mir docking mission ends.

SHARING THE COST

Unfortunately, it could not afford Mir's running costs, estimated at somewhere between $100 million and $250 million a year.

Buran, the Soviet space shuttle, was scrapped, and future Mir development put on hold. NASA, meanwhile, had a Shuttle, but no station. It was financially sensible, and diplomatically advantageous, for Russians and

Americans to pool resources. NASA launched a series of long-term missions on Mir by its astronauts. These were known collectively as Phase One, referring to their value as forerunners of American occupancy of the International Space Station (ISS). The benefits flowed both ways, since Russian cosmonauts were assigned to Space Shuttle missions. And veteran Mir flight engineer Sergei Krikalev will be a member of the first crew to occupy the ISS.

But the new relationship between the former "space race" rivals was sorely tested in 1997, when a sequence of near-disasters threatened Mir. First, in February, a flash fire broke out when an oxygen generator malfunctioned in the Kvant 1 module. With six people on board, including American Jerry Linenger, both of the docked Soyuz craft would have been needed if the order came to abandon ship. But the fire had cut off the route through the station to one of these craft. Fortunately, the fire was contained, but confidence in the aging Mir had been dealt a serious blow.

When Linenger's hazardous stay was over, some U.S. politicians argued that he shouldn't be replaced. But Michael Foale arrived on Mir in May, as scheduled, and reported soon after, "It's a lot easier than I expected." Foale proved to be spectacularly wrong weeks later when a Progress supply vessel slammed into the station, depressurizing his sleeping quarters in the Spektr module. But Foale, his crewmates, and the Phase One project survived.

BUILDING THE ISS

The International Space Station, successor to Mir, is over budget and behind schedule, but given the enormity of the task this is not too surprising. ISS is made up of several modules connected together (and designed to be swapped out and replaced if the need arises). It was projected that over 80 Shuttle and rocket flights would be necessary to construct the complete station, but with the suspension of Shuttle flights and other difficulties it is proving difficult to keep work on the ISS going at all.

The station is currently operating with a reduced crew and is being very slowly expanded. Many more modules were planned and may some day be launched, but at the moment just keeping the station going day to day is proving to be a big job.

WHAT IF...

...SPACE STATIONS WERE BUILT DIFFERENTLY?

Half the size of Manhattan, a 22nd-century space station attracts millions of tourists each year. But tiny, automated stations perform more research.

Some engineers argue that the current ISS plans show how not to build a space station. With so many components—each of them vital—a single launch failure could stall the whole project. It would be much better, they claim, if far fewer but larger sections were launched by really powerful boosters—rockets with the payload capacity of the old Saturn 5 or the abandoned Russian Energia could have lifted most of the space station all at once.

To squeeze more cargo into today's more modest launch packages, engineers could use inflatable structures—orbiting balloons that would accommodate people and equipment inside. Made from multiple layers of Kevlar, the fabric used in bulletproof vests, these inflatables would be more resistant to micrometeor impacts than the thin metal skin of current ISS modules.

The ISS has deliberately been built to make maximum use of weightlessness for research. Future stations could well have a different purpose—as transit stations for interplanetary voyages, say, or as tourist hotels. Artificial gravity would be very desirable, so engineers might well construct such a station in the form of a large spinning wheel. Centrifugal force would generate comfortable gravity levels, but the station would have to be big and spin slowly. A faster-spinning, smaller station could produce the same g forces, but would have violently unpleasant side effects on its occupants.

A system of tethers could also provide spin gravity. Modules would be attached by long cables to a central spinning hub, with no need for an expensive wheel. But it would be difficult for astronauts to move from one module to another, and the tethers would have to be strong enough to withstand impacts from small meteoroids—a broken tether would hurl modules off into deep space on unpredictable trajectories.

Huge stations might turn out to be an expensive mistake, certainly from a scientific viewpoint. Much research is best performed in small, automated stations. In a crewed station, there are always people moving, life-support systems whirring and jolts from frequent spacecraft docking. All these activities cause wobbles that disturb delicate zero-g experiments. Without people, the work would go more smoothly, and scientists could monitor it in comfort from the Earth. Of course, they would miss the excitement of being in space.

ISS COMPONENTS

NAME	PURPOSE	LAUNCH DATE
ZARYA	EARLY ORBIT CONTROL & POWER	NOVEMBER 1998
UNITY	CONNECTOR MODULE	DECEMBER 1998
ZVEZDA	SERVICE MODULE (LIVING QUARTERS, ORBIT CONTROL)	JULY 2000
PHOTOVOLTAIC MODULE	SOLAR PANELS FOR POWER	OCTOBER 2000
PRESSURISED MATING ADAPTER	DOCKING PORT	
	SUPPORT TRUSS	
DESTINY LABORATORY MODULE	RESEARCH LABORATORY	FEBRUARY 2001

PLUG AND PLAY

Like a giant model kit, the International Space Station is to be assembled piece by piece in Earth orbit. Back in 1973, the U.S. lofted the Skylab station—and everything it needed for its two-year life—on one Saturn 5 rocket. But with today's smaller, cheaper boosters, ISS must be launched a section at a time over several years.

In its original designed form, the ISS was to comprise more than 100 major components ranging from complete research laboratories to radio antennas. The parts, hauled into orbit by U.S. Space Shuttles and Russian boosters, are put together in space by spacewalking astronauts, aided by robot arms controlled from inside the ISS and the Space Shuttle.

It is likely that the design of the ISS will be revised as new launch vehicles become available, though many of the modules were under construction or at least into late design work when shuttle flights were suspended. These modules will likely form part of the final station, whatever its form. The Canadian-built Space Vision System provides computer graphics views of what the robot arms are doing, even when they are out of sight. The robot arms are fitted with grapple devices at both ends so that they lock on to any part of the ISS. The biggest arm, the Space Station Remote Manipulator System (SSRMS), will eventually be attached to a mobile platform that can travel the whole length of the station's structure.

Dressed in either the Russian Orlan-M spacesuit or an upgraded Shuttle spacesuit, astronauts and cosmonauts are scheduled to make about 160 spacewalks during the construction period—more than have been performed in the entire history of spaceflight. And ISS spacewalks could be tough duty: The space station cannot be turned from shade into sunlight as easily as the Shuttle's cargo bay, and conditions may be much colder or darker than on a Shuttle mission.

The astronauts will move around ISS with the help of extending poles similar to those used on Mir, robot arms and open trolleys that run along the main ISS girderwork structure. As the world's first space construction gang, they will use a host of manual and power tools to connect components and their numerous cables and pipes. A rocket called SAFER is attached to each astronaut's backpack as the equivalent of a lifejacket in water. Astronauts will wear safety tethers, but should a tether break, SAFER has enough fuel to bring them back to the ISS. The astronauts will have robotic assistants for their spacewalks—the NASA AERCam and German Inspector. These devices float nearby and provide TV views for spacewalk controllers.

The work crew will need all the help they can get. Every piece has to fit as planned and the whole ISS has to work the first time: The station cannot be tested before it is built.

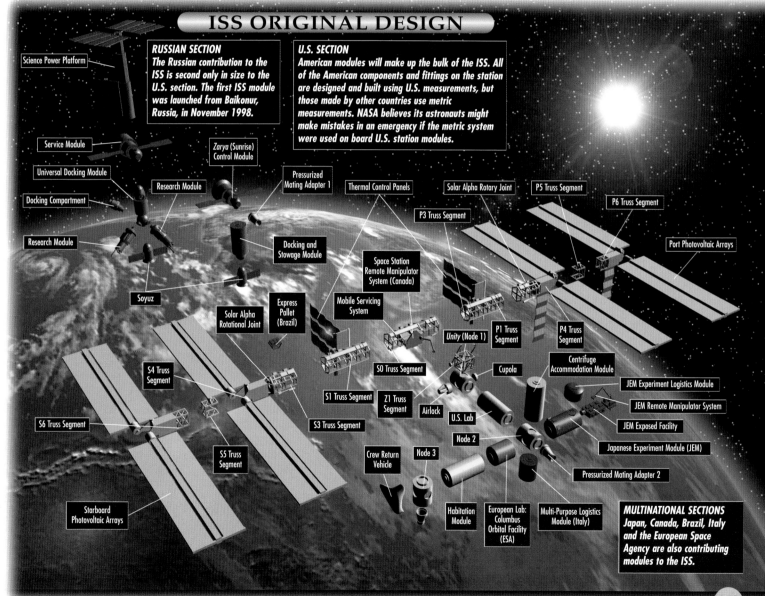

ISS ORIGINAL DESIGN

RUSSIAN SECTION
The Russian contribution to the ISS is second only in size to the U.S. section. The first ISS module was launched from Baikonur, Russia, in November 1998.

U.S. SECTION
American modules will make up the bulk of the ISS. All of the American components and fittings on the station are designed and built using U.S. measurements, but those made by other countries use metric measurements. NASA believes its astronauts might make mistakes in an emergency if the metric system were used on board U.S. station modules.

MULTINATIONAL SECTIONS
Japan, Canada, Brazil, Italy and the European Space Agency are also contributing modules to the ISS.

Science Power Platform
Service Module
Universal Docking Module
Docking Compartment
Research Module
Research Module
Soyuz
Zarya (Sunrise) Control Module
Pressurized Mating Adapter 1
Thermal Control Panels
Solar Alpha Rotary Joint
P5 Truss Segment
P6 Truss Segment
P3 Truss Segment
Port Photovoltaic Arrays
Docking and Stowage Module
Space Station Remote Manipulator System (Canada)
Solar Alpha Rotational Joint
Express Pallet (Brazil)
Mobile Servicing System
P1 Truss Segment
P4 Truss Segment
Centrifuge Accommodation Module
S4 Truss Segment
Unity (Node 1)
SO Truss Segment
Cupola
JEM Experiment Logistics Module
JEM Remote Manipulator System
S1 Truss Segment
Z1 Truss Segment
Airlock
U.S. Lab
JEM Exposed Facility
S6 Truss Segment
S3 Truss Segment
Node 2
Japanese Experiment Module (JEM)
Crew Return Vehicle
Node 3
Pressurized Mating Adapter 2
S5 Truss Segment
Starboard Photovoltaic Arrays
Habitation Module
European Lab: Columbus Orbital Facility (ESA)
Multi-Purpose Logistics Module (Italy)

INTERNATIONAL SPACE STATION

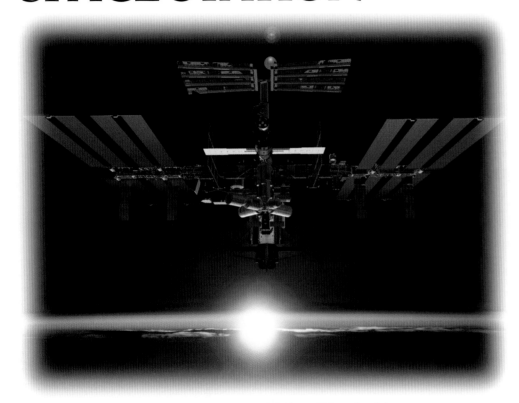

The International Space Station (ISS) will be the biggest space structure ever built. In its low Earth orbit, 220 miles (355 km) up, it should be easy to see in the night sky. The ISS as designed was to be assembled in orbit from components flown up on 33 space shuttle flights and 12 Russian Proton and Soyuz missions. The station will provide living quarters, workshops and laboratories for astronauts from the U.S. and nations around the world throughout its working life, and may be upgraded and adapted with the replacement or addition of modules.

WHAT IF...

...THE ISS FELL TO EARTH?

NASA's biggest struggle so far has been getting the ISS into space, so it has not officially commented on what will happen to the station at the end of its working life. One thing is certain, though—it cannot just be abandoned.

At an altitude of 220 miles (355 km) from Earth, the atmosphere is extremely thin. But there is still enough air to create drag on the orbiting ISS. If the station were to be abandoned, this drag will gradually slow it down over the years until its speed was too low to keep it in orbit. The massive structure would spiral down through the atmosphere toward the Earth. It would start to burn and break up, but it is so big that most of its 475-ton (430-tonne) bulk would remain intact until it hit the Earth. Fragments might come down in the Atlantic or the Pacific, but because the station's orbit takes it across the homes of 95% of the world's population, there is a good chance that some or all of them would slam down onto a town or city and cause serious damage and loss of life.

There are several ways for project managers to avoid such a disaster. For example, the ISS could be boosted into a higher orbit by its on-board engines. To defer the problem for a few decades, it would be enough to move the station into an orbit around 600 miles (950 km) up. During that time, the defunct station could be useful to engineers who want to study the long-

If ISS were abandoned after its nine-year design life, it could be a real hazard to people on Earth. Its orbit would decay after only a few more years, bombarding the planet with massive pieces of debris.

term effects of the space environment on structures.

Another option is to dismantle ISS almost down to the last nut and bolt. Individual components could then be nudged from orbit one at a time, to burn up safely in the atmosphere over an empty ocean area.

Less wastefully, whole modules—or at least any still-functioning equipment—could be used as part of another space station. After all, the most expensive item in any space budget is the thousands of dollars it costs to loft every pound from Earth into orbit; ISS would already be up there, and even as scrap metal for recycling, it would be worth a lot of money.

The best solution might be to keep ISS in service almost indefinitely. Components could be repaired, upgraded or replaced as needed, and the engines refueled whenever it was necessary to adjust a flagging orbit. If the owner nations had no more need for it, the station could even be leased out as a purely private manufacturing and research facility, or even the first orbital hotel. ISS may never wear out.

ISS SPECIFICATIONS

CREW	3 (INITIAL) 7 (FINAL)	PLANNED LIFE	9+ YEARS AFTER FINAL ASSEMBLY
MASS	475 TONS (430 TONNES)	ASSEMBLY COST	$37 BILLION
DIMENSIONS	356 FT X 290 FT (108 X 88 M) (INCLUDING SOLAR ARRAYS)	OPERATING COST	$13 BILLION FOR 9 YEARS OF OPERATIONS
ORBIT	220 MILES (355 KM), INCLINED 51.6° TO EQUATOR	COUNTRIES INVOLVED	U.S., RUSSIA, JAPAN, CANADA, BRAZIL, FRANCE, GERMANY, ITALY, BELGIUM, DENMARK, THE NETHERLANDS, NORWAY, SPAIN, SWEDEN, SWITZERLAND
TIME TO ASSEMBLE	6 YEARS (INCLUDING 930 HRS OF U.S. SPACEWALK TIME)		

SCIENCE IN ORBIT

The crew of a Space Shuttle arriving at the newly completed International Space Station (ISS) would be greeted by a massive structure 356 feet (108 m) long and 290 feet (88 m) wide, nearly the size of two football fields. The station is currently behind schedule and the Shuttle may or not continue flying, but the dream is still alive—ISS has a crew aboard and is carrying out its mission as well as possible under the circumstances.

Clustered at the center of this sprawling space complex are the modules containing laboratories, workshops and the living quarters for up to seven crew members. The work of the crew involves scientific, medical and technological research that can only be carried out in the near-weightless conditions of orbit. It includes studies of the human body to find new ways to prevent and treat diseases, and the development of new types of materials, including semiconductor crystals, plastics and drugs.

The station is also an excellent platform for observing the Earth, because its orbit takes it over 85% of the planet's surface. From their high vantage point, scientists study weather patterns, land usage, the spread of deserts and the destruction of rain forests.

Crews are taken to and from the ISS by U.S. Space Shuttles and Russian Soyuz ferry craft. Supplies and propellants are delivered by Shuttle flights and by unmanned ferries including Russian Progress vehicles, the European ATV and the Japanese HTV.

LIFE-SUPPORT SYSTEMS

On board the station, the life-support systems maintain "shirt-sleeve" conditions for the comfort of the crew. Water is used for drinking and washing and is also electrolyzed to produce oxygen for breathing. This oxygen is mixed with nitrogen—also delivered to the station—to create a fair approximation of the air back home on Earth.

The temperature and humidity of the air are regulated by air conditioners, and the air is circulated around the ISS by fans. Molecular

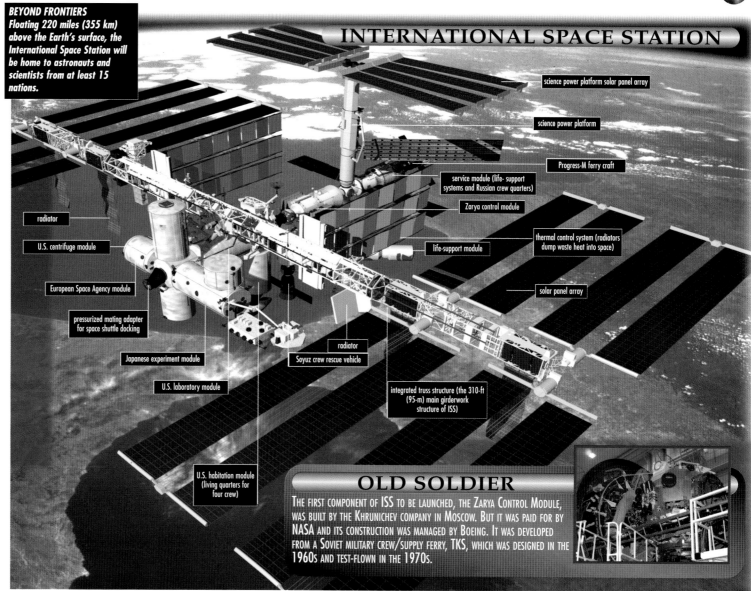

BEYOND FRONTIERS
Floating 220 miles (355 km) above the Earth's surface, the International Space Station will be home to astronauts and scientists from at least 15 nations.

INTERNATIONAL SPACE STATION

- science power platform solar panel array
- science power platform
- Progress-M ferry craft
- service module (life-support systems and Russian crew quarters)
- Zarya control module
- thermal control system (radiators dump waste heat into space)
- life-support module
- solar panel array
- radiator
- U.S. centrifuge module
- European Space Agency module
- pressurized mating adapter for space shuttle docking
- Japanese experiment module
- radiator
- Soyuz crew rescue vehicle
- U.S. laboratory module
- integrated truss structure (the 310-ft (95-m) main girderwork structure of ISS)
- U.S. habitation module (living quarters for four crew)

OLD SOLDIER

THE FIRST COMPONENT OF ISS TO BE LAUNCHED, THE ZARYA CONTROL MODULE, WAS BUILT BY THE KHRUNICHEV COMPANY IN MOSCOW. BUT IT WAS PAID FOR BY NASA AND ITS CONSTRUCTION WAS MANAGED BY BOEING. IT WAS DEVELOPED FROM A SOVIET MILITARY CREW/SUPPLY FERRY, TKS, WHICH WAS DESIGNED IN THE 1960S AND TEST-FLOWN IN THE 1970S.

sieves remove carbon dioxide from it, and activated charcoal filters and catalytic oxidizers scrub away any contaminants. Waste water from the air conditioners, sinks, showers and toilets is recycled for drinking.

The main hazards of life in low Earth orbit are radiation, space debris and micrometeorites. During periods of maximum

solar activity, when solar flares create high levels of radiation, the crews "hide" in the best-shielded parts of the station.

The modules are built to withstand impacts of space debris and meteorite particles up to half an inch in size. The U.S. modules, for instance, are made from 1.25-inch (30-mm) aluminum with layers of Nextel impact protection material

and thermal insulation—making the walls about three inches (75 mm) thick in total. Particles larger than four inches (100 mm) across can be tracked from Earth, and, given enough warning, the ISS can maneuver to avoid a collision. But any particles between half an inch (12 mm) and four inches (100 mm) in size are potentially dangerous.

SERVICING HUBBLE

On December 27, 1999, the Space Shuttle Discovery completed a mission to repair the ailing Hubble Space Telescope—the third Hubble servicing flight since the telescope was launched into Earth orbit in 1990. The first, in 1993, corrected a serious flaw in the telescope's ability to focus, and the second, in 1997, saw the replacement of the telescope's scientific instruments and the installation of a new computer. Two more servicing missions are planned during Hubble's projected life but the curtailment of the Shuttle programme caused their cancellation.

WHAT IF...

...HUBBLE NEEDS MORE WORK?

Hubble's third servicing mission had originally been scheduled for 2000. The flight was not only to repair the faltering gyroscopes and install new electronics, but also to change some of the instruments aboard Hubble. When Hubble's third gyroscope failed in April 1999, the telescope was perilously close to becoming inoperable—if just one more gyroscope failed, Hubble would have to be shut down.

NASA decided to bring the servicing mission forward. Unfortunately, NASA's new Hubble Advanced Camera, scheduled to be installed on Servicing Mission 3, would not be ready for the early launch. Unwilling to have the telescope out of action for a long period of time, NASA decided to split the servicing mission in two.

3A, lauched in December 1999, was successful in replacing the faulty gyroscopes and upgrading the HST's computers, communications gear and sensors as well as conducting routine but necessary maintenance. 3B, carried out in March 2002, replaced the power control unit and added new solar panels as well as intalling a new digital camera. The mission also boosted Hubble into a slightly higher orbit and repaired the HST's protective Multi-Layer Insulation, which prevents damage by micrometeorites.

In January 2004, NASA announced that the Hubble would not receive any more service missions even if the Shuttle returns to full operation. The reasons included budget, crew safety and a commitment to the International Space Station that consumes all available Shuttle resources. Various ideas for alternative service missions have been vetoed.

At present the HST is still operational. Its instruments are capable of continuing for many years to come, but problems with the Hubble's systems – particularly its gyros – may cripple the telescope early. Further failures of the HST's systems, or problems with its instrumentation, may mean the tough decision to launch a new service mission or shut down the telescope.

Even just decommissioning the HST will require a mission of some kind; it has no thrust capability of its own so cannot be deorbited remotely as Mir was. Neither can it be left in a slowly decaying orbit indefinitely. It will be necessary to attach a booster to either push HST into a higher (and safer) orbit or to cause it to enter the atmosphere and burn up in a controlled manner.

It might also be possible to bring Hubble 'home' aboard a Shuttle for display or study – few other objects have been in space for as long as Hubble and the long-term effects of orbital flight could be investigated on the ground long after the telescope's operational life is over.

HST SERVICE HISTORY

APRIL 1990	LAUNCH OF HUBBLE SPACE TELESCOPE: HUBBLE IS PLACED INTO A 380-MILE-HIGH ORBIT BY THE SPACE SHUTTLE DISCOVERY	**2001**	SERVICING MISSION 3B: REPLACEMENT OF SCIENTIFIC INSTRUMENTS, INSTALLATION OF NEW SOLAR ARRAYS, GENERAL HOUSEKEEPING
DECEMBER 1993	SERVICING MISSION 1: CORRECTION TO TELESCOPE'S FOCUS, REPLACEMENT OF SOLAR ARRAYS, UPGRADE OF INSTRUMENTS	**2003**	SERVICING MISSION 4: TWO INSTRUMENTS TO BE REMOVED TO MAKE ROOM FOR NEW INSTRUMENTATION, REFURBISHMENT OF HUBBLE'S POINTING SYSTEM
FEBRUARY 1997	SERVICING MISSION 2: REPLACEMENT OF TWO INSTRUMENTS, INSTALLATION OF NEW COMPUTERS AND MAGNETOMETERS	**2010**	CLOSE-OUT MISSION: HUBBLE TO BE SWITCHED OFF AND EITHER BOOSTED TO A HIGHER ORBIT OR BROUGHT BACK TO EARTH
DECEMBER 1999	SERVICING MISSION 3A: REPLACEMENT OF FAULTY GYROSCOPES, INSTALLATION OF NEW ELECTRONICS AND COMPUTER SYSTEM		

ORBITAL MECHANICS

Astronomers applauded the launch of the Hubble Space Telescope in April 1990 and impatiently awaited its first images. The euphoria didn't last long, though. Instead of showing the crisp points of light expected, Hubble's first pictures were blurred. A tiny error in the machine used to grind the mirror had left the telescope out of focus.

Hubble was designed to be serviced, and NASA planned that five Shuttle missions would visit the telescope to install new instruments and replace out-of-date systems during its 20-year lifetime. But with the mirror's design flaw, the first servicing mission became a rescue bid to save the troubled craft and restore NASA's badly damaged public image.

On December 2, 1993, Space Shuttle Endeavour set off on Servicing Mission 1, and during its 11-day flight, its crew made five spacewalks. On each of these, a pair of astronauts performed the delicate operations needed to bring Hubble back to full fitness. New solar arrays were installed, as were new scientific instruments and sensors to help point the telescope accurately. And with TV news coverage showing every detail of the mission, the credibility of NASA rested on the installation of a single unit: COSTAR, the Corrective Optics Space Telescope Axial Replacement unit. This had been designed to correct the problem with Hubble's mirror, and after it was installed, Hubble's pictures were clear and sharp and far surpassed those available from ground-based telescopes. NASA—and the world's astronomers—breathed a huge sigh of relief.

UPGRADES AND CLOSURE

For over three years, Hubble photographed stunning views of distant stars and galaxies and revealed new clues about the origins of the universe, but by February 1997, a little housekeeping was necessary. Unlike the first Hubble servicing flight, which had been crucial to the survival of the telescope, the second mission was more routine. The aim

SERVICING MISSION 3A

SENSOR CHANGE
On the second of three spacewalks during Servicing Mission 3A, Michael Foale (far left) and Claude Nicollier replace one of Hubble's two fine guidance sensors. These optical sensors help Hubble to locate and point to the objects it is to study.

NEW GYROS
Astronaut Steve Smith installs a replacement rate sensor unit (RSU). Hubble has three RSUs, each containing two gyroscopes, which control and stabilize the telescope.

During Servicing Mission 3A, astronauts Claude Nicollier (on the end of the Shuttle's robotic arm) and Michael Foale work on the Hubble Space Telescope, which has been secured to the Shuttle's cargo bay.

MIRROR IMAGE

When details of Hubble's mirror problem were released in 1990, NASA was heavily criticized and its public image suffered badly. NASA needed Hubble's first servicing mission (above) to be a success, and feared that a failure could result in the agency being scaled down or even scrapped. The flight was deemed so important that much of the mission was shown live on CNN.

was to upgrade many of Hubble's systems, which were based on technologies dating back to the 1970s. These out-of-date systems were limiting the science that Hubble could perform, and the replacement units vastly improved the capability of the telescope.

The next servicing mission was to have been in 2000, but premature failure of some of the gyroscopes used to point the telescope forced NASA to shut the telescope down in November 1999 and schedule an earlier flight. Discovery was launched on December 19, 1999, to bring Hubble back to life by installing new gyroscopes plus an advanced computer and other electronics.

Two further Hubble servicing missions will be made to Hubble. Then, in 2010, the last visit by a Shuttle will switch Hubble off, bringing to a close the remarkable life of the world's first orbital telescope.

MISSION DIARY: SERVICE MISSION 3A

December 19, 1999 The Space Shuttle Discovery is launched to begin Shuttle mission STS-103, Hubble Servicing Mission 3A. The launch comes after nine delays due to bad weather and technical faults. The seven astronauts on board have a combined experience of almost a full year in space.

December 21 French astronaut Jean-François Clervoy of the European Space Agency uses the Shuttle's 50-foot (15-m) robotic arm to capture Hubble (right) and latch it

to a fixture in the Shuttle's cargo bay.

December 22 During the mission's first spacewalk, or extravehicular activity (EVA), Steve Smith and John Grunsfield replace Hubble's three rate sensor units and install voltage/temperature improvement kits on the telescope's six batteries.

December 23 On the second EVA, Michael Foale and Claude Nicollier replace Hubble's central computer and

one of its fine guidance sensors.

December 24 Steve Smith and John Grunsfield, on the third EVA, install a transmitter and a data recorder.

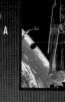

December 25 For only the second time on board a U.S. spacecraft, a crew spends Christmas Day in orbit (left).

December 26 Hubble is gently released back into orbit (right).

December 27 Discovery returns to Earth after a mission that lasted 7 days 23 hr 10 min 47 sec.

GLENN ON DISCOVERY

...GLENN PAVED THE WAY FOR OTHER SENIOR ASTRONAUTS?

John Glenn works out on the ergometer device onboard the Space Shuttle Discovery. At the age of 77 years, Glenn had to put in a lot of exercise to counteract the affects of zero gravity, many of which are similar to the effects of aging.

Through sheer determination to get back into orbit, John Glenn proved to NASA that there is no reason why energetic older astronauts can't fly in space. But any future seniors who wants to reach orbit will have to prove they are as fit as Glenn. A human who ventures into space has to endure physical stresses to the body during and after spaceflight. During liftoff and reentry, astronauts have to withstand high acceleration forces—g forces. When they land, they have to be able to readjust to a 1-g environment after their bones and muscles have wasted in space. They have to be fit enough to cope with any emergency situation—such as tackling an onboard fire—and they have to be able to cope with a rapid evacuation. Glenn found it relatively easy to prove his physical ability, since NASA had medical records on the aging space hero that went back further than any other astronaut's. Even after he left NASA in 1964, Glenn continued to report to the Agency for medical checkups, developing an enormous amount of evidence supporting his good health. And as if to prove the point, at the age of 77, Glenn was still speedwalking two miles each day and lifting weights.

But despite the rigorous physical criteria, it may be that—for scientific reasons—NASA will choose to send other elderly astronauts into space in the future. Some researchers have criticized the scientific basis of Glenn's mission on the grounds that very little can be learned from studying just one person. The answer is to send more than one older person. If NASA decides to adopt the idea, they are unlikely to have any trouble finding recruits. Many veteran astronauts from the Apollo and pre-Apollo era would leap at the opportunity to return to space. Jim Lovell, Commander of Apollo 13, is just one of those who gave his enthusiastic support to Glenn. In fact, he was so eager that he volunteered to become Glenn's backup. At just a few years younger than Glenn, Lovell may yet get his chance.

STS-95 was a routine Shuttle flight, except for one fact—a true American hero, veteran astronaut John Glenn, was on board. In 1962, Glenn had been the first American to orbit the Earth. Now, at the age of 77, he was to fly his second mission—and become the oldest man in space. As Payload Specialist on the Shuttle Discovery, Glenn would receive no special treatment. He had to train as hard as his six crewmates. But it was worth it. On October 29, 1998, he was blasted back into space and fulfilled a lifelong dream.

STS-95 EXPERIMENTS

PROTEIN TURNOVER EXPERIMENT	PEDRO DUQUE'S AND GLENN'S BLOOD SAMPLES MEASURE MUSCLE ATROPHY
HOLTER MONITOR ELECTRODES AND DATA RECORDER	WORN TO MONITOR HEART BEAT
SLEEP EXPERIMENTS	ELECTRODE SENSORS WORN AT NIGHT, AND QUESTIONNAIRES
BACK PAIN QUESTIONNAIRE	CURTIS BROWN, STEVE LINDSEY, STEPHEN ROBINSON AND GLENN ALL ANSWERED THE DAILY QUESTIONNAIRE
FOOD CONSUMPTION RECORDS	CHAIKI NAITO-MUKAI AND GLENN MADE A NOTE OF EVERYTHING THEY ATE
OSTEO BONE CELL GROWTH EXPERIMENT	TO DEVELOP ANTI-TUMOR DRUGS
ADVANCED ORGANIC SEPARATIONS (ADSEP) EXPERIMENT	DESIGNED TO FIND THE BEST WAY TO SEPARATE AND PURIFY BIOLOGICAL MATERIALS IN MICROGRAVITY
MICROGRAVITY ENCAPSULATION PROCESS (MEPS) EXPERIMENT	AIMED AT DEVELOPING ANTI-TUMOR DRUGS
ASTROCULTURE PLANT GROWING EXPERIMENT	ZERO-G HORTICULTURE

SENIOR SPACEMAN

Until John Glenn—aging space hero and U.S. Senator of Ohio—marched into NASA Administrator Dan Goldin's office, no one had seriously considered sending an older person into space. Spaceflight was risky, and you had to be very fit. But Glenn convinced Goldin that there could be some real scientific benefits in finding out how well a senior astronaut adapted to life in space.

Many of the ailments that afflict astronauts are similar to the effects of aging. Microgravity affects balance and perception, causes bone and muscle loss, and weakens the immune system. It upsets the astronaut's metabolism, blood flow and sleep patterns. These symptoms are also common in elderly people. So an older astronaut might be less affected by spaceflight—or perhaps more so. With the support of the National Institute on Aging, it was decided that the question was worth serious investigation. Glenn was added to the crew of STS-95.

Throwing himself into training—much of which involved emergency evacuations—Glenn was spared none of the exercises. He was soon vaulting out of emergency hatches, sliding down ropes, and diving out of airlocks headfirst. On October 29, he climbed into the Shuttle's lower deck for real and was strapped into place. The Discovery launch came nearly on schedule, and nine minutes later, Glenn was in orbit, admiring the blue Earth as it rolled past 340 miles (547 km) below. Zero g took a bit longer to get used to, though. For the first day Glenn floundered around, grabbing at handrails. But he soon found his space legs and began moving around easily.

EMERGING

AFTER THE MERCURY MISSION, PRESIDENT KENNEDY INSISTED THAT JOHN GLENN SHOULD NEVER FLY AGAIN: THE SPACE HERO HAD BECOME TOO POPULAR FOR HIS LIFE TO BE RISKED ON ANOTHER MISSION. BUT WHEN THE GOING GETS TOUGH, THE TOUGH GET GOING. GLENN (ABOVE) ENTERED THE HARD WORLD OF POLITICS IN 1974, RUNNING ON THE DEMOCRATIC PARTY TICKET HE BECAME A SENATOR FOR OHIO. HE HAS EVEN RUN FOR PRESIDENT.

PRESSURE TEST
Glenn (center) undergoes preflight g-force tests. Although he had been weightless during the 1962 mission, he had spent the entire flight strapped in his seat in the tiny Mercury capsule. On Discovery, he would be floating freely.

GANG'S ALL HERE
The crew of STS-95 poses for an onboard photograph. The crew (clockwise from top left) are: Mission Specialist Scott Parazynski, Payload Specialist John Glenn, Mission Commander Curtis Brown, Pilot Steven Lindsey, Mission Specialist Stephen Robinson, Mission Specialist and ESA astronaut Pedro Duque (the first Spanish astronaut in space) and Payload Specialist Chiaki Naito-Mukai.

DRACULA
Less than two days into the mission, Glenn gives the first of 10 blood samples. These were designed to measure any weakening in his muscles. Whenever Mission Specialist Scott Parazynski (left) arrived to take more blood, Glenn would declare, "Here comes Dracula."

BED TIME
Glenn, as well as Payload Specialist Chiaki Mukai, each put electrode caps and sleep nets on their heads and wore special sleep suits. The devices were designed to monitor brain waves, eye and body movements, muscle tension and respiration.

ALL SMILES
There was no doubt that Glenn was happy to be back in space. The crew reported that he was wearing a "grin from ear to ear" that refused to go away.

BATTERY OF MEDS

Soon Glenn was undergoing a battery of medical tests. He gave a total of 16 urine samples and wore electrodes on his chest to record his heart rhythm. In an attempt to establish how far muscle, intervertebral discs and bone marrow change in zero g, he and his colleagues filled in a daily back-pain questionnaire. Glenn's reaction time, short-term memory, hand-eye coordination and other functions were all measured on a laptop computer.

But not all of the more than 80 experiments performed during the mission were on Glenn. On the fourth day, the crew deployed the Spartan 201 solar physics satellite to observe the Sun. It studied the solar corona—the Sun's atmosphere—and the solar wind that accelerates in the corona. The satellite was recaptured two days later.

Glenn also helped to grow plants, feed bone-cell cultures and operate an experiment aimed at developing anti-tumor drugs. This work was no less important for science than the experiments on aging.

On the tenth day, Discovery touched down smoothly at Kennedy Space Center. Back on solid ground, and with his fourth and last term as Senator at an end, perhaps Glenn will settle into a contented, Earth-bound retirement—but nothing can be ruled out.

MISSION DIARY: GLENNS'S MISSION ON STS-95

OCTOBER 29, 1998, 2:19 P.M. EST STS-95 LAUNCHES AFTER A SHORT DELAY, CARRYING JOHN GLENN AND SIX OTHER CREW MEMBERS. 2:28 P.M. SHUTTLE REACHES ORBIT. CREW BEGINS SETTING UP EXPERIMENTS. 5:29 P.M. GLENN TALKS TO EARTH FROM THE SHUTTLE FOR THE FIRST TIME. SEVERAL MORE TALKS WILL FOLLOW IN THE NEXT FEW DAYS. OCTOBER 30 GLENN ACTIVATES THE MICROGRAVITY ENCAPSULATION PROCESS EXPERIMENT FOR DEVELOPING ANTI-TUMOR DRUGS. THE CREW

DEPLOYS PANSAT, A SMALL, NONRETURNABLE COMMUNICATIONS SATELLITE. NOVEMBER 1, 1:59 P.M. DEPLOYMENT OF THE SPARTAN SOLAR PHYSICS SATELLITE. GLENN BEGINS FEEDING BONE-CELL CULTURES AS PART OF THE OSTEO EXPERIMENT, AND STARTS WORK ON THE ADVANCED ORGANIC SEPARATIONS (ADSEP) EXPERIMENT. GLENN AND FELLOW PAYLOAD SPECIALIST CHIAKI MUKAI BEGIN WEARING A

COLLECTION OF SENSORS AT NIGHT TO MONITOR SLEEP PATTERNS. NOVEMBER 3 SPARTAN IS RETRIEVED. NOVEMBER 5 THE CREW BEGINS TO SHUT DOWN SOME OF THE MORE THAN 80 EXPERIMENTS ONBOARD IN PREPARATION FOR THE RETURN TO EARTH. NOVEMBER 6 THE CREW PREPARES FOR LANDING. NOVEMBER 7, 12:04 P.M. DISCOVERY LANDS SAFELY AT KENNEDY SPACE CENTER IN CAPE CANAVERAL, FLORIDA.

COLUMBIA DISASTER

T he launch of the Shuttle Columbia on its 28th mission on January 16 2003 appeared to be entirely routine and successful as the crew of seven astronauts began a two-week scientific mission designated STS-107, the 113th flight of a Space Shuttle orbiter. Less than two minutes after launch, a piece of insulating foam from the huge external fuel tank, barely noticed at the time, fell off and struck the Shuttle's left wing and set in train a sequence of events leading to the break-up of Columbia on re-entry 15 days later.

STS-107 FACTFILE

S hortly after her last mission, India renamed its first weather satellite Kalpana-1 in her honor.

The STS-107 mission was delayed 13 times over a two-year period. When it launched it became the 113 Shuttle flight.

It was the 88th mission since the 1986 explosion of Challenger. For Columbia, the oldest orbiter in NASA's fleet, this was its 28th flight.

All three US spaceflight disasters occurred on around the same date. Challenger was lost on 28 January 1986 and Apollo I on 27 January 1967.

Pieces of wreckage were recovered from a wide area of Texas and Louisiana. Investigators laid them out in the shape of the Shuttle to aid their research into the cause of the diaster.

Several of the student and commercial experiments survived the disaster and data was recovered from them. About 30 per cent of the experiment data had been relayed to Earth during the flight.

Three of the crew were not wearing their gloves during re-entry and one was not wearing a helmet. This was against procedure but would not have affected crew survival.

THE CREW OF STS-107

THE CREW OF COLUMBIA CONSISTED OF SEVEN ASTRONAUTS FROM THREE NATIONS.

RICK HUSBAND, MISSION COMMANDER
U.S. AIR FORCE COLONEL RICK HUSBAND, BORN 1957, HAD MADE ONE PREVIOUS FLIGHT, PILOTING STS-96 IN 1999. HE WAS CHIEF OF SAFETY IN NASA'S ASTRONAUT OFFICE.

WILLIAM McCOOL, PILOT
US NAVY COMMANDER WILLIAM McCOOL, BORN 1961, WAS MAKING HIS FIRST SHUTTLE FLIGHT.

MICHAEL ANDERSON, PAYLOAD COMMANDER
A USAF LIEUTENANT COLONEL BORN IN 1959, MIKE ANDERSON HAD FLOWN ON ENDEAVOUR IN 1998

ILLAN RAMON, PAYLOAD SPECIALIST
BORN 1954, ILLAN RAMON WAS THE FIRST ISRAELI TO FLY IN SPACE. HE WAS A COLONEL IN THE ISRAELI AIR FORCE AND TOOK PART IN THE 1981 ATTACK ON IRAQ'S OSIRAK NUCLEAR REACTOR.

KALPANA CHAWLA, MISSION SPECIALIST
BORN IN INDIA IN 1961, KALPANA CHALWA WAS AN AEROSPACE ENGINEER WHO HAD FLOWN ON COLUMBIA IN 1996. SHE WAS THE SECOND INDIAN TO FLY IN SPACE.

DAVID BROWN, MISSION SPECIALIST
NAVY CAPTAIN, AVIATOR AND FLIGHT SURGEON DAVID BROWN, BORN 1954, WAS ON HIS FIRST SHUTTLE MISSION.

LAUREL CLARK, MISSION SPECIALIST
BORN IN 1961, LAUREL CLARK WAS A NAVY COMMANDER AND FLIGHT SURGEON, ALSO MAKING HER FIRST SPACE FLIGHT.

DOOMED BY DEBRIS

The impact of the foam caused a breach in the shuttle's Thermal Protection System, allowing superheated air to penetrate the leading edge insulation and progressively melt the aluminium structure of the left wing, followed by destruction of the orbiter. The piece of foam weighed only 1.67lb (750 g) but was travelling at approximately 750 feet per second (228 m/s), or 511mph (822 km/h) when it struck the orbiter at an estimated impact angle of less than 20 degrees.

From the moment the foam debris struck the wing, Columbia was doomed, but there was a slim possibility the crew could have been saved by a rescue mission if the extent of damage was understood. A review of launch film on the day of launch did not pick up the debris strike and when it was spotted the following day it was found that no camera position covered the underside of the Shuttle. Various "Debris Assessment Team" meetings and requests for imagery from military sensors achieved little and the crew were not informed of the ground engineers' concerns, the assumption being that the (unseen) damage couldn't be too bad.

After a successful mission lasting 15 days, 22 hours, 20 minutes and 32 seconds during which the crew performed over 80 experiments testing applications of microgravity. The shuttle was travelling at approximately 12,500mph (20,000 km/h or Mach 18) when it broke up over Texas at an altitude of 207,000 feet (63,000 m), spreading debris over a wide area of Texas and into Louisiana.

In August 2003 the accident investigation board delivered a 248-page report, which criticised NASA's safety culture and the report warned that "the scene is set for another accident" without sweeping changes. Recommendations include a redesigned external tank, better imagery of launches, using an extended robot arm with a camera to examine the underside of the Shuttle in orbit and carrying a kit to repair broken thermal tiles with a spacewalk. Shuttle launches were suspended until July 2005.

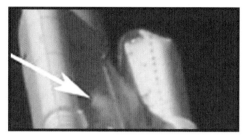

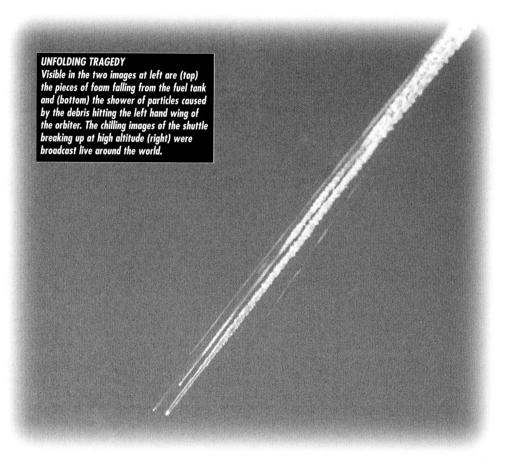

UNFOLDING TRAGEDY
Visible in the two images at left are (top) the pieces of foam falling from the fuel tank and (bottom) the shower of particles caused by the debris hitting the left hand wing of the orbiter. The chilling images of the shuttle breaking up at high altitude (right) were broadcast live around the world.

MEMORIALS

SEVEN ASTEROIDS ORBITING THE SUN BETWEEN MARS AND JUPITER WERE NAMED AFTER THE COLUMBIA CREW. THE ASTEROIDS WERE DISCOVERED AT THE PALOMAR OBSERVATORY NEAR SAN DIEGO IN JULY 2001 BY ASTRONOMER ELEANOR F. HELIN, WHO RETIRED IN JULY 2002.

THE SEVEN ASTEROIDS RANGE IN DIAMETER FROM 3.1 TO 4.3 MILES (5 TO 7 KM). THE NAMES, PROPOSED BY NASA'S JET PROPULSION LABORATORY (JPL), WERE APPROVED BY THE INTERNATIONAL ASTRONOMICAL UNION IN AUGUST 2003.

MISSION DIARY: STS-107 COLUMBIA

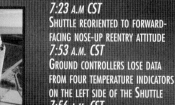

JANUARY 16, 2003, 9:39 A.M. CST
LAUNCH FROM KENNEDY SPACE CENTER.
9:40:20 A.M. CST
EIGHTY-ONE SECONDS AFTER LAUNCH A PIECE OF INSULATING FOAM FALLS FROM THE EXTERNAL FUEL TANK AND PUNCTURES THE UNDERSIDE OF THE LEFT WING LEADING EDGE
FEBRUARY 1, 2003, 7:15 A.M CST
DEORBIT BURN PROCEDURE BEGUN TO BRING COLUMBIA OUT OF ORBIT FOR LANDING

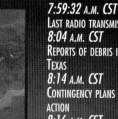

7:23 A.M CST
SHUTTLE REORIENTED TO FORWARD-FACING NOSE-UP REENTRY ATTITUDE
7:53 A.M. CST
GROUND CONTROLLERS LOSE DATA FROM FOUR TEMPERATURE INDICATORS ON THE LEFT SIDE OF THE SHUTTLE
7:56 A.M. CST
SENSORS DETECT RISE IN TEMPERATURE TYRE PRESSURE ON THE LEFT-SIDE LANDING GEAR
7:58 A.M. CST
DATA IS LOST FROM THREE TEMPERATURE SENSORS EMBEDDED IN THE SHUTTLE'S LEFT WING

7:59:32 A.M. CST
LAST RADIO TRANSMISSION
8:04 A.M. CST
REPORTS OF DEBRIS IN SKY OVER TEXAS
8:14 A.M. CST
CONTINGENCY PLANS ORDERED INTO ACTION
8:16 A.M. CST
SCHEDULED LANDING TIME AT KENNEDY SPACE CENTER
3:05 P.M. CST
PRESIDENT BUSH SPEAKS TO THE NATION: "THE COLUMBIA IS LOST. THERE ARE NO SURVIVORS."

TO THE MOON

J ules Verne imagined the first Moon voyagers would arrive by cannon shell, but is sketchy on how they might have returned. Getting to the Moon may have seemed the easy part, but the first attempts by the U.S.A. and U.S.S.R. to send probes to impact the moon in the late 1950s ended ignominiously when the launchers failed. A manned orbiter—Apollo 8—was not to come until a decade later. At the same time the Russians were restricting themselves to unmanned orbiters, landers, and rovers. Their unmanned missions achieved just one sample return to Earth, while the Apollo program delivered a dozen men to the Moon, returning them and hundreds of pounds of lunar rock and soil to Earth.

 Getting two men and some equipment to the Moon required one of the most complicated machines ever built, the Saturn V launcher and associated Command Module, Lunar Module, and Service Module, built by a variety of contractors. Over 90 per cent of these expensive components were burned up, sent into eternal orbit, or abandoned on the moon, with only a cramped capsule returning to splash down. Any future moon mission, if not as efficient as Verne's one-way projectile, will have to be less wasteful, but no less an adventure.

Astronaut Edwin E."Buzz" Aldrin Jr. is photographed on the moon during the Apollo 11 mission: in the center background is the United States flag; in the left background is the black and white lunar surface television camera; in the far right background is the Lunar Module "Eagle."

SOVIET LUNAR PROGRAMME

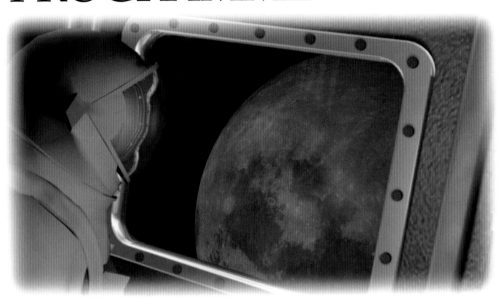

When President John F. Kennedy declared in 1961 that the United States was going to put a man on the Moon by the end of the decade, NASA burst into activity—ready to meet the challenge. But NASA was not alone in their efforts. The Soviet Union had decided that it was going to beat America to the Moon. Thus began a frantic race between the two superpowers to develop the necessary technology and expertise to put a man on the Moon. It was a race the Soviets did not plan to lose.

WHERE ARE THEY NOW?

THE SOVIET UNION'S MOONSHIPS

Had the N-1 been able to successfully launch a payload, it would have been renamed—perhaps as the "Lenin" or "Kommunism" booster. Instead, it disappeared almost without a trace. Ten operational and two mock-up N-1s were built. Four were lost in launch attempts; the remaining eight were dismantled and destroyed.

Scavenged pieces of the boosters ended up being used as carports, hangars and storage sheds all over the Baikonur Cosmodrome. The large stage-one bottom support plate of one booster, complete with 30 engine mounting holes, ended up as the roof of a gazebo in a park. The remaining four LK landers, and one LOK lunar orbiter, survived intact. These vehicles ended up in museums and space engineering institutes, where tourists and students can marvel at what almost was. An LK mock-up was even displayed at the EuroDisney amusement park in the late 1990s.

In the end, the Soviet lunar program spent just $4.5 billion dollars over 15 years, compared to the $24 billion spent on Apollo. Though under-investment was undoubtedly one reason for the Soviet lunar failure, the program also suffered from competition between the design bureaus and poor management of the 26 government bureaus and 500 enterprises that built the N-1 boosters. Perhaps, too, the Soviets underestimated the technical difficulties of sending humans to the Moon.

Kistler Aerospace Corporation is designing the first fully reusable vehicles for carrying payloads into space. Their K1 rocket (illustrations right) will use engines designed and built for the N-1.

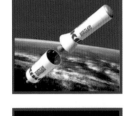

The N-1 did leave a space legacy, though. The launch pads and vehicle integration buildings were converted for use with the Energia superbooster—which was used to launch the Soviet space shuttle, Buran. And the 150 NK-33, NK-39 and NK-43 engines built for the rockets were cocooned and secretly stored, despite orders to destroy them. In 1996, they were sold to the American company Kistler Aerospace, for use as the first-stage engines on the reusable K1 spacecraft.

And, in 1994, the engine designed for the lunar module was sold to India for use in the Indian space program. The remnants of the Soviet crewed lunar program may make it into orbit after all.

SOVIET STEPS TO THE MOON

JANUARY 30, 1956	POLITBURO HEARS PLANS FOR LANDING A COSMONAUT ON THE MOON	FEBRUARY 21, 1969	FIRST N-1 LAUNCH ATTEMPT ENDS IN FAILURE
MAY 1962	N-1 VEHICLE DESIGN COMPLETE	JULY 3, 1969	SECOND N-1 LAUNCH ATTEMPT ENDS IN FAILURE
SEPTEMBER 24, 1962	POLITBURO AUTHORIZES CONSTRUCTION OF N-1	JUNE 27, 1971	THIRD N-1 LAUNCH ATTEMPT ENDS IN FAILURE
AUGUST 3, 1964	CENTRAL COMMITTEE ISSUES DECREE TO BEAT THE U.S. TO THE MOON	NOVEMBER 23, 1972	FOURTH N-1 LAUNCH ATTEMPT ENDS IN FAILURE
SEPTEMBER 1965	ZOND 5 ORBITS MOON		
JANUARY 14, 1966	SERGEI KOROLEV DIES, REPLACED BY VASILI MISHIN	MAY 1974	MISHIN REPLACED BY VALENTIN GLUSHKO, WHO ENDS LUNAR PROGRAM

NEVER THE MOON

Though most people think the Moon race began with President Kennedy's famous 1961 speech, the Soviet Union had already been planning a crewed lunar landing for quite some time. The government's cabinet or Politburo had heard plans for such a mission on January 30, 1956. But the Soviets wasted valuable time. It was not until 1964—when they realized that the U.S. was serious about its lunar ambitions—that the Central Committee issued a decree that would place a Soviet cosmonaut on the Moon before the Americans got there.

Unlike the United States, the Soviet Union did not have a single organization for space exploration. Premier Nikita Khrushchev believed that competition between rival bodies would create better designs. Three design bureaus had been working on plans to land on the Moon: OKB 1, run by Sergei Korolev, the man behind the triumphs of Sputnik and Yuri Gagarin; OKB 586, run by Michael Yangel; and OKB 52 run by Vladimir Chelomei. Each proposed, designed and began building different vehicles for lunar missions. In the end, Korolev's Nositel ("carrier") 1 rocket, or N-1, was selected as the booster and his LOK as the orbiter, with Yangel's LK lander chosen for the descent to the surface. The first landing was planned for 1968.

The plans did not go smoothly. The original flight profile was to have several rockets lift different modules to rendezvous in low Earth orbit. The units would dock, transfer fuel and crew, then boost for the Moon. But this plan soon had to be modified.

DEATH AND DOOM

Korolev, who became the project leader in late 1965, was convinced that the multiple-module approach was too difficult, and did not favor Chelomei's designs. Instead, the design for the N-1 was changed, upgrading its payload capacity. The flight profile was changed to two launches, one for the lunar craft and another for the crew. Korolev was sure this plan would beat the Americans—but disaster struck.

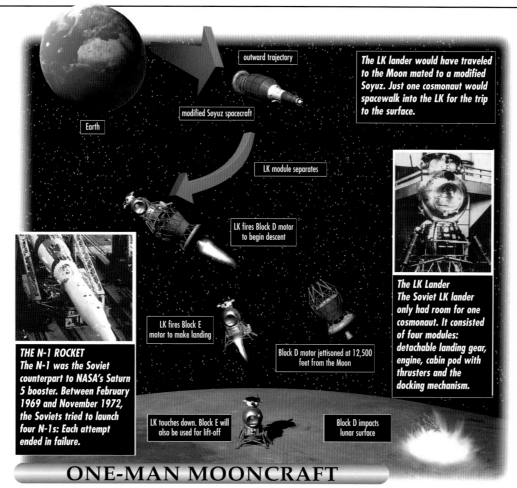

outward trajectory

The LK lander would have traveled to the Moon mated to a modified Soyuz. Just one cosmonaut would spacewalk into the LK for the trip to the surface.

Earth

modified Soyuz spacecraft

LK module separates

LK fires Block D motor to begin descent

THE N-1 ROCKET
The N-1 was the Soviet counterpart to NASA's Saturn 5 booster. Between February 1969 and November 1972, the Soviets tried to launch four N-1s: Each attempt ended in failure.

LK fires Block E motor to make landing

Block D motor jettisoned at 12,500 feet from the Moon

The LK Lander
The Soviet LK lander only had room for one cosmonaut. It consisted of four modules: detachable landing gear, engine, cabin pod with thrusters and the docking mechanism.

LK touches down. Block E will also be used for lift-off

Block D impacts lunar surface

ONE-MAN MOONCRAFT

HASTY

On March 24, 1964, rocket designer Sergei Korolev met with Soviet premier Nikita Krushchev to advocate his plans for lunar exploration. Krushchev expressed interest in the program and Korolev wrote a letter to Leonid Brezhnev (right), then in charge of missile development, citing Krushchev's approval and complaining of a lack of money. The letter may have hurt Korolev's standing with Brezhnev after Brezhnev became premier in 1965.

On January 14, 1966, Korolev died unexpectedly during surgery. Without his genius behind the program, Soviet hopes were lost. Development of the N-1 booster was bogged down in technical problems and redesigns, taking time the Soviets could ill afford.

The N-1 finally lifted off from Baikonur on February 21, 1969—and exploded 66 seconds later. Soviet engineers made modifications to a second N-1 and quickly prepared it for launch. The Soviet lunar lander was not ready, but the OKB 1 scientists hoped to get a lunar flyaround under their belt. On July 3 the rocket exploded at launch. The blast obliterated the launch pad. Thirteen days later NASA launched Apollo 11. The race was over.

MISSION DIARY: THE STORY OF THE N-1

April 12, 1961 Yuri Gagarin becomes the first person in space in a Vostok capsule (Vostok replica, right).
1966 Start of lunar cosmonaut training.
November 1966 The first N-1 Moon-rocket parts arrive for assembly at Baikonur Cosmodrome in Soviet Kazakhstan. The first N-1 launch is set for the third quarter of 1968.
May 7, 1968 The first N-1 Moon rocket (4L) is erected on the launch pad at Baikonur.
June 6, 1968 The maiden flight of the N-1 is postponed

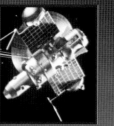

after cracks are found in its first stage.
September 1968 Zond 5 (right), a test for the crewed mission, successfully completes circumlunar navigation.
February 21, 1969 Second N-1 rocket (3L) launched. It crashes to the ground 66 seconds after launch.
July 3, 1969 N-1-5L is launched. The whole rocket falls back onto the launch pad, explodes and destroys the pad (right, pad after crash).
July 20, 1969 Apollo 11 lands on Moon.

1970 Lunar cosmonaut teams disbanded.
June 27, 1971 The next attempt at an N-1 launch (rocket 6L). The rocket rolls out of control and falls apart 48 seconds after launch.
November 23, 1972 Final launch of the N-1 rocket (7L). It reaches an altitude of 25 miles (40 km), but an engine pipeline fire causes engine shutdown 107 seconds after launch.
1974 Soviet Science Commission ends lunar program.

LUNA 1, 2 AND 3

A t the beginning of 1959, the space race was less than two years old—with the Soviet Union, thanks to Sputnik 1, firmly in the lead. Now the Soviets were about to surge even further ahead. In the course of the year, they sent three spacecraft to the Moon. The Soviet Luna probes were relatively crude, but Luna 2 became the first human-made object to reach another celestial body. And Luna 3 sent back the first photographs of the Moon's far side, making Soviet technology the envy of the world.

WHAT IF...

...WE COULD RECOVER THE LUNA PROBES?

W hen the Luna probes were sent into space, their designers never seriously considered their future recovery. Even if they were located and retrieved, there would be little data on board to interest today's scientists. So would it be worth rescuing any of them?

As the first human-made objects to leave the Earth forever, the three Luna probes certainly have a high sentimental value. Any recovery, though, would take much time, money and effort. For a start, no one knows where Luna 1 is. A few days after it passed by the Moon, Soviet controllers lost contact with the spacecraft. It went into solar orbit and by now it is probably somewhere between the Earth and Mars—a vague location that would make the probe difficult, if not completely impossible, to find. New reconnaissance craft would have to be sent out to look for it, and such a search would take years.

Luna 2 would be easier to track down. Its impact on the Moon's surface was exactly where its designers had projected—1° W and 30° N, between the craters of Archimedes, Aristillus and Autolycus. Even if future space archeologists were to excavate the wreckage, Russia would most likely have a political objection to the idea of bringing it back to Earth. In among the probe's electronics, Soviet engineers placed two commemorative metal spheres. Embossed with Soviet insignia and

Soviet designers found room aboard the Luna probes for two metal globes, one of them is shown far right at close to actual size. The globes were embossed with Soviet insignia, the initials C.C.C.P.—Russian for U.S.S.R.—and the date (closeups, near right).

virtually indestructible, the balls are sure to be somewhere in or around Luna 2's impact crater. The the Soviet Union and its symbols may have vanished from history, but Russian pride would still insist that these two tiny artifacts remained on the Moon.

Luna 3's elongated orbit took it completely around both the Moon and the Earth, and it may have burned up in our planet's atmosphere. But if it is still intact, the probe has on board one of the most interesting relics of the early space age. Although much better pictures have subsequently been taken, the original 1959 film of the far side of the Moon would have great historical worth.

Like the program that sent the Luna modules up, any decision to bring them back down would have more to do with politics and public relations than hard science. Replicas of the spacecraft have been made for display in museums. To place the originals next to them would be a costly exercise in nostalgia.

LUNA PROBE FACTS

	LUNA 1	LUNA 2	LUNA 3
MISSION	LUNAR IMPACT ATTEMPT	FIRST LUNAR IMPACT OF LUNAR FAR SIDE	PHOTOGRAPHED 70%
LAUNCH DATE	JANUARY 2, 1959	SEPTEMBER 12, 1959	OCTOBER 4, 1959
WEIGHT	797 LB (361 KG)	860 LB (390 KG)	614 LB (278 KG)
SUMMARY	PASSES WITHIN 3,100 MILES (5,000 KM) OF THE MOON 34 HOURS AFTER LAUNCH. ACHIEVES SOLAR ORBIT.	IMPACTS WITH MOON AFTER 33.5 HOURS OF FLIGHT	AFTER 11 ORBITS AND 177 DAYS IN SPACE, LUNA 3'S CONTROLLERS SWITCH THE PROBE OFF

THE FAR SIDE

Just 15 months after the successful Sputnik 1 mission, the Soviet Union began its Luna program—an ambitious project to send the first spacecraft to the Moon. The Luna probes would provide new information on the Earth's closest celestial neighbor—and Soviet space science would demonstrate to an enthralled world that the Soviet Union was well ahead of its American rivals.

All three Luna missions went astonishingly well. Luna 1, launched on January 2, 1959, became the first artificial object to exceed escape velocity and leave the Earth forever. Designed to impact on the Moon, the probe missed by a mere 3,100 miles (5,000 km)—but sent back data on the Moon's gravity and magnetic field all the same.

Luna 2, launched on September 12, was dead on target. The probe took just 33.5 hours to reach the Moon. En route, Luna 2 released a cloud of light-reflecting sodium vapor that allowed astronomers to track it visually all the way until—at a speed of two miles per second—the spacecraft plowed into the lunar surface.

The abrupt silence from its transmitters was cheered by Soviet controllers. Before impact, data from the module had confirmed the absence of any strong magnetic fields or radiation belts around the Moon.

Luna 3, launched on October 4, gave the Soviet Union even more reason to cheer. The probe's ambitious figure-eight trajectory took it right around the Moon. As it hurtled past, just 41,500 miles (66,787 km) from the surface, onboard cameras took the first pictures of a landscape previously unseen: the far side of the Moon. A course correction nudged Luna 3 into a barycentric orbit—that is, an orbit around the gravitational center of the Earth and Moon system. As the probe approached Earth, its transmitters began to send back the images. This was Luna 3's last duty. Its film exhausted but its mission accomplished, the probe made 11 orbits and spent 177 days in space before its controllers finally switched it off.

Luna 3's fuzzy pictures of the Moon's far side were a major event in space exploration,

and at that time Luna 3 was the most impressive achievement of either the U.S. or Soviet space programs. But the success of the Luna program was measured not only in scientific discoveries—it was also a political coup, and the Soviet Union was jubilant.

HARD EVIDENCE

THE HIDDEN HEMISPHERE
Just as the Earth rotates on its axis, so does the Moon. But the Moon rotates at exactly the same rate as it orbits the Earth—a phenomenon caused by the Earth's gravity and known as synchronous rotation. As a result, only one lunar hemisphere can be seen from the Earth. The far side always remains hidden.

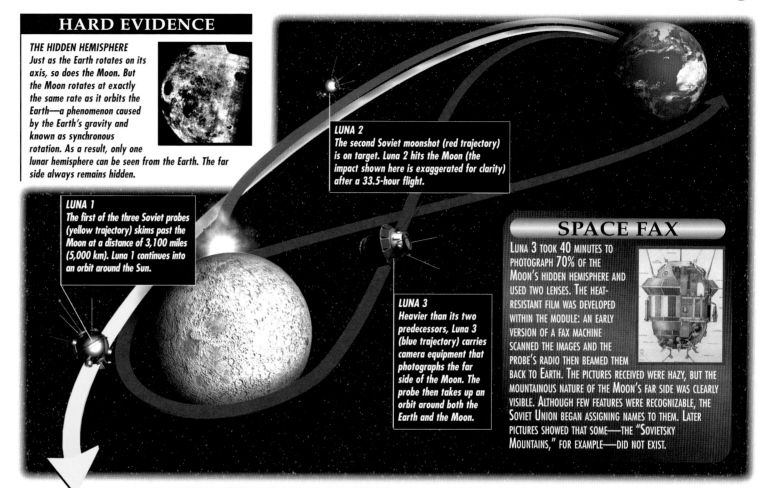

LUNA 1
The first of the three Soviet probes (yellow trajectory) skims past the Moon at a distance of 3,100 miles (5,000 km). Luna 1 continues into an orbit around the Sun.

LUNA 2
The second Soviet moonshot (red trajectory) is on target. Luna 2 hits the Moon (the impact shown here is exaggerated for clarity) after a 33.5-hour flight.

LUNA 3
Heavier than its two predecessors, Luna 3 (blue trajectory) carries camera equipment that photographs the far side of the Moon. The probe then takes up an orbit around both the Earth and the Moon.

SPACE FAX

LUNA 3 TOOK 40 MINUTES TO PHOTOGRAPH 70% OF THE MOON'S HIDDEN HEMISPHERE AND USED TWO LENSES. THE HEAT-RESISTANT FILM WAS DEVELOPED WITHIN THE MODULE: AN EARLY VERSION OF A FAX MACHINE SCANNED THE IMAGES AND THE PROBE'S RADIO THEN BEAMED THEM BACK TO EARTH. THE PICTURES RECEIVED WERE HAZY, BUT THE MOUNTAINOUS NATURE OF THE MOON'S FAR SIDE WAS CLEARLY VISIBLE. ALTHOUGH FEW FEATURES WERE RECOGNIZABLE, THE SOVIET UNION BEGAN ASSIGNING NAMES TO THEM. LATER PICTURES SHOWED THAT SOME—THE "SOVIETSKY MOUNTAINS," FOR EXAMPLE—DID NOT EXIST.

MISSION DIARY: LUNA 1, 2 AND 3

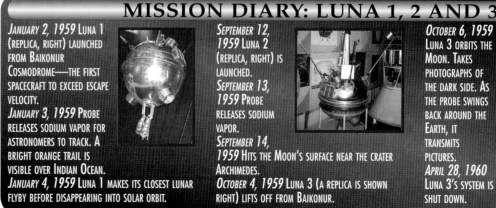

JANUARY 2, 1959 LUNA 1 (REPLICA, RIGHT) LAUNCHED FROM BAIKONUR COSMODROME—THE FIRST SPACECRAFT TO EXCEED ESCAPE VELOCITY.
JANUARY 3, 1959 PROBE RELEASES SODIUM VAPOR FOR ASTRONOMERS TO TRACK. A BRIGHT ORANGE TRAIL IS VISIBLE OVER INDIAN OCEAN.
JANUARY 4, 1959 LUNA 1 MAKES ITS CLOSEST LUNAR FLYBY BEFORE DISAPPEARING INTO SOLAR ORBIT.

SEPTEMBER 12, 1959 LUNA 2 (REPLICA, RIGHT) IS LAUNCHED.
SEPTEMBER 13, 1959 PROBE RELEASES SODIUM VAPOR.
SEPTEMBER 14, 1959 HITS THE MOON'S SURFACE NEAR THE CRATER ARCHIMEDES.
OCTOBER 4, 1959 LUNA 3 (A REPLICA IS SHOWN RIGHT) LIFTS OFF FROM BAIKONUR.

OCTOBER 6, 1959 LUNA 3 ORBITS THE MOON. TAKES PHOTOGRAPHS OF THE DARK SIDE. AS THE PROBE SWINGS BACK AROUND THE EARTH, IT TRANSMITS PICTURES.
APRIL 28, 1960 LUNA 3'S SYSTEM IS SHUT DOWN.

LUNA 10–12

In 1966, the race for the Moon was still an open contest between superpowers America and Russia. Both had successfully placed soft landers on the lunar surface, proving that machinery—and therefore humans—could stand there. The next hurdle was to place a satellite in lunar orbit. As well as commanding considerable prestige, such a mission could gather more information about the Moon's alien environment and collect pictures of potential landing sites. The Russians led the way with Lunas 10, 11 and 12.

WHAT IF...

...THE RUSSIANS RETURNED TO THE MOON?

The world has changed dramatically since the 1960s, and the heady days of the space race are long gone. The contest to be first to the Moon was won by America in 1969, and the U.S.S.R. is no more. So as we enter the 21st century, what future does the Moon hold for the former Soviet states? Will they ever return to the Moon?

Since the breakup of the Soviet Union, Russia itself has concentrated on crewed spaceflight, spearheaded by the Mir space station. Funding for exploration beyond Earth orbit has all but dried up. International cooperation—ironically, mainly with the U.S.—has been the savior of the Russian space program, so any future lunar missions involving Russia or other ex-Soviet states are likely to be joint ventures with foreign partners.

While America and the West have the edge in technology, notably rocket fuels and microelectronics, former Soviet space scientists can bring over 40 years' worth of practical experience to bear. They also have a healthy "make do or improvise" attitude, born out years of having to design their way around obsolete technologies and poor-quality manufacturing processes.

The Russians—or at least Russian-engineered spacecraft—could also make a return to the Moon through private funding. As launching payloads into orbit becomes more economical, space exploration comes within

Privately funded lunar rovers roam the Moon's surface, providing thrills and spills for their drivers back on Earth. The idea has already been taken up commercially, but stands to benefit greatly from the practical experience of former Soviet scientists.

reach of an increasing number of large commercial consortiums.

Already a number of corporations in the U.S. are exploring the idea of landing rovers on the Moon. These rovers could fulfill various roles, including gathering samples and collecting data. But one of the more imaginative uses suggested for the rovers is as vehicles for the ultimate virtual reality experience—paid "drives" across the lunar surface, during which customers pilot the rovers by remote control from the safety of their own homes.

Russian experience of long stays in space gained on Mir could also prove invaluable in founding any kind of colony on the Moon, where conditions are likely to be harsh. Russian astronauts and space scientists have always been renowned for their toughness. The chances are that they will continue to build on that reputation in years to come.

LUNA 10–12 STATS

	LUNA 10	LUNA 11	LUNA 12
MASS AT LAUNCH	3,480 LB (1,578 KG)	3,608 LB (1,636 KG)	3,564 LB (1,616 KG)
MASS IN LUNAR ORBIT	539 LB (244 LB)	2,420 LB (1,097 KG)	2,500 LB (1,133 KG)
ORBIT TIME (APPROX.)	3 HOURS	3 HOURS	3 HOURS
LUNAR ORBIT INCLINATION	71.9°	27.0°	10.0°
NUMBER OF LUNAR ORBITS	460	277	602
NUMBER OF TRANSMISSIONS	219	137	302

LUNAR LONERS

By the mid-1960s, both America and the Soviet Union had soft-landed craft on the lunar surface and were now rehearsing—in Earth orbit—the maneuvers needed to land a crewed spacecraft on the Moon. The next step was to place a satellite in lunar orbit, in order to provide more information about the lunar environment and to search for potential landing sites.

The Soviets were in the lead. In 1959, Luna 2 became the first human-made object to hit the Moon, and in the same year Luna 3 transmitted the first images of the Moon's far side. The U.S.S.R. was also the first to achieve a soft landing with Luna 9.

April 3, 1966, marked yet another Soviet triumph, as Luna 10 became the first spacecraft to orbit the Moon. Where Luna 9 was aimed directly at the landing site, using springs and airbags to cushion the impact, Luna 10 made the journey attached to a larger parent craft, and on approaching the Moon, fired a retro-rocket that slowed it enough to be grasped by the Moon's gravity. Twenty minutes later, the orbiter was detached from the parent craft and released into orbit.

Five feet (1.5 m) long and around 18 inches (50 cm) in diameter, Lunar 10 spun at two revolutions per minute and completed one orbit every 178 minutes. Its limited payload implied that science took second place to national pride: There were devices for measuring electrical, magnetic and radiation fields, but no camera. After 56 days in orbit, Luna 10's batteries died and the craft fell silent.

PARTY PIECE

Luna 10 made history by becoming the first satellite to broadcast a tape recording. At the 23rd session of the Congress of the Soviet Union Communist Party, Luna 10 transmitted the "Internationale," the socialist anthem, while orbiting the Moon. On hearing the triumphal broadcast, Party delegates rose to their feet and saluted before joining a red army choir (above) in singing the song themselves.

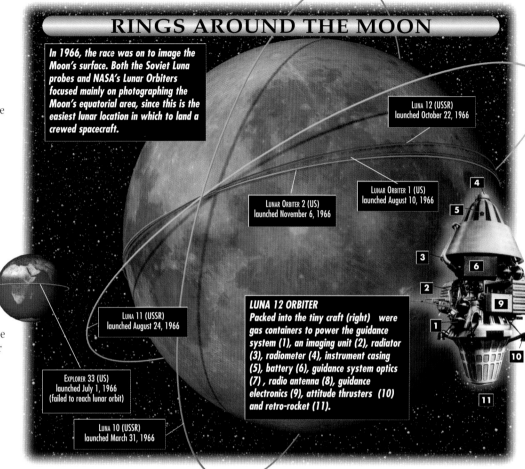

RINGS AROUND THE MOON

In 1966, the race was on to image the Moon's surface. Both the Soviet Luna probes and NASA's Lunar Orbiters focused mainly on photographing the Moon's equatorial area, since this is the easiest lunar location in which to land a crewed spacecraft.

Luna 12 (USSR) launched October 22, 1966

Lunar Orbiter 2 (US) launched November 6, 1966

Lunar Orbiter 1 (US) launched August 10, 1966

Luna 11 (USSR) launched August 24, 1966

Explorer 33 (US) launched July 1, 1966 (failed to reach lunar orbit)

Luna 10 (USSR) launched March 31, 1966

LUNA 12 ORBITER
Packed into the tiny craft (right) were gas containers to power the guidance system (1), an imaging unit (2), radiator (3), radiometer (4), instrument casing (5), battery (6), guidance system optics (7), radio antenna (8), guidance electronics (9), attitude thrusters (10) and retro-rocket (11).

MISSION DIARY

March 31, 1966 Luna 10 launched from Baikonur Cosmodrome; within one hour, the parent craft is on course for the Moon.
April 1, 1966 Luna 10's engine fires to make a mid-course correction en route to the Moon.
April 3, 1966 Luna 10 enters lunar orbit and begins collecting data.
May 30, 1966 The batteries aboard the Luna 10 orbiter (right) fail; its transmitters are switched off.
August 24, 1966 Luna 11 launched from Baikonur Cosmodrome.
August 28, 1966 Luna 11 enters lunar orbit and returns further information on field fluxes.

October 1, 1966 Contact is lost with Luna 11 after its batteries fail.
October 22, 1966 Luna 12 launched from Baikonur Cosmodrome.
October 23, 1966 Luna 12 makes similar mid-course correction to those made by Lunas 10 and 11.
October 25, 1966 Luna 12 enters orbit around the Moon.
October 27, 1966 Luna 12 sends back first images of lunar surface.
January 19, 1967 Last signal from Luna 12 is received as its batteries fail.

PRESERVING THE IMAGE

By the time of the launch of Luna 11, five months later, the U.S. had streaked ahead in the Moon race with Lunar Orbiter 1, which snapped 211 high-quality images of the Moon's surface. Despite Soviet scientists' best efforts, Luna 11 did not match the U.S. probe's success. After entering lunar orbit on August 28, the spacecraft appeared to vanish. The Soviets' official story was that although Luna 11 carried an imaging camera, its mission was simply to test spacecraft systems in lunar orbit, not to take any actual photographs. Russian sources have since revealed that in fact, Luna 11 suffered attitude control problems: It did return images, but they were of empty space.

In October 1966, the Soviet news agency Tass announced that Luna 12 would be launched to photograph the Moon's surface from lunar orbit. Externally similar to Luna 11, it carried a much larger payload than earlier Luna craft, including a camera system with a built-in "dark room" that developed, fixed and dried the film. The developed pictures were then scanned electronically and transmitted back to Earth in a process similar to that used by fax machines.

Luna 12 achieved lunar orbit as planned on October 25, 1966. To make the most of its battery power, it began collecting images of the Moon's equatorial regions almost immediately.

To the delight of Soviet space scientists and Communist Party officials, Luna 12's faxing system worked perfectly and images of the lunar surface were promptly beamed across the nation on Soviet television. But the tide of events had already turned in favor of the U.S. space program. The images provided by Luna 12 were inferior to those of Lunar Orbiter 1, and on January 19, 1967 the spacecraft ceased transmission. American technology was finally beginning to outpace that of its superpower rival. After Luna 12, the Soviet race to the Moon ran out of steam.

RANGER AND SUREYOR

Back in May 1961, when President John F. Kennedy gave the go-ahead to land a man on the Moon, no one knew much about the lunar surface. Was it solid rock or was it covered in treacherous moondust, a quicksand that could swallow an entire spacecraft and its crew? NASA had to know. From 1961 through 1968, the robot Ranger and Surveyor missions were launched to find the answers to the questions that perplexed Apollo planners. These little spacecraft blazed the trail for the first humans on the Moon.

INSIDE STORY

THE RANGER AND SURVEYOR PROBES

A total of 16 spacecraft were launched in the Ranger and Surveyor programs, which began in 1961 and ended in 1968. Of the nine Rangers, 1 and 2 burned up on reentry after failing to leave Earth orbit, while 3 and 5 missed the Moon and went into orbit around the Sun. All of the rest, plus all of the Surveyor craft, are still on the Moon.

The Ranger spacecraft that made it to the Moon arrived at speeds of several thousand miles per hour, and their impacts punched out the craters in the lunar surface that now mark their graves. During 1972, the Apollo 16 orbital camera located the 46-foot (14-m) deep holes made by Rangers 7 and 9. Some day in the future astronauts may return to unearth fragments of these historic spacecraft, but little of them is likely to remain intact.

The Rangers did their job merely by getting to the Moon, but the seven Surveyor craft were designed to land softly. Surveyors 2 and 4 crashed and were probably wrecked, but the others lie intact and lifeless on the lunar surface. Each Surveyor carried silver-zinc batteries that were charged by current from solar panels. But the long, cold lunar night lasts for about two weeks and brings temperatures as low as −256°F (-160°C), which relentlessly drained the spacecraft of their power. Most of the Surveyors succumbed to the bitter cold after just one lunar night and then fell silent.

Apollo 12 astronauts Pete Conrad and Alan Bean photographed Surveyor 3 in 1969, 31 months after the probe's soft landing. The Apollo 12 Lunar Module can be seen in the background of this photograph.

Surveyor 3 is the only one of these craft to have been visited since its arrival on the Moon. In November 1969, the Apollo 12 Lunar Module landed just 520 feet (160 m) to the northwest of it, on the rim of what is now called Surveyor Crater. The astronauts photographed the spacecraft and removed its soil scoop and camera system. These were returned to Earth for examination, along with a sample of its electrical cable. Engineers found that the cable was brittle and that white paint on the scoop had tarnished in the harsh environment.

If humans eventually colonize the Moon, the landers and impact craters are likely to become lunar sites of historic interest. Astronauts not yet born may one day visit these relics and reflect on the pioneering work of the scientists and engineers involved in the Ranger and Surveyor missions.

LUNA LANDING SITES

Mission	Latitude	Longitude	Mission	Latitude	Longitude
Ranger 4	15.5° S	130.5° W	Surveyor 1	2.4° S	43.3° W
Ranger 6	9.39° N	21.51° E	Surveyor 2	5.5° N	12.0° W
Ranger 7	10.35° S	20.58° W	Surveyor 3	3.0° N	23.3° W
Ranger 8	2.6° N	24.7° E	Surveyor 4	0.4° N	1.33° W
Ranger 9	13.3° S	3.0° W	Surveyor 5	1.5° N	23.2° E
			Surveyor 6	0.53° N	1.4° W
			Surveyor 7	40.9° S	11.5° W

MOON PROBED

During the Mercury flights of the early 1960s, when U.S. astronauts first ventured into space, NASA was already aiming for the Moon. Its Ranger spacecraft were designed to crash headlong into our nearest neighbor, beaming back progressively better pictures as they hurtled toward the surface. But the early lunar missions were far less successful than Project Mercury. The first two Rangers, launched in 1961, failed to leave Earth orbit. The following year, Rangers 3, 4 and 5 also ended in failure. Rangers 3 and 5 missed the Moon completely and Ranger 4 tumbled uselessly onto its surface. Ranger 6 reached the Moon successfully in 1964, but sent back no images because its cameras were faulty.

Success finally came later that year with Ranger 7, which collected pictures of the northern basin of the Sea of Clouds during the last 20 minutes before impact. It returned 4,316 images, and the last of them, taken just before the craft struck the Moon at around 5,800 mph (9,344 km/h), were 1,000 times better than any that had been taken from Earth. The views showed the maria (seas) of the Moon are relatively smooth and similar to lava flows on Earth.

Two further missions, Rangers 8 and 9 in 1965, also made precision flights and sent back thousands more close-up images. Ranger 8 aimed for the Sea of Tranquillity, later the site of the first human steps on the Moon, while Ranger 9 headed for the hills. It relayed images of the Alphonsus crater in the southern highlands, an area thought to have once been volcanically active. Ranger 9's impact on the lunar surface brought the series to a successful close.

MOON BUGS

During the Apollo 12 mission in November 1969, astronauts Pete Conrad and Alan Bean visited Surveyor 3 and removed some pieces of it for analysis. Back on Earth, researchers found Earth bacteria in a piece of plastic foam from inside Surveyor's TV camera. These were thought to have been in the foam since before Surveyor was launched; they survived for almost three years on the lunar surface.

RANGER AND SURVEYOR

HARD PUNCHER
Ranger 7 carried six TV cameras, two with 25-mm wide-angle lenses and the other four with 75-mm lenses. It weighed nearly 800 lb (363 kg), the span of its solar panels was about 15 ft (4.6 m) and it was nearly 12 ft (3.7 m) high.

SOFT LANDER
Surveyor 3, shown here, weighed about 665 lb (302 kg) and its equipment included a TV camera and soil sampler. The Surveyor craft were all of the same basic design, but the scientific instruments they carried varied from one mission to another.

THE FINAL COUNTDOWN
Ranger 8 beamed back these pictures (right) in the last 9 sec before it hit the surface of the Moon at almost 6,000 mph (10,000 km/h).

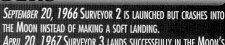

T–9 SECONDS
12 miles (19 km) above the lunar surface.

T–5 SECONDS
6.8 miles (11 km) above the lunar surface.

T–0.7 SECONDS
Just under 5,000 ft (1,524 m) to final impact.

CLOSE-UP VIEW

The next objective was to soft-land on the Moon and take close-up images of the lunar surface. The first craft sent to do this, Surveyor 1, landed gently only 8.7 miles off-target in June 1966 and sent back panoramic views of the landscape. Over the next two years, it was followed by six more Surveyors, not all of which were as successful.

Mission control lost contact with Surveyor 2, when a thruster failure caused it to tumble out of control onto the Moon, and with Surveyor 4, just minutes before touchdown. But Surveyors 3, 5, 6 and 7 all made it safely to the lunar surface. They provided a mass of information, both from their cameras and from on-board instruments that sampled and analyzed the lunar surface. Among their detailed findings was the reassuring confirmation that the surface of the Moon was similar to that of fine-grained soil on Earth and would pose no problem to crewed Moon landings.

MISSION DIARY: NASA RANGERS

August 23, 1961 Ranger 1 fails to leave Earth orbit.
November 18, 1961 Ranger 2 fails to leave Earth orbit.
January 26, 1962 Ranger 3 is launched but misses the Moon by more than 22,800 miles (36,700 km). As with all the Ranger series, the launch vehicle is an Atlas Agena rocket.
April 23, 1962 Ranger 4 is launched but fails to return any data.
October 18, 1962 Ranger 5 is launched but misses the Moon by 450 miles (720 km).
January 30, 1964 Ranger 6 is launched but fails to transmit any images.
July 28, 1964 Ranger 7 is launched. It is the first U.S.

probe to photograph the Moon (right).
February 17, 1965 Ranger 8 launched; 3 days later, it sends its last picture from a height of 2,100 feet (640 m).
March 21, 1965 Ranger 9 sets off for the crater Alphonsus, where it impacts successfully.
May 30, 1966 Surveyor 1 is launched on a successful soft-landing mission. On June 2, the probe settles gently on the floor of the crater Flamsteed.

September 20, 1966 Surveyor 2 is launched but crashes into the Moon instead of making a soft landing.
April 20, 1967 Surveyor 3 lands successfully in the Moon's Ocean of Storms.
July 14, 1967 Surveyor 4 is launched but radio contact is lost en route to the Moon.
September 11, 1967 Surveyor 5 lands in the Sea of Tranquility.
November 10, 1967 Surveyor 6 lands safely—then "hops" 25 yards (23 m) on its thrusters.
January 10, 1968 Surveyor 7 lands in the crater Tycho, far from the potential Apollo sites.

APOLLO TEST FLIGHTS

The mandate of the Apollo program—to put a man on the Moon by the end of the 1960s—demanded a leap forward in spacecraft technology. To meet the goal, NASA devised the Saturn rocket family. Uncrewed missions were carried out to test Apollo before the U.S. made a final commitment to sending astronauts toward the lunar target. On the whole, the missions were successful: Mission Control was able to finesse some useful data out of technical glitches. But on the fourth mission, the glitches turned deadly.

WHAT IF...

...WE COULD LEARN MORE FROM THE APOLLO TEST FLIGHTS?

Uncrewed test vehicles played a unique role in the Apollo program—and they have been obsolete ever since. The complexity of the lunar missions required practice and experience. But the Saturn 5 launch vehicles were essentially ballistic missiles designed to disintegrate on the way to their target: outer space. The so-called "all-up" testing approach used in the Apollo test flights—in which all the pieces of a mission are tested at once—was expensive and has become unnecessary with the rise of the reusable Space Shuttle.

Unlike Saturn 5 rockets, most of the Space Shuttle's testing occurred on the ground, using models. The only real Shuttle flight testing vehicle was Enterprise, flown off the back of a Boeing 747 during approach and landing tests in California in 1977. The Shuttle program also benefited from years of experience of flying rocket research aircraft.

The data keeps flowing in with tests of the new X-38 International Space Station Crew Rescue Vehicle. As single-stage-to-orbit (SSTO) spacecraft are developed and perfected, and as lunar and Mars spacecraft components are constructed in Earth orbit, the importance of atmospheric testing wanes.

The remaining items of Apollo test hardware are museum pieces—and they are

The crucial third stage from an unused Saturn 5 rocket (above) is exhibited at the Kennedy Space Center in Florida. Today, few if any of the Apollo test modules are actively studied: Some are in museums, but many others remain in storage.

overshadowed by relics that actually went to the Moon, some of which have also been installed in museums. CM002, the first production Command Module capsule, is displayed at the Cradle of Aviation Museum, Mitchell Field, New York. Apollo 4 is displayed at the Stennis Space Center, St. Louis, Missouri, and Apollo 6 is at the Fernbank Science Center, Atlanta, Georgia. CM098 is at the Academy of Sciences in Moscow, Russia.

Only the unflown Saturn rocket stages and Lunar Modules are available for exhibition. The rest either remained in space or burned up at the end of their planned missions. But the information gleaned from the test missions lives on in the complex rules of NASA's space program protocol.

APOLLO TEST ROCKET SPECS

Launch Vehicle	Saturn 1	Saturn 1B	Saturn 5
Total Launches	10	4	2
Stages	2	2	3
Height	190 feet (60 m)	224 feet (68 m)	363 feet (110 m)
Launch Engines			
Stage I	8 x H-1	8 x H-1B	5 x F-1
Stage II	6 x RL-10	1 x J-2	5 x J-2
Stage III			1 x J-2

TRIED & TESTED

The six test flights for the Apollo mission examined the hardware for the main stages of the Apollo lunar mission, from launch to stage separation to space maneuvers. The first three missions—AS-201, 202 and 203—launched in 1966 on a Saturn 1B rocket. These missions also checked the separation of the stages of the launcher, their electrical systems and parts of the spacecraft, including the Command Module (CM) for launch and reentry, the Service Module (SM) for orbiting around the Moon, and the Lunar Module (LM).

Planned as the first of a series of crewed test flights in Earth orbit, Apollo 1 ended in tragedy on the launchpad in January, 1967. A flash fire in the Apollo 1 capsule killed the three astronauts and grounded the Apollo crewed missions for 21 months.

The rest of the test flights were uncrewed. Ten months after Apollo 1, the Apollo 4 Command and Service Module (CSM) was sent into Earth orbit on the new, huge Saturn 5 rocket. Apollo 4 reached 11,000 miles (17,700 km) high and reentered the atmosphere at 24,900 mph (40,000 km/h).

Apollo's final Saturn 1B mission in late January 1968 was the first—and only—test launch of the Lunar Module. Once in orbit, the Apollo 5 Lunar Module separated from the second stage of the Saturn launcher. After checking the onboard systems, the descent

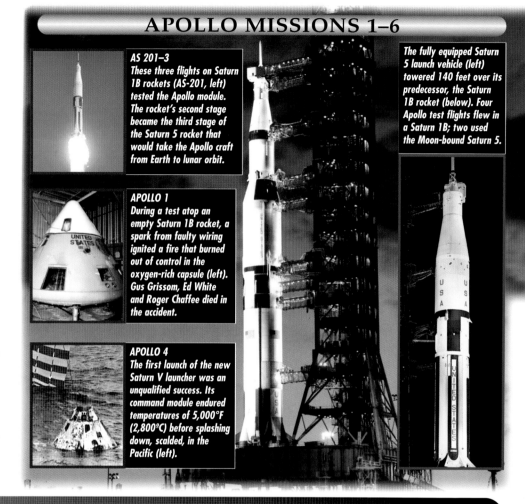

APOLLO MISSIONS 1–6

AS 201–3
These three flights on Saturn 1B rockets (AS-201, left) tested the Apollo module. The rocket's second stage became the third stage of the Saturn 5 rocket that would take the Apollo craft from Earth to lunar orbit.

APOLLO 1
During a test atop an empty Saturn 1B rocket, a spark from faulty wiring ignited a fire that burned out of control in the oxygen-rich capsule (left). Gus Grissom, Ed White and Roger Chaffee died in the accident.

APOLLO 4
The first launch of the new Saturn V launcher was an unqualified success. Its command module endured temperatures of 5,000°F (2,800°C) before splashing down, scalded, in the Pacific (left).

The fully equipped Saturn 5 launch vehicle (left) towered 140 feet over its predecessor, the Saturn 1B rocket (below). Four Apollo test flights flew in a Saturn 1B; two used the Moon-bound Saturn 5.

APOLLO 5
The last Saturn 1B rocket used in an Apollo mission had an oddly shaped cargo (left). This mission tested only the ascent and descent stages of the Lunar Module without a Command/Service Module.

SONIC BOOM

ON NOVEMBER 9, 1967, THE FIRST UNMANNED SATURN 5 LAUNCHED (RIGHT) WITH A DEAFENING ROAR. EYEWITNESSES STOOD MORE THAN THREE MILES (5 KM) AWAY—ANY CLOSER AND THEIR EARDRUMS WOULD HAVE BURST. AT THE PRESS SITE, THE PLATE-GLASS WINDOW OF THE CBS NEWS BOOTH SHOOK SO HARD THAT WALTER CRONKITE AND COLLEAGUES HAD TO HOLD IT IN PLACE DURING HIS LIVE REPORT.

engine was fired up for a mock lunar landing. Four seconds later, onboard systems cut the engine off. A built-in safety system had detected that the engine was not firing powerfully enough, aborted the descent and decoupled the ascent stage of the Lunar Module.

ON TO PLAN B

Although NASA engineers did not obtain data on the Lunar Module's descent engine, they did learn that their backup safety systems worked flawlessly.

The last Apollo test flight, in April, 1968, did not go so smoothly. The first stage was wrenched back and forth by sloshing rocket fuel. Then two of the second-stage engines shut down, forcing the Service Module to ignite to accelerate the spacecraft to escape velocity. Although the spacecraft did not have enough fuel to reach the planned 7-mile-per-second (11 km/s) reentry speed, it did allow NASA to test its contingency procedures. Now the ground was cleared for crewed tests.

MISSION DIARY: APOLLO TEST FLIGHTS

JULY 28, 1960 PROJECT APOLLO ANNOUNCED BY NASA.
MAY 25, 1961 PRESIDENT KENNEDY COMMITS TO A MANNED LUNAR LANDING WITHIN THE DECADE.
NOVEMBER 7, 1963 FIRST OF SIX FLIGHT TESTS OF APOLLO LAUNCH ESCAPE SYSTEM ON TEST ROCKET.
JANUARY 29, 1964 FIRST ORBITAL FLIGHT OF SATURN 1.
MAY 28, 1964 FIRST FLIGHT OF CM TO PROVE COMPATIBILITY WITH SATURN 1.
SEPTEMBER 18 1964 THIRD ORBITAL FLIGHT OF SATURN 1 ROCKET, PREDECESSOR OF SATURN 1B.
JANUARY 20, 1966 FIRST FLIGHT-RATED APOLLO CAPSULE ON SIXTH AND FINAL LITTLE JOE ABORT TEST.

FEBRUARY 26, 1966 FIRST SUB-ORBITAL FLIGHT OF SATURN 1B WITH CSM INSTALLED.
JULY 5, 1966 FIRST ORBITAL FLIGHT OF SATURN 1B.
JANUARY 27, 1967 APOLLO 1 PAD FIRE KILLS THREE ASTRONAUTS BEFORE THE FIRST PLANNED MANNED ORBITAL TEST OF APOLLO CSM ON SATURN 1B IN FEBRUARY.
NOVEMBER 9, 1967 FIRST UNMANNED LAUNCH OF SATURN 5 ROCKET—THE APOLLO 4 MISSION—WITH MOCK LUNAR MODULE. THE MODULE TAKES PHOTOGRAPHS OF THE EARTH FROM 11,000 MILES (17,000 KM).
JANUARY 22, 1968 FIRST UNMANNED LAUNCH OF LM ON SATURN 1B ROCKET.

APRIL 4, 1968 SECOND UNMANNED LAUNCH OF SATURN 5: APOLLO 6. THE CAPSULE SPLASHES DOWN SAFELY (RIGHT).
OCTOBER 1968 FIRST MANNED APOLLO CSM FLIGHT: THE APOLLO 7 MISSION.
DECEMBER 1968 FIRST MANNED SATURN 5 FLIGHT: APOLLO 8.
MARCH 1969 FIRST APOLLO LAUNCH OF SATURN 5 COMPLETE WITH LM (APOLLO 9).
JULY 1969 LUNAR LANDING: APOLLO 11.

APOLLO 1 DISASTER

"Fire, I smell fire." Delivered almost casually over the radio at 6:31 p.m. on January 27, 1967, Roger Chaffee's words heralded the end of Apollo 1's crewed flight simulation. Sixteen seconds later, the crew capsule was split by the intense heat. Fueled by the capsule's pure oxygen atmosphere, fire and smoke had quickly overwhelmed the three men on board and then beat back rescue workers. As America mourned, the Apollo program all but ground to a halt.

WHAT IF...

...IT HADN'T HAPPENED?

The first Apollo launch was intended merely as a test. The capsule would orbit the Earth for approximately two weeks before coming home. In Grissom's words, the mission was "primarily concerned with checking out the spacecraft's systems and seeing whether it was both flyable and livable." If all went well with this and later flights, Apollo would then fly to the Moon. America, for the first time, would have an edge over the Soviet Union.

Unfortunately, the U.S. was too eager to get ahead in the space race. In its enthusiasm, NASA ignored a host of warning signals. The Saturn 1B rocket that would launch the capsule had passed all tests with flying colors during 1966, but the command module itself was beset with problems. As early as April, the environmental control unit had caught fire during a bench test. A transistor failed during the capsule's first simulation. In the second, the environmental unit broke down. Then the fuel tank ruptured. A replacement environmental unit leaked, spilling coolant over the wiring. As engineers worked to overcome these and a host of other ailments, the original launch date of December 1966 was delayed until February 1967 at the earliest.

Apollo's shortcomings were fully documented in a 3,000-page analysis after the tragedy. An electrical arc in the environmental control unit was believed to have sparked the fire. Subsequent inspection of the wiring revealed "numerous examples of poor installation, design

Once the Command Module burst into flames, Ed White struggled desperately to open the hatch. But he didn't have a prayer—hot gases sealed the door shut with thousands of pounds of force.

and workmanship." The hatches, for example, took at least 90 seconds to open—but Apollo 1's crew was overcome by fumes within 16 seconds. The report also noted that a socket wrench had been abandoned deep in the spacecraft's wiring.

The risks inherent in an oxygen-rich atmosphere were well attested. The previous April, the environmental control unit had burst into flames in exactly the same conditions as those aboard Apollo 1. Wernher von Braun, designer of the Saturn rocket, went on record as saying that "pure oxygen in connection with inflammable material can cause a fire in the spacecraft." In the case of Apollo 1, holes burned in aluminum tubing showed that temperatures had exceeded 1,400°F (760°C).

Gus Grissom had perhaps envisioned such a disaster. "There will be risks," he wrote, "as there are in any experimental program, and sooner or later, inevitably, we're going to run head-on into the law of averages and lose somebody. I hope this never happens...but if it does I hope the American people won't feel it's too high a price to pay for our space program." They didn't. Two years later a much improved Apollo spacecraft was sitting on the pad, ready to go to the Moon.

APOLLO 1 CREW

COMMAND PILOT	SENIOR PILOT	PILOT
VIRGIL IVAN "GUS" GRISSOM; LIEUTENANT COLONEL, U.S. AIR FORCE	EDWARD HIGGINS WHITE, II; LIEUTENANT COLONEL, U.S. AIR FORCE	ROGER BRUCE CHAFFEE; LIEUTENANT COMMANDER, U.S. NAVY
DATE & PLACE OF BIRTH	DATE & PLACE OF BIRTH	DATE & PLACE OF BIRTH
APRIL 3, 1926; MITCHELL, INDIANA	NOVEMBER 14, 1930; SAN ANTONIO, TEXAS	FEBRUARY 15, 1935; GRAND RAPIDS, MICHIGAN
PREVIOUS MISSIONS	PREVIOUS MISSIONS	PREVIOUS MISSIONS
MERCURY 4, JULY 21, 1961; GEMINI 3, MARCH 23, 1965	GEMINI 4, JUNE 3, 1965	NONE

FATAL FURNACE

Shortly before 1:00 p.m. on January 27, 1967, three white-suited astronauts walked to launch complex 34 at Cape Canaveral. They were Virgil "Gus" Grissom and Ed White—both old hands from the Gemini program—and rookie Roger Chaffee. Their job was to put Apollo through a routine ground test.

The simulation was to resemble spaceflight as closely as possible. But the capsule in which the men sat contained pure oxygen at a higher-than-normal pressure of 17 pounds per square inch (1.1 bar), in order to protect the spacecraft's electronics from inflows of humid Florida air. The hatches were sealed and locked. Communication would be by radio only. The only deviation from a real launch setup was that the Saturn 1B rocket beneath the capsule had empty fuel tanks. Nothing unusual was expected, and fire crews were ordered on standby rather than full alert.

The three astronauts took their places awkwardly and without much enthusiasm. Chaffee occupied the center couch, with Grissom, the commander, on his left and White, on his right. It was a frustrating period, dogged by constant holdups. By 6:30 p.m. the men had been aboard for five-and-a-half hours and were looking forward to the end of the test. The countdown stood at T minus 10 minutes.

Then, at 6:31 p.m., a brief power surge was picked up in the capsule's 12 miles (19 km) of electrical wiring. It went unnoticed at the time. Ten seconds later, Chaffee reported a fire. White began to wrestle with the hatch's six retaining bolts.

A CLOSE SHAVE

AT ONE STAGE IN THE SIMULATION, GRISSOM BECAME SO IRRITATED BY CONSTANT RADIO PROBLEMS THAT HE COMPLAINED, "IF I CAN'T TALK WITH YOU ONLY FIVE MILES AWAY, HOW CAN WE TALK TO YOU FROM THE MOON?" HE WAS SO CONCERNED THAT HE WANTED TO CHANGE SEATS WITH DEKE SLAYTON (RIGHT), DIRECTOR OF FLIGHT CREW OPERATIONS, SO THAT SLAYTON COULD WITNESS THE GLITCHES FOR HIMSELF.

APOLLO 1 COMMAND MODULE

The Command Module—code-named "spacecraft 012"—sat atop the Saturn 1B rocket on the launchpad. Once the two-piece hatch was sealed, the cabin was pressurized with pure oxygen at 17 psi (1.1 bar).

CREW ACCESS HATCH
Apollo 1 had a two-hatch access door, for maximum protection during spaceflight. But it took more than 90 seconds to open, and required a special tool – used from the outside. Later Command Modules had a single-hatch system with emergency-opening features.

rendezvous window

couch attenuation strut

crew couches

attitude control thrusters for correcting course before reentry

WHITE
Selected as an astronaut in 1962, on June 3, 1965, White lifted off with the four-day Gemini 4 mission. During the flight, he became the first American to walk in space.

GRISSOM
Grissom was selected as one of the seven Mercury astronauts in 1959. He made the second sub-orbital Mercury flight on July 21, 1961, and in 1965 commanded Gemini 3.

CHAFFEE
Apollo 1 was to have been Chaffee's first mission. He became an astronaut in October 1963, after flying reconnaissance missions during the Cuba crisis of 1962.

OUT OF CONTROL

The flames spread rapidly in the oxygen-rich environment. Soon the capsule was filled with smoke. Chaffee's final message was, "We've got a bad fire... We're burning up here!" Then came only unintelligible shouts and the noise of frantic pounding. Finally, silence. From start to finish, the disaster had taken just 16 seconds.

Technicians struggled to open the hatch, unable to see more than six inches ahead through the blinding smoke, their white coats already peppered with burn holes from flying debris. When the first firefighter arrived four-and-a-half minutes later, it was too late. The three astronauts were dead. Grissom lay with his feet on the left-hand couch and his head below the center one, as if he had been trying to find shelter. Chaffee was still strapped in at his post. White had died struggling to undo the hatch.

An autopsy concluded that the men had died of asphyxiation: Once the fire had burned through their spacesuits' supply hoses, they were left breathing the choking fumes in the cabin. Gruesomely, intense heat had melted the suits to their seats. It was seven hours before the bodies could be recovered.

MISSION DIARY: APOLLO 1

MARCH 21, 1966 GRISSOM, WHITE AND CHAFFEE (RIGHT, AT THE LAUNCH SITE) ARE NAMED AS THE APOLLO 1 CREW. THEIR BACKUPS ARE DAVID SCOTT, RUSSELL SCHWEICKART AND JAMES MCDIVITT.
OCTOBER 18, 1966 APOLLO CAPSULE'S FIRST CREWED SIMULATION TEST IN THE KENNEDY ALTITUDE CHAMBER IS TERMINATED BY TRANSISTOR FAILURE.
OCTOBER 21, 1966 A SECOND TEST IS HALTED DUE TO A BROKEN OXYGEN REGULATOR.
OCTOBER 27, 1966 ENTIRE ENVIRONMENTAL UNIT IS REMOVED AFTER FURTHER PROBLEMS EMERGE.
NOVEMBER, 1966 WALTER SCHIRRA, DONN EISELE AND WALTER

CUNNINGHAM ARE APPOINTED AS THE NEW BACKUP TEAM.
DECEMBER 5, 1966 REPLACEMENT ENVIRONMENTAL UNIT BREAKS DOWN AND IS SENT BACK FOR REPAIRS.
JANUARY 6, 1967 APOLLO IS FINALLY MATED TO ITS LAUNCH VEHICLE.
1:00 P.M. JANUARY 27, 1967 GRISSOM, WHITE AND CHAFFEE BEGIN THE FINAL CREWED SIMULATION.
1:20 P.M. TEST IS HALTED BECAUSE OF STRANGE SMELL IN THE OXYGEN SUPPLY. COUNTDOWN RESTARTS AT 2:42 P.M.

5:40 P.M. ANOTHER HALT IS CALLED TO CHECK OUT THE FAULTY RADIO COMMUNICATION SYSTEM.
6:20 P.M. COUNTDOWN IS HALTED DUE TO FAULTY AUDIO CIRCUITS.
6:31 P.M. CAPSULE IGNITES, KILLING ALL ABOARD.
12:30 A.M. JANUARY 28, 1967 RESCUERS BEGIN THE JOB OF REMOVING THE ASTRONAUTS' BODIES.
JANUARY 30, 1967 MEMORIAL SERVICE IN HOUSTON. PRESIDENT JOHNSON LATER ATTENDS GRISSOM'S FUNERAL AT ARLINGTON NATIONAL CEMETERY (LEFT).
FEBRUARY 1967 APOLLO ACCIDENT REVIEWS BEGIN (ABOVE).

APOLLO COMMAND MODULE

Apollo's Command Module took men to the Moon—and brought them back again. Built to house the crew during launch, provide them with air, food and water for the journey, and serve as a base of operations during lunar landings, the tiny craft also had to survive the rigors of reentry into the Earth's atmosphere with little more than a heat shield for brakes. The demands on it were extraordinary. Yet over nine lunar voyages and two Earth orbital missions, the Command Module never failed its passengers.

WHAT HAPPENED...

...TO THE COMMAND MODULES SINCE APOLLO?

Between 1968 and 1972, nine Apollo Command Modules (CMs) journeyed to the Moon, and two underwent test flight missions in Earth orbit. After the lunar landing program was completed, NASA wanted to keep the Apollo hardware flying until the late 1970s for initial space station experiments. But these ambitious plans were scaled down as a result of budget cuts.

Yet the CM did fly four more missions between 1973 and 1975. Three crews visited Skylab—NASA's first experimental space station, adapted from a hollow Saturn rocket stage—and a fourth team made a friendly orbital rendezvous and crew exchange with a Russian Soyuz capsule. This cooperative mission signaled the end of a highly competitive space race between the two superpowers.

The 15 CM capsules used for all these missions are displayed in museums around the world. The CM from the most famous mission, the historic Apollo 11 first lunar touchdown mission in July 1969, takes center stage at the Smithsonian Air and Space Museum in Washington, D.C. The Smithsonian also boasts a factory-fresh set of Apollo hardware, originally built for a canceled lunar mission. The CM can be viewed in pre-flight condition, with its heat shield and glossy foil covering intact.

A reconstructed Apollo CM control panel (above) can be found in the Smithsonian Air and Space Museum in Washington, D.C., alongside the famous Apollo 11 capsule.

The Apollo 13 CM, on the other hand, was not brought out to be displayed for some years. After splashdown and recovery, engineers stripped away most of its equipment as part of their investigation into the accident that had endangered the mission. The tattered capsule was then shipped off to a museum in France.

In 1985, Max Ary, founder of a space museum in Kansas, asked NASA if he could borrow Apollo 13's hollowed-out CM and restore the interior. Permission was granted, and Ary began a painstaking 12-year search for 80,000 missing instruments and sub-system assemblies, locked away and forgotten in various NASA warehouses scattered around the U.S. Today the fully-restored capsule is on display at the Kansas Cosmodrome. For those who are too far to visit, Ary and his team also built a detailed replica of the CM for the 1995 movie Apollo 13.

APOLLO CM SPECS

PRIME CONTRACTOR	NORTH AMERICAN AVIATION	PROPELLANTS	175 LB (79 KG)
CREW	3	ELECTRICAL EQUIPMENT	1,550 LB (703 KG)
HABITABLE VOLUME	210 CU FT (5.9 M³)	COMMUNICATIONS SYSTEMS	225 LB (102 KG)
LENGTH	10 FT 6 IN (3.2 M)	NAVIGATION EQUIPMENT	1,100 LB (499 KG)
DIAMETER	12 FEET 10 INCHES (3.9 M)	ENVIRONMENTAL CONTROLS	450 LB (204 KG)
OVERALL MASS	12,800 LB (5,800 KG)	CREW SEATS AND PROVISIONS	550 LB (249 KG)
HEAT SHIELD	1,900 LB (861 KG)	RECOVERY EQUIPMENT	550 LB (249 KG)
THRUSTER SYSTEM	900 POUNDS (408 KG)	MISCELLANEOUS CONTINGENCY	450 LB (204 KG)

SPACE WOMB

Constructed from 2 million separate components, the 3-person Apollo Command Module (CM) was an incredibly complex machine by any standards—let alone those of the 1960s. Its task was to act as mother ship during the Moon landing missions. One astronaut stayed in the CM during lunar orbit; the other two traveled to the surface in a separate landing craft.

The CM's cramped interior included instruments, navigation computers and life-support equipment designed with the experience gained from the earlier Mercury and Gemini programs. However, the outer shape of the craft required a totally radical approach.

For safety's sake, NASA planners insisted that a CM returning from the moon had to be capable of surviving reentry without using a rocket engine to slow it down beforehand. This was in case something went wrong with Apollo's propulsion systems during the mission. Because of its return trajectory across a quarter of a million miles of space, the CM would hit the Earth's atmosphere at a colossal 25,000 mph (40,000 km/h). This was 7,000 mph (11,000 km/h) faster than previous Mercury or Gemini craft returning from missions in Earth orbit.

Apollo CM designers Maxime Faget and Caldwell Johnson created a cone-shaped craft with a blunt heat shield, similar in some ways to earlier capsules, but with a smoother exterior. Then a lucky accident helped them improve the shape. The diameter of the upper stage of the Saturn 5 rocket under the CM was cut by two inches (5 cm).

HARD FALL

THE APOLLO TEAM CONSIDERED LANDING THE COMMAND MODULE ON SOLID GROUND. BUT TEST IMPACTS (RIGHT) SO BADLY DAMAGED THE MODULE THAT IN 1964, ENGINEERS RECOMMENDED HARD LANDINGS ONLY IN AN EMERGENCY. THE SOVIET UNION, INTERESTINGLY, MADE THE OPPOSITE DECISION. BECAUSE OF THEIR COMPULSIVE SECRECY, THEIR LARGE LAND MASS, AND THEIR LACK OF A GLOBAL NAVY, THE U.S.S.R. TRIED HARD TO LAND ALL THEIR SPACECRAFT ON SOVIET SOIL.

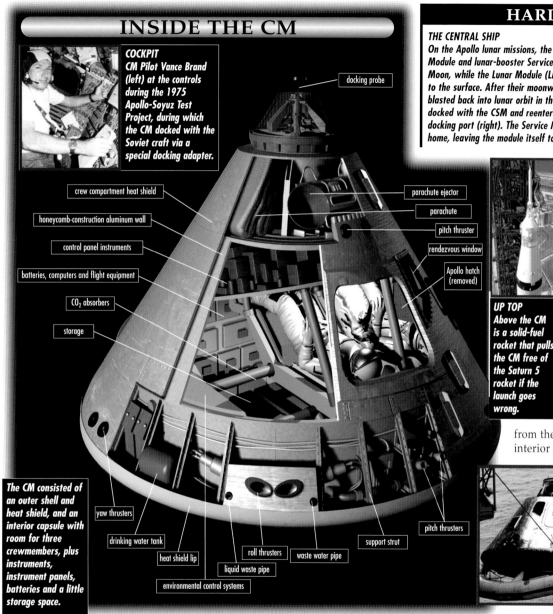

INSIDE THE CM

COCKPIT
CM Pilot Vance Brand (left) at the controls during the 1975 Apollo-Soyuz Test Project, during which the CM docked with the Soviet craft via a special docking adapter.

- crew compartment heat shield
- honeycomb-construction aluminum wall
- control panel instruments
- batteries, computers and flight equipment
- CO₂ absorbers
- storage

- docking probe
- parachute ejector
- parachute
- pitch thruster
- rendezvous window
- Apollo hatch (removed)

The CM consisted of an outer shell and heat shield, and an interior capsule with room for three crewmembers, plus instruments, instrument panels, batteries and a little storage space.

- yaw thrusters
- drinking water tank
- heat shield lip
- roll thrusters
- waste water pipe
- liquid waste pipe
- environmental control systems
- support strut
- pitch thrusters

SMOOTH SHAPE

Faget and Johnson decided to round off the edge of the blunt heat shield at the base, so that the CM would still fit on top of the rocket stage. The result was more aerodynamic than expected, and this new shape turned out to be a crucial benefit.

The CM had to be controllable in air as well as in space. As the module plunged back into

HARD EVIDENCE

THE CENTRAL SHIP
On the Apollo lunar missions, the combined Command Module and lunar-booster Service Module (CSM) orbited the Moon, while the Lunar Module (LM) detached and descended to the surface. After their moonwalk, the landing crew blasted back into lunar orbit in the upper stage of the LM, docked with the CSM and reentered the through the CM's docking port (right). The Service Module's engine was then fired for the return home, leaving the module itself to be jettisoned shortly before reentry.

the atmosphere, air resistance pushed up against the heat shield. If the angle of entry was incorrect by a fraction of a degree, the craft would bounce off the atmosphere and back into space, like a flat stone skimming across the surface of a pond. But if the entry was too steep, the CM would burn to a crisp. The rounded edge allowed the CM to behave like a fat aircraft wing. The astronauts could control the exact angle of approach with 12 tiny thrusters.

In 1970, the CM's engineering saved the day. When Apollo 13's oxygen tank exploded, the craft lost almost all its power. In the chill of interstellar space, sweat from the astronauts soon condensed on interior surfaces such as the instrument panel.

UP TOP
Above the CM is a solid-fuel rocket that pulls the CM free of the Saturn 5 rocket if the launch goes wrong.

TRIAL BY FIRE
The CM was enveloped in an asbestos/epoxy resin heat shield. The mixture was squeezed into a honeycomb layer on the CM's skin, where it hardened like glue and was covered with foil. The shield slowly burned away during reentry, staining the skin (left).

But when the astronauts turned on the power for crucial last maneuvers, there were no electrical shorts that could have stranded them. The CM brought the astronauts back to Earth.

APOLLO LUNAR MODULE

The Apollo Lunar Module was the first vehicle designed to operate purely in the vacuum of space. It was never intended to survive reentry into the earth's atmosphere. Unlike the main Apollo crew capsule, it had no need for an aerodynamic shape or heat shielding. Most of its outer skin was nothing more than lightweight metallic foil, and even the small two-person crew compartment could have been punctured easily. This fragile, bug-like machine was created for a very specialized task—landing astronauts on the Moon.

WHAT IF...

...WE COULD TRACK DOWN THE OLD LUNAR MODULES?

All the Command Modules used during the Apollo missions returned safely to Earth. Today they are exhibited in museums. None of the LMs came home—they were never meant to—but six descent stages still stand on the lunar surface, perfectly preserved.

One day, these historic relics will be visited by lunar tourists. Nearby, they will find the Apollo astronauts' original footprints in the dust, looking almost as fresh as the day they were imprinted. There is no wind on the Moon, no rain, and almost no erosion to sweep away the traces. The footprints may well survive for thousands of years, although micrometeorite impacts, solar radiation, and the stresses of heat and cold in the harsh lunar environment will eventually wear them away. But as long as souvenir hunters leave them intact, the LM descent stages should last a million years or more—perhaps longer than the human race itself.

Further afield, more adventurous tourists will be able to follow the wheel marks made by battery-powered Lunar Rover vehicles. These were carried on the side of the LM descent stages during the last four Apollo lunar missions and were used to transport astronauts across the surface. These rovers, also perfectly preserved, are parked alongside the descent stages, but their tracks, crisscrossing back and forth from the landing sites, stretch over several miles of terrain.

The LM ascent stages are not in such good condition. When Apollo 9 tested an LM in

After an oxygen tank explosion disrupted the Apollo 13 mission, its Lunar Module was jettisoned before reentry. The LM rapidly burned up in the Earth's atmosphere.

Earth orbit and then discarded it, the vehicle assumed a heliocentric orbit, and it presumably remains in orbit today. And then there was the LM of Apollo 13 in April, 1970. After an explosion knocked out power systems in the Command-Service Module on the way to the Moon, the LM served as a lifeboat. Its landing engine pushed the CSM out of lunar orbit and back to Earth.

Six LM ascent stages were discarded in lunar orbit after redocking with their CSMs according to plan. Then they were deliberately allowed to drift off on a crash course with the Moon. The shock waves from the impact reverberated through the Moon, and "moonquake" seismometers left on the lunar surface picked up the vibrations. These experiments gave scientists on Earth new data on the Moon's internal composition.

LUNAR LANDER

WEIGHT	9,180 LB (4,163 KG) DRY; 32,331 LB (14,665 KG) FUELED	WIDTH	31 FT 2 IN (9.5 M) (LEGS EXTENDED)
		DESCENT STAGE ENGINE	1,050–9,870 LB (5-43 KN)
HEIGHT	22 FT 11 IN (7 M)	BURN TIME	15 MIN 10 SEC
WIDTH	14 FT 1 IN (4.2 M)	ASCENT STAGE ENGINE	3,500 LB (15 KN)
	(LEGS STOWED POSITION)	BURN TIME	7 MIN 40 SEC

LUNAR DELIVERY

President John F. Kennedy pledged to the nation in 1961 that a lunar landing would be undertaken within the next 10 years. NASA mission planners, influenced by the ideas of the German-born rocket engineer Wernher von Braun, conceived a spacecraft called "Apollo" to touch down tail-first on the Moon and then blast off back to Earth at the end of its mission.

Other designers, led by NASA engineer John Houbolt, thought this was the wrong approach. If Apollo had to carry fuel for its return to Earth, plus reentry heat shielding and parachutes for splashdown, why must it take all this bulky equipment to the lunar surface? It would just have to be lifted off again, imposing a serious weight penalty on the spacecraft's design.

NASA agreed, and in 1962 the emerging Apollo concept was split into two distinct components: a Command-Service Module (CSM) and a detachable Lunar Module (LM). The LM had to be light enough to be launched from Earth aboard the same Saturn V rocket that carried the CSM. Saving weight proved to be a very difficult challenge for the LM's designers.

Once the spacecraft was successfully in orbit around the Moon, two astronauts clambered aboard the LM for the landing. The third remained aboard the CSM to monitor systems and rendezvous with the LM when it returned from the lunar surface.

The LM consisted of two sections. On completion of lunar surface operations, the landing legs, rocket engine and empty fuel tanks in the lower descent stage were abandoned to save weight. Only the compact crew module, the ascent stage, lifted off for the return trip, using the main fuel tank and engine. After making a rendezvous with the CSM in lunar orbit, the ascent stage was thrown away altogether, saving yet more weight, and leaving just the CSM for the voyage back to earth.

The greatest danger was that the two astronauts in the ascent stage might not be able to locate the CSM in the vastness of space after their ascent from the Moon. Scientists at the Massachusetts Institute of Technology (MIT) designed a computerized navigation system with a radar altimeter. This sophisticated system solved the problem with pinpoint accuracy.

Between 1969 and 1972, the LM performed flawlessly. It carried nine crewed missions, and accomplished an Earth orbital test, a lunar practice descent, six touchdowns and one amazing rescue—Apollo 13.

FLYING LLRV

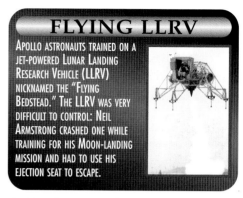

APOLLO ASTRONAUTS TRAINED ON A JET-POWERED LUNAR LANDING RESEARCH VEHICLE (LLRV) NICKNAMED THE "FLYING BEDSTEAD." THE LLRV WAS VERY DIFFICULT TO CONTROL: NEIL ARMSTRONG CRASHED ONE WHILE TRAINING FOR HIS MOON-LANDING MISSION AND HAD TO USE HIS EJECTION SEAT TO ESCAPE.

CROWDED

THE LM CABIN'S INTERIOR WAS SO CRAMPED THAT THE ASTRONAUTS HAD TO STAND UP DURING FLIGHT. DURING REST PERIODS ON THE MOON, THEY FOUND IT IMPOSSIBLE TO LIE DOWN COMFORTABLY, BECAUSE THE ASCENT ENGINE COVER OCCUPIED MOST OF THE FLOOR.

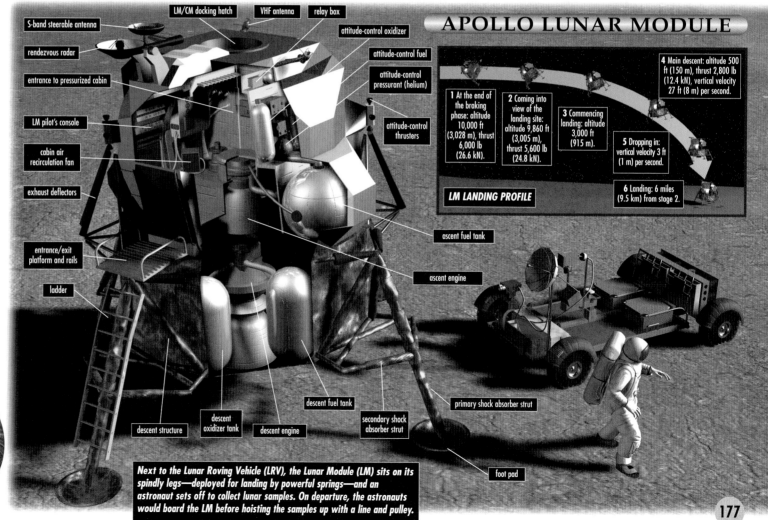

APOLLO LUNAR MODULE

- S-band steerable antenna
- rendezvous radar
- entrance to pressurized cabin
- LM pilot's console
- cabin air recirculation fan
- exhaust deflectors
- entrance/exit platform and rails
- ladder
- LM/CM docking hatch
- VHF antenna
- relay box
- attitude-control oxidizer
- attitude-control fuel
- attitude-control pressurant (helium)
- attitude-control thrusters
- ascent fuel tank
- ascent engine
- descent structure
- descent oxidizer tank
- descent engine
- descent fuel tank
- secondary shock absorber strut
- primary shock absorber strut
- foot pad

LM LANDING PROFILE

1 At the end of the braking phase: altitude 10,000 ft (3,028 m), thrust 6,000 lb (26.6 kN).

2 Coming into view of the landing site: altitude 9,860 ft (3,005 m), thrust 5,600 lb (24.8 kN).

3 Commencing landing: altitude 3,000 ft (915 m).

4 Main descent: altitude 500 ft (150 m), thrust 2,800 lb (12.4 kN), vertical velocity 27 ft (8 m) per second.

5 Dropping in: vertical velocity 3 ft (1 m) per second.

6 Landing: 6 miles (9.5 km) from stage 2.

Next to the Lunar Roving Vehicle (LRV), the Lunar Module (LM) sits on its spindly legs—deployed for landing by powerful springs—and an astronaut sets off to collect lunar samples. On departure, the astronauts would board the LM before hoisting the samples up with a line and pulley.

APOLLO 8

T he crew of Apollo 8 were the first three humans to break the bonds of Earth's gravity and travel to another world. Their 6-day journey included 20 hours in orbit around the Moon. For NASA, the successful mission represented another step toward a lunar landing before 1970. But as the astronauts circled the Moon on Christmas Eve 1968, transmitting their descriptions of the lunar surface to a fascinated television and radio audience around the world, the Apollo 8 mission made history in its own right.

WHAT IF...

...WE COULD SEE THEM NOW?

V isitors to the Museum of Science and Industry in Chicago can see the Apollo 8 command module, scarred by extreme heat when it reentered the Earth's atmosphere but otherwise intact. The first craft to travel to the Moon and back is there on loan from the Smithsonian National Air and Space Museum. But for the men who flew the Apollo 8 capsule, life has gone on.

Frank Borman retired from NASA after the flight and also from the U.S. Air Force, with the rank of colonel. A year after his half-a-million-mile round-trip journey to the Moon, he began a career in more conventional aviation, as special adviser to Eastern Airlines. By 1975, he was Chief Executive Officer of Eastern, and he stayed on until 1986. Two years later, he joined Patlex, a laser technology company. He lives in New Mexico with his wife Susan.

For Bill Anders, the excitement he felt when he was told he would circumnavigate the Moon on Apollo 8 was tinged with disappointment. He had been in friendly rivalry with Neil Armstrong, learning to fly the Lunar Landing Training Vehicle, in the hope and expectation of flying the real thing to the surface of the Moon. But after the LM-less mission of Apollo 8, he thought he would be passed over for future landings. Anders served as backup command module pilot for Apollo

The crew of the Apollo 8 mission (from left to right, James Lovell, William Anders and Frank Borman) have led varied careers after returning to Earth.

11, before quitting the program.

Anders spent the next eight years working for the U.S. government, first as executive secretary of the National Aeronautics and Space Council, and then on the Nuclear Regulatory Commission. He was appointed ambassador to Norway before leaving public service in 1977 for a highly successful career in the business world.

At the time of Apollo 8, Jim Lovell was already a veteran of two Gemini spaceflights, and was looking for his own mission to command. When he was named backup to Neil Armstrong, Apollo 11 commander, he knew it was only a matter of time before he got his own lunar landing.

In April 1970, Lovell and his Apollo 13 crew lifted off for what would become known as "the successful failure." A near-catastrophic explosion in an oxygen tank meant that Lovell never did get to land on the Moon, and found himself staring wistfully at the surface from the window of a crippled spacecraft. But he and his crew lived to tell the tale, and his book about the experience, *Lost Moon*, was adapted for the movie *Apollo 13*.

APOLLO 8 FACTS

CREW	FRANK BORMAN, COMMANDER	EARTH ORBITAL ALTITUDE	115 MILES
	JAMES A. LOVELL, LUNAR MODULE PILOT (NAVIGATOR)	TRANSLUNAR INJECTION BURN	DECEMBER 21, 10:42 A.M. EST
	WILLIAM A. ANDERS, COMMAND MODULE PILOT (FLIGHT ENGINEER)	LUNAR ORBIT INSERTION	DECEMBER 24, 4:59 A.M. EST
LAUNCH	7:51 A.M. EST, DECEMBER 21, 1968, KENNEDY SPACE CENTER	MAXIMUM ALTITUDE ABOVE EARTH	235,000 MILES (378,000 KM)
		CLOSEST APPROACH TO THE MOON	69 MILES (111 KM)
VEHICLE	SATURN 5 ROCKET; FIRST CREWED LAUNCH OF A SATURN 5	DURATION OF FLIGHT	6 DAYS, 3 HOURS, 1 MINUTE

LUNARWATCHER

Astronaut Frank Borman and his crew, Jim Lovell and Bill Anders, were training for an Apollo flight test in Earth orbit in August 1968. But then CIA intelligence revealed that the Soviets were preparing to send a Soyuz spacecraft on a trip around the Moon.

NASA was determined not to be beaten to the Moon. They knew the lunar module (LM) remained in production. But the Saturn vehicle and the command module would be tested in orbit in October, and if all went well, Apollo 8 could orbit the Moon in December, without the LM. Borman told his crewmates they had new orders: They were going to the Moon for Christmas.

The watching world was captivated as the first flight to the Moon lifted off on December 22. After two Earth orbits, Capcom Michael Collins gave the historic order: "Apollo 8, you are Go for TLI." Translunar injection, or TLI, was the engine burn that sent Apollo to the Moon.

CHRISTMAS PRESENT

In the early hours of Christmas Eve, after nearly three days in space, Apollo 8 fired its engine to decelerate, and the Moon's gravity pulled the spacecraft into its orbit. As soon as they could, the crew radioed their impressions of the surface. "Essentially gray, no color…Looks like plaster of Paris," said Lovell. Anders chipped in, "Or sort of a grayish beach sand." They described and photographed the features of the far side of the Moon, never before seen by human eyes.

On the fourth revolution, the astronauts were startled by the beautiful sight of Earth rising above the lunar horizon. And on the ninth orbit, Apollo 8 made its famous Christmas Eve television broadcast to the people of Earth, the planet Lovell called "a grand oasis in the big vastness of space." Twenty hours after reaching lunar orbit, Apollo 8 headed back to Earth.

AROUND THE MOON AND BACK

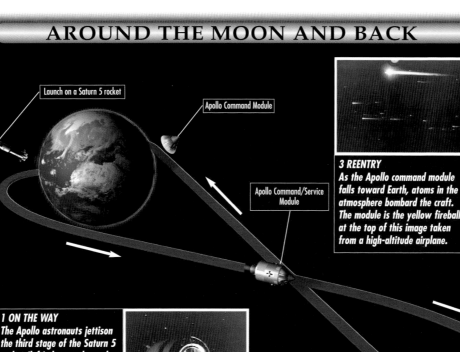

Launch on a Saturn 5 rocket

Apollo Command Module

Apollo Command/Service Module

3 REENTRY
As the Apollo command module falls toward Earth, atoms in the atmosphere bombard the craft. The module is the yellow fireball at the top of this image taken from a high-altitude airplane.

1 ON THE WAY
The Apollo astronauts jettison the third stage of the Saturn 5 rocket (left) that accelerated the Apollo craft from Earth orbit into lunar orbit. The bright balls in the image are droplets of fuel.

The Apollo 8 mission sent a Command/Service Module into Earth orbit, then into some 10 lunar orbits, and then back to Earth. It splashed down in the Pacific Ocean a week later.

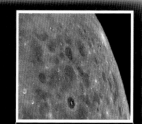

2 THE FAR SIDE
The Apollo 8 crew took many photographs of the far side of the Moon (above), which, unlike the near face's plentiful dark seas, is mostly rocky highlands.

ON THE AIR

NASA'S PUBLIC AFFAIRS OFFICE ASKED MISSION COMMANDER FRANK BORMAN TO "SAY SOMETHING APPROPRIATE" DURING THE CHRISTMAS EVE BROADCAST FROM LUNAR ORBIT. THE TELECAST WOULD BE RELAYED TO MISSION CONTROL (ABOVE) AND OUT TO HALF A BILLION PEOPLE. BORMAN DECIDED ON THE CREATION OF HEAVEN AND EARTH FROM THE BIBLE. THE ASTRONAUTS READ IN TURNS AND SIGNED OFF WITH "A MERRY CHRISTMAS, AND GOD BLESS ALL OF YOU, ALL OF YOU ON THE GOOD EARTH."

MISSION DIARY: APOLLO 8

DECEMBER 21, 1968, 2:36 A.M. EST ASTRONAUTS BORMAN, LOVELL AND ANDERS (RIGHT) ARE AWAKENED TO BEGIN FINAL PREPARATIONS FOR THE FIRST FLIGHT FROM THE EARTH TO THE MOON. *7:51 A.M.* APOLLO 8 LAUNCHES. THE SATURN 5 ROCKET, CARRYING ITS FIRST CREW, CLEARS THE LAUNCH TOWER BURNING 15 TONS OF FUEL PER SECOND. *10:42 A.M.* LOVELL PUSHES THE TRANSLUNAR

INJECTION BUTTON, WHICH IGNITES THE SATURN 5'S THIRD STAGE ENGINE AND SENDS APOLLO 8 ON ITS WAY TO THE MOON. *11:12 A.M.* THE SPENT THIRD STAGE SEPARATES FROM THE COMMAND AND SERVICE MODULE. *DECEMBER 22, 3:06 P.M.* APPROXIMATELY HALFWAY BETWEEN THE EARTH AND THE MOON, THE CREW OF APOLLO 8 MAKES ITS FIRST TELEVISION BROADCAST, BUT HAS DIFFICULTY TRANSMITTING A CLEAR PICTURE OF THE EARTH. THEY STILL MANAGE TO TAKE PHOTOGRAPHS. *DECEMBER 23, 3:00 P.M.* WITH PICTURE PROBLEMS CORRECTED, APOLLO 8 GIVES THE PEOPLE

OF EARTH THEIR FIRST LOOK AT THEIR WHOLE PLANET. *3:29 P.M.* AS APOLLO 8 CROSSES THE POINT WHERE THE MOON'S GRAVITY EXERTS MORE PULL THAN THE EARTH'S, THE SPACECRAFT BEGINS TO PICK UP SPEED. *DECEMBER 24, 4:59 A.M.* LOVELL STARTS THE LUNAR ORBIT INSERTION BURN. THE SERVICE PROPULSION ENGINE BURNS FOR 4 MINUTES TO SLOW APOLLO 8 AND SEND IT IN AN ELLIPTICAL ORBIT AROUND THE MOON. *9:34 P.M.* BORMAN, LOVELL AND ANDERS BROADCAST THEIR PERSONAL IMPRESSIONS OF THEIR

LUNAR FLYBY TO HALF A BILLION LISTENERS. *CHRISTMAS DAY, 1:08 A.M.* APOLLO 8 FIRES ITS ENGINE ONE MORE TIME, LEAVING ITS LUNAR ORBIT ON A TRAJECTORY FOR EARTH. *DECEMBER 27, 10:52 A.M.* SPLASHDOWN IN THE PACIFIC OCEAN (ABOVE). CREW ARE ON BOARD THE U.S.S. YORKTOWN 90 MINUTES LATER.

APOLLO 11

O n July 20, 1969, as the whole world held its breath, Apollo 11's lunar lander, the Eagle, touched down on the Moon. Later, after a few hours rest, the mission commander Neil Armstrong descended the exit ladder followed by Edwin "Buzz" Aldrin and planted the Stars and Stripes in the dust of the Sea of Tranquillity. It was the culmination of a $25 billion program started by President John F. Kennedy in 1961. Armstrong and Aldrin became the first human beings to set foot on another world.

WHAT HAPPENED...

...TO THE CREW OF APOLLO 11?

Been there, done that—the crew of Apollo 11 as they are today. Left to right: Armstrong, Collins and Aldrin.

B orn August 5, 1930, Neil Armstrong was a U.S. Navy pilot and worked as test pilot on the X-15 rocket plane before he joined the space program. He believes that it would be wrong to bask in the glory of being the first man on the Moon and that "he was just doing his job." In Armstrong's view, the main achievement of Apollo 11 was the landing, not the moonwalk; the crew were simply the figureheads of a mission that involved thousands of other Americans. He also points out that the landing could have been made by Apollo 10 or Apollo 12—it just so happened that Apollo 11 was assigned the mission. After leaving NASA, Armstrong went on to become president of a computing systems company. He worked on Presidential Commissions relating to space matters, hosted a TV documentary on aviation, and still gives talks and presentations. He lives on a large farm in Ohio and enjoys playing golf.

Michael Collins was born October 31, 1930. As pilot of the command module, the affable

Left to right: Neil Armstrong, Michael Collins and "Buzz" Aldrin. Not surprisingly, flying the first crewed mission to the Moon proved to be a hard act to follow.

Collins is the forgotten man of the Apollo 11 mission. He calls himself "99% anonymous—until there is an Apollo 11 anniversary!" Collins was offered the commander's job on Apollo 17, but instead left NASA and was instrumental in the establishment of the famed National Air and Space Museum in Washington, D.C. He also wrote *Carrying the Fire*, regarded by many as the best of the astronaut biographies. Collins continued to work in the space industry and set up his own aerospace consultancy. He retired from the Air Force Reserve as a Major General.

Edwin "Buzz" Aldrin was born January 20, 1930. A retired Air Force Colonel, Aldrin is active in the promotion and publicity business. He also gives lectures on space and is a successful author. Aldrin had a difficult time after the Apollo 11 mission: He found the Moon landing a hard act to follow and fell into depression and alcoholism. Aldrin later described his problems in a moving and candid biography, *Return to Earth*, which was subsequently made into a movie starring Cliff Robertson. Today, Aldrin is fully recovered.

APOLLO 11 STATS

CREWED SPACEFLIGHT	NUMBER 33	MASS OF LUNAR SAMPLES RETURNED	48.5 LB (22 KG)
AMERICAN CREWED SPACEFLIGHT	NUMBER 21		
CREWED FLIGHT TO THE MOON	NUMBER 3	PRIOR FLIGHTS OF CREW	ARMSTRONG: GEMINI 8
PREVIOUS CREWED FLIGHTS TO THE MOON	APOLLO 8 AND 10 TEST MISSIONS		COLLINS: GEMINI 10
			ALDRIN: GEMINI 12
FLIGHT OF SATURN 5 ROCKET	NUMBER 5	TRANQUILLITY BASE LOCATION	MOON COORDINATES 0.68°N
FLIGHT TIME	8 DAYS, 3 HR, 18 MIN, 35 SEC		23.43°E
TIME ON MOON	21 HR, 36 MIN		

ONE GIANT LEAP

Apollo 11 carried three astronauts, two of whom—Neil Armstrong and "Buzz" Aldrin—would be the first to set foot on the Moon. The third, Michael Collins, was assigned to remain in the command module, Columbia, taking care of business while it orbited the Moon. The mission began on July 16, 1969, with Armstrong, Aldrin and Collins perched on top of a 363-foot (110-m) Saturn 5 rocket as 7.5 million pounds (33,350 kN) of thrust blasted them into space. Once in Earth orbit, the third and final stage of the Saturn 5 shut down. Apollo 11 then swung around the Earth and the third stage reignited to propel the craft on its 3-day journey to the Moon.

On arrival in lunar orbit, Armstrong and Aldrin crawled into the landing craft, Eagle, which had been tucked into the top of the third stage to protect it during launch. The two spacecraft separated. Then the Eagle's descent engine fired to propel it toward the intended landing site in the Sea of Tranquillity. As the Eagle made its final approach, Armstrong spotted that it was overshooting the landing site and prepared to abort. Mission Control in Houston ordered him to continue.

GOOD JUDGMENT

As the Eagle prepared to land, Armstrong found that the onboard computer was steering it into a rocky crater, with potentially disastrous results. He immediately seized

control and flew over the crater with less than 30 seconds of fuel left, leaving Mission Control powerless to do anything other than watch and trust in his judgment. With about 20 seconds of fuel left, the craft touched down. "Houston, Tranquillity Base here," said Armstrong. "The Eagle has landed."

Later, Armstrong emerged from the Eagle's hatch and climbed down the exit ladder, watched by a huge worldwide TV audience. As he set foot on the surface, he uttered the famous words: "That's one small step for man, one giant leap for mankind" (actually a mistake: he meant to say "...one small step for a man...").

Aldrin joined Armstrong on the surface, where they spent about two hours gathering rocks and deploying experiments. Then they returned to the Eagle and lifted off. Finally, after a delicate docking maneuver with Columbia, Collins fired the rocket engine that would fly the crew home—and into history.

5 DOCKING
Collins took this remarkable photo of the Eagle ascent stage and Earthrise as the lunar module approached the Columbia command module in lunar orbit.

THE APOLLO 11 MISSION

6 SPLASHDOWN
Columbia splashed down beneath 3 parachutes 825 miles (1,327 km) southwest of Honolulu. The crew was on the deck of the recovery ship, USS Hornet, one hour later to be greeted by President Nixon.

3 LUNAR LANDING
"Picking up some dust," said Aldrin. "30 feet, 2½ down...faint shadow...4 forward...4 forward...drifting to the right a little...contact light...OK engine stop."

4 TAKEOFF
Lifting off at a speed of 80 ft (24 m) per second, the Eagle used its spent descent stage as a launchpad as it cleared the dusty, airless surface of Tranquillity Base.

1 SATURN 5
The rocket that powered the Apollo mission is the most powerful ever built in the U.S. At launch, its five F-1 first-stage engines gulped 5,000 gallons (19,000 l) of liquid oxygen and kerosene a second.

2 ESCAPE FROM EARTH
The Saturn 5's S4B third stage was fired to accelerate Apollo 11 to over 25,000 mph (40,000 km/h)—the escape velocity required to enable the craft to break free of the Earth's gravitational field.

MOON JUNK

THE APOLLO 11 ASTRONAUTS LEFT $800,000 WORTH OF DISCARDED EQUIPMENT ON THE MOON: TWO STILLS CAMERAS, LUNAR BOOTS, PORTABLE LIFE-SUPPORT SYSTEM BACKPACKS, MOON TOOLS—AND THE AMERICAN FLAG ON ITS POLE, WHICH WAS KNOCKED OVER BY THE BLAST OF THE EAGLE'S ENGINE. AS THERE IS NO WIND OR RAIN ON THE MOON, THE FOOTPRINTS LEFT BY THE ASTRONAUTS WILL REMAIN THERE FOR CENTURIES.

MISSION DIARY: APOLLO 11

MAY 1961 PRESIDENT JOHN F. KENNEDY COMMITS THE U.S. TO PUTTING A MAN ON THE MOON "BEFORE THE DECADE IS OUT." THE PROGRAM IS NAMED APOLLO.
JULY 16, 1969, 9:32 A.M. EDT APOLLO 11 IS LAUNCHED (RIGHT).
9:44 A.M. APOLLO 11 ENTERS EARTH ORBIT.
12:22 P.M. THE THIRD STAGE OF THE SATURN 5 ROCKET REIGNITES, BLASTING APOLLO 11 OUT OF EARTH ORBIT AND TOWARD THE MOON.

12:49 P.M. THE APOLLO 11 COMMAND AND SERVICE MODULE SEPARATE, TURN AROUND AND DOCK WITH THE LUNAR MODULE.
JULY 19, 1:28 P.M. APOLLO 11 GOES INTO ORBIT AROUND THE MOON.
JULY 20, 1:46 P.M. THE LUNAR MODULE EAGLE SEPARATES FROM THE ORBITING COMMAND MODULE COLUMBIA.
3:08 P.M. EAGLE BEGINS ITS POWERED DESCENT.
JULY 20, 4:18 P.M. "THE EAGLE HAS LANDED."
10:56 P.M. ARMSTRONG BECOMES THE FIRST PERSON ON THE MOON.

JULY 21, 1:54 P.M. THE EAGLE LEAVES THE SURFACE OF THE MOON.
5:35 P.M. EAGLE DOCKS WITH COLUMBIA; THE CREW TRANSFER TO THE COMMAND MODULE.
JULY 22, 12:56 A.M. COLUMBIA BLASTS OUT OF LUNAR ORBIT BACK TOWARD EARTH.
JULY 24, 12:51 P.M. COLUMBIA AND ITS JUBILANT CREW MEMBERS SPLASH DOWN SAFELY IN THE PACIFIC OCEAN (ABOVE).

APOLLO 12

With the flight of Apollo 11 in July 1969, the United States kept the pledge of President Kennedy to land men on the Moon before the decade was out. Before the deadline expired, two more Apollo astronauts would walk on the lunar surface. November's Apollo 12 mission did much more than replicate the achievements of Neil Armstrong and his crew. Astronauts Conrad and Bean stayed longer, and traveled farther on the surface than their predecessors. The age of lunar exploration was underway.

WHAT HAPPENED TO...

...THE CREW OF APOLLO 12?

None of the Apollo 12 team ever returned to the Moon, but the careers of all three men were warped around their day-long stay on the Moon. Their Apollo 12 command module began a new life as a piece of history: The Yankee Clipper is the centerpiece of the visitor's center exhibit at NASA's Langley Research Center, in Hampton, Virginia.

The man who lived in the command module for the 10-day Apollo 12 flight, pilot Dick Gordon, would not fly in space again. He was slated to walk on the Moon on the Apollo 18 mission, after serving with Vance Brand and Harrison Schmitt as the backup crew on Apollo 15. But when Apollo 18 was canceled, Schmitt moved up to the Apollo 17 crew and Brand was reassigned to the Apollo-Soyuz Test Project. Gordon went on to spend a year overseeing Space Shuttle development and testing for the Astronaut Office.

Gordon retired from NASA in 1972, and watched from the stands that December as the last Saturn 5 lifted off for the Moon. He became Vice President of the New Orleans Saints football team, and was later president of a theme park venture, Space Age America, Inc.

Lunar Module Pilot Alan Bean's second spaceflight was in 1973, as commander of Skylab 3, the second mission aboard the United States' first space station. Bean and his colleagues Owen Garriott and Jack Lousma spent 60 days living on the station, running experiments and observing the Earth and the

Today, the Apollo 12 Command Module can be seen at the NASA Langley Research Center in a visitor center exhibit entitled From the Sea to the Stars. Although it never returned to space, two of the Apollo 12 astronauts did.

Sun. Bean retired from NASA in 1981, after six years as head of the Astronaut Candidate Operations and Training Group. Since then, he has made a career as an artist, creating paintings of the Moon and lunar exploration.

Apollo 12 Commander Charles Conrad went on to command the first crewed Skylab mission. He and his crewmates were forced to make essential repairs before moving in to the orbiting module. The job included a three-and-a-half-hour space walk by Conrad and Joseph Kerwin to free a jammed solar panel. Once on board, the three spent 28 days on Skylab. Conrad later said that he was more proud of this successful Skylab mission than of his moonwalk on Apollo 12.

Conrad retired from NASA and the U.S. Navy in December 1973 to pursue a career in technology and communications. By 1995, he was running his own company, Universal Space Lines, developing and testing prototypes of commercial spacecraft. His death in a motorcycle accident in July 1999 deprived the world of spaceflight of one of its heroes.

APOLLO 12 STATS

CREW	COMMANDER CHARLES CONRAD JR.	LUNAR LANDING	NOVEMBER 19, 1969, 1:54 A.M. EST, OCEAN OF STORMS
	COMMAND MODULE PILOT RICHARD F. GORDON JR.		
	LUNAR MODULE PILOT ALAN L. BEAN	LUNAR SURFACE STAY	1 DAY, 7 HOURS, 31 MINUTES
LAUNCH	NOVEMBER 14, 1969, 11:22 A.M. EST	TOTAL EVA TIME	7 HOURS, 45 MINUTES
CALL SIGNS	YANKEE CLIPPER (COMMAND MODULE)	MISSION DURATION	10 DAYS, 4 HOURS, 36 MINUTES
	INTREPID (LUNAR MODULE)	SPLASHDOWN	NOVEMBER 24, 3:58 P.M. EST, PACIFIC OCEAN

BACK SO SOON

Four months had passed since U.S. President Richard Nixon hailed the success of the first lunar landing and made his historic telephone call from the White House to the Moon. Now, on a gray November morning, Nixon took his seat in the stands to watch the next lunar mission take off. Apollo 12 would be the only lunar mission launch cheered on by a serving president. But the weather over Florida was ominous, and as the Saturn 5 rocket cleared the launch tower it was twice struck by lightning, causing electrical surges.

After flicking a few switches, the astronauts put this early disturbance behind them. The rest of Apollo 12's journey to the Moon went without a hitch. Only one mid-course correction was necessary during the 3-day coast to lunar orbit. When the time came for Pete Conrad and Alan Bean to transfer to the lunar module Intrepid, Command Module Pilot Dick Gordon could only manage to say, "I guess I've gotta close the hatch now." Almost three hours after separating from the command module, Intrepid came to rest in the Moon's Ocean of Storms with a distinct thud.

When Pete Conrad stepped onto the surface, his first words raised a laugh at Mission Control, and won him a $500 bet. One of the shortest astronauts ever, at 5 foot 6 and a half (1.68 m), he said, "Whoopee! Man, that may have been a small one for Neil, but it's a long one for me!" Conrad's humor was a marked contrast to the placid bearing of Apollo 11 commander Neil Armstrong.

DOWN TO BUSINESS

Conrad was a professional as well, and foremost in his mind was the key objective of finding Surveyor 3. Launched to examine the Moon's soil in April 1967, Surveyor had operated on the Ocean of Storms for 15 days before running out of power. Apollo 12 was targeted to land as close as possible to the craft, and salvage some of its parts. Conrad soon located Surveyor and relayed the news—Intrepid had almost landed on top of it. He was ecstatic, but the

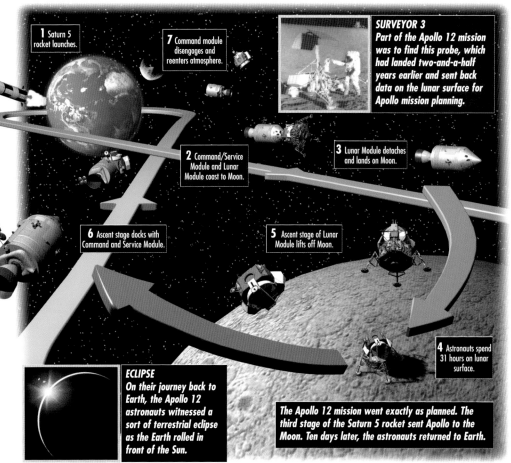

1 Saturn 5 rocket launches.

7 Command module disengages and reenters atmosphere.

SURVEYOR 3
Part of the Apollo 12 mission was to find this probe, which had landed two-and-a-half years earlier and sent back data on the lunar surface for Apollo mission planning.

2 Command/Service Module and Lunar Module coast to Moon.

3 Lunar Module detaches and lands on Moon.

6 Ascent stage docks with Command and Service Module.

5 Ascent stage of Lunar Module lifts off Moon.

4 Astronauts spend 31 hours on lunar surface.

ECLIPSE
On their journey back to Earth, the Apollo 12 astronauts witnessed a sort of terrestrial eclipse as the Earth rolled in front of the Sun.

The Apollo 12 mission went exactly as planned. The third stage of the Saturn 5 rocket sent Apollo to the Moon. Ten days later, the astronauts returned to Earth.

salvage was scheduled for the second of their two moonwalks, and he and Alan Bean concentrated on the job at hand.

The Apollo Lunar Surface Experiment Package was the focus of the first EVA. The job was to deploy a group of instruments—a seismometer, a magnetometer, an atmospheric sensor, an ion detector and a subatomic particle detector—some 600 feet (182 m) from the LM, and to set up the miniature nuclear generator that powered them.

Both EVA missions went smoothly, and on the morning of November 20, Conrad and Bean lifted off from the Moon, after a stay of 31 hours and 31 minutes. They took with them the Surveyor's defunct camera, 76 lb (34.5 kg) of lunar samples, and enough traveler's tales for a lifetime.

MISSION DIARY: APOLLO 12

NOVEMBER 14, 1969 ASTRONAUTS CHARLES CONRAD, RICHARD GORDON AND ALAN BEAN (RIGHT) SUIT UP FOR LAUNCH. **11:22 A.M.** APOLLO 12 LIFTS OFF FROM KENNEDY SPACE CENTER, PAD 39A. LIGHTNING STRIKES THE SATURN 5 ROCKET AT 36 SECONDS AFTER LIFTOFF, AND AGAIN AT 52 SECONDS AFTER LIFTOFF. POWER AND TELEMETRY ARE TEMPORARILY OFFLINE. **2:15 P.M.** AFTER LESS THAN TWO EARTH ORBITS, AND EXTENSIVE SYSTEMS CHECKS, MISSION CONTROL IN HOUSTON GIVES THE GO-AHEAD FOR TRANSLUNAR INJECTION BURN OF THE THIRD-STAGE ENGINES. APOLLO 12 IS ON ITS WAY TO THE MOON.

NOVEMBER 17, 10:47 P.M. APOLLO 12 CRAFT BRAKES INTO AN ELLIPTICAL LUNAR ORBIT THAT MEASURES 194 BY 72 MILES (312 X 115 KM). **NOVEMBER 18, 11:16 P.M.** ABOARD LUNAR MODULE INTREPID, CONRAD AND BEAN SEPARATE THEIR CRAFT FROM COMMAND MODULE YANKEE CLIPPER AND PREPARE FOR POWERED DESCENT TO THE LUNAR SURFACE. **NOVEMBER 19, 1:54 A.M.** INTREPID LANDS ON THE MOON, IN THE OCEAN OF STORMS, JUST 500 FEET (150 M) FROM THE SURVEYOR 3 LUNAR PROBE. APOLLO'S FIRST PINPOINT LANDING IS ACCOMPLISHED. **6:44 A.M.** CHARLES CONRAD BECOMES THE THIRD HUMAN TO SET FOOT ON THE MOON.

10:40 A.M. ASTRONAUTS CONRAD AND BEAN REENTER INTREPID AFTER THEIR FIRST MOONWALK TO DEPLOY EXPERIMENT PACKAGE. **10:55 P.M.** SECOND EXCURSION ONTO THE LUNAR SURFACE. CONRAD AND BEAN RECOVER PARTS OF THE SURVEYOR CRAFT TO RETURN TO EARTH. **NOVEMBER 20, 9:26 A.M.** INTREPID ASCENT STAGE LIFTS OFF FROM THE OCEAN OF STORMS, DOCKING WITH THE COMMAND MODULE IN LUNAR ORBIT THREE-AND-A-HALF HOURS LATER. **NOVEMBER 24, 3:58 P.M.** APOLLO 12 SPLASHES DOWN IN THE PACIFIC. THE ASTRONAUTS SPEND TWO WEEKS IN QUARANTINE (ABOVE).

APOLLO 13

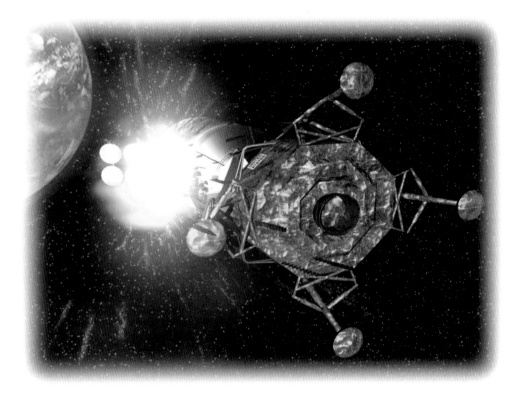

L aunched on April 11, 1970, Apollo 13 was NASA's third attempt at a manned landing on the Moon. The launch itself was a low-key occasion, the press and public having already grown used to the idea of lunar exploration. But then, just over two days into the mission, the world was shocked to attention by the crew's famous radio message: "Houston...we've had a problem here." An onboard explosion had put the lives of the astronauts in serious jeopardy—along with the future of the U.S. space program.

WHAT IF...

...APOLLO 13 HAD SUCCEEDED?

B ecause the Apollo program was largely motivated by a political desire to "beat the Russians in space," it began to run out of steam after the success of Apollo 11, when Neil Armstrong and "Buzz" Aldrin made the first Moon landing. There was a sense of anticlimax, budgets were cut and the public lost interest.

The Moon landings, which were to have continued to Apollo 20, were cut by three missions to finish with Apollo 17. Had Apollo 13 been successful, it is likely that there would have been calls to end the program even earlier, possibly after Apollo 15. Alternatively, had the Apollo 13 crew been killed, cancelation of the entire program would have been the most likely result.

After the safe return of Apollo 13, NASA was given clearance for the 1971 launch of Apollo 14. This was aimed at the same destination—the Fra Mauro Hills in the Ocean of Storms—but with a modified fuel cell

Had Apollo 13 succeeded, Jim Lovell and Fred Haise would have become the fifth and sixth people to walk on the Moon. But those honors fell to Apollo 14 astronauts Alan Shepard (seen here stepping onto the Moon) and Ed Mitchell.

system and other improvements. The agency knew that any failure would mean cancelation of the program. Fortunately, Apollo 14 was a success, and was followed by Apollos 15 (in 1971) and 16 and 17 (in 1972), all of which made use of the Lunar Roving Vehicle.

Although technically successful, the later Apollo missions were regarded by many as an extravagance. So it was that in December 1972, an astronaut left the Moon for the last time in the 20th century.

Even so, Apollo 13 reminded people that flying into space is a dangerous exercise that demands exceptional courage. The fact that the mission snatched triumph from the jaws of disaster also underlined perhaps the greatest achievement of the Apollo program—namely, that it was a triumph of teamwork involving thousands of men and women in hundreds of companies and organizations across the U.S. who together showed the world what the nation could do.

Safely back on Earth: LM Pilot Fred Haise (left), Commander James Lovell (center), CM Pilot Jack Swigert (right).

THE APOLLO 13 CREW

JAMES LOVELL, MISSION COMMANDER
U.S. NAVY CAPTAIN JAMES LOVELL, BORN MARCH 25, 1928, WAS A VETERAN OF THREE SPACEFLIGHTS AND ON APOLLO 13 BECAME THE FIRST PERSON TO MAKE FOUR. HE TOLD OFFICIALS BEFORE THE LAUNCH THAT APOLLO 13 WOULD BE HIS LAST MISSION. LOVELL WROTE THE BOOK APOLLO 13, ON WHICH THE MOVIE WAS BASED.

JOHN SWIGERT, JR., CM PILOT
BORN AUGUST 30, 1931, JACK SWIGERT WAS ASSIGNED TO APOLLO 13 WHEN KEN MATTINGLY WAS DROPPED FOR MEDICAL REASONS. IN 1982, HE WAS ELECTED A REPUBLICAN CONGRESSMAN BUT DIED OF CANCER BEFORE HE ENTERED OFFICE.

FRED HAISE, LM PILOT
FRED HAISE, BORN NOVEMBER 14, 1933, SERVED AS A BACKUP LM PILOT FOR APOLLOS 8 AND 11. AFTER APOLLO 13, HAISE WAS BACKUP COMMANDER OF APOLLO 16 AND DUE TO COMMAND THE CANCELED APOLLO 19.

"HOUSTON, WE'VE HAD A PROBLEM HERE"

As their mighty Saturn 5 launch vehicle thundered into the afternoon skies above the Kennedy Space Center, Apollo 13 astronauts James Lovell, Fred Haise and Jack Swigert began to look forward to their long journey to the Moon. But inside bay 4 of the Service Module, a fault in the number 2 oxygen tank had already turned their spacecraft into a bomb, primed and ready to blow apart.

The tank was part of the Apollo craft's fuel cell system, which produced electricity and water from hydrogen and oxygen. It contained a stirring fan, a heating element and two thermostatic control switches. These switches were designed to operate at 28 volts, but the spacecraft's power supply had been upgraded to 65 volts. As a result, during tests in the weeks before the launch, the higher voltage caused arcing that welded the switches shut. This somehow went unnoticed. During later testing, the faulty switches allowed the temperature of the tank assembly to reach over 1,000°F (540°C), which damaged the insulation of the fan wiring.

The tank finally exploded just under 56 hours into the mission, when Apollo 13 was 205,000 miles (330,000 km) from Earth and the crew was increasing the hydrogen and oxygen pressures to keep the fuel cells functioning properly.

THE JOURNEY HOME

When the accident happened, there was a loud bang and the crew felt the craft shudder. The damaged fan wiring had shorted out, leading to a violent tank explosion that ripped a 13-ft-by-6-ft (4 m x 1.8 m) panel out of the Service Module. Soon, the Command Module was effectively without power, oxygen and water and the main engine, part of the Service Module, was immobilized.

The crew transferred from the Command Module to the Lunar Module, Aquarius. This tiny two-man craft became a "lifeboat," providing essential life-support systems. It also provided propulsion from its descent engine,

which was fired several times to send the crew around the Moon and back to Earth. On the journey home, Lovell, Swigert and Haise huddled in the cold Lunar Module, desperately conserving oxygen, water and power. As Apollo 13 plunged back to Earth, Aquarius was discarded and the crew transferred back to the Command Module. This splashed down safely in the Pacific Ocean, just 4 miles (6.5 km) from the recovery ship *Iwo Jima*.

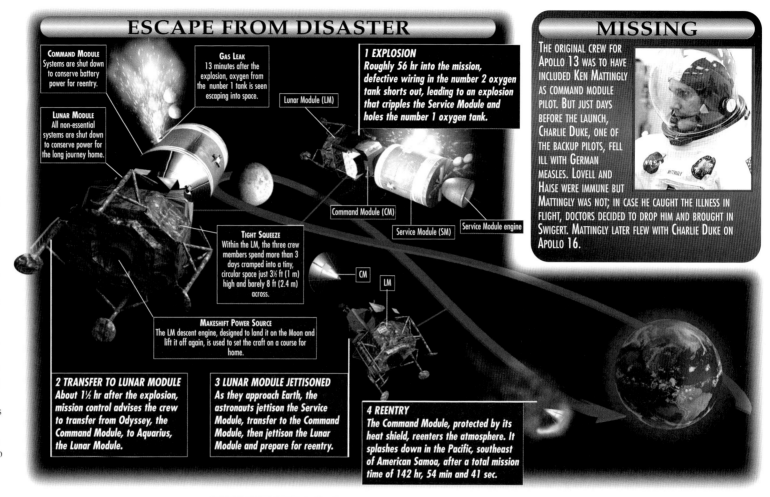

ESCAPE FROM DISASTER

COMMAND MODULE
Systems are shut down to conserve battery power for reentry.

LUNAR MODULE
All non-essential systems are shut down to conserve power for the long journey home.

GAS LEAK
13 minutes after the explosion, oxygen from the number 1 tank is seen escaping into space.

Lunar Module (LM)

1 EXPLOSION
Roughly 56 hr into the mission, defective wiring in the number 2 oxygen tank shorts out, leading to an explosion that cripples the Service Module and holes the number 1 oxygen tank.

Command Module (CM)

Service Module (SM)

Service Module engine

TIGHT SQUEEZE
Within the LM, the three crew members spend more than 3 days cramped into a tiny, circular space just 3½ ft (1 m) high and barely 8 ft (2.4 m) across.

CM

LM

MAKESHIFT POWER SOURCE
The LM descent engine, designed to land it on the Moon and lift it off again, is used to set the craft on a course for home.

2 TRANSFER TO LUNAR MODULE
About 1½ hr after the explosion, mission control advises the crew to transfer from Odyssey, the Command Module, to Aquarius, the Lunar Module.

3 LUNAR MODULE JETTISONED
As they approach Earth, the astronauts jettison the Service Module, transfer to the Command Module, then jettison the Lunar Module and prepare for reentry.

4 REENTRY
The Command Module, protected by its heat shield, reenters the atmosphere. It splashes down in the Pacific, southeast of American Samoa, after a total mission time of 142 hr, 54 min and 41 sec.

MISSING

THE ORIGINAL CREW FOR APOLLO 13 WAS TO HAVE INCLUDED KEN MATTINGLY AS COMMAND MODULE PILOT. BUT JUST DAYS BEFORE THE LAUNCH, CHARLIE DUKE, ONE OF THE BACKUP PILOTS, FELL ILL WITH GERMAN MEASLES. LOVELL AND HAISE WERE IMMUNE BUT MATTINGLY WAS NOT; IN CASE HE CAUGHT THE ILLNESS IN FLIGHT, DOCTORS DECIDED TO DROP HIM AND BROUGHT IN SWIGERT. MATTINGLY LATER FLEW WITH CHARLIE DUKE ON APOLLO 16.

MISSION DIARY: APOLLO 13

APRIL 11, 1970, 2:13 P.M. LAUNCH FROM KENNEDY SPACE CENTER.
4:48 P.M. THE THIRD STAGE OF THE SATURN 5 ROCKET CARRIES THE CRAFT OUT OF EARTH ORBIT TOWARD THE MOON.
6:14 P.M. THE COMMAND MODULE IS TURNED TO DOCK WITH THE LUNAR MODULE. *APRIL 12, 8:53 P.M.* A MID-COURSE MANEUVER PUTS THE CRAFT ON COURSE FOR THE MOON.
APRIL 13, 10:07 P.M. THE NUMBER 1 OXYGEN TANK EXPLODES.
APRIL 14, 3:43 A.M. FIRST LM ENGINE BURN TO SEND CRAFT AROUND THE MOON AND BACK TO EARTH.
9:40 P.M. SECOND LM ENGINE BURN TO CUT JOURNEY TIME TO EARTH.

APRIL 17, 8:14 A.M. SERVICE MODULE IS JETTISONED.
11:43 A.M. LM IS JETTISONED.
1:07 P.M. SPLASHDOWN (LEFT).

APOLLO 14

THE APOLLO 14 CREW

Apollo 14's commander, Alan Shepard, was born in 1923. He graduated from Annapolis in 1944 and served on a destroyer in the Pacific. After World War II he qualified as a pilot and went on to become a test pilot instructor. He was one of the first seven Mercury astronauts and was chosen for the first flight, in 1961, during which he became the first American to go into space. After being diagnosed with an ear infection, Shepard was grounded—and ended up spending six years as chief of the NASA Astronaut Office. But in 1968, an operation cured his ear problem. He went back on the astronaut list and later became commander of Apollo 14.

Shepard retired from the Navy in 1974, with the rank of rear admiral, and became chairman of a construction company and president emeritus of the Astronaut Scholarship Foundation, which raises funds for scholarships for science and engineering students. He died in 1998, after a protracted battle with leukemia.

Stuart Roosa was born in 1933, and was a smoke jumper before he joined the U.S. Air Force at the age of 20. He became an experimental test pilot at Edwards Air Force Base and joined NASA in 1966. Following Apollo 14, Roosa served as backup command pilot for Apollos 16 and 17. He was assigned to the Space Shuttle program until his retirement from the Air Force, as a colonel, in 1976. He pursued various business interests before becoming president and owner of Gulf

The Apollo 14 astronauts, pictured in front of their mission emblem in 1970. Command Module Pilot Stuart Roosa is on the left, Mission Commander Alan Shepard is in the center, and Lunar Module Pilot Ed Mitchell is on the right.

Coast Coors. He died in 1994 of viral pneumonia and is buried in Arlington National Cemetery.

Ed Mitchell was born in 1930. He joined the U.S. Navy after college in 1952. He flew as a carrier pilot and a research pilot, and when the space program began, he was determined to join it—earning a BS in aeronautical engineering from the U.S. Naval Postgraduate School in 1961, and a doctorate in aeronautics/astronautics from the Massachusetts Institute of Technology in 1964.

When Mitchell joined NASA in 1966, he chose to specialize in the Lunar Module. Following the Apollo program, he retired from NASA and from the Navy (as a captain) and founded the Institute of Noetic Sciences in an effort to integrate various scientific disciplines into the study of human consciousness. He has written several books, including the The Way of the Explorer, which addresses the latest research in this field.

Apollo 14 began the heavy-duty exploration of the Moon. Edgar Mitchell and Alan Shepard, America's first man in space, spent two days on the lunar surface, pushing the limits of what could safely be done. By the launch of the mission, on January 31, 1971, the Apollo program was a smoothly running machine. The Apollo 14 spacecraft had been modified to correct the failures of Apollo 13 and to extend their capabilities. But, as all engineers know, anything that can go wrong, will go wrong.

MISSION DATA

LAUNCH	JANUARY 31, 1971, 5:03 P.M. EST	TIME ON MOON	33.5 HOURS
CREW	MISSION COMMANDER ALAN SHEPARD	LUNAR TOUCHDOWN	FEBRUARY 5, 1971, 4:18 A.M. EST
	COMMAND MODULE PILOT STUART ROOSA	LUNAR LIFTOFF	FEBRUARY 6, 1:48 P.M EST
	LUNAR MODULE PILOT EDGAR MITCHELL	SPLASHDOWN	FEBRUARY 9, 1971, 4:05 P.M. EST
TIME IN LUNAR ORBIT	67 HOURS (34 ORBITS)	FRA MAURO BASE	3° 40' 24" SOUTH,
MISSION DURATION	9 DAYS 2 MIN		17° 27' 55" WEST.

LUCKY FOURTEEN

Apollo 14 blasted off from Cape Canaveral into a cloudy and rainy afternoon sky on January 31, 1971, to become a textbook example of how to cope with minor problems. Commander and former Mercury astronaut Alan Shepard, Command Module Pilot Stuart Roosa and Lunar Module Pilot Ed Mitchell had inherited the mission and target base of the failed Apollo 13, with extra objectives and new space techniques to try out.

They were bound for Fra Mauro, a cratered highland area that geologists hoped would provide samples of the earliest bedrock of the Moon and give clues to the early history of the Earth. They would stay longer, conduct more tests, and test the endurance of their spacesuits and themselves more than anyone had done before.

The problems that they were to face would be ironed out by the astronauts and by the well-oiled organization behind them at Mission Control. They had already modified their orbit to make up for the 40-minute launch delay caused by the bad weather.

After the spacecraft had moved into lunar orbit, Roosa maneuvered the Command Module Kitty Hawk to dock with the Lunar Module Antares, still attached to the third stage of the booster. The tiny teeth on the Lunar Module failed to engage. Two more tries were unsuccessful. By this time, engineers at Mission Control had dragged in an identical mechanism to try to find the

MOON TREES

Stuart Roosa, once a Forestry Service smoke jumper, carried over 400 seeds with him in orbit around the Moon. The seeds—of loblolly pine, sycamore, sweet gum, redwood and Douglas fir—were later distributed across the U.S. and to several other countries. Trees grown from them (right) serve as a memorial to the Moon landings and to Roosa.

MOON TRACKS

Sunlight glints on tracks leading across the Moon's Fra Mauro highlands from the Lunar Module Antares. The tracks were made by Apollo 14's Modularized Equipment Transporter during Shepard and Mitchell's first trip away from Antares.

LOST
The rolling highlands of Fra Mauro made navigation difficult. Shepard and Mitchell (above, consulting a map) often lost sight of each other and had difficulty keeping a fix on landmarks.

THE FLYING RICKSHAW
An artist's impression (left) shows Shepard and Mitchell setting out on their first EVA. Shepard, on the right, is pulling the Modularized Equipment Transporter (MET), otherwise known as the "Flying Rickshaw." This 2-wheeled buggy held up to 360 lb (163 kg) of equipment, but the two astronauts sometimes found it easier to carry it than to pull it through the lunar dust.

problem. The fault remains a mystery, but on his sixth try, Roosa locked on. The mission continued, and Apollo 14 entered lunar orbit on February 4.

LUNAR GOLF

As Antares prepared to land, Shepard and Mitchell had to reprogram its control computer while they went through their descent preparations, because the module's "Abort" button was malfunctioning. To everyone's relief, their reprogramming was successful and the module touched down safely on a 7° slope only 175 feet (53 m) from its target. Shepard and Mitchell now began setting up an automated scientific laboratory. This included an instrument to measure lunar seismic activity; a series of experiments to measure charged particles near the surface; a small nuclear generator; and a station to transmit data to earth. On their return to the Lunar Module, they collected some Moon rocks, but they picked up most of their geological samples during their second extravehicular activity (EVA).

Back at the Lunar Module, Shepard produced a golf ball and an improvised club. On his second swing, he claimed that the ball had gone for "miles and miles and miles." It wasn't true, but he had become the first lunar golfer. Shepard later confessed that he had tears in his eyes as he first stood on the Moon.

MISSION DIARY: APOLLO 14

January 31, 1971 Apollo 14 crew suits up (right) for launch. Eight minutes before the launch is due, it is delayed for 40 minutes 2 seconds due to heavy clouds.
January 31, 1971, 4:03 p.m. EST Apollo 14 lifts off, carried by a Saturn 5 launcher.
February 5, 1971, 11:50 p.m. Lunar Module Antares, carrying Alan Shepard and Ed Mitchell, separates from Command Module Kitty Hawk, piloted by Stuart Roosa.

February 5, 1971, 4:18 a.m. Antares (right) touches down on a gentle slope in the highlands near the Moon's Fra Mauro crater. The landing site, at latitude 3° 40' 24" south and longitude 17° 27' 55" west, is only 175 feet from the planned touchdown position.
February 5, 1971, 9:42 a.m. Alan Shepard and Ed Mitchell begin their first moonwalk, or extravehicular activity (EVA), lasting 4 hours 49 minutes.
February 6, 1971, 5:11 a.m. Shepard and Mitchell set

off on their second and final EVA, which lasts for 4 hours 35 minutes.
February 6, 1971, 1:48 p.m. After spending 33.5 hours on the lunar surface, Antares lifts off from the Moon for a rendezvous with Kitty Hawk, carrying 94 lb (43 kg) of Moon samples.
February 9, 1971, 4:05 p.m. Splashdown after 9 days 2 minutes in space. Shepard, Mitchell and Roosa go into the quarantine cabin (above) on U.S.S. New Orleans.

APOLLO 15

Apollo 15 astronauts David Scott and James Irwin spent three days on the surface of the Moon and opened up a new age in lunar exploration. As Scott and Irwin explored the surface, the mission's third crewmember, Alfred Worden, collected other science data from lunar orbit. Apollo 15 was the first truly scientific Apollo mission, and Scott appropriately summed up its objectives as he took his first steps in the lunar dust: "Man must explore. And this is exploration at its greatest."

WHAT HAPPENED...

...TO THE APOLLO 15 CREW?

Apollo 15 was to be the final spaceflight for its three crewmembers. On their return from the Moon in August 1971, the three astronauts were immediately assigned as the backup crew for Apollo 17, the final lunar landing mission. It was a dead-end job for an astronaut, because NASA would probably have delayed the flight rather than replace any of its crew members. But as it turned out, Scott, Irwin and Worden did not even get to stay on the Apollo program as backups. Following an investigation into the sale of the postage stamps that they had taken to the Moon and back—without permission—they were replaced as backups by John Young, Charlie Duke and Ken Mattingly.

Apollo 15 commander David Scott was transferred to a desk job as technical assistant for Apollo. Following the conclusion of the Apollo program, Scott assisted in the training of the crew for the Apollo-Soyuz Test Project, visiting the Soviet Union as part of his work. He was later appointed deputy director and then director of NASA's Dryden Flight Research Center in California. He remained there until October 1977, when he retired from NASA to enter private business, concentrating on areas related to space technology. In recent years, he was technical consultant for the movie Apollo 13 and the television miniseries *From The Earth To The Moon*.

James Irwin, the Lunar Module Pilot,

Posing for the official portrait of the Apollo 15 crew, Mission Commander David Scott stands on the left, Command Module Pilot Alfred Worden sits at center, and Lunar Module Pilot James Irwin is on the right.

resigned from NASA as soon as the stamp scandal broke, and went on to establish the High Flight Foundation, a religious organization that allowed Irwin to preach the gospel through his Apollo 15 experiences. Apollo 15 had changed Irwin spiritually and the remainder of his life focused on sharing his faith with others. But the "Moon Missionary" suffered several heart attacks following Apollo 15 and eventually died of one in 1991, aged 61. He was the first of the 12 moonwalkers to die.

In September 1972, Apollo 15 Command Module Pilot Alfred Worden was assigned to NASA Ames Research Center in California, where he served as senior aerospace scientist and chief of the systems studies division. He resigned from NASA in 1975 and has been involved in the High Flight Foundation, founded by Apollo 15 colleague James Irwin, as well as serving as a senior executive of several technology firms, including the aerospace division of B. F. Goodrich.

The Command Module Endeavour is currently on display in the U.S. Air Force Museum at Wright-Patterson Air Force Base in Dayton, Ohio.

APOLLO 15 MISSION DATA

MISSION	43RD CREWED SPACEFLIGHT, 25TH U.S. CREWED SPACEFLIGHT, 4TH LUNAR LANDING	LUNAR MODULE PILOT	JAMES B. IRWIN
		MISSION DURATION	295 HR 11 MIN 53 SEC
LAUNCH	JULY 26, 1971, KENNEDY SPACE CENTER, LAUNCH COMPLEX 39A	DURATION OF LUNAR LANDING	66 HR 54 MIN 53 SEC; 19 HR 8 MIN SPENT IN EVA
SPACECRAFT	SATURN 5 LAUNCHER, CSM ENDEAVOUR, LM FALCON	DISTANCE TRAVELLED IN LRV	17.5 MILES (28 KM); TOP SPEED OF 8.7 MPH (14 KM/H)
COMMANDER	DAVID R. SCOTT	MASS OF LUNAR SAMPLES	171 LB (77.5 KG); LARGEST ROCK WEIGHED 21 LB (9.5 KG)
COMMAND MODULE PILOT	ALFRED J. WORDEN		

FALCON IN SPACE

Four days after a perfect liftoff from the Kennedy Space Center, Apollo 15's Mission Commander David Scott set the Lunar Module Falcon down within half a mile (800 m) of the planned landing site. The site—between the towering Apennine Mountains and the canyon-like Hadley Rille—was chosen because it gave the astronauts the opportunity to study and take rock samples from three distinctive lunar regions all fairly close to each other.

After a good night's sleep, Scott and Irwin began their first moonwalk. They set up a science station and put the Lunar Roving Vehicle (LRV)—in use for the first time— through its paces on a trip to Hadley Rille and Elbow Crater. While Scott and Irwin worked on the surface, the third member of the crew, Alfred Worden, was keeping himself busy in lunar orbit flying the Command Module and operating scientific equipment located in the Service Module. Worden reported that he could see the Lunar Module on the surface as he passed over the landing site.

On their second moonwalk, Scott and Irwin traveled on the LRV to the Apennine foothills, where they collected 24 bags of rock and soil samples. The moonwalk ended with the astronauts setting up an American flag next to the Lunar Module. The third and final surface activity was a trip to the edge of Hadley Rille, where the astronauts were able to study exposed bedrock and layers of lava flows. Scott capped off this final moonwalk by finding a suitable parking place for the LRV, one from which its television camera could watch Falcon lift off.

LAST-MINUTE TASKS

Falcon blasted off from the lunar surface on August 2, 1971. When the two astronauts were safely aboard the Command Module, the crew deliberately crashed Falcon onto the surface to test seismometers left by previous Apollo crews, and released a small scientific satellite into lunar orbit. The last big event before splashdown was a spacewalk by Worden, who floated between the Command

MISSION SCIENCE

STUDY OF THE SUN
The astronauts took pictures of the solar corona (right), and drove core tubes into the Moon's surface to obtain samples of lunar soil dating back millions of years. Solar particles trapped in these samples provided scientists back on Earth with details of the Sun's history, and this information has increased our understanding of how the Sun affects our climate.

GALILEO'S EXPERIMENT
Scott used his geological hammer and a falcon feather to give a practical demonstration of Galileo's theory that falling objects of different weights, if unhindered by air, drop at the same speed. Both objects fell to the surface together in 1.3 seconds.

LUNAR ECLIPSE
During an eclipse of the Moon a day before splashdown, Scott aimed the crew's TV camera at it and described the scene: "Houston, the Moon is a dull orange ball with a sort of gray area in the center and on one side." This type of eclipse takes place when the Earth's shadow passes across the Moon.

ROCK SAMPLES
The astronauts brought back more than 350 individual samples of rock and soil with a mass of 171 lb (77.5 kg). These samples included a crystalline rock nicknamed the "Genesis Rock" because it was thought to be a remnant of the original lunar crust and as old as the solar system itself. The rocks were studied at the Lunar Receiving Laboratory at the Johnson Space Center in Houston.

MEMORIAL

MISSION COMMANDER DAVID SCOTT PLACED A SMALL MEMORIAL ON THE LUNAR SURFACE TO THE NASA ASTRONAUTS AND SOVIET COSMONAUTS WHO HAD DIED PURSUING THEIR NATIONS' GOALS IN SPACE. THE MEMORIAL (ABOVE) CONSISTED OF A SMALL PLAQUE LISTING THE NAMES OF THE 14 DECEASED MEN—EIGHT AMERICANS AND SIX SOVIETS—ALONG WITH A SMALL ALUMINUM FIGURE THAT REPRESENTED A FALLEN SPACEFARER. IT COULD STILL BE THERE MILLIONS OF YEARS HENCE.

and Service modules to pick up film canisters.

Apollo 15, the most productive lunar mission to date, was hailed as "man's greatest hours in the field of exploration." But there was trouble ahead for the mission's crew. At the end of his final moonwalk, Scott canceled special stamps the crew had brought with them—an unauthorized action that would get them into trouble back on Earth. After an investigation into the sale of the stamps and envelopes carried to the Moon by the crew, all three astronauts were reprimanded and removed from active astronaut status.

Some years later, in 1983, NASA decided to sell commemorative stamps flown aboard the Space Shuttle Challenger. Worden sued the space agency and in a settlement received back the Apollo 15 stamps that NASA had confiscated in 1971.

MISSION DIARY: APOLLO 15

JULY 26, 1971, 9:34 A.M. EDT APOLLO 15 LIFTS OFF ABOARD A SATURN 5 ROCKET FROM THE KENNEDY SPACE CENTER IN FLORIDA.

JULY 29, 1971 APOLLO 15 FIRES ITS MAIN ENGINE TO SLOW DOWN AND EASE ITSELF INTO LUNAR ORBIT.

JULY 30, 1971 THE LUNAR MODULE FALCON DESCENDS STEEPLY OVER TOWERING MOUNTAINS TO LAND IN THE HADLEY-APENNINE REGION OF THE MOON.

JULY 31, 1971 SCOTT AND IRWIN BEGIN THE FIRST MOTOR TRIP ON THE MOON, DRIVING THE LUNAR ROVING VEHICLE TO THE RIM

OF A DEEP CANYON CALLED HADLEY RILLE.

AUGUST 1, 1971 THE ASTRONAUTS MAKE THEIR SECOND MOONWALK, EXPLORING THE FOOTHILLS OF THE APENNINE MOUNTAINS AND DEPLOYING AN AMERICAN FLAG (LEFT).

AUGUST 2, 1971 THE ASTRONAUTS EXPLORE HADLEY RILLE.

AUGUST 2, 1971, 1:11 P.M. FALCON'S LIFTOFF FROM THE MOON IS TELEVISED, USING PICTURES FROM THE LRV'S TV CAMERA.

AUGUST 2, 1971 3:10 P.M. FALCON DOCKS WITH ENDEAVOUR IN LUNAR ORBIT.

AUGUST 4, 1971 AFTER LAUNCHING A SMALL LUNAR SATELLITE, ENDEAVOUR FIRES ITS MAIN ENGINE TO LEAVE LUNAR ORBIT.

AUGUST 5, 1971 WORDEN CARRIES OUT AN EVA (LEFT) TO COLLECT FILM CASSETTES FROM THE PANORAMIC CAMERA AND MAPPING CAMERA, WHICH ARE MOUNTED ON THE SERVICE MODULE, AS ENDEAVOUR HURTLES TOWARD EARTH AT 3,000 MPH (4,800 KM/H).

AUGUST 7, 1971 APOLLO 15 RETURNS TO EARTH SLIGHTLY FASTER THAN PLANNED WHEN ONE OF ITS THREE MAIN PARACHUTES COLLAPSES JUST BEFORE SPLASHDOWN IN THE MID-PACIFIC. THE ASTRONAUTS ARE UNHURT AND ARE PICKED UP BY THE U.S.S. OKINAWA.

APOLLO 16

Fifth of the six successful Moon-landing flights, Apollo 16 had a scientific mission all its own. On April 20, 1972, Apollo 16's Lunar Module made the first touchdown in the lunar highlands. While Command Module Pilot Ken Mattingly made a series of observations from orbit, Mission Commander John Young and Lunar Module Pilot Charlie Duke explored the rugged terrain of the Descartes region on the ground. Scientists had expected signs of volcanism. But the astronauts' fieldwork overturned existing theories.

WHAT HAPPENED TO...

...THE APOLLO 16 CREW?

Despite their close cooperation during the mission's 11-day spaceflight, astronauts Charlie Duke, Ken Mattingly and John Young have gradually gone their own ways in the three decades since they returned.

Lunar Module pilot Charlie Duke was backup LM pilot for the last mission to the Moon—Apollo 17—eight months after Apollo 16. Then he worked on Space Shuttle development until retiring from NASA in 1975 to become a Coors Beer distributor in San Antonio, Texas. In another dramatic career change three years later, Duke became a lay minister. Today, he works in business and continues to preach.

A decade after Apollo 16, Apollo Command Module pilot Ken Mattingly went on to command Space Shuttle mission STS-4 in 1982. The Department of Defense mission was so secret he had to communicate with mission control in code. And his final mission, STS-51C, was another secret: The exact contents of its payload (what appears to have been a spy satellite) were classified. Mattingly retired from the Navy in 1990 and now works for the California-based Rocket Development Company.

Commander John Young—one of NASA's brightest lights—went on to command the first Space Shuttle mission in 1981 and another two years later. He was slated to command STS-31, the Hubble Space Telescope mission, but never flew it. His outspoken

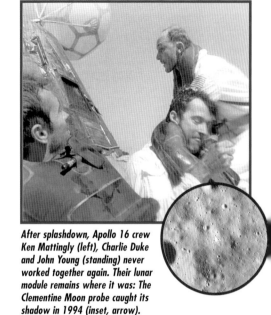

After splashdown, Apollo 16 crew Ken Mattingly (left), Charlie Duke and John Young (standing) never worked together again. Their lunar module remains where it was: The Clementine Moon probe caught its shadow in 1994 (inset, arrow).

criticism of space program safety around the time of the Challenger accident in 1986 knocked him out of the running. But he is still one of the world's most experienced astronauts. He is a safety and engineering advisor at Johnson Space Center, and retains active astronaut status.

Their Command Module, Casper, continues to work for the organization as a display at the U.S. Space and Rocket Center in Huntsville, Alabama, home of the Space Camp school.

APOLLO 16 STATS

CREW COMMANDER JOHN W. YOUNG	**DURATION OF LUNAR LANDING** 71 HOURS; 20 HOURS
COMMAND MODULE PILOT THOMAS "KEN" MATTINGLY II	14 MINUTES SPENT IN EVA
LUNAR MODULE PILOT CHARLES M. DUKE JR.	**TOTAL LUNAR DISTANCE TRAVELED IN LUNAR ROVER**
CRAFT CSM CASPER AND LM ORION ON TOP OF SATURN 5	17 MILES (27 KM); TOP SPEED OF 10.5 MPH (16 KM/H)
LAUNCH VEHICLE	**MASS OF LUNAR EQUIPMENT** 1,240 LB (562 KG)
MISSION DURATION 265 HOURS 51 MINUTES 5 SECONDS	**MASS OF LUNAR SAMPLES** 207 LB (94 KG); LARGEST ROCK WEIGHED 25 LB (11 KG)

HIGHLAND FLING

The first three Apollo landings were highly experimental. The limitations of untried equipment—and NASA's professional caution—meant that astronauts spent very little time on the lunar surface. But from Apollo 15 onward, improved technology and growing confidence allowed NASA to add an increasing science and exploration element to mission profiles.

Apollo 16 was the second of these so-called J-series missions. Its major objective was the investigation of the lunar highlands around Crater Descartes—a much tougher target than the relatively flat lunar maria where other Apollos had landed.

With John Young as Mission Commander, Ken Mattingly as Command Module Pilot and Charlie Duke as Lunar Module Pilot, Apollo 16 made a perfect liftoff from Kennedy Space Center at 12:54 p.m. EST on April 16, 1972. The crew was well aware that because of budget cuts, there would be only one more Moon mission to come—and they were determined to be a hard act to follow.

Just under four days later, the Lunar Module (LM) landed as planned near Crater Descartes. Young and Duke curbed their impatience and slept before their hard work began.

HOME BASE
Despite the rugged terrain, LM pilot Duke set his craft down on a smooth patch of ground just 300 yards (274 m) from the planned landing site. After a few hours' sleep, the astronauts unloaded their Lunar Roving Vehicle and set it up for action.

EVA 1
Charles Duke stands near the 30-ft (9-m) deep Plum Crater, with the Lunar Rover in the background. Most of the first EVA was devoted to setting up the Apollo Lunar Surface Experiment Package (ALSEP)—a nuclear-powered science station that would work long after the astronauts departed.

EVA 2
Duke heads back to the buggy near Stone Mountain. He and commander Young had just deployed a gnomon—a reference marker that would indicate the Sun angle, scale and lunar color for photographic purposes.

With the help of their Lunar Rover, Apollo 16's crew traveled almost 17 miles (27 km) across the lunar surface in three separate excursions—Extravehicular Activities, or EVAs in NASA jargon. Their route is shown at left.

EVA 3
His suit grubby with moondust after three days on the surface, Young carries a sample bag from North Crater toward the waiting moon buggy. Apollo 16's rock samples had some surprises for scientists back on Earth.

RECORD BREAKERS

Fully rested, they unpacked and assembled their Lunar Roving Vehicle (LRV) and began the first of three long EVAs. That first working day, most of their time was spent deploying an automated scientific station—Apollo Lunar Surface Experiment Package (ALSEP).

On the second and third EVAs, Young and Duke checked out geology and gathered samples in selected areas near the the landing site. On the second EVA, the astronauts traveled south-southeast toward Cinco Crater on Stone Mountain. The third and final EVA was to North Ray Crater, for a total distance of around 17 miles (27 km).

By the time Mattingly, Young and Duke returned to Earth, their flight duration of 265 hours, 51 minutes and 5 seconds made them the longest Apollo mission so far—and included a record 20 hours 14 minutes of moonwalking. Their 207 pounds (94 kg) of rock samples set another record. In less than seven months, Apollo 17 would have its chance to make the record books, too.

PROBLEM

A GLITCH IN THE PROPULSION SYSTEM OF THE COMMAND MODULE (SEEN HERE IN LUNAR ORBIT) ALMOST CAUSED A LAST-MINUTE ABORT OF THE APOLLO 16 LANDING. WITH THE CSM ENGINES OUT OF ACTION, THE LANDER'S MOTORS WOULD HAVE BEEN THE ASTRONAUTS' ONLY HOPE OF A SAFE RETURN TO EARTH. BUT AFTER NEARLY SIX HOURS OF SYSTEM CHECKS, MISSION CONTROL GAVE THE GO-AHEAD.

MISSION DIARY: APOLLO 16

APRIL 16, 1972, 12:54 P.M. EST SATURN 5 ROCKET LIFTS APOLLO 16 OFF PAD 39A AT THE KENNEDY SPACE CENTER.

1:06 P.M. THE APOLLO CAPSULE REACHES EARTH ORBIT.

3:27 P.M. THE THIRD STAGE OF THE SATURN 5 ROCKET FIRES AND APOLLO BEGINS ITS TRIP TO THE MOON.

4:16 P.M. THE LM IS UNSTOWED AND DOCKS WITH THE APOLLO CSM.

APRIL 20, 3:28 P.M. APOLLO ENTERS LUNAR ORBIT.

1:08 P.M. THE ORION LUNAR MODULE SEPARATES FROM THE COMMAND/SERVICE MODULE. 10:00 P.M. PROBLEM WITH FLIGHT

CONTROLS OF LM DISCOVERED. AFTER SEVERAL HOURS OF DELIBERATION, MISSION CONTROL (LEFT) DOES NOT ABORT LANDING.

APRIL 21, 2:23 A.M. JOHN YOUNG AND CHARLIE DUKE LAND THE LUNAR MODULE ON THE DESCARTES HIGHLANDS ON THE FAR SIDE OF THE MOON, COORDINATES 8.99° S, 15.49° E.

APRIL 21 DURING THE FIRST EVA, THE TWO MOONWALKERS DEPLOY THE APOLLO LUNAR SURFACE EXPERIMENT PACKAGE (ALSEP).

APRIL 22–3 DURING THE SECOND AND THIRD EVA, THE TWO ASTRONAUTS EXPLORE THE SURFACE AND GATHER SAMPLES OF MOON ROCK.

APRIL 23, 8:26 P.M. YOUNG AND DUKE LIFT OFF IN THE ASCENT STAGE OF THE LUNAR MODULE (LEFT, AS SEEN FROM THE LUNAR

ROVER'S TELEVISION CAMERA).

10:35 P.M. THE LM DOCKS WITH THE CSM IN LUNAR ORBIT.

APRIL 24 THE APOLLO ASTRONAUTS LAUNCH A SUBSATELLITE INTO LUNAR ORBIT.

9:15 P.M. THE SM'S ROCKET IGNITES TO BEGIN THE JOURNEY BACK TO EARTH.

APRIL 25, 3:43 P.M. THOMAS MATTINGLY SPACEWALKS TO RECOVER FILM ON THE OUTSIDE OF THE APOLLO CSM.

APRIL 27, 2:45 P.M. THE CM (ABOVE) SPLASHES DOWN IN THE PACIFIC OCEAN.

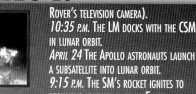

APOLLO 17

The final pair of Apollo moonwalkers, Gene Cernan and Harrison Schmitt, finished the program in style. They stayed longer on the Moon, traveled farther in the lunar rover, and collected more samples than any other mission. Meanwhile, pilot Ron Evans was making the longest ever lunar orbital flight in the Command Module America. But in spite of all its achievements, the mission was overshadowed by reality—NASA had shut down the Apollo program, making Apollo 17 the last voyage of its kind.

WHAT IF...

...THERE HAD BEEN APOLLOS 18, 19 AND 20?

Apollo 20, the last of the scheduled lunar landings, was canceled in January 1970. Later that year, Congress reduced spending on the space program by more than half a billion dollars, to $3.27 billion. The American public had been losing interest in space since the euphoria of the "one giant leap" that had seen Neil Armstrong set foot on the Moon. And in April 1970, after Apollo 13's brush with death, the program began to seem reckless as well as wasteful. So when Thomas Paine, outgoing NASA administrator, announced the agency's plans for 1971, it was no surprise that Apollos 18, 19 and 20 were to be scrubbed.

Harrison Schmitt, geology specialist, had been scheduled to fly on Apollo 18. Schmitt had hoped to head for the Moon's far side in Apollo 18 and to land in the basin of Tsiolkovsky crater. This would have meant putting communication satellites in lunar orbit to allow the far-side explorers to contact mission control.

The cancellation of Apollo 18 forced NASA to make a tough decision. Joe Engle, the scheduled Lunar Module pilot, was pulled from Apollo 17 to make way for Schmitt, breaking up a team that had trained long and hard together. But the change of plan was a consolation to lunar scientists—at last, one of their own would be visiting the Moon.

For Apollo 19, the geologists had hoped for a landing at the lunar north pole, to search for frozen volcanic material. Alternative mission

Space exploration wasn't a priority for President Richard Nixon (left). Lacking his support, NASA had to cancel the final missions of the Apollo program. Fred Haise (below) was one of the astronauts deprived of a lunar landing. The unlucky Haise was also aboard Apollo 13, which famously failed to reach the Moon in 1970.

plans included a look for small volcanoes in the Marius Hills, or landings at great craters such as Tycho and Copernicus. But NASA was listening to Congress, not to the geologists. And they were already looking beyond Apollo, to the Skylab space station project and the Space Shuttle.

One of the three Saturn 5 rockets left idle by the end of Apollo did make it into space, launching Skylab on May 14, 1973. The other two were put on display at the Johnson Space Center in Houston, and the Kennedy Space Center in Florida.

APOLLO 17 MISSION DATA

LAUNCH	12:33 A.M., DECEMBER 7, 1972	DISTANCE COVERED IN LUNAR ROVER	22 MILES (35 KM)
CREW	EUGENE A. CERNAN, COMMANDER		
	RONALD E. EVANS, COMMAND MODULE PILOT	TOTAL WEIGHT OF SAMPLES COLLECTED	243.1 LB (110 KG)
	HARRISON H. SCHMITT, LUNAR MODULE PILOT		
TIME IN LUNAR ORBIT	6 DAYS, 3 HOURS, 48 MINUTES	MISSION DURATION	12 DAYS, 13 HOURS, 51 MINUTES
TIME ON LUNAR SURFACE	3 DAYS, 2 HOURS, 59 MINUTES	SPLASHDOWN	2:25 P.M., DECEMBER 19, 1972

FAREWELL MOON

After a two-hour delay, Apollo 17 lifted off into the Florida night on December 7, 1972. This was the first night launch of the Saturn 5 rocket, and it would be the last, since budget cuts and policy reviews had forced NASA to cancel future missions. As a result, geologist Harrison Schmitt, who had been slated for a seat on Apollo 18, was moved up to the last Apollo crew. The first scientist to go to the Moon, Schmitt knew he had his work cut out for him—so many questions about the Moon and its history remained unanswered, and this mission was the last for the foreseeable future.

As a veteran of Apollo 10, the dress rehearsal for the first landing, mission commander Gene Cernan had been to the Moon before. Cernan and Schmitt had come to this mission by different routes, but for three demanding days on the lunar surface, they worked closely as a unit. The landing site at Taurus-Littrow, a valley deeper than the Grand Canyon, was chosen to provide maximum interest and variety for geological studies. In three seven-hour moonwalks, the astronaut-geologists pushed themselves and their lunar rover to the limit, collecting a record number of lunar samples. But it wasn't all work. Exhilarated and excited, Schmitt and Cernan hopped and skipped and joked their way around the Moon. They left behind a plaque bearing a solemn message: "Here man completed his first explorations of the Moon, December 1972 A.D. May the spirit of peace in which we came be reflected in the lives of all mankind." But Schmitt and Cernan were simply thrilled to be there.

LUNAR ROVING
The Lunar Roving Vehicle makes its way across the Moon's surface, with Gene Cernan in the driving seat. As commander of the mission, Cernan thought it important that his crew didn't just come home with a pile of rocks—he wanted to make sure they made the most of an experience they would remember for the rest of their lives.

ROCKS AND ROVERS
Harrison Schmitt examines a large boulder, making full use of his geological expertise. During this third EVA, Schmitt and Cernan collected samples at the foot of the Taurus mountains.

APOLLO'S SWAN SONG

LEAPS AND BOUNDS
Harrison Schmitt bounds across the Moon in the Taurus-Littrow region. Confidence in the Apollo space suits had grown with each successful mission: Schmitt felt free to leap and sing his way across the lunar surface.

OLD CHARLIE

THE SPECTACULAR LAUNCH OF APOLLO 17 LEFT AT LEAST ONE OF THE KENNEDY SPACE CENTER'S VIP GUESTS UNMOVED. OLDEST LIVING AMERICAN CHARLIE SMITH, WITH 130 YEARS OF EXPERIENCE BEHIND HIM, WAS SKEPTICAL ABOUT THE SATURN 5 ROCKET'S CHANCES OF GETTING CERNAN, EVANS AND SCHMITT TO THE MOON. "I SEE THAT'S A ROCKET," HE SAID TO HIS 70-YEAR-OLD SON CHESTER, "BUT THERE AIN'T NOBODY GOIN' TO NO MOON—ME, YOU, OR ANYBODY ELSE."

MISSION DIARY: APOLLO 17

DECEMBER 7, 12:33 A.M. EST After weeks of preparation and training (RIGHT), APOLLO lifts off from CAPE KENNEDY. DECEMBER 10, 2:47 P.M. LUNAR ORBIT INSERTION maneuver puts APOLLO 17 INTO ELLIPTICAL ORBIT. 7:07 P.M. LUNAR MODULE (LM) SEPARATES FROM COMMAND AND SERVICE MODULE (CSM). DECEMBER 11, 2:55 P.M. LM CHALLENGER LANDS IN TAURUS-LITTROW VALLEY.

6:55 P.M. FIRST EVA, LASTING 7 HOURS 12 MINUTES. CERNAN (RIGHT, WITH EVANS IN CSM) AND SCHMITT DRIVE TO THE CHOSEN SITE AND SET UP THE APOLLO LUNAR SURFACE EXPERIMENT PACKAGE. THEY TAKE 30 LB (13 KG) OF DEEP CORE AND SURFACE SAMPLES. DECEMBER 12, 6:28 P.M. SECOND EVA, A FIVE-MILE DRIVE TO THE SOUTH MASSIF, A MOUNTAIN IN THE TAURUS RANGE. VISITS FOLLOW TO CRATERS LARA AND SHORTY. ORANGE SOIL FOUND AT SHORTY; LATER IDENTIFIED AS

GLASS BEADS CREATED BY VOLCANIC ACTIVITY. EVA LASTS 7 HOURS 37 MINUTES, AND 75 POUNDS OF MATERIAL IS COLLECTED. DECEMBER 13, 5:26 P.M. THIRD EVA. EXAMINATION OF LARGE BOULDER AT NORTH MASSIF AND 145.5 LB (66 KG) OF SAMPLES TAKEN. COMMEMORATIVE PLAQUE ON LANDING GEAR UNVEILED. DECEMBER 14, 5:55 P.M. CHALLENGER (ABOVE) LIFTS OFF THE MOON AND DOCKS WITH THE CSM.

DECEMBER 15, 1:31 A.M. LUNAR MODULE ascent stage jettisoned. DECEMBER 17, 3:27 P.M. COMMAND MODULE pilot RON EVANS LEAVES THE CSM FOR A 67-MINUTE SPACEWALK. DECEMBER 19, 2:25 P.M. SPLASHDOWN OF THE APOLLO 17 COMMAND MODULE IN MID-PACIFIC. CREW PICKED UP BY RECOVERY SHIP U.S.S. TICONDEROGA (ABOVE).

LUNAR ROVERS

The strange looking contraption that Apollo 15 astronauts Dave Scott and Jim Irwin unfolded on the Moon looked like a dune buggy stripped for space. But the Lunar Roving Vehicle—affectionately known as "Rover"—performed almost flawlessly on its maiden mission. Trundling at a top speed of 8 mph (14 km/h), it traversed more of the lunar landscape with the two men and their equipment than was covered by any previous mission. There was one slight disadvantage, though: The ride was rough.

WHAT IF...

...THE ROVER BROKE DOWN?

Given that the Rovers were the most expensive automobiles of all time—the program cost almost $40 million, not including delivery charges to the lunar surface—astronauts had a right to expect reliable transport. But Apollo engineers and managers knew that the Moon missions were pushing technology to its limits, and they were far too cautious to take any chances that they could avoid. So they had to assume that a Rover might break down, or simply have an accident that immobilized it, and make plans to allow for it.

That meant that astronauts were never allowed to drive farther in a Rover than they could reasonably expect to walk back. And the limit to such a safe walk-back distance was not decided by the distance that an astronaut could hope to cover on foot in the one-sixth gravity of the Moon. After all, the astronauts themselves were all trained to a peak of physical fitness. Instead, the limits were set by their life-support systems, which not only supplied their wearers with oxygen but also kept them cool beneath the blazing Sun. Each moonwalker had an emergency oxygen pack in addition to the standard unit. And suit designers had arranged that two astronauts could plug their suits together to share a single cooling unit, should one of them fail. But two astronauts with only one cooling pack between them could not hope to survive for long in the hostile lunar environment.

So individual Rover excursions were

A tow truck arrives to give assistance to a Lunar Rover in trouble. The picture is fanciful, of course: Not even NASA could afford to send tow trucks to the Moon. But fortunately none of the Rovers ever broke down.

limited to a very few miles. Even so, during all of the Rover-equipped missions ground engineers had to live with the fear of a multiple failure: a Rover breakdown coupled with a suit fault that could leave them with at least one dead astronaut on the Moon.

In the end, the technology that NASA had spent so much money on served the agency and its astronauts well. The Rovers performed perfectly on all three of their missions—Apollos 15, 16 and 17. Although each surface excursion was shorter than the Rover's maximum capability, the ultra-expensive vehicles still allowed the astronauts to cover far more ground than they could have on foot. The astronauts never crashed a Rover in a crater, or dumped one into the depths of a hidden crevasse. And their suits held up, too.

LUNAR ROVER SPECS

Length	10.2 feet (3.1 m)	Power	Two 36-volt batteries
Width	6 feet (1.8 m)	Traction drive	4-wheel drive
Ground clearance	14 inches (35 cm)	Last resting place	
Earth weight	462 pounds (209 kg)	Apollo 15 Rover:	Plain near Hadley Rille
Moon weight	77 pounds (35 kg)	Apollo 16 Rover:	Descartes highlands
Top speed	8.7 mph (14 km/h)	Apollo 17 Rover:	Taurus-Littrow

MOON BUGGY

Soon after Apollo 15's lunar module, Falcon, landed on the Moon on July 30, 1971, astronauts Dave Scott and Jim Irwin began to unload the Lunar Roving Vehicle (LRV) that lay folded inside the storage bay. It was to be the first U.S. wheeled carrier on the Moon.

Over the next three days, the Rover transported the two astronauts and their equipment faster and 10 times farther than any previous Apollo moonwalkers. Later models would go on to carry the astronauts of Apollos 16 and 17 over similar distances. The advantages of such increased mobility were immense. The astronauts were forever racing against the clock and their limited oxygen supplies. The more ground that could be covered in the time available, the greater the chance of finding interesting Moon rocks. The Moon wagon helped to make the three last Apollo missions famous for their scientific achievements.

The Rovers were built by Boeing. Made out of lightweight aluminum, they weighed just 462 lb (209 kg) and ran on two 36-volt batteries. But they were sturdy enough to cope with rough lunar terrain, vacuum conditions and extremes of temperature. On board, there was plenty of storage space for basic geology tools, maps and moon rocks, and folding seats for two astronauts.

HOT ROD

The Rover carried a full load of technology. Among this was a navigation system that greatly helped the astronauts to find their way around the lunar surface: Previous moonwalkers had found it extremely difficult to orient themselves. Then there was the communications equipment. Two antennas maintained contact with mission control. Radio signals could be sent continuously, but for TV pictures, the astronauts had to point the high gain antenna toward Earth by hand.

On its first lunar outing, the Apollo 15 Rover averaged a mere 5 to 7 mph (8 to 11 km/h). But, for Scott and Irwin, strapped in their seats, it seemed a lot faster than that. In the light lunar gravity, bouncing over humps and craters, the Rover was positively speeding. Every time the vehicle went over a bump, its wheels came off the ground and it launched into space. At mission control, all they could hear was Scott's occasional, "Hang on!" followed by Irwin's laughter, as they hit a bumpy patch and the Rover took flight.

Arriving within sight of Hadley Rille, the Apollo 15 moonwalkers switched on the Rover's TV camera. Back at mission control, the geologists huddled around the monitors were rewarded with the clearest color pictures ever transmitted from the Moon. At other times, the cameras would be operated remotely from Houston, recording the astronauts as they worked.

Eventually, after three days' hard work, it was time for the Apollo astronauts to leave. But the Rover had to stay behind. Scott parked it on a small rise a short distance from the lunar module, where its TV camera would perform a final service in filming the liftoff.

Two more Rovers followed on the last two Apollo missions in April and December 1972. Both were identical to their pioneering brother. And both met the same fate: Like their predecessor, they, too, were abandoned on the Moon.

LUNAR ROVING VEHICLE

high-gain antenna
low-gain antenna
TV camera
16-mm camera
communications relay unit
steering handle
sample collection store
equipment store
seat
foot rest
under-seat storage
wire wheel
fender

ON TEST
NASA engineers prepare an LRV for loading on the Apollo 15 lunar module, visible in the background. Apollo 15 was the first of three missions to carry wheeled transport moonward.

TUCKED IN
Astronaut David Scott checks out the first LRV, attached to the lunar module, at the Kennedy Space Center's Spacecraft's Operations Building. Its four wheels are folded over the chassis.

UNLOADING THE LRV

1 QUICK FLIP
The LRV is swung out from its storage bay in the lunar module, where it was stowed nose down.

2 WHEELS OUT
On its way out, the rear of the chassis unfolds and the rear wheels spring out into position.

3 ON THE MOON
Next, the front of the chassis unfolds, and the front wheels snap out, ready to hit the ground.

4 ALL SET
With the seats and footrests unfolded and the equipment loaded, the LRV is ready for a ride.

HELD BACK
During the ride on August 1, 1971, while the LRV was parked on an incline, it began to slide away. Astronaut Irwin managed to hold it back (right), because in lunar gravity, it only weighed one-sixth of its Earth weight.

WELL SPRUNG

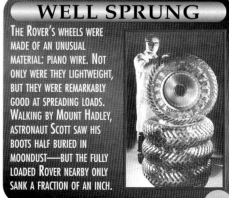

THE ROVER'S WHEELS WERE MADE OF AN UNUSUAL MATERIAL: PIANO WIRE. NOT ONLY WERE THEY LIGHTWEIGHT, BUT THEY WERE REMARKABLY GOOD AT SPREADING LOADS. WALKING BY MOUNT HADLEY, ASTRONAUT SCOTT SAW HIS BOOTS HALF BURIED IN MOONDUST—BUT THE FULLY LOADED ROVER NEARBY ONLY SANK A FRACTION OF AN INCH.

LUNOKHOD ROVERS

T he United States was not the only country to land explorers on the Moon in the early 1970s. While NASA sent teams of Apollo astronauts on dangerous missions to the Moon, the U.S.S.R. launched, landed and operated two radio-controlled Moon buggies at a much lower cost and without putting any people at risk. These so-called Lunokhod ("moon-vehicle" in Russian) rovers trundled for miles across the surface of the Moon, studying and photographing two different regions of the lunar surface.

WHAT IF...

...ROVERS EXPLORED SPACE?

R obot landers and rovers are the quickest and most cost-effective way to explore the surface of the Moon and planets. In 1997, NASA sent the Sojourner rover to Mars as part of the Mars Pathfinder mission. The rover was developed and sent to Mars in only a few years and for relatively little money. The rover returned valuable data without risking human life. Perhaps these remote-controlled cars are the best way to explore space.

Recognizing rovers' potential for exploration, scientists continue to test modern rover prototypes in experimental sites around the world. A rover called Nomad set a distance record exploring the Atacama Desert in Chile in 1997. Controlled from the safety of a NASA center in California, a Russian Marskhod rover explored Kilauea's volcano on Hawaii.

Modern-day technology remains insufficient for such large-scale missions. The next generation of rovers will need to have a range of hundreds of miles, 360-degree vision, high-resolution stereo cameras and equipment for geological work.

But most important, with advanced computers, these roving robots should be able to work independently. The key to successful rover exploration is self-control rather than remote control. The vehicle needs to make decisions

Although rovers would not replace human explorers, they could be used for dangerous, difficult or routine tasks such as primary reconnaissance or mining.

without the constant attention of its controllers.

But even 35 years after the first Moon rovers, artificial intelligence technology remains poor. Although the Sojourner rover was more independent than the Lunokhod rovers, it was hardly smart. If the rover ran into an obstacle, it backed off and waited for instructions. Sojourner's course was actually planned on Earth, and commands were transmitted to it throughout its mission. It could not navigate by itself.

Rovers are less interactive—and less productive—than human explorers. Humans can manipulate tools and samples of soil and rocks more quickly, and with greater dexterity. They can pick up a rock, look at it, manipulate it, all while appraising and reappraising its nature.

Robot rovers are not likely to replace humans in the exploration of the Moon and the planets. Rovers may instead be used for reconnaissance and later may even accompany humans as new worlds are explored. Both humans and machines will have their own roles to play in the future of our solar system.

SOVIET ROVER SPECS

LUNOKHOD 1		LUNOKHOD 2	
LAUNCH DATE	NOVEMBER 10, 1970	LAUNCH DATE	JANUARY 8, 1973
LANDING DATE	NOVEMBER 17, 1970	LANDING DATE	JANUARY 15, 1973
LANDING VEHICLE	LUNA 17	LANDING VEHICLE	LUNA 21
LANDING SITE	MARE IMBRIUM	LANDING SITE	LE MONNIER CRATER
PERIOD OF OPERATION	11 MONTHS	PERIOD OF OPERATION	4 MONTHS
DISTANCE TRAVELED	6.5 MILES (10.4 KM)	DISTANCE TRAVELED	23 MILES (37 KG)
WEIGHT	1,650 POUNDS (748 KG)	WEIGHT	2,000 POUNDS (907 KG)
TV PICTURES TAKEN	20,000	TV PICTURES TAKEN	80,000
PANORAMA PICTURES	206	PANORAMA PICTURES	86
SOIL TESTS	500	SOIL TESTS	740
END OF MISSION	OCTOBER 4, 1971	END OF MISSION	MAY 9, 1973

MOON ROVING

Soviet spacecraft Luna 17 quietly landed on the Moon in November 1970, 16 months after the first men walked on the lunar surface. Within hours of the landing, two ramps attached to the lander extended, and a strange, bathtub-shaped vehicle rolled down to the surface of the Moon. Under the remote control of a Russian control center, the first Lunokhod rover was off.

The portly rover had a rounded body mounted on eight wheels. On a sunny lunar day (one Earth month), the rover's round top opened, and it ran off solar cells. At night, the lid shut to conserve heat.

During its first days on the moon, Lunokhod 1 drove across the pockmarked surface of the Moon, stopping occasionally to perform soil tests. The rover even had a stomping device to measure the strength of the lunar soil. It also measured X-rays from outside our galaxy and the distance between the Earth and the Moon.

LONG-DISTANCE DRIVING

Lunokhod's earthbound drivers peered through a front-facing TV camera for boulders in the Lunokhod's path. But because of the 240,000-mile (386,000 km) distance between Earth and Moon, the images lagged significantly. The drivers had to remember that Lunokhod was several yards ahead of the TV picture on the monitor.

The second rover—which blasted off on the spacecraft Luna 21 two years later—was much improved. It could travel at twice the speed of Lunokhod 1, and had another TV camera to give

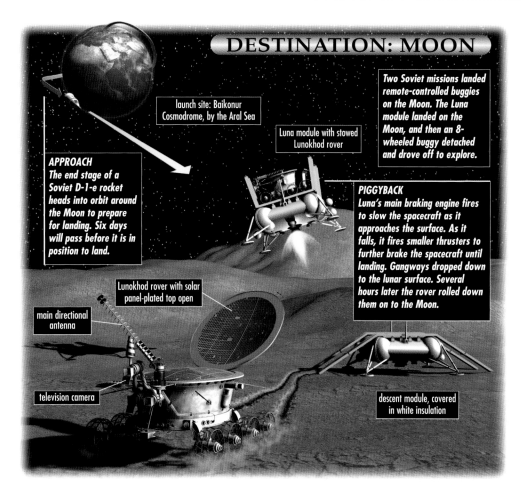

DESTINATION: MOON

launch site: Baikonur Cosmodrome, by the Aral Sea

Luna module with stowed Lunokhod rover

Two Soviet missions landed remote-controlled buggies on the Moon. The Luna module landed on the Moon, and then an 8-wheeled buggy detached and drove off to explore.

APPROACH
The end stage of a Soviet D-1-e rocket heads into orbit around the Moon to prepare for landing. Six days will pass before it is in position to land.

PIGGYBACK
Luna's main braking engine fires to slow the spacecraft as it approaches the surface. As it falls, it fires smaller thrusters to further brake the spacecraft until landing. Gangways dropped down to the lunar surface. Several hours later the rover rolled down them on to the Moon.

Lunokhod rover with solar panel-plated top open

main directional antenna

television camera

descent module, covered in white insulation

LEFT BEHIND

The Lunokhod spacecraft, Luna 17 and 21, carried men to the Moon—in a sense. Like all other Soviet probes of this period, commemorative metal plaques had been attached to the spacecraft. One depicted Soviet revolutionary Vladimir Lenin (above), another the Soviet coat of arms, and a third the Soviet flag. Since the Lunokhod craft were never recovered, the plaque of Lenin is to this day the only image of a political figure on the Moon.

controllers a better view.

Soviet operators sent the "turbo" Lunokhod 2 over a rugged moonscape to the foothills of the Taurus Mountains where, despite having to dodge yard-size boulders, the rover explored a 10-mile (16-km) long crevice. And it made an unexpected discovery.

One of its detectors found that the so-called daytime lunar sky was actually brighter than Earth's night sky—which dampened the enthusiasm of astronomers hoping to set up a lunar observatory. After traveling some 23 miles (37 km), the rover suddenly gave out—as did the project. Despite having built a third, even more sophisticated rover, the Lunokhod mission was judged too expensive to continue.

MOON LICENSE

Five men drove the Lunokhod vehicle from a base in the Soviet Union's Deep Space Communications Center (right). In comparison, more than 20 were present at Mission Control in Houston during an Apollo lunar mission. The Lunokhod team was made up of a commander, driver, engineer, navigator and radio operator. All had to use the rover's cameras as their own eyes.

MISSION DIARY: LUNOKHOD 1

LUNOKHOD 1
November 10, 1970 Luna 17 is launched from Baikonur Cosmodrome, U.S.S.R., carrying Lunokhod 1 payload.
November 17, 1970 Luna 17 lands in the Moon's Mare Imbrium. Lunokhod 1 rolls onto the surface and photographs a hillside (above).
December 22, 1970 Rover parks one mile (1.6 km) from the landing stage.
January 18, 1971 Lunokhod returns to Luna 17 lander and photographs its tracks in the lunar surface (bottom).
Mid-April 1971 While traveling through a boulder field,

Lunokhod's wheels sink in eight inches (20-cm) of dust.
Mid-July 1971 During its ninth lunar day, Lunokhod crosses a 650-foot (200-m) wide crater.
Early September 1971 Lunokhod's equipment begins to wear out.
October 4, 1971 14 years after launch of Sputnik, Lunokhod's mission is over.

LUNOKHOD 2
January 8, 1973 Spacecraft Luna 21, carrying Lunokhod 2, launches from Baikonur Cosmodrome, U.S.S.R.
January 15, 1973 Luna 21 landing stage (right) touches down in the Le Monnier crater. Lunokhod 2 rolls onto the surface.

Late January 1973 Lunokhod nearly collides with the Luna 21 spacecraft. The rover manages to stop only 12 feet (3.6 m) from the lander.
Mid-February 1973 Lunokhod 2 zigzags up slopes toward the Taurus Mountains.
March–April 1973 The Lunokhod rover drives along the rim of a lunar crevice, 165 feet (50 m) deep and 1,300 feet (400 m) wide.
June 9, 1973 Soviets announce that the Lunokhod 2 research program has been completed.

LUNAR PROSPECTOR

NASA's first lunar mission in more than two decades mapped the chemical composition of the Moon's surface, measured the strength of its magnetic and gravitational fields, and found compelling evidence of ice at its poles—all for about $70 million. Called Lunar Prospector, it was developed under the "faster, better, cheaper" concept of NASA's Discovery program. Lunar Prospector orbited the Moon for 18 months before it was deliberately crashed into a crater near the south pole in one last look for ice.

WHAT HAPPENED TO...

...LUNAR PROSPECTOR

In July 1999, Lunar Prospector became the latest piece of space hardware to be abandoned on the surface of the Moon. It joined a variety of Soviet and U.S. probes and Moon rovers, and the descent stages of six Apollo missions.

Future vacationers may tour some of the most important sites in the history of lunar exploration. They will see the bottom half of the Apollo 11 Lunar Module, Eagle, with Neil Armstrong's historic footprint at its base. They will see the rovers used by three Apollo crews, and the Soviet Union's remote-control Lunokhod rovers. They may even visit a small crater within a crater near the Moon's south pole—the impact site of Lunar Prospector. The craft was deliberately crashed into the lunar surface in a search for ice.

Lunar Prospector found evidence of ice near the north and south poles during its first weeks in lunar orbit. The ice appears to be buried just below the surface, in the floors of craters that never see the light of day. While the evidence of ice was compelling, it was not conclusive—Lunar Prospector did not actually "see" any ice, nor could it. But as the end of its mission neared, scientists at the University of Texas at Austin proposed one final experiment that might allow astronomers to directly detect water or related compounds.

The experiment called for the spacecraft to crash into the shadowed region of a small crater near the Moon's south pole. Lunar Prospector's fuel was almost gone, so it would have crashed soon anyway. Scientists hoped the high-speed impact would blast any buried ice into space, where Hubble Space Telescope and observatories on Earth might see it.

Two days before impact, flight controllers began a series of maneuvers to aim Lunar Prospector precisely at the target crater. Engineers expected it to crash at 3,800 mph (6,115 km/h)—an impact speed that would vaporize most of the craft and gouge a small crater in the wall of the larger crater. The impact probably left only a few shards of metal scattered around the crash site.

Unfortunately, there was no big "splash" of lunar water when the spacecraft impacted. Although a half-dozen large telescopes watched the event, they saw no sign of water or ice being blasted up from the surface. Future explorers will have to continue the search for ice buried near the Moon's poles.

LUNAR PROSPECTOR

PROGRAM	DISCOVERY (THIRD MISSION)	**RESULTS**	MAPPED CHEMICAL COMPOSITION OF LUNAR SURFACE; MEASURED LUNAR GRAVITATIONAL AND MAGNETIC FIELDS; DETECTED EVIDENCE OF WATER NEAR NORTH AND SOUTH POLES
PROJECT DIRECTOR	ALAN BINDER		
LAUNCH	9:29 P.M. EST ON JANUARY 6, 1998, FROM SPACEPORT FLORIDA		
LAUNCH VEHICLE	LOCKHEED MARTIN ATHENA 2 (FIRST FLIGHT)	**END OF MISSION**	JULY 31, 1999—SPACECRAFT INTENTIONALLY CRASHED INTO CRATER NEAR LUNAR SOUTH POLE

MAPPING THE MOON

During the 1960s and early 1970s, NASA pursued the Moon like a love-struck suitor. It dispatched more than 30 spacecraft to fly past the Moon, orbit it, or land on its surface, and a dozen astronauts walked on it. After the final Apollo mission, though, the space agency turned its back on the Moon, launching only one small probe in the next quarter-century (and helping the Department of Defense with one more).

NASA finally returned to the Moon in 1998, when it launched Lunar Prospector, one of the first missions chosen under its "faster, better, cheaper" Discovery program. Lunar Prospector was no bigger than the Lunar Orbiter missions of three decades earlier, and didn't even carry a camera. But thanks to advances in computers, communications and instrumentation, it produced a bounty of new findings about the Moon.

Lunar Prospector was launched from Cape Canaveral on January 6, 1998. The craft arrived at the Moon four days later, entering a polar orbit that would allow it to scan almost the entire lunar surface during its mission. A series of brief engine burns eventually dropped it into a circular orbit 62 miles (100 km) high.

During its first year of observations, Lunar Prospector produced a map of the Moon's gravitational field that was the most detailed of any body in the solar system. It also mapped the Moon's magnetic field, which confirmed that the Moon has a small metallic core. And like the 19th-century prospectors of the American West, it found and mapped mineral resources on the Moon's surface.

MOON ICE

Lunar Prospector's most intriguing finding, announced just two months after it entered orbit, was evidence of large amounts of water ice buried in permanently shaded craters at the Moon's north and south poles. Colonists might one day use lunar ice to produce oxygen, drinking water and hydrogen for rocket fuel.

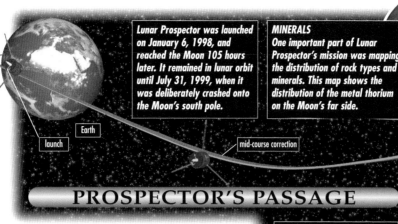

Lunar Prospector was launched on January 6, 1998, and reached the Moon 105 hours later. It remained in lunar orbit until July 31, 1999, when it was deliberately crashed onto the Moon's south pole.

MINERALS
One important part of Lunar Prospector's mission was mapping the distribution of rock types and minerals. This map shows the distribution of the metal thorium on the Moon's far side.

launch

Earth

mid-course correction

ORIGINS

DATA FROM LUNAR PROSPECTOR'S INSTRUMENTS SUPPORTS THE THEORY THAT THE MOON FORMED FROM MATERIAL BLASTED INTO SPACE WHEN A MARS-SIZE BODY RAMMED INTO EARTH SOON AFTER OUR PLANET FORMED. LUNAR ROCKS RESEMBLE THOSE FOUND IN EARTH'S UPPER LAYERS—THE CRUST AND THE MANTLE. THE LUNAR PROSPECTOR PROGRAM WAS CONCEIVED AND MANAGED BY ALAN BINDER (ABOVE).

PROSPECTOR'S PASSAGE

When Lunar Prospector completed its 1-year primary mission, in early 1999, NASA extended the mission by six months, allowing flight controllers to drop the craft's altitude to less than 20 miles (32 km). The orbit change allowed the craft to map finer details in the Moon's gravitational and magnetic fields. As the end of the extended mission neared, Lunar Prospector dropped even lower, skimming just six miles above some mountain peaks. On July 31, it was deliberately crashed into a crater near the south pole in an effort to kick buried ice into space, where telescopes on Earth could see it. Unfortunately, no ice was detected.

GRAVITY
Lunar Prospector produced a highly detailed map of the Moon's gravitational field. This has provided scientists with invaluable data and revealed the presence of gravity anomalies like those shown in red below.

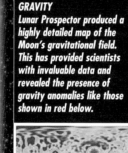

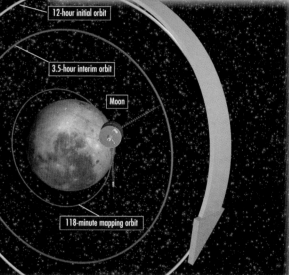

12-hour initial orbit

3.5-hour interim orbit

Moon

118-minute mapping orbit

MISSION DIARY: LUNAR PROSPECTOR

FEBRUARY 28, 1995 NASA SELECTS LUNAR PROSPECTOR (RIGHT) FOR FUNDING UNDER THE DISCOVERY PROGRAM.
SEPTEMBER 1997 THE LAUNCH OF LUNAR PROSPECTOR IS DELAYED WHILE LOCKHEED MARTIN ENGINEERS COMPLETE THE NEW ATHENA 2 BOOSTER.
JANUARY 6, 1998 LUNAR PROSPECTOR IS LAUNCHED AT *9:29 P.M.* EST FROM PAD 46 AT SPACEPORT FLORIDA, WHICH IS A COMMERCIAL LAUNCH SITE AT CAPE CANAVERAL.
JANUARY 11 LUNAR PROSPECTOR ENTERS LUNAR ORBIT.

MARCH 5 SCIENTISTS ANNOUNCE THAT LUNAR PROSPECTOR HAS FOUND EVIDENCE OF WATER ICE BURIED NEAR THE NORTH AND SOUTH POLES.
SEPTEMBER 3 NEW DATA FROM LUNAR PROSPECTOR PLACES THE AMOUNT OF LUNAR WATER AT MORE THAN 6 BILLION TONS (5.4 BILLION TONNES), ABOUT 10 TIMES GREATER THAN EARLIER ESTIMATES.
JANUARY 28, 1999 A SERIES OF ENGINE FIRINGS REDUCE LUNAR PROSPECTOR'S ALTITUDE TO LESS THAN 20 MILES (32 KM). AT THIS LOW ALTITUDE, CAREFUL ATTITUDE MANEUVERS ARE REQUIRED

EVERY FOUR WEEKS TO KEEP THE CRAFT IN ORBIT.
JULY 29 FLIGHT CONTROLLERS INITIATE A SERIES OF ORBITAL MANEUVERS THAT WILL EVENTUALLY BRING LUNAR PROSPECTOR OUT OF ORBIT AND CRASH IT INTO A CRATER NEAR THE MOON'S SOUTH POLE (RIGHT, COLORED BLUE).
JULY 31 LUNAR PROSPECTOR CRASHES, ENDING ITS MAIN MISSION. SCIENTISTS HOPING TO SEE BURIED ICE BLASTED UP INTO SPACE BY THE IMPACT ARE DISAPPOINTED WHEN NONE IS SPOTTED.

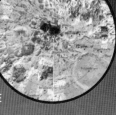

MARS AND VENUS

Earth's nearest neighbours have exerted a fascination on mankind since ancient times. The possibility of intelligent life on either had been ruled out by the time the first probes were sent in the early 1960s. The harsh environment on Venus, with its crushing pressures, howling winds, and extreme temperatures rules out a manned mission, and even unmanned probes last only a short time in the Venusian atmosphere before destruction. Mars is slightly more accommodating, but its constant sub-zero temperatures and frequent dust storms will make human exploration very challenging. The U.S.A.'s Mariner 2 was the first spacecraft to encounter another planet when it flew by Venus in August 1962. A series of attempted and successful flybys and probes flew to Venus before the Soviet Venera 7 landed in December 1970. Magellan, launched from the Space Shuttle in 1989, mapped 98 per cent of Venus using radar.

Mars has seen many successful probes, landers, and rovers, most recently Spirit and Opportunity, which have exceeded all expectations for longevity. The feasibility of a manned Mars mission has been long studied, but with current technology it will be a lengthy mission due to the narrow launch and return windows caused by the relative orbits of Earth and Mars.

The boulder-strewn field of red rocks reaches to the horizon nearly two miles (3 km) from Viking 2 on Mars' Planitia Utopia. Viking 2 landed on Mars on September 3, 1976, some 4,600 miles (7,400 km) from its twin, Viking 1, which touched down on July 20 of the same year.

VENERAS 1–3

In 1960, Venus was an enigma, its face veiled by clouds. And it was not only astronomers who were fascinated by this mysterious planet. The Soviet Union and the U.S. were locked into Cold War rivalries and fears. Three years previous, the Soviets had shocked the Americans by placing the first satellite in orbit. Briefly, the U.S. floundered, while the Soviets thrust ahead, seeking any project that would demonstrate their supremacy: More satellites, a man in orbit, and probes to the Moon and Mars—and to Venus.

WHAT IF...

...VENERA 3 LANDED INTACT?

Venera 3 failed to transmit any data back to Earth when it reached Venus. But in 1970, Venera 7's landing capsule sent back data for 35 minutes as it dropped through the Venusian atmosphere and for 23 minutes after it landed.

The year is 2031, and mission scientists at the Jet Propulsion Laboratory in Pasadena, California, are busy analyzing the massive amount of data that has been sent back to them by Magellan 2. This orbiting probe has been surveying Venus for 10 years, and the highly detailed radar images it has been sending back have led to many startling discoveries about the planet.

But one of its most startling discoveries has nothing to do with its primary radar mapping mission: It is the image of a minute, metallic-looking spot that seemed unlike anything else on Venus's scorched surface. A computer analysis had picked it out as an anomaly. But it was so small—just three feet across—that the mission scientists at the Jet Propulsion Laboratory initially ignored it.

Then one of them related the tiny image to one he had seen in a 1990s history of Soviet space exploration.

"It's a Venera!" he exclaimed.

And indeed, it could have been nothing else. On subsequent passes, Magellan 2 was programmed for close-up inspection of the anomaly, and the data it returned revealed a regular metallic shape. The 60-year-old probe had apparently soft-landed on the surface of Venus back in 1966, as Soviet scientists had

claimed. At the time, this claim had seemed merely a mixture of Soviet propaganda and wishful thinking, because communications had ceased at the moment the probe entered the atmosphere. No remains of the parachute survived—they would have been scorched away by the 900°F (480°C) heat in the intervening decades. And even if the probe itself hadn't been out of commission as it landed, the heat would have melted the insulation of its wiring within hours, and shorted out all its electronic systems.

Yet there it was, glowing but still intact, a find that added a footnote to history. For until that moment, the honor of being the first artificial object to have soft-landed on another planet had belonged to Venera 7, which landed on Venus in December 1970. Venera 3 had its own official record as the first human-made object to reach another planet. Now the record books had to be revised. The Soviet Union had beaten its own record by four years—without even knowing it.

FIRST SOVIET VENUS PROBE

Date	Probe	Outcome		Date	Probe	Outcome
FEBRUARY 4, 1961	SPUTNIK 7	FAILED IN EARTH ORBIT		FEBRUARY 19, 1964	VENERA 1964B	FAILED TO REACH EARTH ORBIT
FEBRUARY 12, 1961	SPUTNIK 8/ VENERA 1	VENUS FLYBY; COMMUNICATIONS FAILURE		MARCH 27, 1964	COSMOS 27	FAILED IN EARTH ORBIT
AUGUST 25, 1962	SPUTNIK 23	FAILED TO LEAVE EARTH ORBIT		APRIL 2, 1964	ZOND 1	VENUS FLYBY; COMMUNICATIONS FAILURE
SEPTEMBER 1, 1962	SPUTNIK 24	FAILED TO LEAVE EARTH ORBIT		NOVEMBER 12, 1965	VENERA 2	VENUS FLYBY; COMMUNICATIONS FAILURE
SEPTEMBER 12, 1962	SPUTNIK 25	FAILED TO LEAVE EARTH ORBIT		NOVEMBER 16, 1965	VENERA 3	HIT VENUS; COMMUNICATIONS FAILURE
FEBRUARY 19, 1964	VENERA 1964A	FAILED TO REACH EARTH ORBIT		NOVEMBER 23, 1965	COSMOS 96	FAILED IN EARTH ORBIT

VOYAGES TO VENUS

In January 1961, when new U.S. president John F. Kennedy announced that "the torch had been passed to a new generation," he took over a nation with no clear sense of mission in space. Later that year, he would dictate an agenda that would take the U.S. to the Moon. But for the moment, the Soviets were reveling in a string of firsts, including the first satellite in orbit and the first probe to photograph the Moon's dark side.

They were also aiming for the planets. Two attempts to launch Mars probes failed, but then Venus swung closer to the Earth, offering a 4-month journey time—half of that to Mars. On February 4, 1961, a 3-stage rocket made it into orbit with a Venus probe in its fourth stage. But when the moment came to ignite the fourth stage and blast it toward Venus, nothing happened. The Soviets canceled the mission, which they called Sputnik 7, after one orbit, and explained away the failure as a successful test of an Earth-orbiting platform from which a planetary probe could be launched.

A week later, Sputnik 8 carried a second probe into orbit and sent it on its way to Venus. This time, all went well, and the Soviets code-named the probe Venera 1. Weighing half a ton, it had two solar panels and instruments to study cosmic radiation, micrometeorites and charged particles. But seven days after launch,

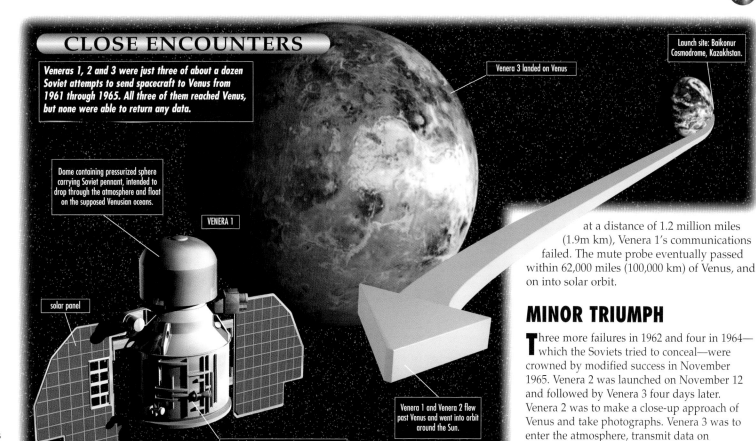

CLOSE ENCOUNTERS

Veneras 1, 2 and 3 were just three of about a dozen Soviet attempts to send spacecraft to Venus from 1961 through 1965. All three of them reached Venus, but none were able to return any data.

Dome containing pressurized sphere carrying Soviet pennant, intended to drop through the atmosphere and float on the supposed Venusian oceans.

VENERA 1

solar panel

probe body containing electronic systems

Launch site: Baikonur Cosmodrome, Kazakhstan.

Venera 3 landed on Venus

Venera 1 and Venera 2 flew past Venus and went into orbit around the Sun.

at a distance of 1.2 million miles (1.9m km), Venera 1's communications failed. The mute probe eventually passed within 62,000 miles (100,000 km) of Venus, and on into solar orbit.

MINOR TRIUMPH

Three more failures in 1962 and four in 1964—which the Soviets tried to conceal—were crowned by modified success in November 1965. Venera 2 was launched on November 12 and followed by Venera 3 four days later. Venera 2 was to make a close-up approach of Venus and take photographs. Venera 3 was to enter the atmosphere, transmit data on temperature and pressure, and then release a descent capsule which would parachute down to the surface.

On February 27, 1966, after more than three months of travel, Venera 2 passed the planet at a distance of 15,000 miles (24,000 km), but again, just as the first pictures should have been sent, the communications system failed. Venera 3, after many mid-course corrections, was perfectly on target. On March 1, it hit Venus as planned but it, too, failed to transmit any information. The Soviets had to be content with a minor triumph: Theirs was the first probe to reach another planet. But Venus would hold her secrets for a few more years.

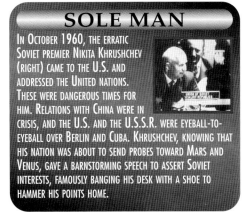
MISSION DIARY: VENERAS 1–3

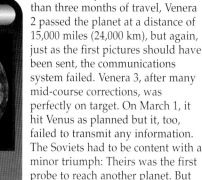

FEBRUARY 4, 1961 Failure of Sputnik 7, the first Soviet attempt to send a probe to Venus.

FEBRUARY 12, 1961 Sputnik 8, put into orbit by a Molniya 8K78 rocket, succeeds in launching Venera 1 toward Venus from Earth orbit.

FEBRUARY 19, 1961 Contact with Venera 1 is lost.

NOVEMBER 12, 1965 Venera 2 is launched from Baikonur Cosmodrome by a Molniya 8K78M (R-7) rocket (above).

NOVEMBER 16 Venera 3 launched. November–December Some

13,000 measurements taken to assess courses of Veneras 2 and 3. A total of 26 communications sessions indicate all is well.

DECEMBER 26 Last course correction places Venera 3 on target to impact Venus (right) three months later.

FEBRUARY 27, 1966 Venera 2 flies past Venus at 15,000 miles (24,000 km), but returns no data.

MARCH 1 Venera 3 impacts Venus 250 miles (400 km) from center of visible face, but it also returns no data.

MARINER TO MARS

T he first close-up images of Mars, sent back by Mariner 4 in 1965, shattered many illusions about the Red Planet. Until then, many people—including scientists—had supposed that Mars was an Earthlike planet, with water and perhaps a breathable atmosphere. There had even been speculation that lines on the surface, visible through telescopes, were canals built by the planet's inhabitants. But the information sent back by Mariner 4 and its sister craft revealed that Mars is a dry, barren, uninhabited world.

WHAT HAPPENED TO...

...THE MARINER MARS PROBES?

U nlike the Pioneer and Voyager probes, which flew past the outer planets Jupiter, Saturn, Uranus and Neptune, the Mariner spacecraft are not destined for the stars. After their close encounter with Mars, Mariners 4, 6 and 7 assumed their own irregular orbits around the Sun and, like comets, became firmly trapped within the solar system. Like comets, too, their orbits can be calculated, so it is possible to estimate where they are, should anyone want to in the future.

NASA did reestablish contact with Mariner 4 in 1967, some two years after its last transmissions from Mars. Engineers were able to detect that the spacecraft had passed through a meteoroid storm on September 15, 1967, when it was 29.6 million miles (47.6m km) from Earth. Some 17 meteoroids hit the craft in a period of 15 minutes, and a drop in on-board temperature showed that they damaged the thermal insulation. In spite of this damage, the craft remained in good shape. The main engine was ignited for 70 seconds, and images 16 and 17 were played back from Mariner's tape recorder memory. When the TV cameras were switched on, they worked but showed only the blackness of space.

As with other NASA spacecraft, the communications link between the Mariner probes and Earth was provided by the Deep Space Network's facilities at Goldstone in the Mojave Desert, California (above), and near Madrid, Spain, and Canberra, Australia.

Today, three decades of interplanetary travel will have taken their toll on the fragile explorers. Exposed to harmful radiation and the extreme temperatures in the vacuum of space, their electronics will have long since perished. The circuits and chips were not designed to last as long as those on probes to the outer solar system, which are shielded from prolonged exposure to radiation and temperature extremes. Further meteoroid strikes are likely to have damaged the delicate solar panels, and any large strike will have caused more serious structural damage.

As for Mariner 9, it is still orbiting Mars. The orbit is designed to last for about 50 years and should not decay until 2021. If humans visit Mars before then and happen to gaze up at the night sky from the Martian plains, they might just be able to make out a tiny speck of light moving rapidly across the sky—an intrepid explorer from the previous century.

MARINER MARS PROBES

Mariner 4			**Dimensions**	10 feet 10 inches (3.2 m) tall, 19 feet ¾ inch (5.8 m) long (with solar panels extended)
Weight at launch	575 lb (261 kg)			
Dimensions	6 feet (1.8 m) tall, 22 feet (6.7 m) long (solar panels extended)		Power	450–500 watts from 4 panels of solar cells
Power	195 watts from 4 panels of solar cells		Data storage	195 Mbits non-reusable tape (Mariners 6 & 7);
Data storage	Magnetic tape			51.22 Kbits reusable magnetic tape (Mariner 9)
Camera	Vidicon TV camera			
Telescope	30.5-millimeter focal length reflector		Cameras	Narrow- and wide-field vidicon TV cameras (508- and 52-millimeter focal length for Mariners 6 & 7; 50-millimeter for Mariner 9)
Mariner 6, 7 and 9				
Weight at launch	910 lb (412 kg) (Mariners 6 &7); 300 lb (136 kg) (Mariner 9)			

FIRST TO THE RED PLANET

NASA's Mariner missions to Mars contributed enormously to our knowledge of the planet, and the program was undeniably a great success despite the loss of two of the six probes shortly after they were launched. The first of these failures came on November 5, 1964, when Mariner 3 was lost after the nose fairing of its Atlas Agena D launch vehicle failed to jettison and free the craft for its journey to the Red Planet. But three weeks later, its sister craft, Mariner 4, was launched successfully. When it swung past Mars at a distance of just 6,116 miles (9,842 km) on July 14, 1965, it sent back the first-ever TV images of the Martian surface.

When scientists received Mariner 4's images and other data from Mars, they were stunned by what they saw. The TV images showed Mars to be a barren, cratered world, very different from the Earthlike planet that many had expected. As Mariner 4 dipped behind the far side of Mars, the planet's atmosphere distorted its radio signals slightly. From this distortion, the atmospheric pressure was calculated to be approximately 5 to 10 millibars—far too low to allow liquid water to exist on the surface. Furthermore, surface temperatures were estimated to be –148°F (-100°C). The findings made depressing reading for those biologists who had hoped that Mars harbored life.

The images from Mariner 4 were limited and of poor quality, but Mars came into sharper focus when the more sophisticated Mariners 6 and 7 flew by in 1969, a memorable year for NASA. Hot on the heels of the historic Apollo 11 Moon mission, Mariners 6 and 7 swooped by Mars at a "grazing" distance of about 2,100 miles (3,370 km). Cameras attached to scanning platforms enabled each craft to collect sharp images of the Martian surface, and filters placed in front of the cameras meant that color photographs could be created. These showed the now-familiar rusty orange-red color of the planet's surface. But by chance, the cameras missed some of the most spectacular features of Mars' ancient terrain—its huge volcanoes, including the largest in the solar system, Olympus Mons.

MARINERS 8 AND 9

To achieve a global survey of Mars and find landing sites for future Viking craft, NASA needed a spacecraft in Martian orbit. Mariners 8 and 9 were built to achieve this, but the initiative got off to a disastrous start when Mariner 8 plunged into the Atlantic Ocean shortly after launch. Mariner 9 was launched successfully, and on November 13, 1971, it became the first artificial satellite of Mars. But the first images showed absolutely nothing—for a planet-wide dust storm was at its height. When the storm subsided, Mariner 9 made many important discoveries, including the Olympus Mons volcano and the 2,450-mile (4,000-km)-long Valles Marineris valley. It sent back a total of 7,329 images and was a fitting end to the Mariner missions.

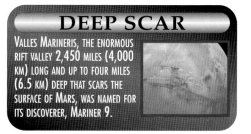

DEEP SCAR

VALLES MARINERIS, THE ENORMOUS RIFT VALLEY 2,450 MILES (4,000 KM) LONG AND UP TO FOUR MILES (6.5 KM) DEEP THAT SCARS THE SURFACE OF MARS, WAS NAMED FOR ITS DISCOVERER, MARINER 9.

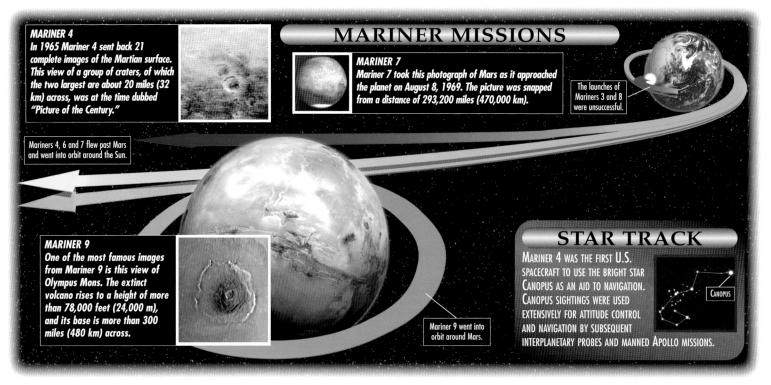

MARINER MISSIONS

MARINER 4
In 1965 Mariner 4 sent back 21 complete images of the Martian surface. This view of a group of craters, of which the two largest are about 20 miles (32 km) across, was at the time dubbed "Picture of the Century."

MARINER 7
Mariner 7 took this photograph of Mars as it approached the planet on August 8, 1969. The picture was snapped from a distance of 293,200 miles (470,000 km).

The launches of Mariners 3 and 8 were unsuccessful.

Mariners 4, 6 and 7 flew past Mars and went into orbit around the Sun.

MARINER 9
One of the most famous images from Mariner 9 is this view of Olympus Mons. The extinct volcano rises to a height of more than 78,000 feet (24,000 m), and its base is more than 300 miles (480 km) across.

Mariner 9 went into orbit around Mars.

STAR TRACK

MARINER 4 WAS THE FIRST U.S. SPACECRAFT TO USE THE BRIGHT STAR CANOPUS AS AN AID TO NAVIGATION. CANOPUS SIGHTINGS WERE USED EXTENSIVELY FOR ATTITUDE CONTROL AND NAVIGATION BY SUBSEQUENT INTERPLANETARY PROBES AND MANNED APOLLO MISSIONS.

CANOPUS

MISSION DIARY: EXPLORING MARS

NOVEMBER 5, 1964 MARINER 3 IS LOST IN AN ACCIDENT SHORTLY AFTER LAUNCH.

NOVEMBER 28, 1964 MARINER 4 IS LAUNCHED SUCCESSFULLY FROM CAPE CANAVERAL, FLORIDA, ON AN ATLAS AGENA ROCKET.

JULY 14, 1965 MARINER 4 PASSES WITHIN 6,100 MILES (9,800 KM) OF MARS AND SENDS BACK THE FIRST CLOSE-UP IMAGE OF THE PLANET.

OCTOBER 1, 1965 THE LAST TELEMETRY FROM MARINER 4 IS RECEIVED WHEN THE SPACECRAFT IS 192 MILLION MILES (308M KM) FROM EARTH. CONTACT IS REESTABLISHED DURING THE FALL AND

WINTER OF 1967.

FEBRUARY 24, 1969 MARINER 6 IS LAUNCHED SUCCESSFULLY.

MARCH 27, 1969 MARINER 7 IS LAUNCHED SUCCESSFULLY.

JULY 28, 1969 THE FIRST MARINER 6 IMAGES ARE TRANSMITTED: 33 PICTURES OF THE FULL MARS GLOBE.

JULY 31, 1969 THE CLOSEST APPROACH OF MARINER 6 ALSO MARKS THE END OF ITS MISSION.

AUGUST 5, 1969 MARINER 7'S CLOSEST APPROACH OCCURS WHEN THE SPACECRAFT FLIES OVER THE SOUTH POLE OF MARS AT A MINIMUM ALTITUDE OF 2,177 MILES (3,500 KM). THE MISSION

FORMALLY ENDS ON THE SAME DAY.

MAY 8, 1971 MARINER 8 IS LOST SHORTLY AFTER LAUNCH.

MAY 30, 1971 MARINER 9 IS LAUNCHED SUCCESSFULLY.

NOVEMBER 13, 1971 MARINER 9 BECOMES THE FIRST ARTIFICIAL SATELLITE OF MARS WHEN IT ENTERS AN ELLIPTICAL ORBIT THAT TAKES IT WITHIN 1,050 MILES (1,689 KM) OF THE SURFACE.

OCTOBER 27, 1972 CONTACT WITH MARINER 9 IS CUT, SIGNIFYING THE END OF THE MARINER MISSIONS TO MARS.

VENERA 9 AND 10

Before the advent of space flight, Venus was an intriguing mystery to scientists. The Moon and Mars had been extensively studied by telescope for centuries, but the surface of Venus was hidden from view by impenetrable cloud cover. In 1961, the Soviets began the Venera program—a long-term attempt to land a probe on the surface of Venus. After many attempts, some of which ended in success and some in failure, Venera 9 and 10 sent back the first black and white pictures of the surface of Venus in October 1975.

WHAT IF...

...THERE WERE MORE MISSIONS TO VENUS?

Engineers knew it would be difficult to design a spacecraft able to withstand the surface conditions of Venus. In the face of temperatures high enough to melt lead, and atmospheric pressure reaching more than half a ton per square inch, the Venera landers were only expected to last 32 minutes. Both craft exceeded this target, but by the time they had been on the surface for an hour, the environment was taking its toll.

The spacecraft that have landed on the Moon and Mars remain as a testament to our first steps into the universe. When astronauts once again visit the Moon, these remnants from a previous era will be the first monuments commemorating space exploration, preserved by the lack of atmosphere. On Mars, landers will gradually erode and wear away as they are blasted by dust whipped up by the strong winds. But on Venus, the Venera landers will be eaten away by the highly corrosive hydrochloric and hydrofluoric acid in the atmosphere.

In such a harsh environment, the next step in exploration—collection of soil samples—is problematic. Apollo astronauts brought back samples of the lunar regolith, and future sample return missions will be launched to Mars. But to return soil samples from Venus, the craft must survive long enough to collect the samples and either return to the orbiter, or analyze the data on the surface and then

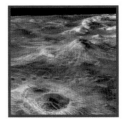

This image, taken by the Magellan probe, gives a 3-dimensional perspective view of the Venusian terrain. The 124-mile (200-km) -wide Nagavonyi Corona, a volcanic feature, can be seen in the foreground.

transmit the analysis to the orbiter. Russian and U.S. spacecraft designers are currently working to overcome some of the problems the mission will present.

The craft will again comprise an orbiter and a lander. Aboard the lander will be a small booster used to loft the sample to Venus orbit. Included as part of the orbiter will be an Earth Return Vehicle (ERV)—a rocket stage to propel the sample back to Earth. The lander will descend to the surface and use a robotic arm to scoop up a sample of the soil. The sample will be deposited inside the booster. Since launching the booster from the surface would require more propellant than the lander could carry, the booster will first be carried to a high altitude using a balloon, and then ignited. It will then dock with the orbiter and transfer the sample to the ERV. When Earth and Venus are correctly aligned for the return trip, the ERV rockets will fire and send the sample on its way.

Scientists hope the new missions will provide even more information about planet formation and the evolution of planetary atmosphere.

VENERA 9 AND 10

	VENERA 9	VENERA 10
LAUNCH DATE	JUNE 8, 1975	JUNE 14, 1975
DATE OF SEPARATION	OCTOBER 20, 1975	OCTOBER 23, 1975
DATE OF LANDING	OCTOBER 22, 1975	OCTOBER 25, 1975
LANDING TIME	5:13 A.M. GMT	5:17 A.M. GMT
LANDING SITE	32° S, 291° E	16° N, 291° E
DURATION OF SURFACE OPERATION	53 MINUTES	65 MINUTES

VENUS BOUND

By 1975, eight Venera spacecraft had left Earth orbit en route to Venus. The Venera series had proved highly successful, exposing many of the planet's mysteries for the first time. Two of the eight craft had descended into the atmosphere, sending back data describing a hostile environment with an atmospheric pressure 90 times that of Earth, surface temperatures over 900°F (480°C) and an atmosphere composed of 97% carbon dioxide.

For the next series of Venera, the Soviet Union developed more advanced craft, capable of returning pictures from the surface for the first time. The Soviets launched two new missions to Venus during the 1975 launch window. Venera 9 lifted off on June 8, 1975, followed six days later by Venera 10. Once placed in orbit by a Proton booster, mission controllers gave them a final check before igniting the rocket stage to propel the craft on their three-and-a-half month journey to Venus.

Two days before arriving at Venus, the lander and the orbiter separated to follow different trajectories. The lander was encased in an 8-foot (2.4-m) diameter spherical capsule which would provide protection during entry. The hermetically sealed sphere distributed the heat load and prevented the craft from imploding due to the immense atmospheric pressure. This protective heat shield was designed to survive temperatures up to 21,630°F (12,000°C). To give it even more protection from the fiery atmosphere, the lander was cooled with refrigerant before it began its descent.

SUCCESSFUL LANDING

Forty miles from the surface, the heat shield was discarded and parachutes were deployed to stabilize the lander and slow descent. Thirty-one miles from the surface, the dense atmosphere allowed aerodynamic braking. A circular collar around the top of the craft generated enough drag to slow the rate of descent. Final touchdown, 75 minutes after entering the atmosphere, was cushioned by a metallic shock-absorbing ring.

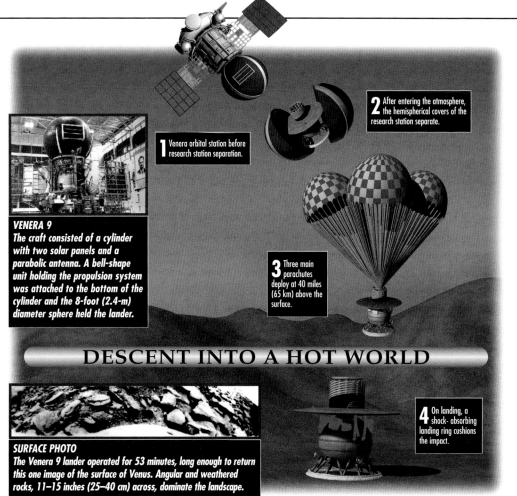

1 Venera orbital station before research station separation.

2 After entering the atmosphere, the hemispherical covers of the research station separate.

3 Three main parachutes deploy at 40 miles (65 km) above the surface.

4 On landing, a shock-absorbing landing ring cushions the impact.

VENERA 9
The craft consisted of a cylinder with two solar panels and a parabolic antenna. A bell-shape unit holding the propulsion system was attached to the bottom of the cylinder and the 8-foot (2.4-m) diameter sphere held the lander.

DESCENT INTO A HOT WORLD

SURFACE PHOTO
The Venera 9 lander operated for 53 minutes, long enough to return this one image of the surface of Venus. Angular and weathered rocks, 11–15 inches (25–40 cm) across, dominate the landscape.

The Venera 9 operated for 53 minutes before failure. Three days later, Venera 10 touched down 1,364 miles (2,195 km) away and survived for 65 minutes. Surface data and video images collected by the probes were transmitted to the orbiter for later relay back to Earth. The probes had also taken measurements of the composition of the atmosphere and structure of the clouds during descent.

The Venera 9 and 10 orbiters relayed the data from the landers back to Earth and studied the upper atmosphere of Venus. But the highlight of the missions was the first photographs of the surface of Venus. The images of the rocky landscape provided the first clues to our understanding of planetary evolution.

MISSION DIARY: VENERA 9–10

JUNE 8, 1975 VENERA 9 IS LAUNCHED INTO ORBIT BY PROTON ROCKET (RIGHT) FROM BAIKONUR COSMODROME IN KAZAKHSTAN, USSR.
AFTER ONE EARTH ORBIT VENERA 9 ROCKET STAGE IGNITES.
JUNE 14 VENERA 10 IS LAUNCHED INTO ORBIT FROM BAIKONUR COSMODROME.
AFTER ONE EARTH ORBIT VENERA 10 ROCKET STAGE IGNITES.
OCTOBER 20 VENERA 9 DESCENT CRAFT SEPARATES FROM ITS ORBITER.
OCTOBER 22 VENERA 9 INSTRUMENT COMPARTMENT COOLED TO 14°F (-10°C).
DESCENT TO 78 MILES (125 KM) ABOVE SURFACE VENERA 9 CAPSULE ENTERS THE VENUSIAN ATMOSPHERE (RIGHT) AT 6.6 MILES (10 KM) PER SECOND. THE TEMPERATURE IS 21,630°F (12,000°C). THE COVERS PROTECTING THE LANDER SEPARATE. AT A VELOCITY OF 820 FEET (250 M) PER SECOND, THE DROGUE PARACHUTE DEPLOYS.
40 MILES (65 KM) FROM THE SURFACE THREE MAIN PARACHUTES DEPLOY.
31 MILES (50 KM) FROM THE SURFACE PARACHUTES JETTISON; DRAG SLOWS CRAFT FURTHER.
OCTOBER 22, 5:13 A.M. GMT VENERA 9 LANDS ON THE SURFACE WITH AN IMPACT VELOCITY OF 20–26 FEET (6–8 M) PER SECOND. THE TV CAMERA COVERS EJECT AND CAMERA AND

INSTRUMENTS SWITCH ON.
OCTOBER 22, 6:06 A.M. GMT VENERA 9 CEASES TO FUNCTION.
OCTOBER 23 VENERA 10 DESCENT CRAFT SEPARATES FROM ITS ORBITER.
OCTOBER 25, 5:17 A.M. GMT VENERA 10 LANDS ON SURFACE OF VENUS. THE FIRST PHOTOS OF THE SURFACE (ABOVE) ARE RELAYED TO THE ORBITER.
6:22 A.M. GMT VENERA 10 CEASES TO FUNCTION.

VIKING TO MARS

After a 10-month journey through interplanetary space, two U.S. spacecraft reached Mars orbit in the summer of 1976. Vikings 1 and 2 were the most sophisticated robot probes yet built, and they had a mission to match their capabilities. While the orbiters mapped the Red Planet from high above its atmosphere, each one sent a Viking lander to a soft touchdown on the surface. The probes beamed back the first images of the rock-strewn landscape, sniffed the Martian air and soil—and searched for signs of life.

WHAT IF...

...VIKING HAD FOUND LIFE ON MARS?

Since the advent of the telescope, the red disk of Mars and its shifting patterns have provoked wonder and fascination. The tantalizing idea that our neighbor planet might be teeming with exotic life-forms has inspired many science fiction classics. But when the Viking landers dropped in on Mars, they found a dry, forbidding landscape, where the temperature could drop as low as –190°F (-125ºC). The two craft sifted the red soil of the planet looking for evidence of microscopic organisms eking out an existence in the soil. They found none. But what if they had? Certainly, the efforts of the Viking scientists would have been vindicated and their billion-dollar budget would have seemed a small price to pay.

But the discovery of life on another world would have had far-reaching consequences. Public reaction might have been one of passionate curiosity coupled with a sense of unease. If there are microbes on Mars, could there be intelligent, even hostile aliens not too much farther away? Humanity has long thought itself one of a kind. Just a few centuries ago, the idea that the Earth travels around the Sun was outrageous. Now Viking would be telling us we were not alone in the universe, that life is abundant and tenacious. It would have been the discovery of the century.

In 1969, NASA and the Space Task Group set up by President Nixon proposed that the U.S. commit itself to an ambitious program of space

Astronauts make a scientific foray from their Closed Environment Life Support System on the Martian surface. Their research mission forms part of the type of Mars program that might have been prioritized by NASA if Vikings 1 and 2 had found traces of organic compounds.

exploration, culminating in the departure of a crewed mission to Mars in 1981. The program was never adopted—the Space Shuttle was the only component to survive budget cuts. But if Viking had found life on Mars in 1976, the space program would have followed a different course. NASA would not have waited 21 years before sending another lander to the planet. Exploratory missions would have been fast-tracked and Mars would have seen a host of robotic exploration missions well before the Pathfinder-Sojourner project of 1997.

Once the danger of contamination from Martian microbes had been dealt with, samples of extraterrestrial life would have been brought back to Earth for study. Perhaps before the end of the century, inspired by the possibility of more exciting discoveries to come, a crew of astronaut explorers might have left Earth for Mars to follow up the momentous discoveries of Vikings 1 and 2.

VIKING PROBE STATISTICS

VIKING 1		VIKING 2	
LAUNCH	AUGUST 20, 1975, KENNEDY SPACE CENTER	LAUNCH	SEPTEMBER 9, 1975, KENNEDY SPACE CENTER
LAUNCH VEHICLE	TITAN 3E-CENTAUR	LAUNCH VEHICLE	TITAN 3E-CENTAUR
TOTAL MASS (UNFUELED)	3,247 POUNDS	TOTAL MASS (UNFUELED)	3,247 LB (1,473 KG)
MARS ORBIT INSERTION	JUNE 19, 1976	MARS ORBIT INSERTION	AUGUST 7, 1976
LANDING	JULY 20, 1976, CHRYSE PLANITIA	LANDING	SEPTEMBER 3, 1976, UTOPIA PLANITIA
ORBITER SHUTDOWN	AUGUST 17, 1980	ORBITER SHUTDOWN	JULY 25, 1978
LOSS OF CONTACT WITH LANDER	NOVEMBER 13, 1982	LOSS OF CONTACT WITH LANDER	APRIL 11, 1980

VIKING EXPLORERS

The Red Planet's first Earth visitor was a Soviet probe that landed in 1971. The probe transmitted TV pictures for 20 seconds and then went silent, possibly because of a radio relay failure—or perhaps as a result of the planet's most violent dust storm in decades. As that storm raged, the U.S. Jet Propulsion Laboratory was steering Mars' first artificial satellite, Mariner 9, into orbit. When the skies cleared, Mariner shot detailed TV pictures of the planet's giant volcanoes and valleys.

The data transmitted by Mariner proved vital in preparations for Viking's journey to Mars four years later. But Viking was a far more ambitious project than Mariner. The Viking team managed to launch two lander-orbiter combination craft within weeks of each other, bringing both Vikings into Martian orbit after a 10-month cruise halfway around the Sun. The mission was one of the most complex ever attempted. It was also the most expensive uncrewed space project to date, costing about a billion dollars. But Viking was worth every penny.

Both landers successfully separated from their orbiters and touched down on the planet's surface to examine Martian biology and sample the chemistry of this distant world. Neither lander found the hoped-for signs of life. Nevertheless, Viking was a great success. The orbiters sent back tens of thousands of images of our neighbor and its moons, Phobos and Deimos. They measured the structure and composition of the atmosphere, and detected water vapor. And the landers kept working for several years, providing data on climate and seismology, as well as vivid panoramas of the reddish landscape.

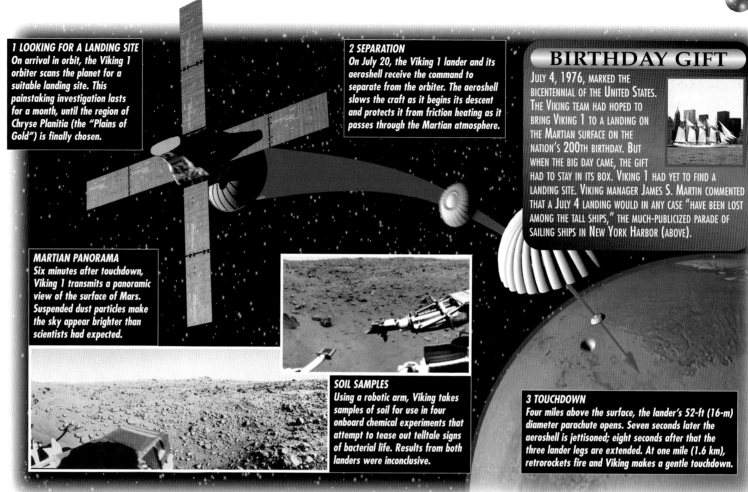

1 LOOKING FOR A LANDING SITE
On arrival in orbit, the Viking 1 orbiter scans the planet for a suitable landing site. This painstaking investigation lasts for a month, until the region of Chryse Planitia (the "Plains of Gold") is finally chosen.

2 SEPARATION
On July 20, the Viking 1 lander and its aeroshell receive the command to separate from the orbiter. The aeroshell slows the craft as it begins its descent and protects it from friction heating as it passes through the Martian atmosphere.

MARTIAN PANORAMA
Six minutes after touchdown, Viking 1 transmits a panoramic view of the surface of Mars. Suspended dust particles make the sky appear brighter than scientists had expected.

SOIL SAMPLES
Using a robotic arm, Viking takes samples of soil for use in four onboard chemical experiments that attempt to tease out telltale signs of bacterial life. Results from both landers were inconclusive.

BIRTHDAY GIFT
JULY 4, 1976, MARKED THE BICENTENNIAL OF THE UNITED STATES. THE VIKING TEAM HAD HOPED TO BRING VIKING 1 TO A LANDING ON THE MARTIAN SURFACE ON THE NATION'S 200TH BIRTHDAY. BUT WHEN THE BIG DAY CAME, THE GIFT HAD TO STAY IN ITS BOX. VIKING 1 HAD YET TO FIND A LANDING SITE. VIKING MANAGER JAMES S. MARTIN COMMENTED THAT A JULY 4 LANDING WOULD IN ANY CASE "HAVE BEEN LOST AMONG THE TALL SHIPS," THE MUCH-PUBLICIZED PARADE OF SAILING SHIPS IN NEW YORK HARBOR (ABOVE).

3 TOUCHDOWN
Four miles above the surface, the lander's 52-ft (16-m) diameter parachute opens. Seven seconds later the aeroshell is jettisoned; eight seconds after that the three lander legs are extended. At one mile (1.6 km), retrorockets fire and Viking makes a gentle touchdown.

MISSION DIARY

NOVEMBER 15, 1968 PROJECT VIKING INITIATED AS A JOINT VENTURE OF THE JET PROPULSION LABORATORY AND NASA'S LANGLEY RESEARCH CENTER. WORK SOON BEGINS ON THE LANDER'S PROTECTIVE AEROSHELL (RIGHT).
AUGUST 20, 1975 AT CAPE CANAVERAL, A TITAN ROCKET LAUNCHES THE VIKING 1 SPACECRAFT ON ITS 62-MILLION-MILE (100M KM) VOYAGE TO MARS.
SEPTEMBER 9, 1975 VIKING 2 LAUNCHED.

JUNE 14, 1976 APPROACHING THE RED PLANET, VIKING 1 CAMERAS COME TO LIFE, TAKING IMAGES OF THE GLOBE OF MARS (RIGHT).
JUNE 19 VIKING 1 BRAKING MANEUVER PUTS THE SPACECRAFT INTO MARTIAN ORBIT.
JUNE 21 AFTER AN ADJUSTMENT TO ITS ORBIT, VIKING 1 SCANS THE PLANET FOR A SUITABLE LANDING SITE.
JULY 20, 3:32 A.M. EST VIKING 1 LANDER SEPARATES FROM THE ORBITER AND BEGINS ITS DESCENT.
JULY 20, 6:53 A.M. VIKING 1 LANDER TOUCHES DOWN IN CHRYSE PLANITIA; 25 SECONDS LATER IT TRANSMITS A PICTURE OF ONE OF ITS OWN FOOTPADS ON FIRM, ROCKY

GROUND (FAR RIGHT).
AUGUST 7 VIKING 2 ARRIVES IN MARTIAN ORBIT.
AUGUST 9 WITH THE HELP OF DATA FROM ITS TWIN, VIKING 2 BEGINS THE SEARCH FOR A LANDING SITE.
SEPTEMBER 3, 5:37 P.M. VIKING 2 LANDER SETTLES ON THE SURFACE THOUSANDS OF MILES FROM VIKING 1, AT UTOPIA PLANITIA.
JULY 25, 1978 VIKING 2 ORBITER IS POWERED DOWN

AFTER A SERIES OF ATTITUDE CONTROL THRUSTER GAS LEAKS. IT HAS MADE 706 ORBITS OF MARS.
APRIL 11, 1980 VIKING 2 LANDER TERMINATED AFTER BATTERY FAILURE.
AUGUST 17, 1980 CONTACT LOST WITH VIKING 1 ORBITER, AFTER OVER 1,400 ORBITS.
NOVEMBER 13, 1982 PROJECT VIKING COMES TO AN END WITH LOSS OF SIGNAL FROM THE VIKING 1 LANDER.

MAGELLAN

The 1970s was NASA's decade. Apollo had landed on the Moon, Voyager was soaring through the solar system, and Viking had landed on Mars. Riding this wave of success was a proposed $800-million mission called Venus Orbiting Imaging Radar (VOIR). This expensive project was canceled in 1982, but good ideas are hard to kill, and in 1983, a scaled-down version called Venus Radar Mapper made it back into the budget. After surviving several cancellation threats, it was renamed Magellan and launched in May 1989.

WHAT IF...

...WE WENT BACK TO VENUS?

Magellan fell into Venus' atmosphere on October 13, 1994. As it fell, the increasing atmospheric pressure and friction destroyed the spacecraft. In all likelihood, the solar panels were the first to be ripped off, followed quickly by the main antenna. Then smaller pieces would have been torn off and burned up by the atmospheric friction. But Magellan entered Venus' atmosphere at a relatively slow velocity, so it is quite possible that the entire vehicle did not burn up in the process. Some larger pieces probably fell gently to the surface, drifting down at an ever-slower rate as the atmospheric pressure increased, and landing like pebbles thrown into a lake. Judging from where the probe's final signals were sent, pieces of Magellan could be resting on the sides of Maxwell Montes, the highest volcano on Venus.

Now that Magellan is gone, what might the next mission to Venus be like? That would depend upon how much the mission was trying to discover. A fairly basic mission would need only a small, relatively simple craft with carefully chosen instruments, and could be put into orbit around Venus for as little as $200 million. A more ambitious mission, needing a larger, more complex craft (possibly carrying landers), would be much costlier.

An orbiter with near-infrared cameras could take 3-dimensional pictures of the clouds of Venus, revealing details of their structure and composition. Entry probes that plunge down into the hostile atmosphere of the planet could provide invaluable data on its chemical makeup as well as information on wind speeds and directions. And landers could not only provide photographs of the ground, but

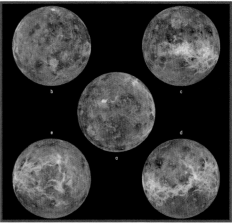

These composite images of Venus were built up from Magellan's complete radar image collection. Venus Express, launched in November 2005, will conduct further investigation of Earth's hostile sister planet.

also carry out chemical tests of the rocks and soil.

There is also the possibility of using balloons to explore Venus. The Soviet Vega missions, in 1985, used balloons to take atmospheric measurements, and an updated version of that idea—using reversible-fluid balloons that can adjust their buoyancy to control their altitude—could be an effective way of studying the planet. These "aerobots" could drop down into the hot lower atmosphere to carry out experiments, take pictures, or possibly even land and drop off instrument packages before zooming back up to cooler altitudes.

Whatever form a future mission to Venus takes, one thing is certain—there is a tremendous amount of knowledge waiting there to be discovered.

SPACECRAFT FACTS

LAUNCH DATE	MAY 4, 1989	POWER	28 V SUPPLIED BY NI-CAD BATTERIES AND 1,200-WATT SOLAR PANELS
DESTRUCTION DATE	OCTOBER 13, 1994		
LENGTH	15.4 FEET (4.6 M)	ONBOARD PROPULSION	ONE 15,000-LB THRUST (66.6 KN)
WIDTH (STOWED)	6.6 FEET (2 M)		SOLID ROCKET 24 LIQUID HYDRAZINE
WIDTH (SOLAR PANELS DEPLOYED)	32.8 FEET (10 M)		THRUSTERS (8 X 100-LB(440 N), 4 X
WEIGHT AT LIFT-OFF	7,612 LB (3,452 KG)		5-LB (25 N), AND 12 X 2-LB (10 N)

MAPPING VENUS

Venus is often called our "sister planet" because it is almost the same size as Earth. But Venus has a thick atmosphere of carbon dioxide with clouds of sulfuric acid, and its surface temperature is about 860°F (460°C). A mission to map the surface of Venus would have to be able to penetrate the clouds with radar to see the surface below—and Magellan did just that.

Magellan was the first planetary exploration spacecraft to be launched by a Space Shuttle. On May 4, 1989, the Shuttle Atlantis lifted the spacecraft into low Earth orbit. Later in the day, the astronauts released Magellan from the Shuttle's cargo bay and a solid-fuel rocket motor, called the Inertial Upper Stage, fired to send the spacecraft on its way. Magellan looped around the Sun one-and-a-half times before arriving at Venus over 15 months later.

As Magellan closed in, Venus' gravity accelerated it from its cruise velocity of 9,990 mph (16,000 km/h) to over 24,340 mph (39,000 km/h). At 8:30 a.m. Pacific Time on August 10, 1990, Magellan dove over Venus' north pole and disappeared behind the planet. Thirty anxious minutes later, scientists confirmed that the spacecraft's Star-408 motor had correctly burned for 84 seconds and put the spacecraft into an elliptical orbit around Venus.

RADAR SCANS

Magellan's standard mapping cycles each took 243 days and 1,852 orbits to complete. Each of these highly elliptical orbits took less than 3.5 hours, and each mapping pass imaged a strip of Venus 16 miles (25 km) wide by 10,000 miles (16,000 km) long. At the low part of each orbit, Magellan scanned the surface of the planet with its radar, and at the high part it transmitted the radar data back to Earth, where it was used to compile detailed images of the planet.

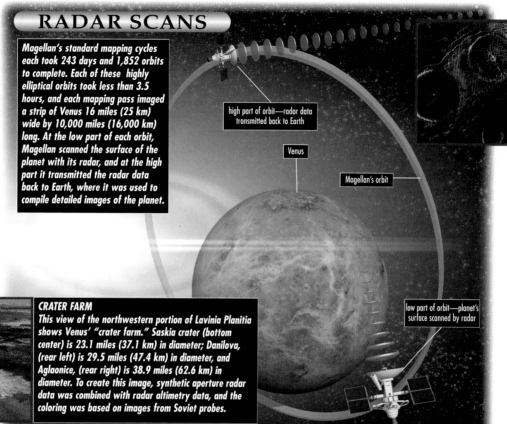

high part of orbit—radar data transmitted back to Earth

Venus

Magellan's orbit

low part of orbit—planet's surface scanned by radar

CRATER FARM
This view of the northwestern portion of Lavinia Planitia shows Venus' "crater farm." Saskia crater (bottom center) is 23.1 miles (37.1 km) in diameter; Danilova, (rear left) is 29.5 miles (47.4 km) in diameter, and Aglaonice, (rear right) is 38.9 miles (62.6 km) in diameter. To create this image, synthetic aperture radar data was combined with radar altimetry data, and the coloring was based on images from Soviet probes.

PANCAKE DOMES
These large circular features in the Eistla region of Venus are volcanic "pancake domes," 39 miles (63 km) in diameter, with broad, flat tops less than 3,000 feet (914 m) high. These unusual pancake-like shapes formed during a volcanic eruption, when an extremely thick, sticky form of lava flowed slowly outward while being flattened by the intense atmospheric pressure.

VENUS STRIPS

For the next five-and-a-half years, Magellan swooped down over Venus, first collecting strips of surface imagery with its synthetic aperture radar, then collecting gravity data on the interior of the planet.

The main part of Magellan's mission was divided into a total of five orbital cycles, each 243 days long, and during each cycle it flew over the entire surface of Venus. Magellan imaged 98% of Venus' surface with a resolution that revealed features down to about 300 feet (90 m) in size, and built gravity maps of 94% of the planet. The spacecraft's images showed a surface covered by volcanic material, with lava channels thousands of miles long and domes made of super-thick lava. There was no evidence of plate tectonics, nor of substantial wind erosion, but plenty of evidence that the surface of Venus is geologically young—less than 800 million years old.

After Magellan's fourth mapping cycle, it used the aerodynamic drag of the atmosphere to slow down and drop into a lower orbit. Finally, on October 11, 1994, the mission flight controllers terminated the mission by commanding Magellan to dive into Venus' atmosphere. Radio contact was lost on October 12, and Magellan was declared lost the following day.

MAGELLAN

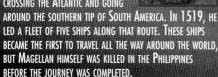

BORN AROUND 1480 IN PORTUGAL, FERDINAND MAGELLAN (RIGHT) WAS ONE OF THE FINEST NAVIGATORS OF HIS TIME. HE BELIEVED HE COULD SAIL FROM PORTUGAL TO INDONESIA BY CROSSING THE ATLANTIC AND GOING AROUND THE SOUTHERN TIP OF SOUTH AMERICA. IN 1519, HE LED A FLEET OF FIVE SHIPS ALONG THAT ROUTE. THESE SHIPS BECAME THE FIRST TO TRAVEL ALL THE WAY AROUND THE WORLD, BUT MAGELLAN HIMSELF WAS KILLED IN THE PHILIPPINES BEFORE THE JOURNEY WAS COMPLETED.

MISSION DIARY: MAGELLAN

MAY 4, 1989, 2:48:59 P.M. EDT SPACE SHUTTLE ATLANTIS LIFTS OFF FROM KENNEDY SPACE CENTER ON MISSION STS-30, WITH MAGELLAN IN ITS CARGO BAY. THE SHUTTLE CARRIES MAGELLAN UP TO A LOW EARTH ORBIT, AT A HEIGHT OF ABOUT 185 MILES (300 KM) ABOVE THE GROUND. *MAY 4, 1989, 9:03 P.M. EDT* MAGELLAN, MOUNTED ON THE INERTIAL UPPER STAGE ROCKET THAT WILL CARRY IT OUT OF EARTH ORBIT (RIGHT), IS RELEASED FROM THE SHUTTLE TO BEGIN ITS 15-

MONTH-LONG JOURNEY TO VENUS. MAGELLAN IS THE FIRST AMERICAN PLANETARY MISSION IN 11 YEARS. *AUGUST 10, 1990* MAGELLAN GOES INTO ORBIT AROUND VENUS. *SEPTEMBER 15, 1990* FIRST ORBITAL CYCLE BEGINS; RADAR MAPPING STARTS. *MAY 15, 1991* SECOND ORBITAL CYCLE BEGINS. *JANUARY 15, 1992* THIRD ORBITAL CYCLE BEGINS. *SEPTEMBER 14, 1992* FOURTH ORBITAL CYCLE

BEGINS; GRAVITY DATA ACQUISITION STARTS. *MAY 24, 1993* MAGELLAN USES AEROBRAKING TO DROP INTO A LOWER, CIRCULAR ORBIT. *AUGUST 3, 1993* FIFTH ORBITAL CYCLE BEGINS. *OCTOBER 11, 1994* MISSION TERMINATED AFTER A TOTAL OF 4,225 USABLE IMAGING ORBITS THAT PRODUCED BETTER VIEWS OF VENUS (ABOVE) THAN HAD EVER BEEN ACHIEVED BEFORE.

MARS GLOBAL SURVEYOR

Mars Global Surveyor (MGS) is a surveyor satellite, weather satellite and communications satellite all rolled into one. In circular polar orbit around Mars since March 1999, its onboard cameras have taken thousands of high-resolution surface photos. It also sends back daily temperature and atmospheric moisture data. And even when its surveying mission ends, Surveyor will still play a crucial role in Martian exploration by relaying signals from future missions back to Earth.

WHAT IF...

...MARS GLOBAL SURVEYOR'S MISSION IS FURTHER EXTENDED?

MGS could be used as a communications and data relay for future Mars missions.

Mars Global Surveyor is already operating outside its original planned lifetime. The necessity of extended aerobraking sessions essentially added an extra year onto the mission. This was necessary to bring the satellite into an orbit where its primary mission could be fulfilled. That mission is still ongoing in late 2005.

Scientists can use the data gathered from Surveyor both to learn about the Earth by comparing it to Mars, and to build a set of comprehensive data to aid in planning future missions. Mars and Earth shared similar conditions billions of years ago, but appear much different today. A comparison of the two planets will allow scientists to understand Earth's history and possibly its future.

There is no limit to the extra work available for MGS beyond its original mission, particularly following the announcement in June 2000 of what look like recently formed gullies down the walls of gullies and craters. One possibility is that these gullies were caused by explosive upwelling of liquid water from beneath the surface in very recent geological times. If this is true—although scientists caution that there are other possible explanations for the existence of the gullies—then it has huge implications for the search for life on Mars. If water still flows beneath the Martian regolith, there may be life beneath the regolith as well.

MGS's Mars Orbital Camera (MOC) is really two cameras, a double fish-eye lens camera for

producing wide-angle panoramas, and a high-power CCD device that can resolve small segments of the surface down to 6 ft (2 m). Once the wide-angle camera finds a feature of interest, the high-resolution camera can zero in on it. In the case of the gullies, the MOC can repeatedly revisit them and use the CCD to check for any alteration—potential evidence of current activity. The extended mission MGS will also help gather data for future Mars missions. Spacecraft operations such as aerobraking and information exchange can be further investigated, while MGS can also help find good landing sites for surface missions.

The lost NASA lander, Mars Polar Lander, probably failed because it fell onto rocks or down into a ravine. Its landing site had to be selected from Viking photographs too low-resolution to show such dangers. MGS's CCD can image potential sites for future lander missions in detail to prevent such a loss in the future. Also, MGS is fitted with a 3-foot (0.9-m)-high antenna built to work as a relay for future landers. MGS would have performed this task for the failed Mars Polar Lander and Deep Space 2 missions, and will be available as a relay for as long as the spacecraft survives.

MARS GLOBAL SURVEYOR

MISSION PHASES			
LAUNCH	NOVEMBER 7 1996	BEGIN MAPPING	APRIL 4, 1999
MARS ARRIVAL	SEPTEMBER 12, 1997	PAYLOAD	MARS ORBITER CAMERA (MOC)
AEROBRAKING 1	NOVEMBER 7, 1997		MARS ORBITER LASER ALTIMETER (MOLA)
SCIENCE	MARCH 27, 1998		THERMAL EMISSION SPECTROMETER PROJECT (TES)
AEROBRAKING 2	SEPTEMBER 23, 1998		MAGNETOMETER AND ELECTRON REFLECTOMETER (MAG/ER)

MAPPING MISSION

Mars Global Surveyor (MGS) began its 466-million mile (750 million km) journey from Cape Canaveral Air Station on November 7, 1996. Once out of Earth orbit, the craft unfurled its twin solar panels, only for mission controllers to find that the latch on one of the panels had cracked, leaving the panel itself stuck at an angle. The panels were designed to provide power as well as to help the craft assume the proper orbit. Normally, this is accomplished by using the rocket engine to slow the craft. But the rocket used to launch MGS lacked the propellant to both lift the craft from Earth and slow it down once in Mars orbit. Instead, the plan was for Surveyor to initially assume a highly elliptical orbit. At the low point of this orbit, the craft would just skim the Martian atmosphere. Friction would slow the craft on each pass, until it finally assumed the correct orbit.

But engineers now worried that the stress of aerobraking could cause the damaged panel to break off entirely. Mission operators rescheduled the aerobraking procedure to place less stress on the damaged panel, revising the original 4-month schedule to a longer, 12-month schedule.

During the longer hiatuses from aerobraking, MGS was able to conduct scientific studies. During one such study, Surveyor's Magnetometer and Electron Reflectometer (MER) detected local "fossil" magnetic fields from Mars' oldest rocks.

On April 4, 1999, Surveyor attained mapping orbit and its 687-day mapping mission finally began.

MAJOR DISCOVERIES

MGS's Mars Orbiter Laser Altimeter (MOLA) uses reflected laser beams to gather topographical details of the Martian surface. MOLA has turned up vast plains in the northern hemisphere, flatter than those found on Earth. These could be sheets of sediment left by evaporating oceans as the planet cooled, or possibly vast frozen lakes left covered in dust.

MARS GLOBAL SURVEYOR

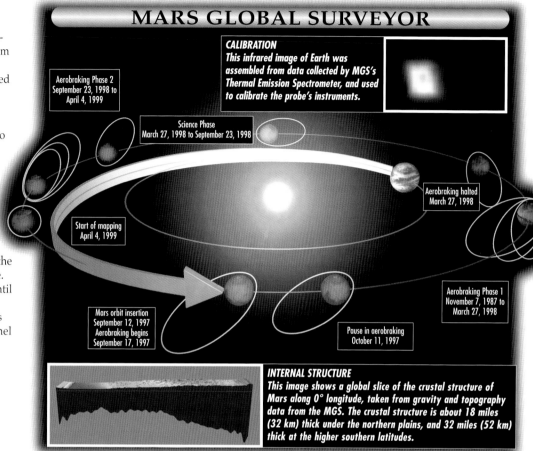

Aerobraking Phase 2
September 23, 1998 to April 4, 1999

CALIBRATION
This infrared image of Earth was assembled from data collected by MGS's Thermal Emission Spectrometer, and used to calibrate the probe's instruments.

Science Phase
March 27, 1998 to September 23, 1998

Aerobraking halted
March 27, 1998

Start of mapping
April 4, 1999

Mars orbit insertion
September 12, 1997
Aerobraking begins
September 17, 1997

Pause in aerobraking
October 11, 1997

Aerobraking Phase 1
November 7, 1987 to
March 27, 1998

INTERNAL STRUCTURE
This image shows a global slice of the crustal structure of Mars along 0° longitude, taken from gravity and topography data from the MGS. The crustal structure is about 18 miles (32 km) thick under the northern plains, and 32 miles (52 km) thick at the higher southern latitudes.

TWIN ROVERS

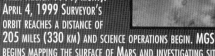

NASA REALLOCATED FUNDING FROM OTHER PROGRAMS TO TAKE ADVANTAGE OF 2003'S EXCELLENT LAUNCH WINDOW. THE RESULTING MISSION SENT TWO ROVERS TO MARS AT ONCE. THE ROVERS (ILLUSTRATION, ABOVE) ARE NAMED SPIRIT AND OPPORTUNITY AND ARE EXACT DUPLICATES. THEY LANDED AT DIFFERENT LOCATIONS. THEIR MISSION, STILL ONGOING IN SEPTEMBER 2005, IS TO GATHER INFORMATION ON ANCIENT WATER AND CLIMATE ON MARS.

The onboard Thermal Emission Spectrometer (TES) charts the temperature and chemical composition of the Martian surface and atmosphere to provide a detailed mineral map of the entire planet.

The most spectacular data are the 25,000 photographs taken by MGS. These reveal a dynamic world of winds and dust dunes. It is also a mysterious world, with numerous landforms scientists cannot explain. The two Martian poles appear completely different—the north is flat and pitted while the south has a series of holes and mesas. More mysterious still are numerous gullies that seem to have been caused by recent liquid water flows—which should be impossible according to conventional views of Martian geology.

MISSION DIARY: MARS GLOBAL SURVEYOR

NOVEMBER 7, 1996 MARS GLOBAL SURVEYOR (RIGHT, IN ASSEMBLY) IS LAUNCHED FROM CAPE CANAVERAL AIR STATION BY DELTA 7925 ROCKET (SECOND RIGHT).
SEPTEMBER 12, 1997 MGS ARRIVES AT MARS. A 22-MINUTE FIRING OF MAIN ROCKET ENGINES PLACES THE SPACECRAFT IN AN ELLIPTICAL ORBIT.
SEPTEMBER 17, 1997 START OF AEROBRAKING. MGS PERFORMS A SERIES OF ORBIT CHANGES TO SKIM THE MARTIAN ATMOSPHERE, USING AIR RESISTANCE TO SLOW DOWN A TINY AMOUNT WITH EVERY ORBIT.

OCTOBER 11, 1997 PAUSE IN AEROBRAKING. TWO OF SURVEYOR'S SOLAR PANELS HAD BENT SLIGHTLY UNDER PRESSURE—AEROBRAKING WAS HALTED TO ALLOW THEM TO RESUME POSITION.
NOVEMBER 7, 1997 RESUMPTION OF AEROBRAKING, AT A SLOWER PACE.
MARCH 27, 1998 AEROBRAKING IS HALTED TO ALLOW SURVEYOR TO DRIFT INTO THE PROPER POSITION WITH RESPECT TO THE SUN. THE HIATUS IS USED TO COLLECT SCIENTIFIC DATA.
SEPTEMBER 23, 1998 AEROBRAKING PHASE 2—RESUMPTION OF AEROBRAKING TO SHRINK THE HIGH POINT OF SURVEYOR'S

ORBIT DOWN TO 205 MILES (330 KM) FROM THE MARTIAN SURFACE (MGS VIEW OF MARS TAKEN MARCH 1999, RIGHT).
APRIL 4, 1999 SURVEYOR'S ORBIT REACHES A DISTANCE OF 205 MILES (330 KM) AND SCIENCE OPERATIONS BEGIN. MGS BEGINS MAPPING THE SURFACE OF MARS AND INVESTIGATING SITES OF INTEREST.
2005 MGS MISSION SEEMS SET TO CONTINUE FOR THE FORESEEABLE FUTURE.
2050 ORBIT WILL COMPLETELY DECAY AND SURVEYOR WILL CRASH INTO MARS.

PATHFINDER TO MARS

On July 4, 1997, an object that looked like a cluster of beachballs hit Mars' rusty surface, bounced and came to rest in an ancient flood channel. Mars Pathfinder had landed. The first spacecraft to visit Mars in 21 years, Pathfinder and its companion rover Sojourner spent almost three months probing the red planet. As planned, they beamed back information about Martian geology and climate. Perhaps more importantly, as the pioneers of NASA's Discovery program, the two little probes inaugurated an exciting new era in space exploration.

WHAT IF...

...WE COULD ALL GO ROVING IN THE SOLAR SYSTEM?

Many years will pass before spaceflight is cheap enough and safe enough to attract vacationers to the Moon. In the meantime, the success of NASA's Mars Rover has inspired a new kind of extraterrestrial entertainment: Remote tourism.

A group of business executives, scientists and former NASA officials has already set up Lunacorp—a corporation that will develop a futuristic lunar rover to be driven by the public. It will be the world's first interactive space exploration event.

Would-be moon cruisers will first have to take a lunar driving test on a simulator. Then, if they make the grade, they will be allowed to take remote control of a rover and drive it live on the Moon. Stop-offs along the way will include Apollo 11's Tranquillity Base, where Neil Armstrong first set foot on the Moon in 1969; remote drivers will have to take care that they do not disturb this or any other historic sites. But a Lunacorp mission will not be all play: The company's lunar rovers will bristle with scientific instruments dedicated to furthering our knowledge of the Moon.

The biggest problem for Earth-based drivers will be the time lag caused by sheer distance. A radio signal takes almost 1.3 seconds to travel the 239,000 miles (385,000 km) from the Earth to the Moon, so quick reactions to a crisis will be impossible.

Guided by remote control from Earth, a Lunacorp rover (top) checks out some of the Moon's attractions. But it will be many years before any kind of tourist can see a Martian sunrise (above).

Roving on Mars would be trickier still. Even when the planet is at its closest to Earth, signals need more than six minutes to go back and forth across the gulf of space. Imagine a car in which the controls for steering, brakes and gas had no effect until six minutes after you used them. Even with a top speed of only a few feet per hour, a Mars Rover would not be for beginners.

PATHFINDER FACTS

Launch Vehicle	Delta 2 7925 with PAM-D upper stage		up to **850 watts** (on Mars)
	Pathfinder Lander	**Camera**	Stereo; 140° field of view
Dimensions	Tetrahedron, 3 feet (0.9 m) tall; with camera, 5 feet (1.5 m) tall	**Sojourner Rover**	
		Dimensions	2 ft (0.6 m) long, 1.5 ft (0.45 m) wide, 10 inches (0.25 m) high
Weight	2,062 lb (935 kg) at launch; 793 lb (340 kg) on Mars; 300 lb (136 kg) in Mars gravity	**Weight**	22 lb (10 kg) on Mars; 8.36 lb (3.79 kg) in Mars gravity
Power	178 watts (during cruise);	**Maximum speed**	2 ft (0.6 m) per minute

LOW BUDGET

The Pathfinder mission was the first in a new era of cost-conscious space exploration, a demonstration that NASA had opted for a "faster, better, cheaper" policy. Each spacecraft in the new Discovery program had to be designed, built and launched within three years. It had to stay within a tight budget of around $150 million. Above all, it had to do the job.

Pathfinder's launch was routine, but its arrival on Mars was not. The probe slammed into the atmosphere at more than 16,500 mph (26,554 km/h). Its heat shield glowed as friction with Mars' thin air robbed it of its speed. At 1,000 mph (1,600 km/h), a parachute opened. The probe still fell rapidly until, only eight seconds before impact, it fired braking rockets and inflated a cocoon of airbags. At 40 mph (64 km/h), Pathfinder hit the rocky plain of Ares Vallis and bounced to a halt.

The airbag technique had never been tried before, but it worked flawlessly. Pathfinder emerged from its protective cocoon and, four hours later, released the Discovery program's next novelty: Sojourner, a tiny, six-wheeled autonomous rover. The first of its kind to see active interplanetary service, Sojourner was only 10 in (25 cm) high; yet it contained an alpha proton X-ray spectrometer and a miniature processing lab that could analyze specimens of Martian rock and soil. As Sojourner toiled over its samples—each analysis required the rover to be stationary for 10 hours—Pathfinder sent home streams of data on surface weather conditions.

The mission blinked out on September 27, 1997, when the last successful transmission passed between Mars and Earth. NASA's new small-scale, elegant approach had been a triumph.

NAMESAKE

APPROPRIATELY, SOJOURNER MEANS "ONE WHO STAYS A WHILE." BUT THE MARS ROVER WAS NAMED FOR SOJOURNER TRUTH (RIGHT), AN AFRICAN-AMERICAN LEADER WHO WAS ACTIVE DURING THE CIVIL WAR. SHE WAS THE SUBJECT OF A WINNING ESSAY IN A NASA SCHOOLS' COMPETITION.

TOUCHDOWN ON MARS

1 DESCENT
The probe's heat shield protects the probe until it slows enough to fire retro-rockets and deploy its parachute.

2 TOUCHDOWN
Airbags inflate to protect Pathfinder from a 40-mph (65-km/h) landing impact.

3 DEPLOYMENT
The lander extends its three solar panels and releases the Sojourner rover from its protective packing.

SOJOURNER
Topped by a flat solar power panel and a transmission antenna, the Sojourner rover resembled a child's toy. But the 10-inch (25-cm) high machine was capable of complex chemical analysis and functioned perfectly for many weeks beyond its design specification.

The fully deployed lander (right) was renamed the Sagan Memorial Station in honor of space scientist Carl Sagan, who died in December 1996 while Pathfinder was in interplanetary space.

WHAT THEY SAW

YOGI ROCK
The Martian boulder that mission controllers dubbed "Yogi" may be the most-photographed rock in the solar system. Situated very close to the Pathfinder landing site, its surprisingly smooth contours indicate that water once flowed on the surface of Mars.

MINI MATTERHORN
This rock was one of many scattered around the Ares Vallis landing site. Planetary geologists chose the area because of its wide assortment of boulders, which may have been left behind by a flood millions of years ago, when Mars had water.

TWIN PEAKS
On the horizon, as seen from Pathfinder, a pair of low hills nicknamed "Twin Peaks" intrigued scientists. Some believe that traces of stratification on the hills is another sign that water once played an important part in the shaping of Mars.

MISSION DIARY: MARS PATHFINDER

1994 PROJECT PLANNING BEGINS.
JUNE 1995 DESCENT SYSTEMS—ROCKETS PLUS AIR BAGS—ARE TESTED. SOJOURNER IS DEVELOPED ON A SIMULATED MARTIAN LANDSCAPE (RIGHT).
JANUARY 31, 1996 PATHFINDER LANDER IS TEAMED WITH SOJOURNER AND TEST-FITTED IN THE CASING (RIGHT) THAT WILL CARRY THEM TO MARS.
AUGUST 14, 1996 COMPLETE MARS PATHFINDER PROBE ARRIVES AT CAPE CANAVERAL TO PREPARE FOR SUBSEQUENT LAUNCH.
DECEMBER 5, 1996 MARS PATHFINDER IS LAUNCHED ON A DELTA 2 (LEFT) FROM CAPE CANAVERAL PAD 17B.
JULY 4, 1997 THE PROBE LANDS ONLY 13 MILES (21 KM) OFF TARGET IN THE ARES VALLIS REGION OF MARS (RIGHT).
JULY 6, 1997 THE SOJOURNER ROVER MAKES ITS FIRST TRIP.
MARCH 18, 1998 MISSION CONCLUDED.

PHOBOS 1 AND 2

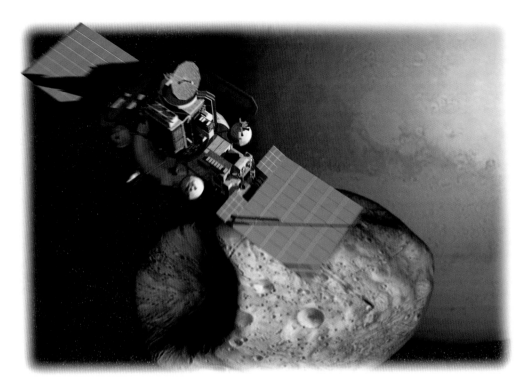

W estern spacewatchers were impressed when, in 1985, the Soviets announced their twin Phobos craft—the most ambitious planetary exploration probes yet unveiled. Named after the Martian moon that they set out to explore, the probes carried an arsenal of scientific instruments to study not just their namesake, but also the Red Planet itself, the Sun and the interplanetary medium. Twelve European nations and the U.S. contributed to the experiments on board. But ultimately, Phobos was a flop.

WHAT IF...

...WE KNEW WHY THE PHOBOS MISSION FAILED?

T he two Phobos craft may have had their share of bad luck, but human error also played a large part in their failure. Their extraordinary scientific capability was scarcely brought to bear on their main objective: the moon, Phobos.

Phobos 1 didn't even reach Mars orbit: It sailed right past into the outer solar system. Phobos 2 failed at the last hurdle while in Mars orbit on its way to the moon.

Controllers lost contact with Phobos 1 on August 31, 1988, after the spacecraft received a faulty command. The set of computers that was used for scientific instruments on board was also used to control the spacecraft—a simplification that saved weight. But an error in an instruction sent to an experiment was misinterpreted by the craft as a command to turn off its attitude control thrusters. Soon, the craft was no longer pointing at the Sun.

Since Phobos relied on its large solar panels to generate electricity—and since the main batteries could provide backup for only 10 hours—the craft began to lose power. By the time the mistake was recognized, power had run out. Mission controllers tried desperately to reactivate the spacecraft from ground stations in the Soviet Union, but nothing worked.

The failure of Phobos 2 was also due to poor design. The Phobos craft relied on its high-gain antenna for communication with mission control—when the antenna pointed

If the Phobos probe had succeeded, it would have orbited around Mars (background), dropped its landers on its namesake moon (foreground), and carried out a range of valuable experiments.

toward Earth. But because the various scientific instruments carried on board were fixed all around the main body, the whole spacecraft had to be turned in order to make use of them. This meant a loss of radio contact with Earth while camera images or other data was gathered. On March 27, 1989, contact was lost after one such period, just before Phobos flew by the Martian moon. Communications were restored for a moment, but it was soon clear that the craft was out of control and power was running low. Shortly afterward, radio contact was lost for good. The computers on board were also misbehaving; the official investigation into the loss of Phobos 2 found many other pieces of hardware that could also have failed.

For a while, the Soviets considered launching yet another Phobos. But the plan was abandoned months later, when the Soviet Union itself went the way of its daring probes—into oblivion.

PHOBOS 1 AND 2

Launch rocket	Proton-K		Phobos 1	Phobos 2
Main engine	KTDU-425A, 4,250 lb (1.927 kg)	Launch dates	July 7, 1988	July 21, 1988
Thrusters	28	Probe launch mass	13,684 lb	13,640 lb
Scientific instruments	22 (about 1,100 lb/500 kg on each)		(6206 kg)	(6,186 kg)
Communication rate to Earth	4,000 bits per second			
Spacecraft computer memory	30 million bits			

PHOBOS, MEET PHOBOS

From the late 1970s, the Soviet Union was seeking complex interplanetary missions that might match the achievements of the U.S. Voyager probes. Planners eventually decided on a program that would send robot craft to Phobos, the larger of the two tiny moons of Mars. Probably a captured asteroid, Phobos is interesting in its own right. It might also be used as a staging post for crewed missions to the planet.

Phobos 1 was launched on July 7, 1988, with Phobos 2 following close behind just two weeks later. The spacecraft were very similar, except for an extra lander carried by Phobos 2. Each had its own Autonomous Propulsion System (APS), which was fired to propel it away from Earth and, later, would be fired again to put the spacecraft in a wide Martian orbit.

Once in orbit, Phobos would jettison the large APS and switch to onboard thrusters. Cameras on board would precisely measure the orbit of the tiny Martian moon and then the craft would drop to within 170 feet of Phobos.

BALLISTIC TESTING

A laser and a krypton "gun" were to fire pulses at the tiny moon, releasing gases that would be analyzed by a spectrometer on board. A radar system would beam radio waves through the surface to help map the layers beneath. As the probe pulled away, it would release a lander that would conduct further experiments.

Then the main spacecraft would settle into a long spell of Mars observation—probing volcanic hotspots with infrared sensors, seeking water with a neutron detector, and training various telescopes on the Sun as it set behind the edge of the planet, to study the Martian atmosphere.

But technical glitches put Phobos 1 out of contact with Earth even before it reached Mars orbit. Phobos 2 did only slightly better: It achieved Mars orbit. But with the target moon in sight, another communications failure left it tumbling, lost forever to its controllers.

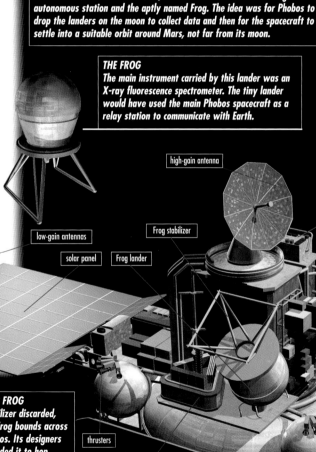

PHOBOS 2

Both Phobos craft carried landers, but Phobos 2 carried two: a long-term autonomous station and the aptly named Frog. The idea was for Phobos to drop the landers on the moon to collect data and then for the spacecraft to settle into a suitable orbit around Mars, not far from its moon.

THE FROG
The main instrument carried by this lander was an X-ray fluorescence spectrometer. The tiny lander would have used the main Phobos spacecraft as a relay station to communicate with Earth.

LONG-TERM AUTONOMOUS STATION
The LAS would have embedded itself in the surface with a harpoon to avoid floating off in Phobos' low gravity. Its transmitter would have sent back data directly to Earth.

high-gain antenna
low-gain antennas
solar panel | Frog lander
Frog stabilizer
instrument section
instrument section
propellant tank
thermal radiator panel
thrusters
LAS lander (folded)
thermal radiator panel | APS | attitude thruster

LEAP FROG
Stabilizer discarded, the Frog bounds across Phobos. Its designers intended it to hop across the little moon on the slender springs that served as its legs, taking samples with its spectrometer until its batteries ran down.

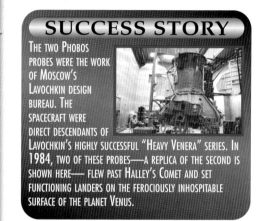

SUCCESS STORY

The two Phobos probes were the work of Moscow's Lavochkin design bureau. The spacecraft were direct descendants of Lavochkin's highly successful "Heavy Venera" series. In 1984, two of these probes—a replica of the second is shown here—flew past Halley's Comet and set functioning landers on the ferociously inhospitable surface of the planet Venus.

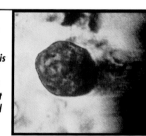

MARS' MOON
The 3-channel Fregat camera aboard Phobos 2 captured this combined optical and infrared image of the Martian moon against the bright background of its parent planet.

MARS SURVEYOR 98

T he planet Mars is scarred by channels carved long ago into its surface by running water. Did oceans once exist on the planet? Today, there is no liquid water on Mars, so where has it all gone? These are among the questions that the two Mars Surveyor 98 spacecraft set out to answer. First up was Mars Climate Orbiter, a Mars weather satellite, closely followed by Mars Polar Lander with its two Deep Space 2 microprobes—targeted for daring touchdowns at the edge of the planet's south polar cap.

WHAT IF...

..MARS MISSIONS CONTINUE?

A s history has shown, humans constantly feel the need to seek out new frontiers, and the exploration of the space frontier is well underway. And despite the failure of the Mars Surveyor 98 mission, the first steps toward the human exploration of Mars have been taken.

Over the next decade, a fleet of robot spacecraft will travel from Earth to Mars to gather information about the Martian environment and test the new technologies that will be needed for a human mission to the Red Planet. The Mars Odyssey and Mars Express missions are currently ongoing. More missions are due soon such as Phoenix in 2007, a low-cost mission searching for evidence of water, and the ambitious Mars Science Laboratory due to launch in late 2009.

These uncrewed missions are designed to increase our knowledge about the planet's climate, geology and water distribution both in the past and at present. As well as preparing the groundwork for the arrival of humans, they will attempt to find out whether life has ever existed on the planet—or still exists there.

Missions to Mars have suffered numerous setbacks. The Russian Mars '96 mission was lost, as was Polar Lander, probably due to difficulties in selecting a suitable landing site before MGS data

was available. The Beagle-2 probe carried by Mars Express was successfully ejected from the orbiter but has not been heard from since.

It is hoped that having satellites in orbit as relays (a sort of 'Interplanetary Internet' as some scientists call it) and improved images of projected landing sites will prevent future losses. New techniques for investigating Mars are under constant development. Balloons and extremely light aircraft may be used by future probes to carry instrument packages, and ever-more-advanced rovers will be delivered.

It is hoped to extract rock samples from the surface of Mars and by drilling and return them to a spacecraft in orbit, which will then bring them to Earth for scientists to examine.

All this investigation into the climate, topography and conditions on Mars will lead to a greater understanding of our own planet, and will pave the way for a manned mission to Mars.

The goal is for humans to reach Mars to carry out on-site science investigations, and that event could take place by as early as 2020. The work being carried out by the robot probes of the Mars Surveyor program will help make a human mission all the easier. Spacecraft such as the Mars Global Surveyor are blazing the trail that one day will lead to human footprints in the red Martian soil.

MARS SURVEYOR 98

MARS CLIMATE ORBITER	
LAUNCH	DECEMBER 11, 1998
LAUNCHER	DELTA 2
MASS	1,387 POUNDS (629 KG)
DIMENSIONS	6.9 FEET (2.1 M) TALL, 5.4 FEET (1.6 M) WIDE, 6.4 FEET (2 M) DEEP
ARRIVAL DATE	SEPTEMBER 23, 1999; SPACECRAFT DESTROYED ON ENTERING MARS ORBIT
SCIENCE INSTRUMENTS	RADIOMETER (PMIRR), IMAGER (MARCI)

MARS POLAR LANDER	
LAUNCH	JANUARY 3, 1999

LAUNCHER	DELTA 2
MASS	1,270 POUNDS
LANDER MASS	639 POUNDS (576 KG)
DIMENSIONS	3.5 FEET (1 M) TALL, 12 FEET (3.6 M) WIDE
ARRIVAL DATE	DECEMBER 3, 1999; CONTACT LOST ON ARRIVAL
SCIENCE INSTRUMENTS	IMAGER, ROBOT ARM, WEATHER INSTRUMENTS, GAS ANALYZER, ICE AND DUST CLOUD DETECTOR, MICROPHONE

DEEP SPACE 2 MICROPROBES	
MASS	5.3 POUNDS (2.4 KG) EACH
DIMENSIONS	8.3 IN (21 CM) TALL, 5.3 IN (13.5 CM) WIDE
LANDING SPEED	400 MPH

MISSING MISSION

The 2-spacecraft Mars Surveyor 98 mission should have added enormously to our knowledge of Mars. The spacecraft—the Mars Climate Orbiter and the Mars Polar Lander, with its two Deep Space 2 microprobes—would have searched for evidence of life on the Red Planet, and sent back valuable data about its atmosphere and climate.

Both spacecraft were launched successfully—the Climate Orbiter on December 11, 1998 and the Polar Lander on January 3, 1999—and they both reached Mars. But navigational errors brought the Climate Orbiter too close to the planet, and on September 23, 1999, it plunged into the Martian atmosphere and was destroyed. Just over two months later, its companion probe was also lost. The Polar Lander apparently reached its destination on December 3, but no signal was ever heard from it.

ORBITER AND LANDER

The Mars Climate Orbiter was intended to study the Martian atmosphere and climate and act as a relay station for the lander. The Mars Polar lander would have used a stereo camera, an instrument package and a robot arm to search for water on the Martian south polar cap.

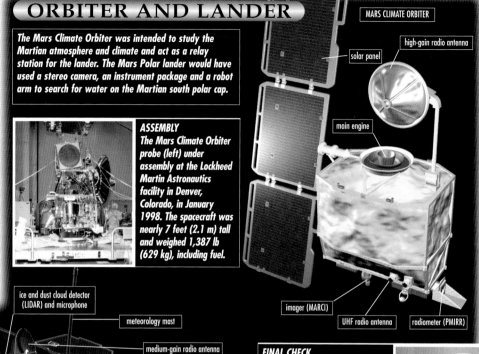

MARS CLIMATE ORBITER

high-gain radio antenna

solar panel

main engine

imager (MARCI)

UHF radio antenna

radiometer (PMIRR)

ASSEMBLY
The Mars Climate Orbiter probe (left) under assembly at the Lockheed Martin Astronautics facility in Denver, Colorado, in January 1998. The spacecraft was nearly 7 feet (2.1 m) tall and weighed 1,387 lb (629 kg), including fuel.

ice and dust cloud detector (LIDAR) and microphone

stereo surface imager

UHF radio antenna

robotic arm

meteorology mast

instrument package (MVACS)

medium-gain radio antenna

robotic arm camera

propellant tank

scoop

meteorology submast

solar panel

auxiliary solar panel

descent imager (MARDI)

MARS POLAR LANDER

FINAL CHECK
At the Kennedy Space Center's Spacecraft Assembly and Encapsulation Facility 2, a technician checks the Mars Polar Lander (right) before it is finally sealed within its protective casing, or backshell.

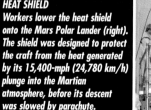

HEAT SHIELD
Workers lower the heat shield onto the Mars Polar Lander (right). The shield was designed to protect the craft from the heat generated by its 15,400-mph (24,780 km/h) plunge into the Martian atmosphere, before its descent was slowed by parachute.

PENETRATORS

JUST BEFORE ITS ENTRY INTO THE MARTIAN ATMOSPHERE, THE POLAR LANDER SHOULD HAVE RELEASED ITS PAIR OF DEEP SPACE 2 MICROPROBES. HOUSED INSIDE BASKETBALL-SIZED CASINGS (RIGHT), THEY WERE DESIGNED TO SLAM INTO THE PLANET AT A SPEED OF ABOUT 400 MPH (650 KM/H) AND BURY THEMSELVES 6 FT (1.8 M) BELOW THE SURFACE. THEY WERE MEANT TO OPERATE FOR ABOUT 50 HOURS, MEASURING THE WATER CONTENT AND TEMPERATURE OF THE MARTIAN SOIL AND TRANSMITTING THE DATA TO THE CLIMATE ORBITER, WHICH WOULD RELAY IT TO EARTH. THE MICROPROBES WERE NAMED AMUNDSEN AND SCOTT, AFTER THE FIRST TWO EXPLORERS TO REACH THE EARTH'S SOUTH POLE.

WIDE COLLABORATION

The Polar Lander's camera system, the Mars Descent Imager (MARDI), was designed to take images of the planet's surface during the final two minutes of the probe's descent through the atmosphere. Once safely on the surface, the lander would have deployed its science instruments and begun its study of the planet. Its principal science instrument package was the Mars Volatiles and Climate Surveyor (MVACS). This consisted of weather instruments, a surface imager, a robotic arm for collecting soil samples, and instruments that could analyze the samples.

The lander also carried Russian-made equipment—a Light Detection and Ranging (LIDAR) instrument for detecting ice and dust clouds. This instrument was provided by the Space Research Institute of the Russian Academy of Sciences, and was the first Russian instrument to be carried on an American planetary space probe. And the Mars Microphone, with which the lander could listen in on the sounds of the planet, was provided by the U.S.-based Planetary Society. It was the first privately funded instrument from a nonprofit organization to be carried on a NASA mission.

SURVEYOR'S SCIENCE

If its arrival at Mars had gone according to plan, the Climate Orbiter would have used the friction of the thin Martian atmosphere to slow itself down and enter a low elliptical orbit around the planet. When safely in orbit, the Climate Orbiter was to have kept a close eye on the Martian weather, using an instrument called the Pressure Modulator Infrared Radiometer (PMIRR) to scan the atmosphere and measure its temperature and the amount of dust and water vapor it holds. The orbiter also carried a camera system—the Mars Color Imager (MARCI)—with which to photograph the planet, and it would have served as a relay station, picking up the signals from the Polar Lander and beaming them back to Earth.

MARS EXPRESS

A t the end of 2003, six months after lifting off from Baikonur Cosmodrome in Kazakhstan, a European spacecraft entered Mars orbit. Its mission was to study the Martian atmosphere and to release a small probe named Beagle 2. The Beagle probe was to drop down to the planet's surface to analyze soil and rock samples. Contact was lost with Beagle 2 after separation from the Mars Express orbiter and was never re-established. Its fate remains unknown.

WHAT IF...

...BEAGLE 2 HAD NOT BEEN LOST?

The Beagle 2 lander carried instruments designed to study the geology of the landing area, to search for 'life signatures' in the form of chemical compounds, and to study the Martian climate. It was self-powered with solar panels to allow a long duration mission.

Beagle 2 was designed to take rock samples with a robotic arm. First its panoramic camera would make a full 360° survey of the landing site. When a suitable nearby rock was located, Beagle-2 was to reach out its robotic arm and use a small drill to remove a sample core from the rock. The core would then be loaded into a combustion chamber within the lander and heated to high temperatures.

Organic matter, such as bacteria or plant material, will burn at 570°F (300°C) and rock will burn at 1,110°F (595°C). Even diamond will burn at higher temperatures. By measuring the amount of carbon released from the sample at each temperature, scientists can determine what type of material is present in the rock.

Previous missions have only taken samples of the surface of Mars by using a scoop at the end of a robotic arm. In addition to the samples of rock, Beagle-2 was to endeavor to take a soil sample from below the surface. It is hoped that water, and possibly life, may still exist below the surface, where water could

The 132-pound (60-kg), 2-foot (60 cm)-diameter Beagle-2 lander was released from the orbiter section of Mars Express five days before the spacecraft entered Mars orbit.

have escaped evaporation by the Sun's heat.

To achieve this goal, Beagle 2 would have deployed a cylindrical "mole" that could crawl across the surface at 2.5 inches (6 cm) per second until it reached a rock. Using the rock to guide it downward, the mole would then tunnel into the ground to a depth of about three feet. Then the mole, with a sample of material in its "mouth," was to be pulled back to the lander where the sample would be analyzed.

The loss of Beagle 2 is a setback to the exploration of Mars, but the data it would have gathered can be obtained by other missions. The orbiter has gathered good data and remains in service, making Europe's first interplanetary probe at least a partial success.

MARS EXPRESS

	MARS EXPRESS ORBITER	BEAGLE-2 LANDER
DRY MASS	1,224 LB (555 KG)	132 POUNDS (60 KG)
PROPELLANT MASS	941 KG (427 KG)	N/A
LANDING MASS	N/A	66 LB (30 KG)
SCIENTIFIC INSTRUMENT MASS	256 LB (116 KG)	33 LB (15 KG)
DIMENSIONS	4.9 BY 5.9 BY 4.6 FEET 1.5 X 1.8 X 1.4 M	2 FEET (60 CM) IN DIAMETER, 5 IN (12.7 CM) HIGH
POWER REQUIREMENTS	500W	58W
LAUNCHER; SOYUZ/FREGAT	LAUNCH MASS: 2,792 LB (1,266 KG)	LAUNCH DATE: JUNE 1, 2003

MARS MISSION

The launch of Mars Express, in June 2003, came just over four years after the mission was announced by the European Space Agency (ESA). Many previous ESA spacecraft took a decade to develop, but for this mission ESA chose to emulate NASA's "better, faster, cheaper" approach to design, manufacture and test a spacecraft in a fraction of that time.

To achieve this, Mars Express borrowed the design of many of its components from other probes instead of developing new technologies. For instance, the main rocket engine has a long history of use in space, the thrusters were designed for the ill-fated Cluster satellite mission, and the software needed to point the spacecraft in the right direction was developed for the Rosetta cometary probe. Even some of the scientific instruments were first used on the Russian Mars 96 probe, which failed to leave Earth orbit due to a launch problem.

Besides being low cost, Mars Express is also one of the smallest spacecraft ever launched by ESA. The entire craft is about the size of an office desk, with two solar panels extending from its sides to produce electrical power. Mars Express continues a series of innovative ESA missions, including the Giotto probe that visited Comet Halley in 1986, and Ulysses, which flew over the poles of the Sun in 1994 and 1995.

THE MISSION

Even though the U.S. and the Soviet Union have had difficulties in exploring Mars,

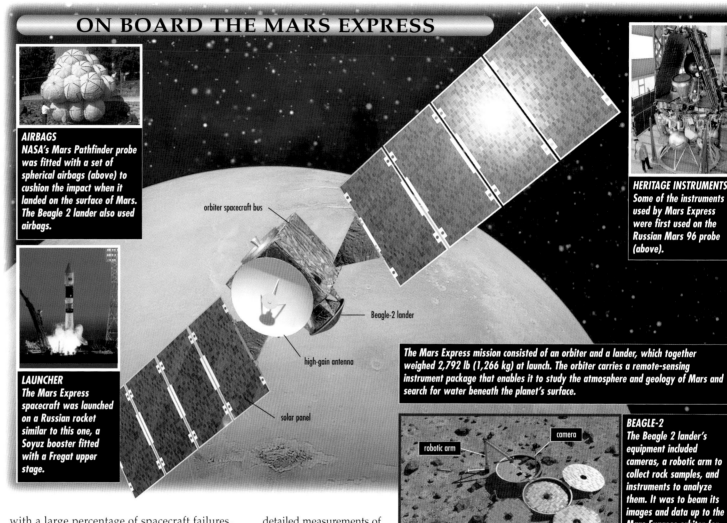

ON BOARD THE MARS EXPRESS

AIRBAGS
NASA's Mars Pathfinder probe was fitted with a set of spherical airbags (above) to cushion the impact when it landed on the surface of Mars. The Beagle 2 lander also used airbags.

LAUNCHER
The Mars Express spacecraft was launched on a Russian rocket similar to this one, a Soyuz booster fitted with a Fregat upper stage.

orbiter spacecraft bus

Beagle-2 lander

high-gain antenna

solar panel

HERITAGE INSTRUMENTS
Some of the instruments used by Mars Express were first used on the Russian Mars 96 probe (above).

The Mars Express mission consisted of an orbiter and a lander, which together weighed 2,792 lb (1,266 kg) at launch. The orbiter carries a remote-sensing instrument package that enables it to study the atmosphere and geology of Mars and search for water beneath the planet's surface.

camera

robotic arm

solar panels

BEAGLE-2
The Beagle 2 lander's equipment included cameras, a robotic arm to collect rock samples, and instruments to analyze them. It was to beam its images and data up to the Mars Express orbiter for transmission back to Earth.

with a large percentage of spacecraft failures, Mars Express is just as ambitious as the previous larger ESA missions. As well as performing its own experiments, the Mars Express orbiter can operate as a communications relay, gathering scientific data from probes launched by other nations and transmitting it to Earth for scientists to study.

After a 6-month journey, Mars Express reached its destination in December 2003. The main spacecraft then entered an elliptical orbit around Mars from where it can train its instruments on the surface. The initial mission specifications required Mars Express to make detailed measurements of Mars' atmosphere and geological structure for a full Martian year—687 Earth days. This mission looks set to be extended after the nominal period.

Five days before the spacecraft arrived in Mars orbit, the Beagle 2 lander was released from the orbiter to follow its own trajectory to the surface of Mars. If all had gone as planned Beagle 2 would have slammed into the Martian atmosphere at 20,000 mph (32,000 km/h).

At first slowed by a protective heat shield, Beagle 2 was then to deploy a parachute to slow it further as it dropped through the planet's thin atmosphere. Finally, a set of airbags would be inflated to give the craft a soft landing. Once settled, Beagle 2 would reach out its robotic arm and release a tunneling "mole" to search for signs of life. However, contact was not established with the probe and its fate remains unknown.

BEAGLE 2

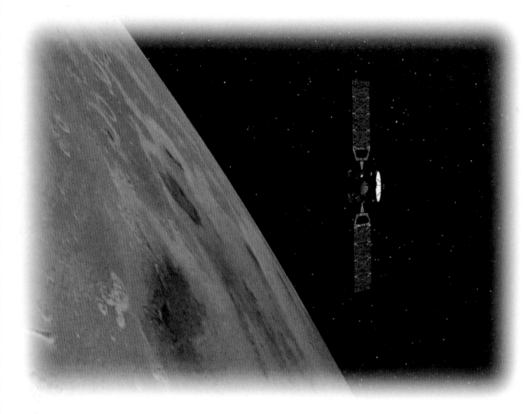

BEAGLE FACTS

Beagle 2 was named after HMS *Beagle*, the ship that took Charles Darwin on the voyage of exploration that led to the writing of *The Origin of Species* and the modern theory of evolution.

At one point during construction of Beagle 2, many of the team caught colds and had to be taken off the project lest they contaminate the lander and thus the samples collected from Mars.

The PAW ready to be delivered to the aseptic assembly facility at The Open University.

Mars missions are fraught with risk. At least one in three probes sent to the Red Planet has been lost without returning much or any data.

Building a small lander to "piggy back" on Mars Express presented many challenges to the Beagle 2 team. Not only did the outer shell have to survive impact on the Martian surface at any angle but also had to offer thermal and radiation exposure protection and not be harmed by the pre-launch sterilisation procedure. It was also very important that the

structure have low mass and low cost. The primary structure consisted of an inner shell comprising carbon-fiber skins on an aluminium honeycomb core.

The British pop band Blur recorded a series of tones to be broadcast to Mars Express and relayed to Earth on landing, but these were never received. The artist Damien Hirst painted a series of colored spots on an instrument to aid calibration of sensors.

High above Mars, a spacecraft jettisoned a tiny lander and sent it towards the surface, beginning the last phase of a mission to seek evidence that there once was, or just possibly still is, some form of life beneath the surface. On Christmas Day, a group of expectant scientists listened and waited for a series of musical tones, confirming a safe landing, but the chances of anything coming from Mars were diminishing by the hour…

MARS EXPRESS DATA

MARS EXPRESS:		ORBIT:	
LAUNCH DATE:	2 Jun, 2003 17:45 UT	# Orbital Inclination:	86.3°
MISSION END:	30 Nov, 2005 (Nominal Mission)	# Pericentre:	160 miles (258 km)
LAUNCH VEHICLE:	Soyuz-Fregat	# Apocentre:	7,183 miles (11,560 km)
LAUNCH MASS:	(2,696 lb (1,223 kg)	# Period:	6 hours 43 minutes
		BEAGLE 2:	
		Diameter	25.5 in (640 mm)
		Height	9 in (230 mm)
		Weight	152 lb (68.8 kg)

CUT-PRICE SCIENCE

Launched as an adjunct to the Mars Express orbiting surveyor mission, Beagle 2 was a small lander craft, about the size and shape of a barbecue, designed to sit on the Martian surface and analyse the soil for evidence of past or present life beneath the surface.

Beagle 2 was a rare British spacecraft, although developed under the auspices of the European Space Agency or ESA at Darmstadt, Germany. It was designed by a team led by Professor Colin Pillinger of the UK's Open University. Built to extremely strict weight and budgetary guidelines, building and testing Beagle 2 was a battle to raise funding and meet objectives with as little delay and expense as possible. This led to some shortcomings in testing features such as the airbags designed to protect the lander on impact. When Beagle 2 was first conceived in 1997 it was envisaged to weigh about 238 lb (108 kg) and to be part of a network providing geophysical survey data. This was shelved when the mass available on Mars Express was revised to only 132 lb (60 kg).

Beagle 2 was equipped with a variety of sensors on a robotic arm. A stereo camera would guide the placement of sample collecting tools. A rock grinder would crush the sample for study by the microscope and chemical analysis by the spectrometer and mass spectrometer. Power was to come from a group of four solar panels charging 42 lithium ion batteries to keep the experiments running during the cold Martian night.

The Russian Soyuz rocket carrying the Mars Express and Beagle 2 was launched from the Baikonur space center in Kazakhstan on 2 June, 2003 and spent the next seven months reaching Mars. On 19 December a pyrotechnic device fired to slowly release a loaded spring, which gently pushed Beagle 2 away from the mother spacecraft (artist's impression, right). A camera on Mars Express captured Beagle 2 floating away. This was the last positive sighting of the lander. Unlike most other landers, there was no datalink to the orbiter during the descent and the first transmission was expected once Beagle 2 had bounced to a landing, unfurled its antenna and was in view of the next orbit of Mars Express.

MARTIAN MYSTERY

Despite many passes by Mars Express and other orbiting craft, no signal was ever received. Imaging of the landing zone by numerous sensors have failed to find any trace of Beagle 2, its airbags, parachute or other equipment. The lack of telemetry during the descent and landing phase has hindered the search for what went wrong. One theory is that a thinner than expected atmosphere increased the descent rate of the lander and the parachute and airbags opened too late to arrest Beagle's descent rate from 3 miles (5 km) per second to one gentle enough for survival. Despite all the missions to Mars over the years, much about the Martian atmosphere remains mysterious, but without a fully-developed parachute Beagle 2 may have hit the ground at over 650 ft (200 m) per second, completely destroying it.

A report into the loss of Beagle 2 criticised the management of the project and the pressure to do it on the cheap, which led to inadequate testing. The Open University team is hoping to hitch-hike a Beagle 3 aboard a 2009 US mission to Mars.

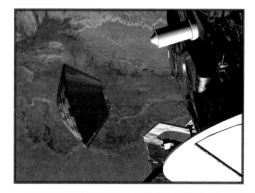

MISSION OBJECTIVE

ARTIST'S IMPRESSION
Had Beagle 2 reached the surface safely, this is how it would have looked, unfolded and with its robotic arm deploying a variety of sensors.

COLIN PILLINGER

Professor Colin Pillinger of the Open University became the public face of Beagle 2, and his optimism and enthusiasm helped raise funds for the mission. Indeed, some critics of the project have pointed to the possibility that Pillinger was too closely focused on fundraising and publicity to manage such a complex project effectively. On the Open University's plans for a Beagle 3, Pillinger is quoted as saying: "It's back to the bottom of the hill to start pushing that boulder back up again."

MISSION DIARY

2 June, 2003
Mars Express launched from the Baikonur launch pad in Kazakhstan onboard a Soyuz Fregat launcher
19 December, 2003 08:31 (GMT):
Beagle 2 separates from mothership and begins descent towards Mars
25 December, 00:20 (GMT)

Beagle 2's timer chip wakes the on-board computer to get ready to control the pre-programmed descent activities
25 December 02:47 (GMT)
Beagle 2 enters the Martian atmosphere
25 December 02:54 GMT
Beagle 2 lands on Mars
25 December 03:00 (GMT)

Mars Express orbital insertion
7 January, 2004
Beagle 2 declared lost
2004–2005
Mars Express continues to orbit Mars and send back useful data

MARS ROVERS: OPPORTUNITY AND SPIRIT

I n a quest to discover once and for all whether water ever existed on Mars, two identical six-wheeled rover vehicles have followed the success of the much smaller Mars Sojourner by exploring further and for longer than any other mission to the Red Planet.

MARS ROVER TOOLS

T he Rock Abrasion Tool (RAT) brushes dust from and then bores into interesting rock surfaces and exposes fresh material for examination by other instruments onboard.

The Miniature Thermal Emission Spectrometer (Mini-TES) measures the thermal signature of the terrain to identify promising rocks and soils for closer examination and for determining the processes that formed Martian rocks. It can also look skyward to provide temperature profiles of the Martian atmosphere.

A panoramic camera is used to determine the mineralogy, texture, and structure of the terrain. Two 'hazcams' give the rover drivers a 3D picture to help them steer around rocks and other hazards.

Magnets are used to collect ferrous dust particles which are then studied close-up by devices called the Mössbauer Spectrometer and the Alpha Particle X-Ray Spectrometer (APXS).

A Microscopic Imager (MI) is used for obtaining close-up, high-resolution images of rocks and soils. Many types of soils and surface were found in the Eagle Crater. NASA scientists named them after ice cream flavours such as "coffee and cream", "vanilla", "black forest" and "Neopolitan".

Although the rovers' motors and electronics lasted longer than predicted, the wheels and other mechanical components suffered from wear over time. To reduce wear on Spirit's front wheels, the rover was driven backwards after several months' use.

MARS ROVER SPECS

MASS AT LAUNCH

Rover	408 lb (185 kg)
Lander	767 lb (348 kg)
Backshell / Parachute	742 lb (209 kg)
Heat Shield	172 lb (78 kg)
Cruise Stage	425 lb (193 kg)
Propellant	110 lb (50 kg)
Total Mass	2,343 lb (1,063 kg)

REMOTE CONTROL

Launched on Delta II rockets, the Mars landers and rovers entered the Martian atmosphere inside protective "aeroshells". Each was slowed in the thin air by a parachute and Rocket Assisted Descent (RAD) motors that fire 30–50 ft (10–15 m) above the surface, bringing the vehicle to a dead stop. At the same time, a six-lobed airbag inflated and cushioned the impact of the craft striking the surface and subsequent bounces.

The two Rovers landed on opposite sides of Mars, Spirit landing in the large shallow Gusev Crater near the south rim of the deeper Bonneville Crater. Its heat shield fell on the north rim and was spotted when Spirit drove up to the crater's edge. Opportunity landed on the Meridiani Planum, an area known to be rich in hematite, a material associated with hot springs or standing pools of water.

After several Martian days (sols), two-way communication was established and tested and the rovers were ready to explore Mars in more detail than had ever been possible before. To communicate with Earth, the rovers relay information by way of orbiting craft; 70 per cent of the mission data is transmitted via the Mars Odyssey Orbiter and 30 per cent by way of the Mars Global Surveyor.

Driving a vehicle by remote control 50 million miles (80 million km) from Earth requires caution. Several cameras are used to select a path, steer the rovers and avoid close obstacles. The rovers are not designed for speed. On a hard, flat surface they could reach 2 inches (5 cm) per second, but were limited by hazard avoidance software to 10-second runs, followed by 20-second pauses to assess the terrain, giving an average speed of ⅓ in (1 cm) per second. Records for distance driving in one day reached 722 ft (220 m) in March 2005. Driving over a surface and then reversing allows the cameras to study the tracks and get an idea of the softness or otherwise of the terrain.

Each future move is discussed and planned in great detail before being commanded. A duplicate rover is tested on Earth on various simulated Martian surfaces, which are often tilted to angle equal to slopes faced by the actual rovers. Occasionally, softer than expected sand caused the rovers to become stuck for long periods. Scientists simulated this with various building and gardening materials and developed techniques for driving out of these "sand traps".

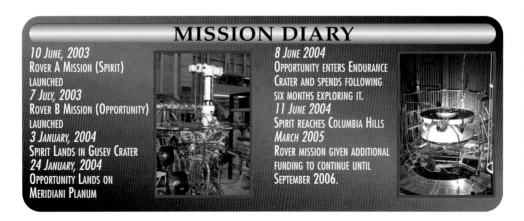

A color image of the martian landscape at Meridiani Planum, where the Mars Exploration Rover Opportunity landed on Saturday 24 January 2004. This is one of the first images beamed back to Earth from the rover shortly after it touched down, captured by the rover's panoramic camera.

After exploring its immediate environment, Spirit then set off towards a feature dubbed the Columbia Hills. Each of the seven peaks is named for one of the astronauts lost in the 2003 Columbia disaster. The hills, rising to 300 ft (90 m) above the plain were estimated as 1 mile (1.6 km) distant, a drive that would take up to 160 sols if the rover lasted that long.

EXTENDED WARRANTY

Opportunity drove up to the stadium-sized Endurance Crater. After much debate it was decided to enter the crater as the possible science benefits outweighed the risk of not being able to drive out again. As it happened, Opportunity later became stuck in a sand trap for nearly five weeks but was successfully extracted.

Near its own heat shield, Spirit discovered an iron meteorite. This was first meteorite discovered on another world.

Many other discoveries pointed to evidence of the former presence of water on Mars, the main scientific target of the mission. These included: spherules; "smile"-shaped marks in rocks and ripple patterns on rock surfaces. Evidence of magnesium sulphates pointed to the evaporation of water in the distant past.

The Rovers were able to observe the sky as well as the Earth, and photographed the Earth and stars from an unusual viewpoint. They also witnessed lunar eclipses where Mars's moons, Phobos and Deimos passed in front of the Sun. A streak in the sky in one photo perplexed scientists, but is now thought to have been the Viking 2 orbiter, which is still circling Mars.

The rovers pleased NASA and the Jet Propulsion Laboratory by remaining in functional condition well past the "warranty" of their 90-sol primary missions, which were completed in April 2004. From then on the crews could take greater risks than before. A software update allowed two hours "autonomous" driving each day after an hour of "blind" driving following a route pre-planned on Earth. Autonomous driving, which in fact meant driving for 6 ft 6 in (2 m) then pausing to look for obstacles, allowed the rovers to cover as much ground in three sols as they had in 70 during early parts of the mission.

Dust accretion on the solar panels caused a reduction in available power from time to time and a partial loss of vision from the cameras, but strong winds later blew the dust clear, giving improved performance. With the rovers still functioning well more than year after landing, mission funding was extended by 18 months to at least September 2006.

DELTA LAUNCH

BOTH ROVERS WERE LAUNCHED ON DELTA II ROCKETS, ALTHOUGH THE ROVER B MISSION NEEDED A HEAVY VERSION OF THE ROCKET WITH MORE FUEL BECAUSE MARS AND EARTH HAD MOVED FURTHER APART IN THE MONTH SINCE THE ROVER A LAUNCH.

MISSION DIARY

10 JUNE, 2003
ROVER A MISSION (SPIRIT) LAUNCHED

7 JULY, 2003
ROVER B MISSION (OPPORTUNITY) LAUNCHED

3 JANUARY, 2004
SPIRIT LANDS IN GUSEV CRATER

24 JANUARY, 2004
OPPORTUNITY LANDS ON MERIDIANI PLANUM

8 JUNE 2004
OPPORTUNITY ENTERS ENDURANCE CRATER AND SPENDS FOLLOWING SIX MONTHS EXPLORING IT.

11 JUNE 2004
SPIRIT REACHES COLUMBIA HILLS

MARCH 2005
ROVER MISSION GIVEN ADDITIONAL FUNDING TO CONTINUE UNTIL SEPTEMBER 2006.

MARS ORBITERS

After the race for the Moon came the race for the planets. In 1971, the Soviet Union and the United States each sent two probes on the way to Mars. The final sprint to be first to orbit another planet was on. Over the coming decades, many more U.S. and Soviet probes would follow in the path of these pioneers. Many never made it. Missions to the Red Planet soon won a reputation, particularly in Soviet circles, for being dogged by bad luck. It seemed Mars was not going to surrender its secrets without a fight.

WHAT IF...

...WE RETURNED TO MARS?

When Mars Global Surveyor and its Pathfinder lander were launched in 1996, they became the first in a long series of NASA missions destined for Mars. In December 1998 and January 1999, the next two in the series, the Mars Climate Orbiter and Mars Polar Lander, were blasted into space.

The Mars Climate Orbiter reached the planet in September 1999, but its braking maneuvers took it far too close to the Martian surface. The navigation error meant that NASA lost all contact with the spacecraft, which apparently crashed. Scientists were vexed at the loss of the data the orbiter would have sent back, but NASA's Mars Global Surveyor has been able to carry on beyond its original mission and take up some of the slack.

While some missions have been a failure, like Polar Lander in 1999 and the Beagle-2 probe, each time lessons have been learned. It is just as important to learn about the technical difficulties of reaching and landing on Mars as it is to discover the secrets of the planet itself.

For humans to be able to explore Mars we will need to know certain critical facts about the planet – how much water is available if any, and also what the radiation environment is like.

Mars' thin atmosphere and weak magnetic field could make it susceptible to high levels of solar radiation. The planet lacks a

The Mars Surveyor Orbiter will carry a new Gamma Ray Spectrometer (GRS) similar to the one lost with Mars Observer. The GRS will analyze the composition of the Martian surface.

protective zone like the Earth's Van Allen belt, so astronauts will probably need substantial protection against radiation.

It is not just NASA whose probes will be peering in on Mars in the coming years. The Japanese space agency, ISAS, successfully launched the Planet-B probe on July 4, 1998. Its mission includes monitoring Mars' magnetic field and atmosphere and investigating suggestions that the planet has a dust ring, similar to the ones around Saturn, along the orbit of its moon Phobos.

The Europeans are also getting in on the act. After the failure of Russia's Mars '96 mission, the European Space Agency (ESA) swiftly put together a new project. Named Mars Express, the probe launched in 2003. One of the principal objectives of the Mars Express mission was to survey the subsurface of the planet for water. Its lander, Beagle 2, was to carry out biological and geochemical investigations on the surface of Mars—hoping to find conclusive evidence of the existence of life, past or present, on the planet. Unfortunately Beagle 2 was lost after separation from the Mars Express probe.

MARS MISSES

Name	Country	Launch Date	Result
Mariner 8	U.S.	May 8, 1971	Failed during launch
Kosmos—419	U.S.S.R.	May 10, 1971	Failed to leave Earth orbit
Mars 4	U.S.S.R.	July 21, 1973	Failed to enter Mars orbit
Mars 5	U.S.S.R.	July 25, 1973	Entered Mars orbit but shut down a few days later
Phobos 1	U.S.S.R.	July 7, 1988	Lost en route in August 1989
Phobos 2	U.S.S.R.	July 12, 1988	Lost near Phobos in March 1989
Mars Observer	U.S.	September 25, 1992	Lost just before Mars arrival on August 21, 1993
Mars 96	Russia	November 16, 1996	Ditched in Atlantic Ocean

MARTIAN PROBES

The race to be the first to orbit Mars began in 1971, when two American and two Soviet spacecraft set off for the Red Planet. Six years earlier, in 1965, the American Mariner 4 had shot past Mars in a high-speed fly-by, without collecting much significant data; Mariner 6 and 7 had followed in 1969 with similar results. A closeup orbital mission was needed, and Mariners 8 and 9 and the Soviet combined orbiter/lander craft Mars 2 and Mars 3 were built for the job.

But Mariner 8 ditched in the ocean after its second rocket stage failed. And on September 22, 1971, as the three remaining craft sped toward Mars, observers on Earth saw a white cloud developing on the planet. It grew into a huge dust storm, engulfing Mars.

Mariner 9 arrived on November 13, just 14 days ahead of Mars 2. Mariner entered orbit successfully, and waited for the dust storm to die down before transmitting data. It sent back more than 7,000 sharp images of huge mountains, giant valleys and—to the astonishment of scientists back on Earth—what looked like riverbeds.

The two Soviet craft were less successful. When they arrived, their landers had to be deployed immediately—the probes had not brought enough fuel to carry them any longer. Both of the landers plummeted into the atmosphere and perished in the dust storm. The orbiters circled the planet and sent back limited data over the next four months.

PAST AND PRESENT

The Soviets launched more probes over the following decades, trying to equal the success of Mariner 9, but it seemed they were not destined to reach Mars again. Two probes were launched for Mars' moon Phobos in 1988. Both were lost in space before completing their missions. And in 1996, the Russian-led Mars '96 mission ditched in the Atlantic Ocean.

Later U.S. probes fared better. In August and September 1975, NASA launched Vikings 1 and

RED PLANET TRAIL BLAZERS

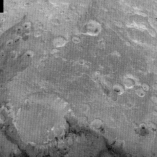

NASA and the European Space Agency (ESA) have scheduled several missions to Mars over the next 20 years. The program may culminate in a crewed mission by the end of the 21st century.

MARINER 9
Mariner 9, the first probe to orbit another planet, spent a year orbiting Mars—and thrilling scientists with its pictures of the planet.

Mars 3
Mars 3 was the most successful of the pair of Soviet probes that chased Mariner 9 to Mars. As well as a few photographs, it sent back infrared, ultraviolet and magnetic field data.

SURVEYOR 96
Entering Mars orbit on September 12, 1997, Mars Global Surveyor 96 is making the most detailed map of Mars yet. Its camera is capable of 5-foot (1.5-m) resolution.

CLIMATE ORBITER
Part of the Surveyor 98 mission, the Climate Orbiter was launched on December 11, 1998. Its will observe Mars' weather over a Martian year.

JINXED

THE RUN OF BAD LUCK CONNECTED WITH MARS MISSIONS HAS LED SOME TO HINT AT FOUL PLAY BY MARTIANS. AS WELL AS THE RUSSIANS' ABYSMAL RECORD, INCLUDING THE LOSS OF MARS '96 (RIGHT), OTHER COUNTRIES HAVE EXPERIENCED THE MARS JINX. NASA'S MARS ORBITER, LAUNCHED IN 1992, CEASED TRANSMITTING THREE DAYS BEFORE REACHING THE PLANET. THE MARS GLOBAL SURVEYOR SUFFERED A BROKEN SOLAR PANEL, AND JAPAN'S NOZOMI PROBE NEARLY FAILED TO ESCAPE EARTH ORBIT.

2. Like the earlier Soviet Mars probes, the Vikings were dual orbiter/lander spacecraft. Public attention focused on what became the first successful Martian landings, but the Viking orbiters' mapping project was equally important. The two orbiters successfully imaged the surface of Mars at a resolution of 300 meters (roughly 1,000 feet), giving us an even more complete map of Mars than the one assembled from Mariner 9's data. The Vikings sent back pictures of vast lava plains, giant volcanoes, deep canyons and distinctive wind- and water-formed features. The two spacecraft took 52,000 photographs and spun around the planet 2,106 times before ceasing transmission in 1978 and 1980.

On November 7, 1996, NASA launched the Mars Global Surveyor (MGS), at present the longest-serving spacecraft to orbit Mars. Surveyor carried much new technology and also tried out an innovative—and hair-raising—method of decelerating and settling into orbit. Instead of using its rockets to slow down, it aerobraked, or skimmed through the upper reaches of the Martian atmosphere, allowing friction to reduce its speed. This meant that the spacecraft needed to carry considerably less fuel, increasing its load capacity for scientific instrumentation. MGS has been so successful that its mission continues to be extended without any end date in sight.

MISSION DIARY

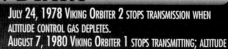

MAY 30, 1971 MARINER 9 LAUNCHED.
NOVEMBER 13, 1971 MARINER 9 ARRIVES AT MARS AND BECOMES THE FIRST SPACECRAFT TO GO INTO ORBIT AROUND ANOTHER PLANET. FINDS APPARENT EVIDENCE THAT THE PLANET ONCE HAD VAST WATER SYSTEMS (RIGHT).
OCTOBER 27, 1972 MARINER 9 DEACTIVATED.
AUGUST 20, 1975 VIKING 1 LAUNCHED.
SEPTEMBER 9, 1975 VIKING 2 LAUNCHED.
JULY 19, 1976 VIKING 1 ENTERS MARS ORBIT.
AUGUST 7, 1976 VIKING 2 ENTERS MARS ORBIT. BOTH ORBITERS EMBARK ON MAPPING MISSION.

JULY 24, 1978 VIKING ORBITER 2 STOPS TRANSMISSION WHEN ALTITUDE CONTROL GAS DEPLETES.
AUGUST 7, 1980 VIKING ORBITER 1 STOPS TRANSMITTING; ALTITUDE CONTROL GAS EXHAUSTED.
NOVEMBER 7, 1996 MARS GLOBAL SURVEYOR LAUNCHED.
SEPTEMBER 12, 1997 MARS GLOBAL SURVEYOR ENTERS ORBIT AROUND MARS.
DECEMBER 11, 1998 MARS CLIMATE ORBITER LAUNCHED.
MARCH 9, 1999 MARS GLOBAL SURVEYOR BEGINS HIGH-RESOLUTION MAPPING OF MARS.
SEPTEMBER 23, 1999 MARS CLIMATE ORBITER ARRIVES AT MARS.

EXPLORING FURTHER

Exploration of the outer planets is a time-consuming business due to the great distances involved. The length of some missions has exceeded the careers of the scientists who devised and nurtured them. Pioneer 10 passed Jupiter in December 1973, and in 1990 it became the first man-made object to leave the Solar System. It was last heard from in 2003 when over 10.6 billion miles (17 billion km) from Earth. The Huygens probe touched down on Saturn's moon Titan in January 2005, making the furthest landing from Earth. The Cassini mothership continues to study Saturn, Titan and Saturn's 32 other moons. Experiments are underway to use solar wind to power future spacecraft, with photons from the sun pushing against a sail made of Mylar or a similar thin, lightweight fabric. Such a craft would accelerate very slowly, but would eventually reach enormous velocities. For example, a future probe to the outer solar system could reach 60,000 mph (100,000 km/h) in three years, enough to reach Pluto in about half the time of a conventionally-powered spacecraft.

Cassini-Huygens is a joint NASA/ESA/ASI unmanned space mission to study Saturn and its moons. This photograph shows scientists performing final checks on the Cassini spacecraft's antennae prior to launch in October 1997.

PIONEER SOLAR MISSIONS

Although the earliest U.S. interplanetary probes lost the race to the Moon, they discovered something more interesting along the way: a torrent of charged particles emanating from the Sun. Instead of jockeying for a piece of lunar territory, the later Pioneer probes were sent into orbit around our nearest star. Weighing in at about 140 pounds (64 kg), these little explorers proved that interplanetary space—previously thought to be empty—is filled with powerful magnetic fields and a strong wind that blows from the Sun.

WHAT IF...

...ALL SATELLITES OPERATED FOR 30 YEARS?

NASA generally expects its probes to operate longer than their scheduled missions. After completing primary tasks, every probe is assigned a provisional extended mission. Pioneer probes 6–9 not only outlasted their primary missions, but also extended their extended missions.

Unfortunately, there comes a time in every probe's life—even if it is still fully operational—when contact with ground control is cut. The most likely reason for abandoning a probe is that the information it is providing is not worth the cost of receiving it.

With a designed operational life of six months, Pioneers 6, 7, 8 and 9 surpassed expectations by more than 30 times. With the launch of Pioneer 9—by which time Pioneer 6 had already operated for six times its scheduled lifetime—four reliable probes were being used as an integrated solar warning station.

But by the end of the decade, NASA was launching more technologically advanced solar and terrestrial probes. Two Orbiting Solar Observatories (OSO 5 and 6) and the last of six Orbiting Geophysical Observatories (OGO 6) were launched in 1969. These new probes were telling scientists far more about the Earth and the Sun than the Pioneer probes ever could. The OGO 6 probe carried with it 26 experiments to study the Earth and its magnetosphere—four times the number of

In 1997–8, NASA spent $234 million on space communications, including the cost of operating the Deep Space Network (right). Space probes past their "expiration dates" are rarely worth tracking.

experiments on board Pioneer 9.

In the same year, 1969, two Intercosmos solar and interplanetary probes were launched by the member states of the Soviet bloc. And with the launch of the European Space Agency's Highly-Eccentric Orbit Satellite, the Pioneer probes quickly lost their edge.

During the decades that followed, science and technology continued to make huge advances while the Pioneer probes became less relevant and more expensive. Receiving transmissions from a space probe is "pay-per-view," and with a sharp increase in the number of interplanetary missions, time on NASA's Deep Space Network became a precious resource.

In 1995, the Deep Space Network made contact with Pioneers 6, 7 and 8 (Pioneer 9 had retired in 1983). But NASA was not after scientific data—it tracked the probes as practice for the Lunar Prospector mission. The deathblow for the Pioneer solar program finally came in November 1997, when contact with Pioneer 6—by then the only Pioneer left operational—was cut, and NASA left its longest-serving probe to explore space alone.

PIONEER SOLAR MISSIONS

Name	Launch Date	Mission Length	Operational Life	Weight	Instruments	Experiments
Pioneer 5	Nov 11, 1960	10 months	29 years	95 lb (43 kg)	4	4
Pioneer 6	Dec 12, 1965	6 months	30 yr 8 mo	141 lb (64 kg)	6	10
Pioneer 7	Aug 17, 1966	6 months	29 yr 6 mo	141 lb (64 kg)	6	8
Pioneer 8	Dec 13, 1967	6 months	29 yr 8 mo	141 lb (64 kg)	6	8
Pioneer 9	Aug 11, 1968	6 months	24 years	141 lb (64 kg)	6	8

SOLAR SUCCESS

The Pioneer program got off to a rocky start. Originally planned to be a series of lunar probes, only one of the first five Pioneer probes, Pioneer 4, made it anywhere near the Moon. Another of the early probes, Pioneer 3, discovered a second belt of trapped radiation around Earth—the first belt was discovered by the first U.S. satellite, Explorer 1, in 1958. But these early Pioneers were largely regarded as failures. With the Moon proving difficult to reach, the next five Pioneer probes were designed to study the space between the Earth and the Sun, and they would do so with spectacular success.

On March 3, 1960, Pioneer 5 was launched from Cape Canaveral. Once out in interplanetary space, the probe—which operated for 106 days— detected complex magnetic patterns. Like the earlier Pioneers, Pioneer 5 carried an instrument to detect the levels of radiation trapped in the Earth's magnetic fields. The probe also carried two other high-energy particle detectors and a magnetic-field detector. It correlated changes in magnetic fields with the eruption of solar flares and proved that the Sun made space a dangerous place for an unprotected astronaut.

The next Pioneer design, which lasted for four missions, was designed to investigate these intriguing electrical and magnetic phenomena in greater detail. After the early failures, the Pioneer program was finally ready to make some important contributions to the emerging field of space science.

LAUNCH

ALTHOUGH LAUNCHED BY THE U.S. AIR FORCE ON OCTOBER 11, 1958 (LEFT), PIONEER 1 WAS THE FIRST SPACECRAFT TO BE CONTROLLED BY THE NEWLY-CREATED NATIONAL AERONAUTICS AND SPACE ADMINISTRATION—NASA. THE PROBE FAILED TO REACH ITS INTENDED DESTINATION, THE MOON, BUT IT DID RETURN SOME USEFUL DATA ABOUT THE MAGNETIC FIELDS THAT SURROUND THE EARTH.

SOLAR PIONEER ORBITS

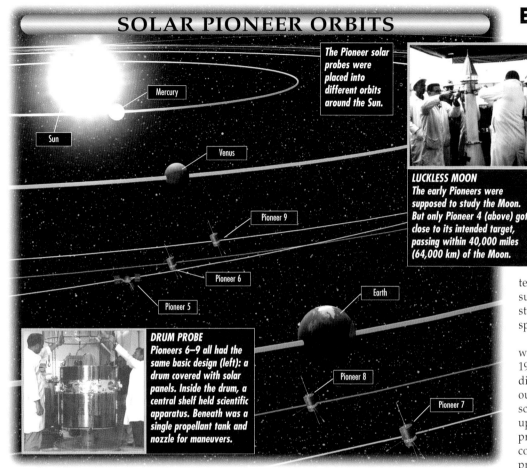

The Pioneer solar probes were placed into different orbits around the Sun.

Mercury

Sun

Venus

Pioneer 9

Pioneer 6

Earth

Pioneer 5

Pioneer 8

Pioneer 7

LUCKLESS MOON
The early Pioneers were supposed to study the Moon. But only Pioneer 4 (above) got close to its intended target, passing within 40,000 miles (64,000 km) of the Moon.

DRUM PROBE
Pioneers 6–9 all had the same basic design (left): a drum covered with solar panels. Inside the drum, a central shelf held scientific apparatus. Beneath was a single propellant tank and nozzle for maneuvers.

MISSION DIARY: PIONEER 6

DECEMBER 16, 1965 PIONEER 6 IS LAUNCHED FROM LAUNCH COMPLEX 17A, CAPE CANAVERAL (RIGHT).
NOVEMBER 1968 PIONEER 6 GOES BEHIND THE SUN. BY MONITORING HOW ITS TRACKING SIGNAL CHANGES, SCIENTISTS LEARN ABOUT THE COMPOSITION OF THE SUN'S CORONA, ITS OUTER ATMOSPHERE.
NOVEMBER 8, 1968 WITH THE LAUNCH OF PIONEER 9, THE PIONEER SOLAR STORM OBSERVATION NETWORK NOW NUMBERS FIVE MEMBERS—FOUR SATELLITES AND THE EARTH.
DECEMBER 15, 1995 PIONEER 6'S PRIMARY TRANSMITTER FAILS.

JULY 11, 1996 NASA'S RADIO TELESCOPE COMMUNICATION SYSTEM—THE DEEP SPACE NETWORK (FAR RIGHT)—TRACKS PIONEER 6 AND SUCCESSFULLY COMMANDS THE PROBE TO SWITCH TO ITS BACK-UP TRANSMITTER. AFTER 30 YEARS AND 8 MONTHS, PIONEER 6 IS STILL WORKING.
OCTOBER 6, 1997 PIONEER 6 IS TRACKED FOR THE LAST TIME BY DEEP SPACE STATION 43, A RADIO TELESCOPE NEAR CANBERRA, AUSTRALIA.
NOVEMBER 1997 NASA FORMALLY ABANDONS PIONEER 6.

EXTRA LONG LIFE

In addition to Pioneer 5's instrumentation, Pioneers 6–9 also had two instruments designed to measure the density of electrons in the solar wind. NASA's ingenious scientists even monitored changes in the probe's tracking signal to glean information about the wind the transmission passed through. All this information allowed a better understanding of the structure and flow of the solar wind.

Among the new technology tested on the Pioneer probes was a gyroscope-type stabilization system. Like a spinning top or a rotating bicycle wheel, a space probe is less likely to wander off course if it is spinning. First used with Pioneer 6, the technique proved so stable—the probe spun successfully at about 60 rpm—that it became standard on all NASA's subsequent deep-space probes.

Each of the next three Pioneers—Pioneer 7 was launched in August 1966, 8 in December 1967 and 9 in November 1968—had a slightly different orbit so that a network of solar outposts began to form. The probes allowed scientists to forecast solar storms accurately up to two weeks in advance—plenty of time to prepare to study the event. The solar network continued to operate until the early 1970s, providing a wealth of information about the complex electrical and magnetic lines that swirl around the Sun.

The satellites themselves operated far longer than the network. The youngest satellite, Pioneer 9, was the first to go—it stopped transmitting in 1983. The other three probes held on for another decade. NASA's Deep Space Network last made contact with Pioneer 7 in 1995, when only one of its instruments was still working. The story was the same for Pioneer 8, contacted a year later. Most surprising of all was the oldest sibling, Pioneer 6, which was still fully functional in the summer of 1996. With more than 30 years of continuous operation, Pioneer 6 made it into the record books as NASA's oldest operational probe.

PIONEER 10 AND 11

Pioneer 10 and 11 are humanity's first emissaries to the galaxy at large. Launched in the early 1970s to investigate Jupiter and Saturn, they were the first spacecraft designed to leave the solar system behind. Now, they are more than 7 billion miles from home and still traveling at more than 27,000 mph (43,500 km/h). Pioneer 11's power ran down in 1995, but Pioneer 10—officially retired—made contact with NASA in 2003, and may still be transmitting. Should either one encounter an alien race in the vastness of interstellar space, each carries a message from humanity.

WHAT WILL HAPPEN TO...

...PIONEER 10 NOW?

The little probe is showing its age. Although most of the damage was sustained in the dusty planetary system of its birth and the ravaging radiation storm of its planetary encounter, 2 million years is a long time. The interstellar medium through which it has traveled is a vacuum a trillion times greater than can be produced on Earth, but it still takes its toll, eroding the space probe's skin at a rate of about one angstrom per year. Despite this, the message etched onto the gold plaque mounted on the spacecraft's body is still readable.

Pioneer's greeting is simple. The plaque shows two upright bipeds, a single hand raised in greeting, standing by an outline of the spacecraft. When it was built, decisions about the racial characteristics of the figures and whether or not to show their sex organs caused fierce debate—but such concerns have been left far behind, long ago. Next to the figures, a starburst-like diagram depicts the position of the solar system from which the probe has trekked in relation to 14 pulsars—stars whose radiation varies, each with a unique and identifiable signature. With this map, it should be possible for an alien race to track and single out our Sun among the sea of stars that makes up the Milky Way.

Arranged along one edge of the plaque is a stylized representation of all the planets of our solar system and their relative distances. A final cryptic figure establishes the scale of all the other measurements relative to the properties of hydrogen, the most common atom in the universe. From this

The Pioneers carried gold plaques designed by astronomers Frank Drake and Carl Sagan. When the design was published, American newspapers were deluged with letters objecting to the nude figures— and the use of tax money "to send smut into space."

battered gold plaque, an alien race can decipher what we look like, and where we come from. Or, 2 million years hence, what we looked like, and where we came from: After all, those "aliens" may turn out to be our own descendants.

Pioneer 10's last transmission was received in January 2003, at which point the spacecraft was 7.6 billion miles (12.2 billion km) from Earth. In the last years of Pioneer 10's life its instruments were only turned on for a few hours each week. In February 2003 when contact was established, the probe's power supply was too weak for its signals to reach Earth – over 20 hours away at the speed of light. Pioneer might still be transmitting but the distance is simply too great for us to pick up its signals.

Pioneer 10 is headed roughly in the direction of Aldebaran in the constellation of Taurus, about 68 light years away. It will pass by in about two million years. Pioneer 11 went out of contact in 1995 when Earth's motion took it out of the field of view of the probe's antenna. Pioneer 11 may or may not be transmitting as it heads out towards the constellation of Aquila, the Eagle. It will pass one of the stars of Aquila in about 4 million years.

PIONEER 10 AND 11

MANUFACTURER	TRW	INSTRUMENTS CARRIED
DESIGN LIFETIME	2.5 YEARS	MAGNETOMETER, PLASMA ANALYZER, CHARGED PARTICLE DETECTOR,
ANTENNA DIAMETER	8 FT 10 IN (2.6 M)	IONIZING DETECTOR, NON-IMAGING TELESCOPES WITH OVERLAPPING
CURRENT SPEED, PIONEER 10	242M MILES/389M KM PER YEAR	FIELDS OF VIEW TO DETECT SUNLIGHT REFLECTED FROM PASSING
CURRENT SPEED, PIONEER 11	232M MILES/372M KM PER YEAR	METEOROIDS, MICROMETEOROID DETECTORS, UV PHOTOMETER,
ONBOARD POWER	RTGS (RADIOISOTOPE THERMONUCLEAR GENERATORS) PROVIDING 155 W	IR RADIOMETER, AND AN IMAGING PHOTOPOLARIMETER. PIONEER 11 ALSO CARRIED A LOW-SENSITIVITY FLUXGATE
COMMUNICATIONS RATE	16–2,048 BPS THROUGH NASA DSN STATIONS	MAGNETOMETER.

DISTANT TRAVELERS

Pioneer 10 left for Jupiter on March 3, 1972. It was the latest in a long line of Pioneer probes designed to explore interplanetary space. At Jupiter's distance from the Sun, Pioneer 10 and its sibling Pioneer 11 could not rely on solar cells for energy. Instead, they carried a nuclear generator that drew power from the heat produced by radioactive plutonium. The most visible feature of both Pioneer craft was an umbrella-like antenna nine feet (2.75 m) across, needed both to transmit data to Earth and to receive instructions from Mission Control. These instructions were vital—1970s computers were too heavy for the probes to carry, so there were no onboard brains. Controllers would have to put up with the fact that the probes would be so far distant that radio signals would take an hour and a half to reach them.

Pioneer 10, spinning five times a minute to stabilize its antenna, plunged into the asteroid belt beyond Mars in mid-July 1972, and emerged unharmed in February 1973. The project scientists had worried that it would be crippled or destroyed by a 30,000-mph (50,000 km/h) collision with an interplanetary pebble, but Pioneer 10's survival gave them the confidence to launch Pioneer 11 in its wake.

Pioneer 10 began its encounter with Jupiter on November 26, 1973, when its instruments registered the presence of the giant planet's stormy magnetosphere. In the 26 days of its flypast, the little probe was battered by the intense radiation belts that surround Jupiter. As Pioneer skimmed past the planet, 81,000 miles (130,000 km) above the cloud tops, its instruments turned on the Jupiter's most

striking feature—the Great Red Spot. Pioneer's pictures confirmed that the Spot was a giant storm. Its encounter over, Pioneer 10 headed out of the solar system at 25,000 mph (40,000 km/h) and passed the orbit of Pluto in 1990.

After Pioneer 10's success, there seemed no need to repeat the same program. Pioneer 11 was given a course correction that pointed it toward Jupiter's south pole rather than its equator, and took the spacecraft much closer to the planet than its predecessor—only 21,000 miles (33,796 km/h) above the clouds. As Pioneer 11 whipped past, Jupiter's gravity accelerated the probe to a speed of 107,373 mph (172,800 km/h) and flung it onward to a second planetary encounter—with Saturn. At Saturn, Pioneer 11 swung past only 1,200 miles

(1,931 km) from the ring system, and 13,000 miles (20,920 km) above the clouds. It discovered that Saturn had radiation belts and a strong magnetic field, found two new rings and a moon, and helped scientists understand much more about the structure and composition of the gas giant. Its mission completed, Pioneer 11 also headed out of the solar system, in the opposite direction to its predecessor.

ONWARD AND OUTWARD

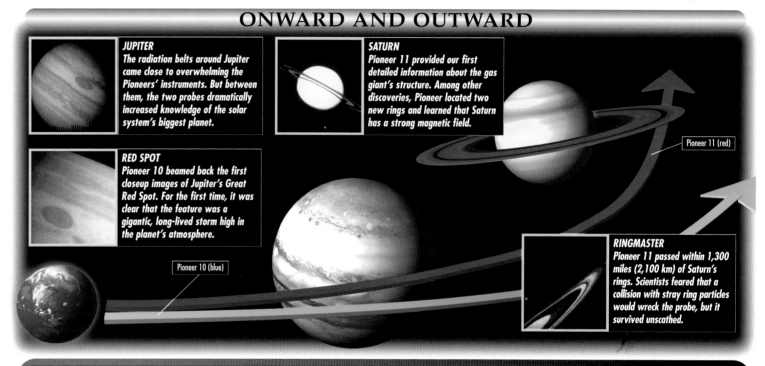

JUPITER
The radiation belts around Jupiter came close to overwhelming the Pioneers' instruments. But between them, the two probes dramatically increased knowledge of the solar system's biggest planet.

RED SPOT
Pioneer 10 beamed back the first closeup images of Jupiter's Great Red Spot. For the first time, it was clear that the feature was a gigantic, long-lived storm high in the planet's atmosphere.

SATURN
Pioneer 11 provided our first detailed information about the gas giant's structure. Among other discoveries, Pioneer located two new rings and learned that Saturn has a strong magnetic field.

Pioneer 11 (red)

Pioneer 10 (blue)

RINGMASTER
Pioneer 11 passed within 1,300 miles (2,100 km) of Saturn's rings. Scientists feared that a collision with stray ring particles would wreck the probe, but it survived unscathed.

SNAPSHOTS

ALTHOUGH THE PIONEER MISSIONS PROVIDED THE FIRST CLOSEUP VIEWS OF JUPITER AND SATURN, THE PROBES WERE NOT EQUIPPED WITH CAMERAS. INSTEAD, THEY SCANNED THEIR TARGETS WITH A DEVICE CALLED A PHOTOPOLARIMETER. BY COMBINING SEVERAL SETS OF RAW DATA (SHOWN ABOVE), MISSION SCIENTISTS WERE ABLE TO CREATE PERFECT FULL-COLOR IMAGES OF THE PLANETS.

MISSION DIARY: PIONEER 10 AND 11

MARCH 3, 1972 PIONEER 10 LAUNCHED FROM CAPE CANAVERAL ON ATLAS-CENTAUR BOOSTER.
JULY 1972–FEBRUARY 1973 PIONEER 10 SUCCESSFULLY TRAVERSES THE ASTEROID BELT BETWEEN MARS AND JUPITER.
APRIL 6, 1973 PIONEER 11 LAUNCHED.
NOVEMBER 26, 1973 PIONEER 10 PASSES WITHIN THE ORBIT OF SINOPE, JUPITER'S OUTERMOST KNOWN MOON.
DECEMBER 4, 1973, 6:26 P.M. PIONEER 10'S CLOSEST APPROACH TO JUPITER, 81,000 MILES (130,000 KM) ABOVE THE PLANET'S CLOUDS.
DECEMBER 3, 1974 PIONEER 11'S JUPITER ENCOUNTER, CLOSEST APPROACH 26,725 MILES (43,000 KM).
SEPTEMBER 1, 1979 PIONEER 11'S CLOSEST APPROACH TO

SATURN, AT 13,000 MILES (20,920 KM) (RIGHT).
JUNE 13, 1983 PIONEER 10 CROSSES THE ORBIT OF NEPTUNE (WHICH WAS THEN THE OUTERMOST PLANET DUE TO THE ECCENTRICITY OF PLUTO'S ORBIT) AND BECOMES THE FIRST CRAFT TO LEAVE THE SOLAR SYSTEM.
FEBRUARY 23, 1990 PIONEER 11 LEAVES THE SOLAR SYSTEM.
SEPTEMBER 22, 1990 PIONEER 10 REACHES 50 TIMES THE EARTH'S DISTANCE FROM THE SUN.
NOVEMBER 1995 LAST COMMUNICATIONS RECEIVED FROM

PIONEER 11.
MARCH 31, 1997 FORMAL END OF PIONEER MISSIONS.
FEBRUARY 17, 1998 VOYAGER 1 SPACECRAFT OVERTAKES PIONEER 10 TO BECOME MOST DISTANT HUMAN OBJECT.
28107 A.D. PIONEER 10 PASSES 6.4 LIGHT YEARS FROM THE STAR PROXIMA CENTAURI.
2 MILLION A.D. PIONEER 10 REACHES THE NEIGHBORHOOD OF THE RED STAR ALDEBARAN.
4 MILLION A.D. PIONEER 11 MAKES A CLOSE APPROACH TO A SMALL STAR IN THE CONSTELLATION AQUILA (ABOVE).

VOYAGER MISSIONS

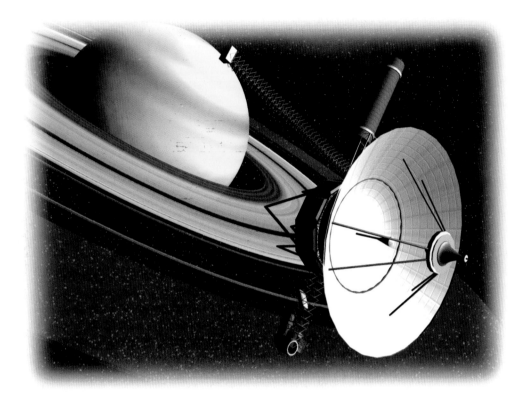

The twin spacecraft Voyagers 1 and 2 have transformed our understanding of the outer solar system. Originally designed to study only Jupiter and Saturn, these two intrepid probes have visited all the gas giants—Jupiter, Saturn, Uranus and Neptune. They have sent back startling images of churning atmospheres, complex ring systems and exotic moons, some of which are large enough to be worlds in their own right. Now, as the Voyagers head for the stars, they continue to report from the edge of the solar system.

WHAT IF...

...THE VOYAGERS REACH ANOTHER SOLAR SYSTEM?

After 20 years and more than 7 billion miles (11 billion km), the Voyager space probes are still in fairly good shape. Both spacecraft continue to return data from their studies of the solar wind—the stream of charged particles emitted by the Sun. Using this data, scientists hope to investigate the heliopause, the distant region where the solar wind finally dies away.

NASA plans to keep in contact with the Voyagers for the next 20 to 30 years, which is when their nuclear power packs are expected to run down. After that, and assuming that nothing collides with them, the probes will continue on their journey to the stars.

Voyager 1 is currently traveling into the outer solar system at a rate of about 324 million miles (520 million km) per year. After both Voyagers break through the heliopause, which

The Voyager probes are already the most distant of all man-made objects and may one day reach another solar system.

is expected to happen in 2014, they will travel for some 24,000 years before they reach the Oort Cloud, the area on the edge of the solar system where comets are thought to originate.

Beyond the Oort Cloud, the Voyagers will enter true interstellar space. But even at speeds of more than 35,000 miles per hour, it will be tens of thousands of years before either craft reaches another star.

About 40,000 years from now, Voyager 1 will pass less than a light-year from AC+79 3888, a star in the constellation of the Little Bear. According to present calculations, Voyager 2 should arrive in the vicinity of the bright star Sirius in the year 296,036. It could be millions of years before either spacecraft drifts into another planetary system and is perhaps picked up by intelligent life.

VOYAGER PROBE FACTS

LAUNCH VEHICLE	TITAN 3E WITH CENTAUR UPPER STAGE	POWER SUPPLY	3 THERMOELECTRIC GENERATORS USING PLUTONIUM 238
MAXIMUM HEIGHT	9.84 FT (3 M)		
BOOM LENGTH	EXTENDABLE TO 42.5 FT (13 M)	HIGH GAIN ANTENNA	12 FT IN DIAMETER (3.6 M)
SPACECRAFT WEIGHT	1,820 LB (825 KG)	TRANSMITTER POWER	23 WATTS
SCIENCE INSTRUMENT WEIGHT	234 LB (106 KG)	CAMERAS	TELEPHOTO 0.4°, WIDE ANGLE 3°
DATA STORAGE CAPABILITY	538 MILLION BITS		

THE GRAND TOUR

Voyagers 1 and 2 set off on their epic journeys of discovery in 1977. Their launches were timed to take advantage of a rare planetary alignment. Every 176 years, the giant planets of the outer solar system are aligned in such a way that a well-aimed spacecraft can use their gravitational fields to slingshot its way from one to the other. After each encounter, the spacecraft picks up speed—enough to have reached Neptune by 1989 in the case of Voyager 2. Without such boosts, the trip would have taken at least 30 years.

First port of call on the "Grand Tour" was Jupiter. Voyager 1 reached there first, in the spring of 1979, followed by its sister ship in July of the same year. The two probes investigated Jupiter's Great Red Spot, found a previously undiscovered ring system and even detected powerful lightning bolts on the planet.

Hurled onward to Saturn by Jupiter's gravity, the Voyagers reached the planet in

orbital plane of the solar system) and onward into interstellar space.

Voyager 2 continued to Uranus and sent back images of a planet that up until then had been little more than a blank disk to earthbound astronomers. At Neptune, the spacecraft passed within 3,000 miles (4,800 km) of the surface, its closest approach to any planet since it had left Earth. Its last encounter was with Neptune's largest moon, Triton. There, almost 3 billion miles (4.8 billion km) from the Sun, Voyager 2 made the totally unexpected discovery of ice volcanoes. It was a fitting end to a great interplanetary adventure.

1981 and beamed back our first detailed pictures of its intricate system of rings and moons. Voyager 1 made a close fly-by of the moon Titan, where it found an atmosphere denser than the Earth's. But the Titan mission was the spacecraft's last: The moon's gravity swung Voyager 1 high above the ecliptic (the

TRAVEL SNAPSHOTS

VOYAGER 1

NEPTUNE

URANUS

VOYAGER 2

SATURN

JUPITER

EARTH

The Voyagers are accelerated toward the outer planets by the gravity of Jupiter and Saturn.

A high-altitude cloud soars above the blue haze of Neptune.

Voyager 2 discovers six new Neptune satellites, including lumpy Proteus.

Night looms over the top right of Uranus. Its far side is always dark.

Uranus is unusual for lying on its side. Its polar region is shown here in red.

Saturn's flattened sphere is obvious as Voyager 2 approaches the planet.

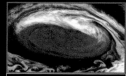

These streaks on Dione were caused by the fracturing of the moon's surface.

The shadow of the moon Ganymede falls on Jupiter. Fiery Io is to the right.

The Great Red Spot churns on Jupiter. It is 15,000 miles (24,000 km) from top to bottom.

MISSION DIARY: VOYAGER 1 AND 2

AUGUST 20, 1977 VOYAGER 2 IS LAUNCHED FROM CAPE CANAVERAL. SEPTEMBER 5, 1977 VOYAGER 1, DESPITE BEING NUMBER ONE, IS LAUNCHED FROM THE CAPE 16 DAYS LATER (RIGHT). MARCH 5, 1979 VOYAGER 1 REACHES JUPITER (SECOND RIGHT).

EN ROUTE, IT HAS OVERTAKEN VOYAGER 2 AND JUSTIFIED ITS DESIGNATION. JULY 9, 1979 VOYAGER 2 MAKES A FLYBY OF JUPITER ON A COURSE THAT HAS BEEN CHOSEN TO COMPLEMENT VOYAGER 1'S. THE PROBE MAKES A CLOSE APPROACH TO THE MOON EUROPA AND INVESTIGATES JUPITER'S SOUTHERN LATITUDES. NOVEMBER 12, 1980 VOYAGER 1 PASSES SATURN.

AFTER A CLOSE ENCOUNTER WITH THE MOON TITAN, IT IS HURLED OFF-COURSE BY TITAN'S GRAVITY. AUGUST 26, 1981 VOYAGER 2 PLUNGES THROUGH THE RINGS OF SATURN (ABOVE) AND PICKS UP ENOUGH SPEED FROM THE PLANET'S GRAVITY TO CARRY IT ONWARD

TO URANUS. JANUARY 24, 1986 VOYAGER 2 SURVEYS URANUS AND ITS MANY SATELLITES. AUGUST 25, 1989 VOYAGER 2 PHOTOGRAPHS THE COLDEST PLACE SO FAR FOUND IN THE SOLAR SYSTEM— THE SURFACE OF NEPTUNE'S LARGEST MOON TRITON (ABOVE).

PIONEER VENUS

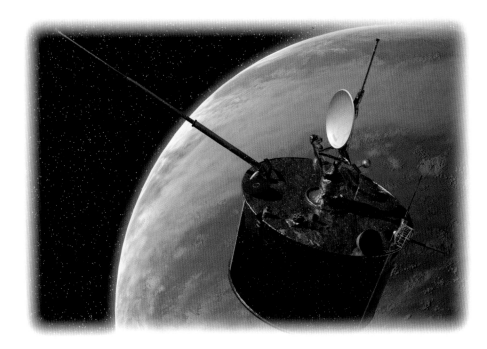

Nobody at NASA expected the Pioneer Venus mission to last for long. After entering Venus' hostile atmosphere, the four probes that made up one-half of the mission were silenced within hours. But they perished in a burst of glory, as their instruments returned important new data on Venus' atmosphere. The second part of the mission stayed away from Venus' punishing extremes in orbit to return data for over a decade. Long and short, these two Pioneer missions have both increased our knowledge of Venus.

WHAT IF...

...WE DESIGNED BETTER VENUS PROBES?

Multiprobes offer some of the best two-for-one deals in space exploration. By equipping a spacecraft with a few detachable extras, scientists get much more data than a single spacecraft could return, at little extra cost.

In view of the success of Pioneer Venus, you might expect NASA to be eager to get other multi-missions off the ground. But no multiprobes have visited Venus since Pioneer, and none are approved for the near future.

Some of NASA's reluctance is due to the fact that Pioneer Venus did such a good job. Thanks to Pioneer, we know a lot more about Venus' atmosphere, and after the mission NASA decided it was time to map out the planet's surface in similar detail. Given that multiprobes cannot see very much of a planet after they have dropped to its surface, the task of mapping was better suited to a single orbiter—in this case, the Magellan Venus Orbiter, launched in 1989.

With Venus' surface mapped, NASA once again turned its attention to what a multiprobe does best: studying the atmosphere. The Venus Multiprobe Mission was proposed for launch as part of the Discovery series in early 1999. It was designed to drop 16 probes. These would measure pressure and temperature on their drop to Venus' surface, while their velocities would tell scientists about the circulation of the atmosphere.

Until the technology exists to toughen up multiprobes, balloons may be our best way of exploring Venus' searing atmosphere. The Vega balloons survived for two days and circled a third of the planet.

But the mission was never approved. Multiprobes have gone as far as they can under current technology. The main problem is building anything to survive Venus' surface, where the pressure is 90 Earth atmospheres and the temperature 900°F (480°C). A titanium coat boosts a lander's chances, but this has to be thick enough not to cave in—a tall order when you want to pack a number of light probes onto one "spacebus."

While multiprobes are on hold, Russian and French scientists have pioneered a new breed of interplanetary probe—a balloon. The 1985 Venus Vega Mission dangled scientific equipment in Venus' atmosphere beneath two teflon-coated balloons. Although the balloons had to contend with clouds of sulfuric acid, the conditions were far less hostile than on Venus' surface. Each one lasted for about 46 hours—longer than any Pioneer probes.

Balloons may steal the limelight now, but multiprobes have not been forgotten. And when technology catches up with the aspirations of mission planners, a new family of probes could be on its way to Venus.

PIONEER VENUS MISSIONS

PIONEER VENUS ORBITER	SPACECRAFT NAME: PIONEER 12	MULTIPROBE	SPACECRAFT NAME: PIONEER 13
LAUNCH	MAY 20, 1978	LAUNCH	AUGUST 8, 1978
SCIENCE INSTRUMENTS	17	SCIENCE INSTRUMENTS	CARRIER: 2, LARGE PROBE: 7, SMALL PROBES: 3 EACH
VENUS ENCOUNTER	DECEMBER 4, 1978		
NUMBER OF ORBITS OF VENUS	OVER 5,000	VENUS ENCOUNTER	DECEMBER 9, 1978
END OF MISSION	OCTOBER 8, 1992	END OF MISSION	DECEMBER 9, 1978

PROBING VENUS

With its dense, poisonous atmosphere and a surface temperature that would melt lead, Venus must be approached with caution. NASA took no chances with the Pioneer Venus mission, which sent an orbiter and a "probe bus" carrying four atmospheric probes to Venus. Each probe was protected by a heat seal of toughened carbon—and with six separate spacecraft involved after the probes were deployed, the chances of success for at least one part of the mission were high.

The orbiter Pioneer 12 was launched three months before the multiprobe, and took a wide-swinging flight path that slowed down its approach to Venus and used less fuel—an important saving that would allow the mission to continue for 14 years. Upon arrival, the orbiter began to log details of everything near Venus from the properties of the solar wind to the planet's gravitational field. It also made radar maps of the planet's surface, which revealed new mountains and rift valleys.

While the orbiter took care of the bigger picture, the multiprobe was designed to study the composition of Venus' violent atmosphere, made of clouds of acid. The multiprobe took the fast route to Venus, and arrived just a few days behind the orbiter. The four probes, one large and three small, separated from the bus 8 million miles (13 million km) from Venus. Each was destined for different parts of the planet, and since none of them would survive for long, it was vital that they arrive in the right place. The large probe entered the atmosphere over Venus' equator on December 9, 1978. It was slowed down by a parachute and returned data all the way to the surface.

A FIERY END

The three smaller probes followed closely behind. These had no parachutes, but were braked by friction on their heat shields. As the probes plunged through the clouds, their temperature sensors shorted out when they came in contact with sulfur particles. Nonetheless, all three returned valuable data on the atmosphere's composition, right down to the size and shape of cloud particles.

THE PIONEERS ARRIVE

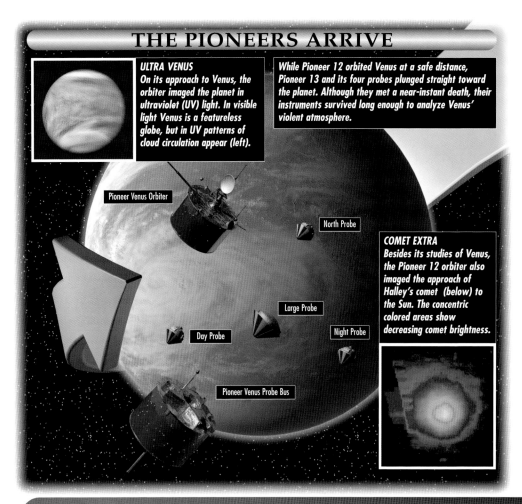

ULTRA VENUS
On its approach to Venus, the orbiter imaged the planet in ultraviolet (UV) light. In visible light Venus is a featureless globe, but in UV patterns of cloud circulation appear (left).

While Pioneer 12 orbited Venus at a safe distance, Pioneer 13 and its four probes plunged straight toward the planet. Although they met a near-instant death, their instruments survived long enough to analyze Venus' violent atmosphere.

Pioneer Venus Orbiter

North Probe

COMET EXTRA
Besides its studies of Venus, the Pioneer 12 orbiter also imaged the approach of Halley's comet (below) to the Sun. The concentric colored areas show decreasing comet brightness.

Large Probe

Day Probe

Night Probe

Pioneer Venus Probe Bus

RADAR MAPS

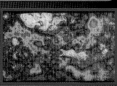

THE PIONEER VENUS ORBITER MAPPED OUT VENUS' SURFACE BY RADAR FOR A TOTAL OF AROUND FOUR YEARS. IT PRODUCED RELIEF MAPS OF GEOLOGICAL FEATURES (ABOVE) THAT INCLUDED MOUNTAINS AND VALLEYS, AND REVEALED THAT VENUS' SURFACE IS SMOOTHER THAN THE EARTH'S. THE RADAR WAS SWITCHED OFF FOR 10 YEARS IN 1981 TO SAVE PRECIOUS FUEL. IT WAS ACTIVATED AGAIN IN 1992 SO THAT PIONEER COULD MAP OUT THE SOUTHERN REGIONS OF VENUS.

On hitting the ground, one of the probes was destroyed instantly, while another lasted for just two seconds. The third transmitted for over an hour before its battery power was exhausted. By then, its internal temperature was a scorching 259°F (126°C). The multiprobe carrier, Pioneer 13, entered the atmosphere later the same day. Without a heat shield, it did not survive for long, and burned up about 70 miles (113 km) above Venus' surface.

The Pioneer 12 orbiter was still going strong. Most of its fuel was spent by 1980, but there was enough in reserve for it to study hitherto unexplored areas of Venus in 1991. In September 1992, Pioneer 12 entered Venus' upper atmosphere. It burned up a month later—the end of a long and unique space odyssey.

MISSION DIARY: PIONEER-VENUS

1974 U.S. CONGRESS APPROVES THE VENUS MISSION AND THE ORBITER'S SCIENCE INSTRUMENTS ARE SELECTED.
1878 THE ORBITER AND MULTIPROBE ARE GIVEN THEIR FINAL PRELAUNCH CHECKS (RIGHT).
MAY 20, 1978 THE PIONEER 12 ORBITER LIFTS OFF FROM CAPE CANAVERAL TO START ITS MISSION TO OUR NEAREST PLANETARY NEIGHBOR. THE ORBITER'S JOURNEY SWINGS IT MORE THAN HALFWAY AROUND THE SUN TO HELP IT SAVE FUEL.
AUGUST 8 PIONEER 13 DEPARTS CAPE CANAVERAL FOR VENUS WITH ITS CARGO OF FOUR ATMOSPHERIC ENTRY PROBES. THE LARGEST PROBE

WEIGHS **695** POUNDS AND CARRIES SEVEN SCIENTIFIC INSTRUMENTS. THE THREE SMALL PROBES WEIGH **165** LB (**75** KG) EACH.
NOVEMBER 16 PIONEER 13'S LARGE PROBE IS RELEASED FROM THE CARRIER SPACECRAFT.
NOVEMBER 20 PIONEER 13'S THREE SMALL PROBES (ABOVE) ARE RELEASED.
DECEMBER 4 PIONEER 12 BECOMES THE FIRST U.S. SPACECRAFT TO ENTER ORBIT AROUND THE PLANET VENUS. IT STARTS TO MAKE RADAR MAPS OF THE PLANET AND STUDIES THE UPPER ATMOSPHERE.

DECEMBER 9 PIONEER 13 AND ITS ARMADA OF FOUR PROBES ENTERS THE VENUSIAN ATMOSPHERE. THE LARGE PROBE ENTERS FIRST, AND A PARACHUTE IS DEPLOYED TO SLOW ITS DESCENT (RIGHT).
MARCH 1981 THE RADAR INSTRUMENT ABOARD THE ORBITER IS SWITCHED OFF.
1991 THE ORBITER'S RADAR MAPPING INSTRUMENT IS TURNED ON TO SCAN SOUTHERN AREAS OF THE PLANET.
OCTOBER 8, 1992 WITH NO FUEL REMAINING, PIONEER 12 SINKS TOWARD VENUS, WHERE IT IS DESTROYED BY THE HEAT OF ATMOSPHERIC DRAG.

VEGA 1 AND 2

Halley's comet headed sunward in 1986, as it does every 76 years—but this time it had company. Twin Soviet probes, Vega 1 and 2, were part of an international flotilla of spacecraft on an intercept course. Vega 1's camera returned the first close-up image of Halley's nucleus. The twins had already proved their worth, though: Their mission explored two alien bodies for the price of one. To reach Halley, the probes swung by Venus, where they deployed landers and balloons into the planet's corrosive atmosphere.

WHAT IF...

...WE COULD SEE THE VEGA PROBES AGAIN?

Mission planners knew they were sending the Vega craft into two very hostile environments. The Venusian atmosphere will crush, melt and corrode a lander all at the same time. And comets are surrounded by a halo of dust known as a coma. At the enormous speeds of the comet flyby, a single dust particle could slam through an aluminum sheet three inches (8 cm) thick.

But the Soviets had more than 15 years' experience designing equipment to survive on Venus. Each Vega lander was held in an insulated sphere until it commenced atmospheric entry. The sphere "hatched" about 40 miles (64 km) above the planet's surface. After atmospheric research balloons were deployed, a pilot parachute system opened to increase stability. The probe also had a ring of shock absorbers to absorb impact.

Once the probe had landed, onboard instruments measured atmospheric constituents and activated a drill to examine soil samples (although the Vega 1 drill failed because it deployed prematurely). Once these jobs were done, the landers' missions were complete. Soon, they began to fail under the heat load of the dense, 900°F (480°C) atmosphere. By now, there can be little left of either but some molten, acid-eaten scrap, possibly buried by windblown sand.

The Teflon-coated balloons were built to float around 33 miles (53 km) up, right in the middle of the sulfuric cloud layer, in

Those were the days: Vega project director V.M. Kovtunenko proudly winds up a triumphant news conference at the mission's end in March 1986. Five years later, the Soviet Union had collapsed.

temperatures and conditions similar to those of Earth's surface. Each balloon lasted around 46 hours and traveled more than a third of the way around the planet, moving from east to west. Onboard instruments measured pressure, temperature and vertical wind speed, and hunted for lightning strikes. What finally destroyed the balloons was traveling from night into the Venusian morning. Solar heating expanded the helium gas inside the balloons until they burst. There may be Teflon scraps yet blowing on the Venusian winds.

As for the two comet interceptors, they were protected from comet dust by multiple layers of super-thin aluminum. This mitigated the worst effects, but dust impacts still caused some equipment failures on the Vega pair, and the power available from the solar cells dropped by 50%. The probes survived the encounter with Halley, though, and remain in solar orbit. For a while it was suggested that they should be reactivated to survey other comets or asteroids. But the Soviet ability to sustain such efforts was failing, and the idea was not taken up. Now, after more than a decade in space, the probes are long past salvage.

VEGA SPECS

COUNTRY	SOVIET UNION	HEIGHT	15 FT 10 IN (4.8 M)
MISSION	VENUS LANDER AND BALLOON, COMET HALLEY FLYBY	CLOSEST APPROACH TO HALLEY NUCLEUS	5,544 MILES/8,922 KM (VEGA 1), 5,018 MILES/8,075 KM (VEGA 2)
LAUNCH VEHICLE	PROTON, FROM BAIKONUR		
SPACECRAFT MASS	10,824 LB (4,900 KG)	FLYBY SPEED	49 MILES/79 KM PER SECOND (VEGA 1),
"WINGSPAN" (SOLAR PANELS)	33 FT 2 IN (10 M)		48 MILES/77 KM PER SECOND (VEGA 2)

DOUBLE DEAL

Together, Vega 1 and 2 made up the last successful Soviet interplanetary mission—and by far the most triumphant. Originally, the twin probes had been intended to be simply the latest in the long-running Venera series of Venus probes. The Soviets had been sending Venera spacecraft into the Venusian atmosphere ever since 1967. Their designs had improved enough for successive landers to penetrate clouds of sulfuric acid and survive, at least briefly, the planet's hellish surface. There, temperatures reach almost 900°F (480°C) —hot enough to melt lead—and pressures are 90 times Earth's.

SCIENTIFIC COCKTAIL

Vega 1 and 2 were originally planned as orbiter-lander combinations in the same sequence of probes, designated Venera 17 and 18. These probes were the first major Soviet mission to invite contributions from foreigners. French scientist Jacques Blamont suggested that the probes could deploy balloon "aerostats" to explore the Venusian atmosphere. And Blamont was also ultimately responsible for the redirection of the mission towards Halley's comet. At a 1980 cocktail party, U.S. space scientist Louis Friedman joked that Venus would be a great place to see Halley's comet when it came around again in 1986, because it would get closer to Venus than it did to the Earth. Blamont considered the feasibility of sending the probes on to Halley by swinging past Venus. The Russians liked the idea, particularly after NASA announced that it would not be sending a Halley probe of its own. A Soviet comet mission would not only be a scientific bonanza, it would be a propaganda coup.

INTERNATIONAL SCIENCE

So the probes were renamed Vega 1 and 2, for "Venera-Galley" (there is no "H" in the Russian alphabet). The lander design was left unchanged, but the orbiter needed adaptation. The spacecraft was protected against dust impacts by aluminum sheeting. A movable instrument platform carried narrow- and wide-angle cameras, a spectrometer and an infrared sounder to measure the nucleus' surface temperature. The spacecraft's computers made sure that the panel pointed constantly at the nucleus. Other instruments on board included magnetometers and cometary dust collectors that would measure particle size. The Vega mission was a fine example of international cooperation. Once the aerostats were released into the Venusian atmosphere, their passage across the planet was tracked by 12 radio observatories in 10 separate countries. And data from Vega's Halley flybys was used to pinpoint the course of another space probe zeroing in on the comet. Thanks to Vega's initial observations, the European Space Agency's Giotto spacecraft came within 372 miles (600 km) of the Halley nucleus.

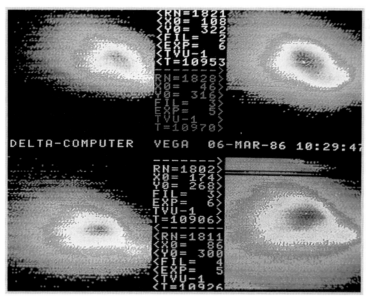

COMET CORE
These false-color images of Comet Halley's nucleus were beamed back by Vega 1 near its closest approach on March 6, 1986. From blue to red, the pictures show the increasing density of the gas around Halley's core.

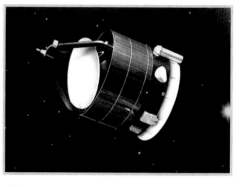

EURO PARTNER
The European Space Agency's Giotto probe (above) made the closest approach—within 400 miles (640 km)—to Comet Halley. But its close pass was only possible thanks to data from Vega 1 and Vega 2.

BACK-DOOR FRIEND

VEGA WAS THE FIRST SOVIET MISSION OPENED UP TO SUBSTANTIAL PARTICIPATION BY THE INTERNATIONAL SPACE SCIENCE COMMUNITY (RIGHT). BUT POLITICAL TENSIONS WERE A PROBLEM. RESEARCHER JOHN SIMPSON FROM THE UNIVERSITY OF CHICAGO CONTRIBUTED COMET-DUST DETECTORS TO THE PROBES, BUT IN ORDER TO AVOID ANY CONTROVERSY HE HAD THE INSTRUMENTS MANUFACTURED IN WEST GERMANY AND PUBLICIZED AS A WEST GERMAN CONTRIBUTION.

MISSION DIARY: VEGA 1 AND 2

1981 SHORT OF FUNDS, NASA CANCELS ITS PROPOSED COMET HALLEY PROBE.
1981 SOVIET UNION UPGRADES TWO PLANNED VENUS PROBES FOR A HALLEY ENCOUNTER.
OCTOBER 16, 1982 MOUNT PALOMAR'S HALE TELESCOPE TRACKS RETURNING COMET HALLEY FOR THE FIRST TIME EVER.
DECEMBER 15, 1984 VEGA 1 IS LAUNCHED FROM BAIKONUR IN THE SOVIET REPUBLIC OF KAZAKHSTAN.
DECEMBER 21 VEGA 2 (ABOVE) UNDERGOES FINAL CHECKS AND IS

LAUNCHED FROM BAIKONUR ON THE SAME TRAJECTORY AS ITS SISTER PROBE.
JUNE 11, 1985 VEGA 1 REACHES VENUS. IT DEPLOYS VENUS LANDER AND RECEIVES A SPEED-BOOSTING GRAVITY ASSIST FROM THE PLANET.
JUNE 15 VEGA 2 VENUS ENCOUNTER AND LANDER DEPLOYMENT. BOTH PROBES DEPLOY BALLOONS IN THE VENUSIAN ATMOSPHERE (ABOVE).
SEPTEMBER 11, 1985 NASA SPACECRAFT INTERNATIONAL COMETARY EXPLORER TRAVERSES THE PLASMA TAIL OF COMET

GIACOBINI-ZINNER, A DRY RUN FOR THE COMING HALLEY FLYBYS.
FEBRUARY 9, 1986 COMET HALLEY PASSES PERIHELION, ITS CLOSEST APPROACH TO THE SUN. BOTH VEGAS NEAR THEIR TARGET.
MARCH 6 VEGA 1 MAKES ITS CLOSEST HALLEY FLYBY (5,544 MILES/8,922 KM).
MARCH 9 VEGA 2 MAKES ITS CLOSEST HALLEY FLYBY (5,018 MILES/8,075 KM). SOVIET SCIENTISTS (ABOVE) SOON BEGIN TO SHARE DATA FROM THE VEGA MISSIONS WITH THE REST OF THE WORLD.
MARCH 14 ESA'S GIOTTO PROBE USES VEGA DATA TO MAKE THE CLOSEST HALLEY FLYBY, AT JUST 372 MILES (600 KM).

GALILEO ORBITER

Galileo is one of the most complex spacecraft ever built, and it orbited Jupiter for four years making detailed measurements of the planet and its moons. While Galileo's orbiter photographed the moons and analyzed Jupiter's strong magnetic field, the probe it had carried descended into the atmosphere on a one-way journey to destruction. Galileo's operations were scheduled to end in January 2000, but the spacecraft was functioning so well that NASA decided to extend its mission.

WHAT IF...

...WE SENT A NEW GALILEO?

By the time Galileo reached Jupiter, it was taking almost an hour for radio signals from Earth to reach the spacecraft. With such a long delay, it was impossible for controllers on Earth to direct its course minute-by-minute. Instead, the spacecraft had to take care of itself, storing all of the navigation information it needed on-board and maneuvering automatically when necessary.

And because Galileo was so far from the Earth, it needed a large antenna—the high-gain antenna—to be able to return all of the data it was due to transmit. The 16-ft (4.9-m) diameter antenna was too large to fit into the payload bay of the Space Shuttle, so it had to be folded like an umbrella and unfurled when out in space.

With such a complex craft, every component had to be thoroughly tested, time after time, to minimize any risk that it would fail. This testing added millions of dollars to the cost of the project. But in spite of this testing, the high-gain antenna failed to open fully. All attempts to release the antenna failed, and the engineers back on Earth had to find a way around the problem. New software that enabled Galileo to compress its data—so that it could be sent back by its low-gain antenna, which transmitted data at a much slower rate—was written and uploaded to the spacecraft. In addition, the ground stations that received the Galileo data were upgraded to make them much more sensitive. These remedies reduced the amount of data that was lost.

The Europa Orbiter, one of the new breed of relatively simple and cheap planetary probes, is scheduled to continue Galileo's observations of Jupiter's moon Europa in about 2006.

Galileo has produced a great wealth of data about Jupiter and its moons, but future probes to Jupiter will have to be much smaller and simpler. NASA can no longer afford projects that consume the kind of time and money it lavished on Galileo, and its philosophy is now "smaller, faster, cheaper."

Spacecraft must now be less expensive, not just to save money up front but also to make failures more readily accepted and cheaper to replace. So it is unlikely that another craft as large and complex as Galileo will be built—but thanks to improvements in technology, there will probably be smaller, quickly built and cheaper spacecraft that will be just as capable.

Using this new approach—already employed on missions such as Mars Pathfinder and Deep Space 1—missions will be launched more frequently. Scientists will explore the solar system and increase our understanding of the universe at a quicker pace and at a fraction of the cost of large complex missions such as Galileo.

GALILEO ORBITER

ORBITER		ATMOSPHERIC PROBE	
LAUNCH MASS	4,691 LB (2,128 KG)	TOTAL MASS	747 LB (339 KG)
DIMENSIONS	20 FT 8 IN (6.3 M) HIGH, MAX. 16 FT (4.9 M) IN DIAMETER	DIMENSIONS	5 FEET (1.5 M) IN DIAMETER
		DESCENT MODULE MASS	266 LB (121 KG)
SCIENCE PAYLOAD MASS	260 LB (117 KG)	SCIENCE PAYLOAD MASS	66 POUNDS (30 KG)
PROPELLANT MASS	2,035 LB (923 KG)		

JUPITER OBSERVER

Galileo was built to give scientists a better understanding of Jupiter, the largest planet in the solar system. The 2.5-ton vehicle was one of the most complex spacecraft ever built and consisted of three sections—a spinning section to provide stability, a non-spinning section that carried the cameras, and a probe that undocked from the main spacecraft to plunge down through the Jovian atmosphere.

The first objective of the Galileo mission was to deliver the probe, which slammed into Jupiter's atmosphere at 106,000 mph (170,500 km/h) on December 7, 1995. After entry, the probe switched on its instruments and measured the composition of Jupiter's atmosphere for 58 minutes before it was destroyed by the building pressure and heat. The information it had gathered was relayed to Earth via the orbiter, traveling high above the planet and well away from its dangerous atmosphere. The orbiter itself went on to complete a long tour of duty studying Jupiter's moons.

Sending the spacecraft 500 million miles across space had presented its designers with a number of problems. For instance, it had to carry a lot of fuel for course corrections and orbital maneuvers, and this fuel amounted to about 40% of its takeoff weight. And at Jupiter the sunlight

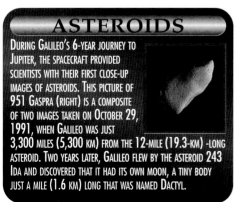

ASTEROIDS

DURING GALILEO'S 6-YEAR JOURNEY TO JUPITER, THE SPACECRAFT PROVIDED SCIENTISTS WITH THEIR FIRST CLOSE-UP IMAGES OF ASTEROIDS. THIS PICTURE OF 951 GASPRA (RIGHT) IS A COMPOSITE OF TWO IMAGES TAKEN ON OCTOBER 29, 1991, WHEN GALILEO WAS JUST 3,300 MILES (5,300 KM) FROM THE 12-MILE (19.3-KM) -LONG ASTEROID. TWO YEARS LATER, GALILEO FLEW BY THE ASTEROID 243 IDA AND DISCOVERED THAT IT HAD ITS OWN MOON, A TINY BODY JUST A MILE (1.6 KM) LONG THAT WAS NAMED DACTYL.

is only 4% the strength it is at Earth, so it could not use solar panels to generate its electricity. Instead, Galileo had to use small nuclear power sources—radioisotope thermoelectric generators (RTGs).

ORBITER IN ACTION

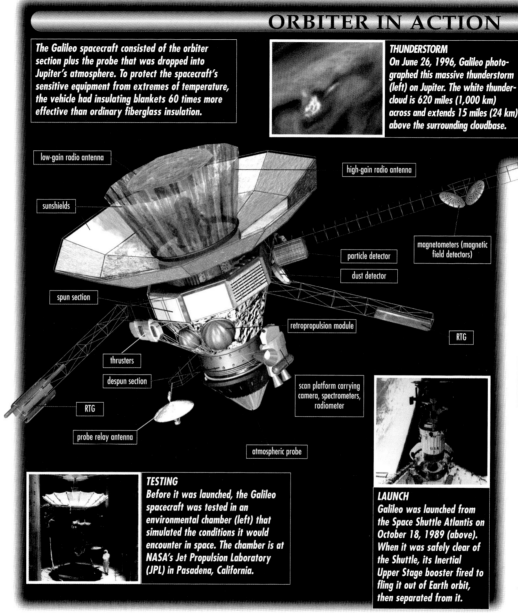

The Galileo spacecraft consisted of the orbiter section plus the probe that was dropped into Jupiter's atmosphere. To protect the spacecraft's sensitive equipment from extremes of temperature, the vehicle had insulating blankets 60 times more effective than ordinary fiberglass insulation.

THUNDERSTORM
On June 26, 1996, Galileo photographed this massive thunderstorm (left) on Jupiter. The white thundercloud is 620 miles (1,000 km) across and extends 15 miles (24 km) above the surrounding cloudbase.

low-gain radio antenna
sunshields
high-gain radio antenna
plasma wave detector
magnetometer (magnetic field detector)
magnetometer boom
particle detector
magnetometers (magnetic field detectors)
dust detector
spun section
retropropulsion module
RTG
thrusters
despun section
scan platform carrying camera, spectrometers, radiometer
RTG
probe relay antenna
atmospheric probe

TESTING
Before it was launched, the Galileo spacecraft was tested in an environmental chamber (left) that simulated the conditions it would encounter in space. The chamber is at NASA's Jet Propulsion Laboratory (JPL) in Pasadena, California.

LAUNCH
Galileo was launched from the Space Shuttle Atlantis on October 18, 1989 (above). When it was safely clear of the Shuttle, its Inertial Upper Stage booster fired to fling it out of Earth orbit, then separated from it.

HARD EVIDENCE

COMET STRIKE
In July 1994, when Comet Shoemaker-Levy plunged into Jupiter, the planet's immense gravitational pull broke it into 20 large pieces. On impact, these fragments together released more energy into

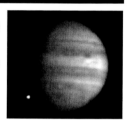

the planet's atmosphere than would result from setting off the Earth's entire nuclear arsenal. Images from the Hubble Space Telescope showed some of the collisions, but only Galileo, 148 million miles (238m km) from the planet, was able to witness the full 6-day onslaught and destructive power of the impacts. The bright point in this Galileo image (above) is an impact on the dark side of Jupiter.

busy for years to come. Its science instruments were divided between the rotating upper section of the craft and the "despun" lower section.

The upper section turned at about 3 rpm and housed instruments that needed to sweep around to gather information. These included sensors to detect and measure charged particles, cosmic and planetary dust and the magnetic fields of Jupiter and its moons. It also carried the spacecraft's power supply, propulsion system and most of its electronics. The lower section carried instruments that needed to point at specific targets, such as the camera system and spectrometers for the analysis of gases and surface chemistry.

SPINNING SCIENCE

During its long operational life, the Galileo orbiter generated a massive amount of data that will keep scientists

241

GALILEO PROBE

After a journey of nearly six years, the Galileo spacecraft's Jupiter probe parted company with its mother craft to begin a 50-million-mile (80-million-km) descent toward the giant planet. Four-and-a-half months later, it slammed into the Jovian atmosphere at 106,000 mph (170,500 km/h). Aboard the tiny probe was a series of experiments designed to unravel the mysteries of Jupiter's turbulent atmospheric shroud. They had less than one hour to gather data before the probe disintegrated under a combination of unimaginable heat and crushing pressure.

WHAT IF...

...WE KNEW WHAT HAPPENED TO THE GALILEO PROBE?

Although contact with the Galileo probe was lost after 125 miles (200 km) of its descent through Jupiter's atmosphere, the tiny craft almost certainly continued on its journey for some time afterward. It is highly unlikely that it met any solid resistance during its plunge into the gas giant—but as it fell farther toward the core, temperatures and pressures would have risen inexorably.

Around 200 miles (322 km) into the atmosphere, two hours after the probe's initial entry, the surface temperature would have reached 500°F (260°C)—enough to melt the plastic parachute, causing the probe to pick up speed. Just 40 minutes later, the craft would have reached a depth of 400 miles (645 km). At this point, surface temperatures would have risen to 1,200°F (650°C) and the pressure bearing down on the craft would have been 280 times the pressure on the Earth's surface.

Under such conditions, those parts of the craft made of aluminum would have begun to melt and fall like raindrops. Meanwhile, the descent module's hardier titanium casing would have continued to fall, getting hotter all the time. After six hours, at a depth of around 800 miles (1,300 km), the temperature would have reached a scorching 3,100°F (1,700°C) and casing, too, would have vaporized—along with everything inside it.

Could future probes to the gas giants do better? Perhaps. The problem of withstanding Jupiter's massive atmospheric pressure could be overcome by using a balloon—that is, by varying the internal pressure of the craft to

The Galileo probe continued on its plunge through Jupiter's atmosphere long after contact was lost. But eventually, the buildup of heat and pressure was enough to annihilate the robust little craft.

cope with the massive force squeezing it from the outside. But to send a balloon to enter Jupiter's atmosphere at the speed of the Galileo space probe is impractical. So any future probe to the gas giant would need a seriously powerful rocket engine that would slow it down before its atmosphere encounter.

Unfortunately, there would be a tougher problem to contend with: Jupiter's weather. A balloon probe descending through the atmosphere would be a little like a stray party balloon in the middle of a hurricane—and would need a way to respond accordingly. And since meteorologists on Earth have enough trouble coping with the vagaries of atmospheric conditions on our own planet, it may be a long time before we learn more about Jupiter than the remarkable amount Galileo and its probe have already told us.

GALILEO PROBE SPECS

PROBE MASS	747 LB (339 KG)	ENTRY SPEED INTO JOVIAN ATMOSPHERE	106,000 MPH
HEAT SHIELD MASS	335 LB (152 KG)		(170,500 KM/H)
PROBE DIAMETER	5 FT (1.5 M)	MAX. DECELERATION FORCE	230 G
PARACHUTE DIAMETER	8 FT (2.4 M)	MAX. TEMPERATURE DURING ENTRY	28,000°F (15,500°C)

HEROIC SACRIFICE

The Galileo Jupiter probe consisted of two modules: an outer deceleration module, encased in heat shields, and an inner descent module that carried scientific instruments and communications systems. The probe was carried aboard the Galileo spacecraft, which was launched by the Space Shuttle Atlantis in October 1989.

After a 6-year journey that included gravity-assist loops around Venus and the Earth to help it gain speed, Galileo reached Jupiter in December 1995. By that time, the probe had already detached itself from its mother craft and begun its perilous descent toward the largest planet in the solar system.

The release needed to be executed with pinpoint precision, since the probe possessed no steering capability of its own. The craft had to enter the Jovian atmosphere at an angle of precisely 8.5°. Just 1.5° less, and it would have ricocheted off the outer atmosphere into deep space; 1.5° more, and it would have plunged through the atmosphere too fast, causing it to vaporize like a shooting star.

Like the Space Shuttle during reentry, friction between the probe and Jupiter's atmosphere generated massive amounts of heat that caused temperatures on the surface of the craft to rise sharply. But the Galileo probe was traveling some five times faster than a returning Shuttle. As a result, the friction was so great that the spacecraft's heat shields almost instantaneously began to glow white-hot. The moment of entry was probably the most critical of the probe's long journey.

HARD EVIDENCE

A CLEAR DAY
Although scientists calculated every inch of the Galileo probe's flight in advance, one factor that they could not allow for was the weather. Just like the Earth, the Jovian atmosphere contains clouds of water vapor (left). But after its 6-year journey, the Galileo probe just happened to arrive on a day when there was barely a cloud in the sky. As a result, it recorded a far lower hydrogen and oxygen content than expected.

GALILEO PROBE

The Galileo probe—the fastest artificial object of its day—was carefully designed to withstand the tremendous forces exerted on it by Jupiter's massive gravity. The scientific instruments aboard proved more than equal to the task and returned vast amounts of invaluable data.

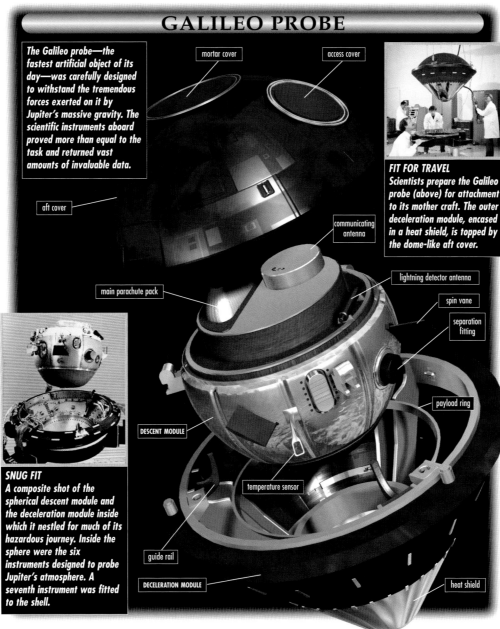

mortar cover

access cover

aft cover

communicating antenna

main parachute pack

lightning detector antenna

spin vane

separation fitting

DESCENT MODULE

payload ring

temperature sensor

guide rail

DECELERATION MODULE

heat shield

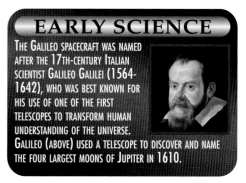

FIT FOR TRAVEL
Scientists prepare the Galileo probe (above) for attachment to its mother craft. The outer deceleration module, encased in a heat shield, is topped by the dome-like aft cover.

SNUG FIT
A composite shot of the spherical descent module and the deceleration module inside which it nestled for much of its hazardous journey. Inside the sphere were the six instruments designed to probe Jupiter's atmosphere. A seventh instrument was fitted to the shell.

NOT SO FAST

During the first two minutes, as the density of the atmosphere increased, the probe began to slow down. A drogue parachute was deployed to slow the craft even further. This reduced the friction enough for the heat shields to be discarded. As the aft cover was ejected, the main parachute unfurled and the inner descent module separated from its the outer shielding.

Four minutes later, the descent module had slowed down to a modest 250 mph (400 km/h). At one point, the force of deceleration reached 230 g—230 times the Earth's gravity. If a person weighing 100 lb (45 kg) had been aboard, she would have felt as if she weighed 11½ tons (10½ tonnes).

To the probe's mission controllers, 500 million miles (800 million km) away, it came as a great relief when the six analysis instruments sparked into life and worked flawlessly after their 6-year hibernation. But as the descent module hurtled downward, the mounting temperature and pressure threatened to ruin 15 years of careful and detailed planning.

Four of the instruments aboard the probe busied themselves analyzing the pressure, temperature and composition of the Jovian atmosphere, while a fifth studied the clouds and a sixth searched for lightning. A seventh experiment, activated a few minutes before arrival, measured the strength of Jupiter's magnetic field.

Unable to communicate directly with mission control, the probe radioed its readings to the Galileo mother craft, which by now was in orbit around Jupiter. This left mission control with less than an hour in which to take measurements. For 58 minutes, the orbiter received and transmitted data—then contact was lost.

But the Galileo probe's short-lived efforts were not in vain: Its fiery sacrificial journey revealed a wealth of information about Jupiter, which in turn has greatly enhanced our understanding of the solar system's largest planet.

GIOTTO

Every 76 years, Halley's Comet swings back into the inner solar system and passes close enough to the Earth to be seen with the naked eye. Halley's last return, in 1986, was different from all its previous visits because a tiny European spacecraft called Giotto flew out to study it at close range. Giotto met up with Halley in March 1986. Despite taking a battering from cometary debris, it shot some remarkable close-up photographs of the comet, and its instruments recorded a wealth of valuable data.

WHAT IF...

...WE SEND MORE PROBES TO VISIT COMETS?

When Giotto's encounter with Comet Grigg-Skjellerup was completed, the spacecraft had only a few pounds of propellant left in its tanks—just enough to change its orbit for a final Earth flyby. For seven years, the spacecraft silently traveled through space, its orbit bringing it closer to home for one final Earth flyby.

On July 1, 1999, Giotto came home for the last time. The little spacecraft sped by halfway between the Earth and the Moon, approaching from the south and passing over Antarctica and South America before returning once again to deep space. The spacecraft had surpassed all expectations and provided scientists with much more information about comets than they could have expected when the project was first proposed.

Giotto was Europe's first comet probe. Its next, called Rosetta, is a much more advanced spacecraft that will catch up with Comet 46 P/Wirtanen in 2011, as it makes one of its frequent visits to our part of the solar system. Rosetta will go into orbit around the comet, map it in detail and land an instrument package on its surface.

NASA also has a comet-visiting program, which includes the Stardust spacecraft that will collect samples of cometary dust and bring them back to Earth. Stardust was launched in February 1999, and in 2004, at a distance of 155 million miles (249m km) from Earth, it flew within 150 miles (240 km) of the surface of Comet Wild 2. It is currently en route to Earth with its cometary dust samples, and parachute them down to the ground in early 2006.

Giotto's historic encounter with Halley (left) was the first time a space probe had made a close encounter with a comet, but it will certainly not be the last.

In 2002, while Stardust was still on its way to Comet Wild 2, NASA's Contour (Comet Nucleus Tour) probe was launched. If all had gone as planned, Contour would have become the first spacecraft to encounter three comets. In 2003, it was to fly by Comet Encke, then continue for a 2006 encounter with Comet Schwassmann-Wachmann 3 and finally a 2008 rendezvous with Comet d'Arrest. At each encounter, Contour would take pictures and analyze the comet's nucleus and the dust flowing from it. In the event, Contour was lost as it maneuvered to leave Earth orbit.

In 2005, another NASA mission will took the exploration of comets to a higher level when the Deep Impact comet probe met Comet Tempel 1. The Deep Impact mission is actually two spacecraft. One is an instrument platform that slowly flies past the comet to record information, while the second is an "impactor" that crashes into the comet to help scientists learn more about the physical composition of comets. The future of close-up comet exploration looks promising.

Deep Impact was a big success, allowing scientists to learn a lot more about comet composition from the formation of the impact crater and the material ejected.

GIOTTO MISSION FACTS

LAUNCH	JULY 2, 1985, 11.35 GMT ABOARD AN ARIANE 1 ROCKET	NUMBER OF SCIENTIFIC INSTRUMENTS	10
LAUNCH SITE	KOUROU, FRENCH GUIANA (ELA 1)	TARGETS	COMETS HALLEY AND GRIGG-SKJELLERUP
SPACECRAFT MASS	2,117 LB (960 KG)	CLOSEST APPROACH TO HALLEY	370 MILES (595 KM)
SPACECRAFT DIMENSIONS	6 FEET (1.8 M) IN DIAMETER, 9 FEET (2.7 M) HIGH	CLOSEST APPROACH TO GRIGG-SKJELLERUP	125 MILES (201 KM)
		ORIGINAL MISSION (HALLEY)	END DATE APRIL 2, 1986
		EXTENDED MISSION	END DATE JULY 23, 1992

HALLEY HUNTER

Giotto was blasted into space aboard an Ariane 1 rocket launched from Kourou, French Guiana, on July 2, 1985. It arrived at Halley's Comet on the night of March 13–14, 1986, and during the early evening it crossed the "shock front" that formed the boundary between the solar wind and the outer regions of the comet's dusty atmosphere. The spacecraft's camera was switched on and immediately began to transmit the first fuzzy, close-up images of Halley back to the mission controllers on Earth. The pictures showed the comet to be a dark, peanut-shape body, blacker than coal, with bright jets of material spouting from its surface.

To protect Giotto and its payload of instruments from the hazardous dust around the comet, the spacecraft was fitted with a tough, 2-layer aluminum and Kevlar dust shield. The payload, tucked away safely behind the shield, included a camera, instruments to measure the composition of the comet's gas and dust, a dust impact detector, experiments to study the solar wind and charged particles, and an instrument to study changes in the magnetic field. There was also an optical probe that could study the brightness of the comet's atmosphere and a radio science experiment to detect electrons. Giotto's electrical power was supplied by a solar array wrapped around the outside of its cylindrical body, and at the opposite end from the dust shield was the main communication dish that beamed images and data back to Earth.

EXTENDED MISSION

Giotto's dust shield began to prove its worth about two hours before the spacecraft's closest approach to Halley, when the first of 12,000 impacts of cometary dust were recorded. The impact rate increased as Giotto closed in on Halley and passed through the jets of material, heated by the Sun, that were being blasted from the comet's surface. Then, only 14 seconds before its closest approach, Giotto was sent into a spin when it was hit by a particle thought to be about the size of a grain of rice and weighing

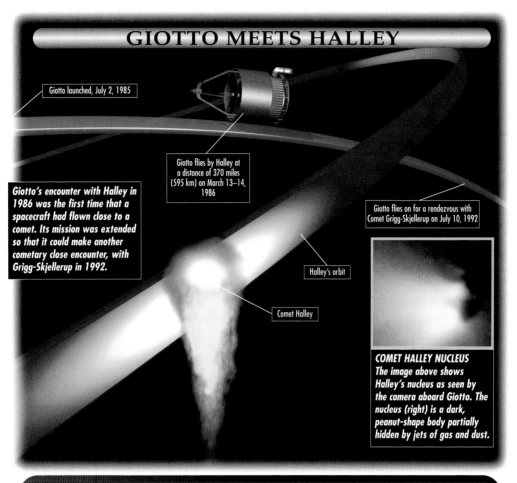

GIOTTO MEETS HALLEY

Giotto launched, July 2, 1985

Giotto flies by Halley at a distance of 370 miles (595 km) on March 13–14, 1986

Giotto's encounter with Halley in 1986 was the first time that a spacecraft had flown close to a comet. Its mission was extended so that it could make another cometary close encounter, with Grigg-Skjellerup in 1992.

Giotto flies on for a rendezvous with Comet Grigg-Skjellerup on July 10, 1992

Halley's orbit

Comet Halley

COMET HALLEY NUCLEUS
The image above shows Halley's nucleus as seen by the camera aboard Giotto. The nucleus (right) is a dark, peanut-shape body partially hidden by jets of gas and dust.

MISSION DIARY: GIOTTO

1979 Studies begin on the design of a joint U.S.-European spacecraft to rendezvous with Halley's Comet.
1980 The U.S. withdraws from the project due to the high costs. The European Space Agency (ESA) decides to proceed alone.
1982 Construction of Giotto begins (above).
July 2, 1985 Giotto is launched by an Ariane 1 rocket from the ESA launch site at Kourou, French Guiana.
March 13–14, 1986 Giotto has its close encounter with Halley's Comet, passing it at a distance of about 370 miles (595 km).
April 2, 1986 Placed in "hibernation."
February 1990 Giotto is "awakened" by ground controllers.
July 2, 1990 Swings past Earth to get on course for a second encounter.
July 23, 1990 Begins a second period of hibernation.
May 4, 1992 Awakened again.
July 10, 1992 Passes within 125 miles (201 km) of Comet Grigg-Skjellerup.
July 23, 1992 Operations end.
July 1, 1999 The sleeping Giotto flies past Earth, but is not awakened.

DARMSTADT

GIOTTO, EUROPE'S VERY FIRST INTERPLANETARY SPACE MISSION, WAS MONITORED AND CONTROLLED FROM THE EUROPEAN SPACE OPERATIONS CENTER (ESOC). LOCATED IN THE CITY OF DARMSTADT, GERMANY, ESOC IS THE EUROPEAN SPACE AGENCY'S PRIMARY SPACE MISSION CONTROL CENTER. IT MONITORS THE LAUNCH AND OPERATION OF EUROPEAN SATELLITES AND SPACE PROBES USING ESTRACK, ITS GLOBAL NETWORK OF TRACKING STATIONS.

less than 1/30th of an ounce.

At Giotto's mission control—the European Space Operations Center in Darmstadt, Germany—computer screens went blank and contact with the spacecraft was lost. But to everyone's amazement and relief, the tough little spacecraft soon began to send bursts of data back to Earth, and after half an hour it settled down to normal transmissions. Giotto continued its up-close survey of Halley's Comet for nearly a day. Its instruments were turned off in the early hours of March 15 as the spacecraft and the comet went their separate ways.

No one had expected Giotto to survive its bruising encounter with Halley, but although several instruments were damaged, the spacecraft was still operational and still had fuel on board. Sending Giotto to study a second comet was now a real possibility. So Giotto was put into a "hibernation" mode, and four years later its course was altered: It skimmed past Earth on its way to its second target—Comet Grigg-Skjellerup.

Giotto met up with the much smaller and less active Grigg-Skjellerup on July 10, 1992, passing only 125 miles (201 km) from the comet's surface—the closest any spacecraft had been to a comet. Giotto's camera was out of action, but eight of its science instruments returned useful information. Giotto's science instruments were turned off for the last time on July 11, and on July 23, the spacecraft was returned to its deep-sleep mode, from which it would never be reawakened.

DEEP SPACE 1

Deep Space 1 is a revolutionary new spacecraft. It is powered by a new, super-efficient engine and carries an intelligent computer that pilots the craft with minimal instructions from ground control. The mission of this little probe is to test out these and other new technologies in space. It is the first spaceflight NASA has launched purely to test innovations. It is part of NASA's New Millennium Program, which will help shape the spacecraft of the future.

WHAT IF...

...ALL SPACECRAFT HAD ION ENGINES?

Science fiction starships have had ion drives for decades. NASA actually built its first ion drive at the Lewis Research Center as long ago as 1960. During the 1960s and '70s, Soviet space probes tested ion thrusters and NASA tried out the drives in the lab, on test flights and on satellites in Earth orbit. But Deep Space 1 is the first craft to fly beyond Earth's orbit powered solely by one of these devices. If all goes well with DS1, the next New Millennium Program missions will probe Mars and map the Earth's surface—all under ion propulsion. The final phase, Deep Space 4, will carry four ion engines of the same design as DS1's single engine and use them to fly in formation with a comet while taking pictures and perhaps even sampling the comet nucleus.

Ion drives apply a gentle, steady acceleration rather than the hefty kick of a chemical rocket, so they are best suited to missions that can afford the time to accelerate for days or weeks on end. But despite this slow build-up, an ESA (European Space Agency) ion-propelled craft called "Smart-1" was launched in 2003 on a flyby mission to the Moon.

Deep Space 1's descendants will have larger solar arrays to generate more electrical power and will carry more propellant to take them farther. Ion drives will make it possible to send a crewed mission to Mars—a feat impractical for a conventional craft due to the huge store of propellant required. The

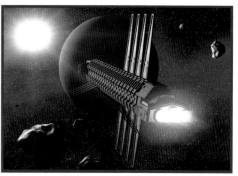

Crewed spacecraft powered by high-efficiency ion drives hold the key to future exploration throughout the solar system and beyond.

continual thrust provided by an ion drive is ideal for certain types of transfer orbits to Mars. And because the ion engine produces constant but gentle acceleration, it provides micro-gravity inside the craft, making conditions more comfortable than in the zero gravity of a chemical rocket, which would coast to the Red Planet.

The ion drive could also be used for a mission to Pluto. But this type of mission presents special problems, since the sunlight in the outer reaches of the solar system is far too dim to power the solar arrays that generate electric power for the drive. One approach is to accelerate in the glare of the inner solar system, and coast onward into the frozen darkness. An alternative is to carry another source of power, such as a nuclear generator, or—if those science fiction starships are ever built—a fusion reactor or antimatter generator.

DS1 SPECIFICATIONS

TOTAL COST	$152.3M (FY95-99)	HIGH GAIN ANTENNA DIAMETER	11 IN (28 CM)
LAUNCH DATE	OCTOBER 24, 1998	COMMUNICATIONS FREQUENCIES	X, KA
LAUNCH SITE	CAPE CANAVERAL, FLORIDA	MAXIMUM DATA RATE	20 KILOBITS PER SECOND
END OF MISSION DATE	SEPTEMBER 1999	MAXIMUM POWER	2500W (2100W USED TO
LAUNCH MASS (SPACECRAFT AND PROPELLANTS) 1072.13 LB			POWER ION ENGINE)
(486.3 KG)			

SPACE GUINEA PIG

When it blasted into orbit aboard a Delta rocket on October 28th, 1998, Deep Space 1 (DS1) carried on board no fewer than 12 new technologies for testing. Some will make the spacecraft of the future smaller and cheaper; others aim to increase the precision of space astronomy. But the key innovations aboard DS1 are its engine and its control system.

DS1 is propelled by an ion engine. In a chemical rocket like the Space Shuttle, a continuous controlled explosion hurls burning gas out of the rocket nozzle, driving the vehicle onward and upward. An ion engine is somewhat more sedate. Instead of a chemical reaction, it uses electric power to accelerate charged particles of gas out of the engine nozzle. Instead of thundering sheets of flame, the ion engine produces an eerie blue glow. And instead of hundreds of tons of thrust, the ion engine produces a thrust of just one-third of an ounce (0.275 N)—or about one-tenth of the weight of an apple.

But appearances can be misleading. Ion engines are deceptively powerful and extremely efficient. They can go on producing thrust continuously for months on end. Deep Space 1's supply of 180 pounds (82 kg) of xenon gas is enough to thrust continuously for 20 months, during which time the engine will gradually accelerate the spacecraft up to speeds of 10,000 mph (16,000 km/h).

CLEAN FUEL

THE ION DRIVE POWERING DS1 IS THE MOST EFFICIENT ENGINE EVER FLOWN IN SPACE——MANY HUNDREDS OF TIMES MORE SO THAN THE FLAME-BELCHING SPACE SHUTTLE MAIN ENGINES, WHICH REQUIRE HUGE RESERVES OF FUEL. BUT WHERE EACH OF THOSE MONSTERS PRODUCES AROUND 200 TONS OF THRUST, THE THRUST FROM DS1'S ION ENGINE IS LITTLE MORE THAN THE WEIGHT OF A SINGLE SHEET OF PAPER.

SOLAR-POWERED IONS

The ion engine is powered by an array of solar cells. At full throttle, the engine consumes 2.5 kilowatts of power—about the same as a large electric heater. This is a great deal of power to generate using solar cells, so Deep Space 1 is testing a new type of high-powered solar array. The spacecraft has two "wings," each 14 ft 9 in (4.5 m) by 5 ft 3 in (1.6 m) in size and composed of 360 silicon lenses that focus sunlight onto 1,800 solar cells. These "solar concentrator arrays" yield up to 20% more power than the best existing solar cell designs.

Deep Space 1's sophisticated automatic pilot system is just as ground-breaking as its revolutionary engine. It makes DS1 virtually independent of NASA's tracking network and ground controllers. The system has two main components. The first, AutoNav, can determine exactly where DS1 is in the solar system so that the probe can fine-tune its own flight path. To do this, it carries a database of the orbits of 250 asteroids and the positions of 250,000 background stars. By regularly taking pictures of asteroids and comparing the images to its stored data, DS1 can calculate its own position and adjust the thrust of its ion engine as required. The second component of the control system is a piece of software called "Remote Agent." NASA ground controllers feed only very general instructions into Remote Agent. The software then calculates not only how to carry out the orders, but also the best sequence in which to execute them.

If these major new technologies work successfully, and all the signs from the mission suggest that they are doing so, then Deep Space 1 may spawn a new generation of space probes and spark a golden age of solar system exploration.

ON ITS OWN

WHICH ARE GUIDED FROM THE GROUND, DS1 CARRIES NEW AUTOMATED NAVIGATION SOFTWARE. WITHOUT ANY HELP FROM MISSION CONTROL, THE PROBE'S TEST TRACK TOOK IT TO WITHIN JUST A FEW MILES OF THE SURFACE OF ASTEROID 1992KD—THE CLOSEST ASTEROID FLYBY EVER.

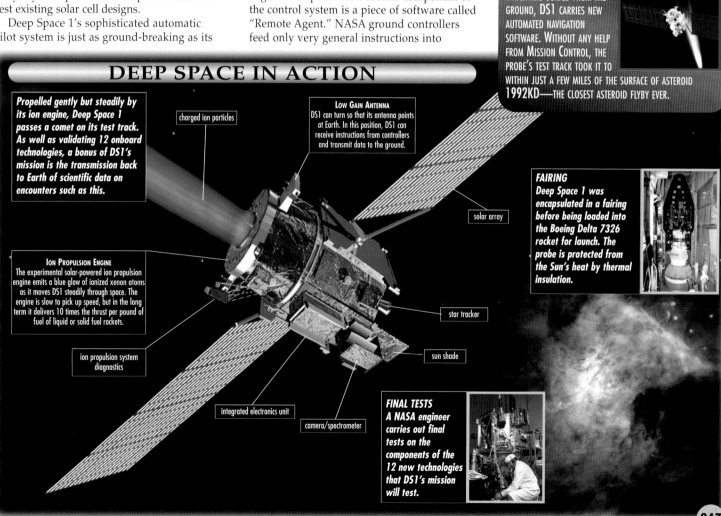

DEEP SPACE IN ACTION

Propelled gently but steadily by its ion engine, Deep Space 1 passes a comet on its test track. As well as validating 12 onboard technologies, a bonus of DS1's mission is the transmission back to Earth of scientific data on encounters such as this.

charged ion particles

LOW GAIN ANTENNA
DS1 can turn so that its antenna points at Earth. In this position, DS1 can receive instructions from controllers and transmit data to the ground.

solar array

ION PROPULSION ENGINE
The experimental solar-powered ion propulsion engine emits a blue glow of ionized xenon atoms as it moves DS1 steadily through space. The engine is slow to pick up speed, but in the long term it delivers 10 times the thrust per pound of fuel of liquid or solid fuel rockets.

ion propulsion system diagnostics

integrated electronics unit

camera/spectrometer

star tracker

sun shade

FAIRING
Deep Space 1 was encapsulated in a fairing before being loaded into the Boeing Delta 7326 rocket for launch. The probe is protected from the Sun's heat by thermal insulation.

FINAL TESTS
A NASA engineer carries out final tests on the components of the 12 new technologies that DS1's mission will test.

CASSINI-HUYGENS

A quarter of a century after Pioneer 11 took the first close-up pictures of Saturn, the ringed planet came under the gaze of a new NASA craft. The massive Cassini-Huygens spacecraft reached Jupiter in 2000. In December 2004, the Huygens probe—released from Cassini—plunged into the thick atmosphere of Saturn's moon Titan, sending back enough data to keep scientists busy for decades. The $3.25 billion project is set to be the last big-budget probe mission for some time. And if all goes according to plan, it should also be one of the longest.

CASSINI-HUYGENS PHOTO FILE

The narrow angle camera onboard Cassini took a series of exposures of Saturn and its rings and moons on February 9, 2004, which were composited to create this stunning color image. At the time, Cassini was 43 million miles (69 million km) from Saturn, less than half the distance from Earth to the Sun.

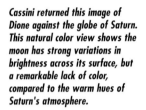

Cassini returned this image of Dione against the globe of Saturn. This natural color view shows the moon has strong variations in brightness across its surface, but a remarkable lack of color, compared to the warm hues of Saturn's atmosphere.

Cassini captured this revealing view of Saturn's cloud-tops, which shows that Saturn's hydrogen- and helium-rich atmosphere is a dynamic place, filled with spots, ovals and swirling vortices and filaments of gas. The image has been highly processed to show detail.

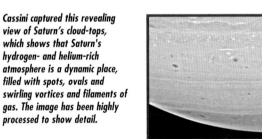

This is the first colored view of Titan's surface. The two rock-like objects just below the middle of the image are about 6 in (15 cm) (left) and 1.5 in (4 cm) (centre) across respectively, at a distance of about 2 ft 9 in (85 cm) from Huygens.

CASSINI-HUYGENS STATS

	CASSINI ORBITER	HUYGENS PROBE
NUMBER OF INSTRUMENTS	18	7
POWER GENERATION	PLUTONIUM THERMOELECTRIC GENERATORS	LITHIUM SULFUR DIOXIDE BATTERIES
UNFUELED NAVIGATION	GRAVITY ASSISTS	GRAVITY, PARACHUTES
FUELED NAVIGATION	TWO 100-POUND-FORCE (440 N) THRUSTERS	THREE 112-POUND-FORCE (498 N) SPRINGS
PROPELLANT	MONO-METHYL-HYDRAZINE, NITROGEN TETROXIDE	NONE
UNFUELED WEIGHT	4,750 LB (2,154 KG)	770 LB (349 KG)
WEIGHT OF FUEL	6,905 LB (3,132 KG)	NONE
DIMENSIONS	22 FT (6.7 M) HIGH, 13 FT (4 M) WIDE	9 FT (2.7 M) IN DIAMETER
DURATION OF MISSION	43 MONTHS	3 HOURS

SATURN SAILOR

After its long journey from Earth, Cassini made its closest approach to Jupiter on 30 December 2000, and began taking measurements and photographs. In all, 26,000 images were taken, allowing the creation of the most detailed global portrait ever made of the planet.

One area of new study was the rings of Jupiter, barely visible from Earth. Cassini showed that they were made of irregular, rather than spherical particles, suggesting they were ejected from Jupiter's moons Metis and Adrastea by micrometeorite impacts.

The scientific discoveries of Cassini are almost too numerous to describe. Among them are two new moons of Saturn, spotted in June 2004. These bodies are very small, but have been named Methone and Pallene.

Direct observation of the planets and moons of the outer Solar system was only one area of study. The Cassini science team took the opportunity to put of Einstein's theory of general relativity to the test. They experimented with radio signals from Cassini to prove that a massive object like the Sun causes space-time to curve, and a beam of radio waves or light that passes by the Sun has to travel further because of the curvature and the delay in reaching Earth can be used to measure the amount of curve. Results from Cassini have not conclusively proved the theory, but have increased scientific confidence in it greatly.

The most spectacular experiment conducted by Cassini was the despatch of the

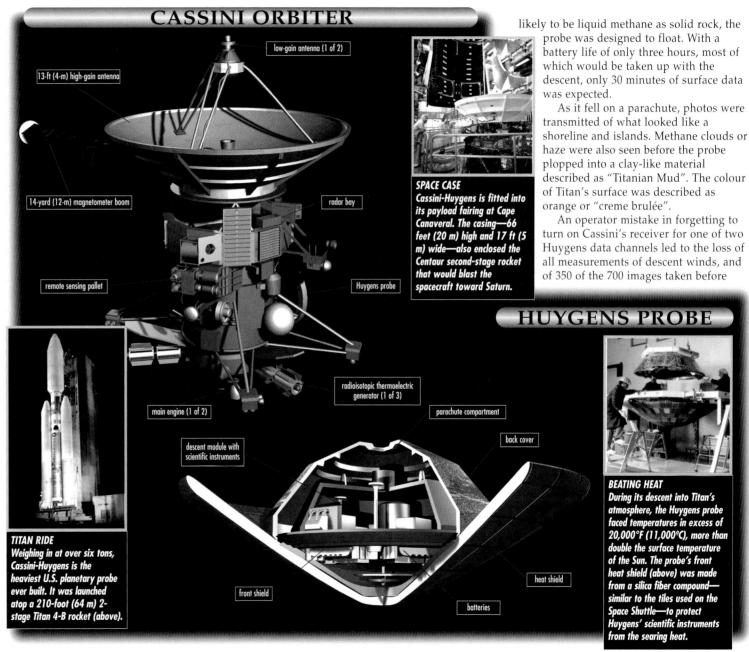

CASSINI ORBITER

low-gain antenna (1 of 2)

13-ft (4-m) high-gain antenna

14-yard (12-m) magnetometer boom

remote sensing pallet

radar bay

Huygens probe

main engine (1 of 2)

radioisotopic thermoelectric generator (1 of 3)

parachute compartment

back cover

descent module with scientific instruments

front shield

heat shield

batteries

TITAN RIDE
Weighing in at over six tons, Cassini-Huygens is the heaviest U.S. planetary probe ever built. It was launched atop a 210-foot (64 m) 2-stage Titan 4-B rocket (above).

SPACE CASE
Cassini-Huygens is fitted into its payload fairing at Cape Canaveral. The casing—66 feet (20 m) high and 17 ft (5 m) wide—also enclosed the Centaur second-stage rocket that would blast the spacecraft toward Saturn.

HUYGENS PROBE

BEATING HEAT
During its descent into Titan's atmosphere, the Huygens probe faced temperatures in excess of 20,000°F (11,000°C), more than double the surface temperature of the Sun. The probe's front heat shield (above) was made from a silica fiber compound—similar to the tiles used on the Space Shuttle—to protect Huygens' scientific instruments from the searing heat.

likely to be liquid methane as solid rock, the probe was designed to float. With a battery life of only three hours, most of which would be taken up with the descent, only 30 minutes of surface data was expected.

As it fell on a parachute, photos were transmitted of what looked like a shoreline and islands. Methane clouds or haze were also seen before the probe plopped into a clay-like material described as "Titanian Mud". The colour of Titan's surface was described as orange or "creme brulée".

An operator mistake in forgetting to turn on Cassini's receiver for one of two Huygens data channels led to the loss of all measurements of descent winds, and of 350 of the 700 images taken before

RINGED THINGS

DUTCH ASTRONOMER CHRISTIAAN HUYGENS (1629–93) DISCOVERED TITAN IN 1655, AND WAS THE FIRST TO ARGUE THAT SATURN'S UNUSUAL SHAPE COULD BE EXPLAINED BY RINGS. TWENTY YEARS LATER, FRENCH ASTRONOMER GIOVANNI DOMENICO CASSINI (1625–1712; LEFT) DISCOVERED FOUR MORE SATURNIAN MOONS, AS WELL AS THE GAP BETWEEN TWO OF THE RINGS.

Huygens probe to the surface of Titan. After a large number of preliminary flybys, Cassini ejected Huygens towards Titan on 25 December 2004, although it didn't arrive for a

further three weeks.

As Huygens entered Titan's atmosphere it sent data and images to Earth, relayed via Cassini. Betting that the landing site was as

landing. Nonetheless, the probe survived on the surface and transmitted for over an hour and 12 minutes and is regarded as a great success.

GENESIS SOLAR WIND SAMPLER

The Genesis Solar Wind Sampler mission was launched to answer such questions as: what is the Sun made of? What makes the Earth different from the Sun? These and other questions about the origins of the solar system could possibly be answered by studying material travelling away from the Sun, borne by the so-called "solar winds".

THE LANDING PLAN

FISHING FOR SPACECRAFT

Although it ended with a bump on the desert floor, the plan to retrieve the Genesis Sample Return Capsule (SRC) was worked out to the finest detail. The 452 lb (205 kg) SRC was to enter the atmosphere at 410,000 ft (125 km) over northwestern Oregon at a speed of 6.8 miles (11 km) per second. At 98,430 ft (30 km) altitude a drogue chute was to deploy, followed by the main chute (actually a parafoil) at 20,000ft (6100m). A nearby USAF radar station would guide a primary and a backup helicopter to intercept the capsule about 4,500ft (1,370 m) above the ground and catch it with a hook on a long line. If the first helicopter failed, the backup had enough time for several attempts to hook the SRC. From re-entry to capture was expected to take about 80 seconds, but sooner than expected the SRC went past with no parafoil to catch.

What should have happened. Catching materials from space with aircraft was a method used to retrieve capsules containing photographs from early spy satellites.

LAST WORDS

These are some of the messages from mission control to the helicopter pilots as the Sample Return Capsule fell to Earth.

"We are able to identify a tumbling object"

"Be advised, from our vantage point we do not see a drogue chute. Negative drogue"

"The track there is two zero zero, nine miles"

"It looks like we have a no chute sir. Vector two zero zero, eight miles. Look for an impact on the surface."

"Impact at 5:58 AM. Impacted vehicle on the surface"

HERE COMES THE SUN

Solar wind particles are believed to consist of the same material from which the planets formed, being atoms, ions or high-energy particles, but cannot be detected on Earth or by craft in orbit.

These elements were formed more than five billion years ago, when the Sun was formed from the collapse of a cloud of gas and dust called a solar nebula. The outer layer of the Sun contains the remnants of this nebula, which are constantly being propelled outwards by its heat. Enthusing about the Genesis mission, one physicist said: "It's effectively like dipping your spoon into the Sun and being able to analyse that, almost like you would a sample of seawater."

To collect these tiny and lightweight particles, a series of collector arrays made of extremely pure sapphire, silicon, gold and diamond were put into the core of the Genesis probe. These would only be exposed to the outside when the probe was in the optimum location.

Launched in August 2001, Genesis was despatched on a mission to a point in space known as LaGrange 1 or L1, where the gravitational pull of the Earth and Sun is equal, allowing a craft to remain in the same relative position between the two bodies, almost as though it could 'hover' on the spot. Once at L1 it collected samples for 884 days and travelled about 1.8 billion miles (2.8 billion km) relative to the Sun. Three types of solar wind were collected, "fast", "slow", and the massive expulsion of plasmas known as coronal mass ejections. One

problem noted by mission control was overheating batteries, which were running at 104°F (40°C) rather than the optimum 74°F (23°C). This was not thought to be a critical problem at L1 because the batteries' main function was to steer the craft during re-entry and open its parachute.

SALT LAKE PITY

Once the material had been gathered, and the collector plates had been sealed, the most dangerous phase of the mission began, returning the spacecraft and its delicate cargo back to Earth without damaging the valuable

GENESIS SPACECRAFT

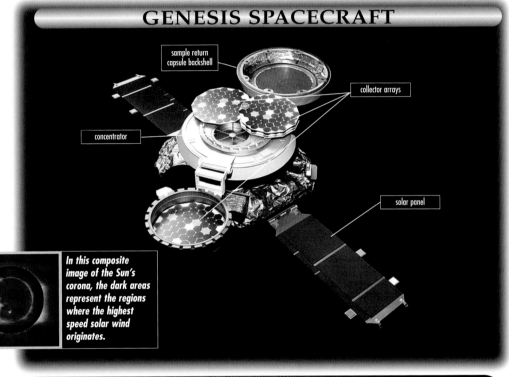

- sample return capsule backshell
- collector arrays
- concentrator
- solar panel

In this composite image of the Sun's corona, the dark areas represent the regions where the highest speed solar wind originates.

SUPER CLEAN

THIS WAS THE FIRST SPACE PROJECT TO REQUIRE A CLASS 10 CLEANROOM, IN WHICH THERE ARE NO MORE THAN 10 PARTICLES OF CONTAMINANT PER 35 CU FT (1 M³).

samples. The method chosen was to fly Eurocopter Squirrel helicopters under the path of the Sample Return Capsule, which was jettisoned above and catch it with a hook before it reached the ground. The capture area was chosen as a salt lake at the US Army's Dugway Test Range in Utah because it allowed for a very large landing footprint, of about 18 by 52 miles (30 km by 84 km). The helicopters were flown by experienced Hollywood stunt pilots who were well practiced and in position, but the parachute failed to deploy at 20,000 ft (6,100 m) as planned, and the capsule, wobbling like a dinner plate as it fell plunged into the salt flats at a speed measured at 193 mph (311 km/h). Embarrassingly for NASA, this occurred on live TV. Long-range images and helicopter flybys showed the capsule half-buried on its edge in the ground and it looked like the expensive mission had been a total failure, but the wreckage was carefully removed and transferred to a clean environment.

As feared, the delicate collector plates were broken in the impact, but were not contaminated by any Earth materials. In April 2005 scientists disassembled the sample collectors in the Johnson Space Center's Class-10 clean room and announced that there were useable intact samples. In all, Genesis collected about 1020 ions, the equivalent of about 0.4 miligrams of material, including protons, electrons, and ions of heavier elements such as helium and oxygen. Analysis of these samples is scheduled to continue into 2007.

SAMPLES

GENESIS WAS THE FIRST CRAFT TO BRING BACK EXTRATERRESTRIAL SAMPLES BACK TO EARTH SINCE THE LAST APOLLO MISSIONS IN THE EARLY 1970s

MISSION DIARY

8 AUGUST, 2001
LAUNCHED

16 NOVEMBER, 2001
HALO ORBIT INSERTION

12 MARCH 2002
START OF SAMPLE COLLECTION

6 APRIL 2004
GENESIS COLLECTS ITS LAST SAMPLE

8 SEPTEMBER 2004
THE SAMPLE RETURN CAPSULE ENTERED EARTH'S ATMOSPHERE AND CRASHES INTO THE UTAH DESERT (RIGHT)

2004-SEPTEMBER 2007
ANALYSIS OF THE SAMPLES

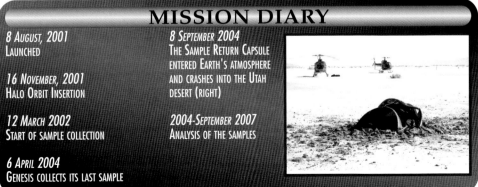

LOST PROBES

Exploration has always had its dangers, and exploring space presents many hazards to the robotic spacecraft sent out to investigate the mysteries of the solar system. These probes have to survive the extremely hostile environment of space, often for long periods of time, and despite their usually reliable technology, they occasionally break down and are never heard from again. But each failure teaches spacecraft designers valuable lessons that will help to improve the success rate of future missions.

WHAT IF...

...PROBES WERE IMPROVED?

The loss of a planetary spacecraft is highly frustrating for those involved in the mission, and knowing that a spacecraft is intact but unable to communicate can be worse than learning that the spacecraft has been destroyed. Desperate efforts are made to try to contact the missing craft, until eventually all hope is lost and attempts cease. And as if losing a spacecraft were not bad enough, planetary missions usually draw the attention of the media. In the glare of the world's media, any loss is highly embarrassing.

One problem faced by all spacecraft as they venture away from Earth is the length of time it takes to send signals to them. For Mars, this time is only 15 minutes, but for a spacecraft at Jupiter or beyond, the signal takes almost an hour to reach it. If a spacecraft out there gets into trouble, it could be lost before its controllers have a chance to send it signals that could help it resolve the problem.

In an effort to avoid this kind of situation, space probes are being made more autonomous. Low-cost spacecraft such as Deep Space 1 are testing the new technologies needed to improve the chances of future missions. Using the stars to navigate, Deep Space 1 carefully made its way to asteroid 1992KD and reported home after its arrival.

This was not the case for Mars Polar Lander. When it reported back to its controllers just before entry into the Martian atmosphere, the craft appeared to be fully functional. At this point communication was intentionally stopped

The Cassini Saturn probe can put its systems into a "safe" mode if something goes wrong, to give its controllers time to find a solution to the problem.

to allow the craft to realign itself for descent to the south pole of Mars. A little while later, controllers listened carefully for the craft's first signal from the surface. But it never arrived. For days, every attempt was made to contact the probe, but no answer was ever received. NASA continued to search for the probe for several weeks, using the Mars Global Surveyor to photograph the landing site, but found no trace of it. Just weeks before the loss of Mars Polar Lander, NASA also lost the Mars Climate Orbiter due to a miscalculation in its trajectory.

Though it always expected that some of these "smaller, faster, cheaper" spacecraft would be lost, NASA is rethinking its plans for the exploration of Mars. Instead of each probe communicating directly with Earth, NASA is now thinking of setting up a network of communications satellites around Mars to relay messages from scientific satellites in orbit and on the surface. Whether or not this improves the success rate of NASA's Mars exploration efforts, spacecraft will still have to be very carefully designed to ensure future missions are not lost.

SPACE CRAFT LOSSES

MOON MISSIONS				MARS 3	U.S.S.R.	VENUS MISSIONS	
		LUNA 6	U.S.S.R.	MARS 6	U.S.S.R.		
LUNA 1	U.S.S.R.	SURVEYOR 4	U.S.			VENERA 1	U.S.S.R.
PIONEER 4	U.S.	LUNA 18	U.S.S.R.	PHOBOS 1	U.S.S.R.	ZOND 1	U.S.S.R.
RANGER 3	U.S.			PHOBOS 2	U.S.S.R.	VENERA 2	U.S.S.R.
RANGER 4	U.S.	MARS MISSIONS		MARS OBSERVER	U.S.	VENERA 3	U.S.S.R.
RANGER 5	U.S.	MARS 1	U.S.S.R.	CLIMATE ORBITER	U.S.		
LUNA 4	U.S.S.R.	ZOND 2	U.S.S.R.	POLAR LANDER	U.S.		
RANGER 6	U.S.	MARS 2	U.S.S.R.	BEAGLE 2	U.K.		

LOST IN SPACE

Since the very earliest days of spaceflight, probes have been launched out into the solar system. The Moon was the first target, since the U.S. and the Soviet Union needed to gather data and accumulate technical experience in preparation for crewed lunar missions. And in the Cold War rivalry between the two superpowers, each tried to prove they had the world's best technology: Missions to Mars and Venus soon followed the first Moonshots.

But in the early 1960s, so little was known of conditions in space that many satellites and probes failed within days of launch. To reach Mars or Venus, a spacecraft had to survive a flight lasting nine months or more, so the probes had to be highly reliable. This was achieved by adding backup systems that would take over if a primary system failed. Backups could be supplied for most of the essential functions of the spacecraft, but there were always a small number of components and systems whose failure would end the mission. Such devices had "single-point criticality."

The largest single-point-critical element in any space mission is the launcher. If the booster fails during ascent to orbit, the mission ends before it has even left the Earth. And even when it has safely entered its interplanetary trajectory, a spacecraft still faces dangers and can malfunction. For instance, a computer or thruster failure can leave the spacecraft facing in the wrong direction, so that its communication antenna loses contact with the Earth.

LEARNING FROM FAILURES

Although many missions have been lost, their scientific instruments have usually managed to return a small amount of information before failure. Using this information, scientists have been able to understand the space environment better, which in turn has helped engineers to improve probe design. As designs have improved, so has the success rate, but failures have by no means been eliminated.

In preparation for the human exploration of Mars, NASA has once again turned its attention to that planet. Two spacecraft are sent during each launch window but problems have plagued almost every flight. During the 1990s, Mars Pathfinder was a success but Mars Orbiter, Mars Climate Orbiter and Mars Polar Lander were all lost. The most recent of these missions, Mars Polar Lander, disappeared in December 1999. It probably reached the surface of the planet, but no signals were ever received from it.

SUCCESSES

THE MOST SUCCESSFUL INTERPLANETARY MISSIONS EVER STAGED WERE THE PIONEER (RIGHT) AND VOYAGER FLIGHTS TO THE OUTER PLANETS. IN SPITE OF COUNTLESS DANGERS, THESE PROBES TO JUPITER AND BEYOND REMAINED IN CONTACT WITH EARTH AND PROVIDED A WEALTH OF SCIENTIFIC DATA ABOUT THE GIANT GAS PLANETS AND INTERPLANETARY SPACE.

PROBES TO MARS AND VENUS

MARS 1
The first mission to fly past Mars was the Soviet Mars 1 probe, launched in 1962. It probably got to within 118,000 miles (191,000 km) of Mars in June 1963, but all contact with it had been lost in March.

Most of the interplanetary probes that have been launched—and most of those that have been lost—were headed for our nearest planetary neighbors, Mars and Venus.

MARS

PHOBOS PROBES
In 1985, the Soviet Union sent a pair of probes to study Phobos, one of the moons of Mars. But contact with Phobos 1 was lost before it reached Mars, and Phobos 2 was lost shortly after it entered Mars orbit.

THE MOON

EARTH

SURVEYOR 98
The ill-fated Mars Surveyor 98 mission consisted of the Mars Climate Orbiter (top) and the Mars Polar Lander (bottom). The Climate Orbiter burned up in the Martian atmosphere September 1999, and all contact with the Polar Lander was lost in December 1999.

VENERAS
The Soviet Venera missions to Venus got off to a bad start in the 1960s with the loss of the first probe in 1961 and the second and third in 1964. But Veneras 4 (in 1967) through 16 (in 1981) were successful.

VENUS

THE LOST ZONDS
The first two of the Soviet Zond series of probes were lost due to communications failure. These were Zond 1, a mission to Venus launched on April 2, 1964, and the Zond 2 Mars probe, launched on November 30, 1964.

DATA BLOCK

SOON AFTER THE 1989 LAUNCH OF GALILEO (RIGHT) TO JUPITER, CONTROLLERS FEARED THEY WOULD LOSE THE SPACECRAFT WHEN ITS MAIN ANTENNA FAILED TO DEPLOY PROPERLY. A LATER PROBLEM WITH A TAPE RECORDER ALSO JEOPARDIZED THE MISSION. BY UPLOADING NEW SOFTWARE, NASA RETRIEVED MOST OF THE CRAFT'S SCIENTIFIC DATA, BUT ITS DATA TRANSMISSION RATE IS NOW LESS THAN A TEN-THOUSANDTH OF ITS PLANNED 134 KILOBITS PER SECOND.

INDEX

Page numbers in *italics* refer to illustrations